LEADING THE WAY: A HISTORY OF JOHNS HOPKINS MEDICINE

Leading the Way:

A History of
Johns Hopkins Medicine

Neil A. Grauer

Johns Hopkins Medicine in association with
The Johns Hopkins University Press
Baltimore, Maryland

Frontispiece: From its founding under "The Four Doctors" (left to right in painting, William H. Welch, William S. Halsted, Sir William Osler and Howard A. Kelly) to today's physicians, scientific researchers and medical students, Johns Hopkins Medicine has been in the forefront of the mission to unlock the mysteries of disease, cure it, and teach others how to improve health throughout the nation and around the world.

Photograph by Claudio Vasquesz

www.hopkinsmedicine.org

Printed in the United States on acid-free paper by
R.R. Donnelley & Sons Company

Johns Hopkins Medicine
Marketing and Communications
901 South Bond Street
Suite 550
Baltimore, Maryland 21231-3359

Distributed by the Johns Hopkins Press
2715 North Charles Street
Baltimore, Maryland 21218-4363

www.press.jhu.edu

Library of Congress Cataloging-in-Publication Data

Grauer, Neil A.
Leading the way: a history of Johns Hopkins Medicine / Neil A. Grauer.
p. ; cm.
Includes bibliographical references and index.

ISBN 978-1-4214-0657-2 (hdbk. : alk. paper) — ISBN 1-4214-0657-8
(hdbk. : alk. paper)
I. Johns Hopkins Medicine. II. Johns Hopkins University. III. Title.
[DNLM: 1. Johns Hopkins Medicine. 2. Schools, Medical—
history—Maryland. 3. History, 19th Century—Maryland. 4. History,
20th Century—Maryland. 5. History, 21st Century—Maryland. W 19]

610.71'1752—dc23
2011046145

Composed in Adobe Garamond Pro (designed by Robert Slimbach, based
on the roman types of Claude Garamond (c. 1480–1561) and the italic types
of Robert Granjon (1513–1589), and Gill Sans MT Standard, based on the
famous set of humanist sans serif fonts originally designed by Eric Gill and
the Monotype Type Drawing Office, first appearing in 1928.

Vice President, Marketing and Communications: Dalal Haldeman

Editor and Production Manager: Patrick Gilbert

Copy Editors: Justin Kovalsky and Eileen O'Brien

Cover photo: Keith Weller

Cover design: Jennifer E. Fairman and David A. Rini. Department of Art
as Applied to Medicine, Johns Hopkins University School of Medicine

Book design: David Dilworth, Johns Hopkins Medicine, Office of
Marketing and Communications

Dedication

To Edward D. Miller,
dean/CEO of Johns Hopkins Medicine and
vice president for medicine of
The Johns Hopkins University,
and Ronald R. Peterson,
president of The Johns Hopkins Hospital,
the Johns Hopkins Health System, and
executive vice president of Johns Hopkins Medicine,
who together have given dynamic life,
purpose and foresight to the once new
and untested but now extraordinary and
expanding entity known as Johns Hopkins Medicine;
and to all the physicians, research scientists,
nurses, residents, interns, fellows,
employees, volunteers, trustees—and,
most important, patients—
of Johns Hopkins Medicine.

Contents

Introduction

Those who work for The Johns Hopkins University, its School of Medicine and The Johns Hopkins Hospital sometimes wonder if Mr. Hopkins could ever have imagined the world-wide impact of these entities at the time he made his 1873 bequest to create them.

The common-sense answer, of course, is no. Yet we hardly can scoff at his inability to foresee the astounding century-and-a-quarter growth in his creations. Indeed, we can empathize with it. The Johns Hopkins Health System and University trustees who devised the unified governance of these institutions under the banner of Johns Hopkins Medicine in 1997 surely can gaze with a measure of what would be Mr. Hopkins' astonishment at their own creation's even more incredible growth, burgeoning influence, and continuing expansion in a mere 15 years.

The last major history of Johns Hopkins Medicine, entitled *A Model of Its Kind*, was published in 1989, the 100th anniversary of the opening of The Johns Hopkins Hospital. The School of Medicine celebrated its centennial in 1993. In the two decades since then, not only has there been a complete change in the way the Hospital and School of Medicine are run, but an entirely new Johns Hopkins Hospital has been built—a billion-dollar, 1.6-million-square-foot, twin-tower complex for adults and children. In addition, the number of School of Medicine departments has grown to 33, and 31 of them have new directors; dozens of multidisciplinary research institutes and centers have been created; an entirely new Medical School curriculum has been formulated; four hospitals—two in Maryland, one in Washington, D.C., and one in Florida—have joined the Hopkins Medicine family; and Johns Hopkins Medicine International has been founded, expanding our influence exponentially.

The history of medicine as taught, advanced through research and practiced at Johns Hopkins is one of constant striving to garner knowledge and use it to improve the health of individuals throughout the world. Like all histories, it is replete with inspiring achievements, as well as undoubted shortcomings.

It also can be a bit confusing—at least in terms of nomenclature. That is because "Johns Hopkins Medicine," as an entity, is a so-called virtual corporation that was devised in mid-1996 and came into full operation in January 1997. However, the term had been around before then. As early as 1989, the celebration of the 100th anniversary of the opening of the Hospital was hailed as the centennial of "Johns Hopkins Medicine." The name stood for several different things before becoming what it is today; and under its umbrella operate a broad array of Hopkins Medicine enterprises.

Although many individuals casually acquainted with Johns Hopkins routinely misstate its name as John Hopkins, without the unfamiliar "s" at the end of Johns, even heads of Hopkins' medical departments may flinch if asked to explain how Johns Hopkins Medicine differs from the Johns Hopkins Health System, The Johns Hopkins Health System Corporation, and the Johns Hopkins Medical Institutions. Here are a few brief explanations:

The Johns Hopkins Medical Institutions is not a formal legal entity but an unofficial title, believed to have been in use since the 1970s to describe the School of Medicine, Hopkins Hospital, and the other schools—the Bloomberg School of Public Health and the School of Nursing—which are located on Hopkins' East Baltimore medical campus. The late Robert Heyssel, who ran Hopkins Hospital from 1972 to 1992, called it "a term of convenience." It endures in many ways, not the least of which is as the e-mail address domain—"jhmi.edu"—of most Hospital and medical school staff.

The Johns Hopkins Health System Corporation is a legal entity, created in 1986 as a private, non-profit corporation to formulate policy among and provide centralized management for its affiliates. JHHSC functions as the parent holding company of its wholly owned affiliates, including The Johns Hopkins Hospital, Johns Hopkins Bayview Medical Center, Howard County General Hospital, Suburban Hospital, Sibley Memorial Hospital, All Children's Hospital, Johns Hopkins Community Physicians, Inc., and Johns Hopkins Employer Health Programs, Inc. These affiliates are often referred to collectively as **The Johns Hopkins Health System**.

Johns Hopkins Medicine, as explained above, is not a legal entity but rather the result of a formal collaboration entered into by The Johns Hopkins University and Johns Hopkins Health System Corporation. Johns Hopkins Medicine (JHM) provides a vehicle for internal operational coordination among JHHSC, the affiliates and the Johns Hopkins University School of Medicine. Johns Hopkins Medicine also provides a united voice for external activities. JHHSC and JHU as distinct yet interdependent corporations trading as JHM now are able to respond in an integrated fashion to opportunities and pressures affecting the medical enterprise. Under the auspices of JHM, JHHSC and JHU jointly own a number of affiliates, including Johns Hopkins HealthCare, Johns Hopkins Home Care Group, and Johns Hopkins Medicine International.

This book endeavors to recount the remarkable accomplishments of Johns Hopkins Medicine over the nearly 140-year period between the 1873 incorporation of The Johns Hopkins University and The Johns Hopkins Hospital to the present day, especially during the two decades not previously chronicled in such a way. Not surprisingly, this required a lengthy manuscript. Too lengthy, in fact, to fit comfortably either between hard covers or in a reader's lap. Significant editing had to be done to make this a book of manageable size. Such editing does not mean, however, that what was removed from this text has been consigned to the cutting room floor. Instead, thanks to the blessings of the Internet, a medium impervious to such constraints as length, we hope, eventually, to place on the Web what had to be edited out. Readers still will find a substantial amount of information in this book, but for now, if they want even more, especially about each clinical and basic science department, go to the Johns Hopkins Medicine website, http://www.hopkinsmedicine.org/ . There you will find a Web page for each department and division that gives full descriptions of all the faculty and the services they provide.

1889–1939:
The First Fifty Years

THE BENEFACTOR

President Charles W. Eliot of Harvard

Severely ill with pneumonia and assuming he was about to die, Johns Hopkins joked about it.

With his nephew Joseph sitting beside his bed, Hopkins observed, "Joe, it is very hard to break up an old habit. I've been living for seventy-eight years now, and I find it hard to make a change in my ways."[1]

Shortly thereafter, at 3:45 a.m., on a frigid, overcast December 24, 1873—Christmas Eve—Hopkins, a multi-millionaire Baltimore businessman, banker and philanthropist, died quietly in his sleep at his Saratoga Street home.[2]

What really changed that day was not so much Mr. Hopkins' mortality, but the future of medical research, education and patient care throughout the United States and around the world.

The front page of the *Baltimore Sun* on Christmas Day featured what then would pass for a banner headline: "Death of Johns Hopkins," with a sub-headline proclaiming "His Benevolent Enterprises, Monuments of Learning and Charity, &c."[3]

In referring to the "beneficence this community is so largely to realize in the future," *The Sun* was not displaying unusual prescience. It was well known that in 1867, Johns Hopkins had asked Maryland's legislature, the General Assembly, to establish two corporations, The Johns Hopkins Hospital and The Johns Hopkins University. It also was well known that he had drawn up a will pledging virtually his entire fortune to them—a then-staggering $7 million bequest. It was the largest gift of its kind ever bestowed in the United States, estimated to be worth up to $11 billion today—more than three times what Harvard's endowment was then and seven times what Princeton and Cornell combined had in their coffers.[4]

Less than a year before his death, in a supplemental letter to the 14 friends he had appointed as the future institutions' trustees, Hopkins outlined a visionary plan for his hospital that would forever link it with his university—in a previously unimagined way—to foster the science of health care. He wrote: "In all your arrangements in relation to this Hospital, you will bear constantly in mind that it is my wish and purpose that the institution shall ultimately form part of the Medical School of that university for which I have made ample provision in my will."[5] This association between the university and hospital was so well known in advance of Hopkins' death that it was mentioned in *The Sun*'s obituary.[6]

"I congratulate you … on the prodigious advancement of medical teaching, which has resulted from the labors of the Johns Hopkins faculty of medicine…. [I]n the development of medical teaching and research… you … here have led the way."

—President Charles W. Eliot of Harvard, 1901

The Johns Hopkins Hospital, shown here at the time of its completion in 1889, was considered a municipal and national marvel when it opened. It was believed to be the largest medical center in the country with 17 buildings, 330 beds, 25 physicians and 200 employees. As a *Baltimore American* headline put it on May 7, 1889, the Hospital's opening day, "Its Aim Is Noble," and its service would be "For the Good of All Who Suffer."

1795
May 19: Johns Hopkins is born at Whitehall, his family's tobacco plantation in Anne Arundel County, Maryland.

1812
Hopkins, at age 17, moves to Baltimore to become an apprentice of his Uncle Gerard, a wholesale grocer.

How or why Hopkins, a Quaker with little formal education, conceived of the then-unique and extraordinarily beneficial collaboration between a hospital and medical school remains a mystery. It was a stroke of genius, however, that led to creation of one of the world's finest centers for medical science and health care, an inspiration both to the ill and to those who treat them.

With the opening of the Hospital in 1889 and the Medical School in 1893, Hopkins' Founding Four physicians—internist William Osler, pathologist William Welch, surgeon William Halsted and gynecologist Howard Kelly—revolutionized the teaching and practice of medicine. Their innovative methods, emulated and enhanced by those who succeeded them, became the benchmark for hospitals and medical schools everywhere—both then and now.

From the teaching of medical students beside the patient's bed, to the placement of diagnostic laboratories adjacent to the clinics, to the emphasis on antiseptic surgery, Hopkins' pioneering physicians set standards that remain hallmarks of Hopkins today.

Similarly, from the birth of cardiac surgery with the "Blue Baby" operation, the creation of CPR (cardiopulmonary resuscitation) and the invention of electronic heart defibrillators, to the development of neuroscience and neurosurgery, fostering the human genome project, and making DNA breakthroughs that provided the foundation of stem cell research, Hopkins indeed has been leading the way in medical research, teaching and practice—just as it was when Harvard's President Charles W. Eliot lavished praise upon it during the University's 25th birthday celebration in 1901.

All of this likely would have flabbergasted Johns Hopkins himself—although we cannot entirely be certain of that. While anecdotes abound about Johns Hopkins the man, not much more really is known about him as a person than there is about how he came up with the idea of affiliating his Hospital with his university's School of Medicine.

Mr. Hopkins the Man

Johns Hopkins was born on May 19, 1795, the second of what would become the 11 offspring of his parents, Samuel and Hannah Janney Hopkins. They were prosperous Quaker owners of the 500-acre tobacco plantation on which their children were born in Anne Arundel County, Maryland. The homestead, known as Whitehall, still stands near what now is Crofton, Maryland.

Young "Johnsie," as he was known to his siblings, came by his unusual first name, Johns, in what, for the Hopkins family, was a perfectly usual manner: tradition. When his great-grandfather Gerard Hopkins married his great grandmother, Margaret Johns, they named their 10th and last child Johns Hopkins to perpetuate his mother's maiden name. This Johns Hopkins was the grandfather of the baby born in 1795. (In like fashion, many of the brothers and sisters of the latter Johns Hopkins had "Janney" as part of their names.)[7]

"Johnsie" was not born poor. The vast acreage of his family's plantation was on land granted by the King of England to an ancestor, William Hopkins, in the seventeenth century. Its soil was tilled, planted and harvested by the family's slaves, and tobacco was a profitable crop.[8]

In 1807, however, the family's circumstances changed dramatically. Leaders of the Society of Friends had begun to preach that slavery was inconsistent with their faith, and Samuel Hopkins freed all of his slaves. Johns—then only 12—was required to leave boarding school, return to the plantation and go to work in the fields.[9]

Although his formal education had ended, Johns Hopkins was encouraged by his book-loving mother to continue his studies as best as he could—at night, in near-darkness, or whenever else possible. He did so and developed a lifelong devotion to books that, combined with an unusually retentive memory, rendered him an exceedingly well-read man.[10]

Young Johns Hopkins also displayed a remarkably sharp mind for figures. When he was 17, his mother observed, in what proved to be a masterful understatement: "Thee has business ability, and thee must go where the money is." The money was in Baltimore, where his father's brother, named Gerard, owned a wholesale grocery business. He hired his nephew as an apprentice.[11]

Johns Hopkins—also something of a master of understatement—later recalled that he "took kindly to merchandizing." By 1814, when he was 19, Hopkins had developed into a promising merchant who had succeeded in increasing his uncle's business considerably.[12]

Hopkins lived in the home of his Uncle Gerard and Aunt Dolly for seven years. As a former university archivist, Kathryn Allamong Jacob put it, "no story would be complete without a sad love affair," and for Johns Hopkins, that reportedly was the love that blossomed between him and one of his uncle and aunt's daughters, Elizabeth Hopkins, then 16. The two wished to marry, but the Society of Friends frowned on marriages between first cousins and Elizabeth's parents forbade it. Johns and Elizabeth vowed never to marry anyone else—a pledge they kept for the rest of their lives. They remained close friends, and in later years, Johns Hopkins bought her a house near his own and bequeathed it to her upon his death. She died there in 1889.[13]

> *"Thee has business ability, and thee must go where the money is."*

■ One of the main foundations of the early fortune of Mr. Johns Hopkins (1795–1873) was moonshine whiskey. As a young Baltimore wholesale merchant, Hopkins agreed to take moonshine in return for goods. He re-bottled it and sold it as "Hopkins' Best." His fellow Quakers disapproved and he temporarily was suspended from the Meeting, but years later, he was pleased to recall that "the first year I was in business, I sold $200,000 worth of goods."

1819
At the age of 24, Hopkins begins his own wholesale business; in 1822, he forms Hopkins Brothers wholesale firm, using three of his brothers as salesmen. The firm grows rapidly, doing business in Maryland, Virginia and North Carolina.

1845
Hopkins retires from wholesaling, a wealthy man at 50. He becomes a prominent Baltimore banker and a major, personal investor in the Baltimore and Ohio Railroad. He becomes a director of the B&O in 1847.

1857
During a nationwide financial panic, Hopkins pledges his personal fortune to ensure the solvency of the B&O Railroad, saving both it and Baltimore from financial ruin.

1861
With the outbreak of the Civil War, Hopkins joins with the B&O president, John Work Garrett, to overcome opposition from Southern-sympathizing board members and puts the B&O at the service of the Union. Hopkins meets and corresponds occasionally with Abraham Lincoln to discuss the war.

His courtship thwarted, Johns Hopkins moved out of his uncle and aunt's home, but apparently it was not the unpleasantness surrounding his rejected marriage proposal that soon led to his also leaving his uncle's business. Johns and Uncle Gerard disagreed on the propriety of peddling whiskey. Uncle Gerard refused to "sell souls into perdition." Johns Hopkins thought receiving moonshine whiskey from farmers in return for wholesale goods, then packaging and selling the whiskey, was a perfectly legitimate undertaking. He and his uncle parted company commercially—and although Uncle Gerard had qualms about personally trading in souls, he evidently had no objection about investing in it. He generously loaned Johns $10,000 to launch his first independent enterprise, and other relatives also invested in the young man with "business ability."[14]

Going into business for himself, Johns enlisted three of his brothers as salesmen and formed a wholesale company called Hopkins Brothers. Soon it was doing extensive business in Maryland, Virginia and North Carolina.[15] A lucrative aspect of their trade was, indeed, whiskey. Hopkins took the moonshine farmers offered in payment and bottled it under the brand name "Hopkins' Best." The Baltimore Society of Friends did not approve and briefly expelled him from the Meeting. Some sources say Hopkins later expressed regret over trading in whiskey (and perhaps souls); others note that years later, he proudly told a cousin that "the first year I was in business, I sold $200,000 worth of goods." (Never a teetotaler, Hopkins enjoyed fine wines and usually served champagne to dinner guests.)[16]

The rapidly growing, multi-state merchandise trade of Hopkins Brothers prospered because of much more than corn liquor. After 25 years at its helm, Johns Hopkins left the firm a wealthy man. He was just 50 years old and his fortune really had only begun its growth. Using his substantial funds and keen business instincts, he quickly became Baltimore's leading finance capitalist.[17]

He was named president of the Merchants National Bank of Baltimore and a director of five other banks. He also became an early and substantial backer of the Baltimore and Ohio Railroad, and soon was its largest stockholder. In 1847 he became a director of the company, and in 1855, he was named chairman of its powerful finance committee.[18]

When the Civil War erupted, Hopkins was instrumental in ensuring that the B&O's trains and tracks were committed to the Union's cause. Hopkins threw his considerable influence behind his close friend, John Work Garrett, president of the B&O, to overcome the Southern sympathizers on the railroad's board and make the B&O a vital resource to the Union armies, despite heavy losses inflicted on it by Southern saboteurs.[19]

A volatile city in a border state, Baltimore was the scene of the Civil War's first bloodshed. On April 19, 1861, a week after Southern forces had fired upon Charleston's Fort Sumter (but inflicted no casualties), a large brick-throwing and gun-toting Baltimore mob attacked federal troops. The object of their fury were soldiers from the Sixth Massachusetts Regiment, responding to Abraham Lincoln's call for volunteers and changing trains in the city as they headed to Washington. Four soldiers and 12 civilians were killed and dozens wounded in the riot.[20]

Located below the Mason-Dixon Line, Baltimore had a population fiercely split between those favoring the North and the South during the Civil War. Many historians consider that conflict's first genuine bloodshed to have been spilt when Southern sympathizers attacked the Sixth Massachusetts Regiment on Pratt Street, beside the inner harbor, on April 19, 1861. The Northern troops were responding to Abraham Lincoln's call for volunteers and had to change trains in Baltimore on their way to Washington. Four soldiers and 12 civilians were killed and dozens wounded in the riot. Baltimore remained under the control of federal forces for the rest of the war.

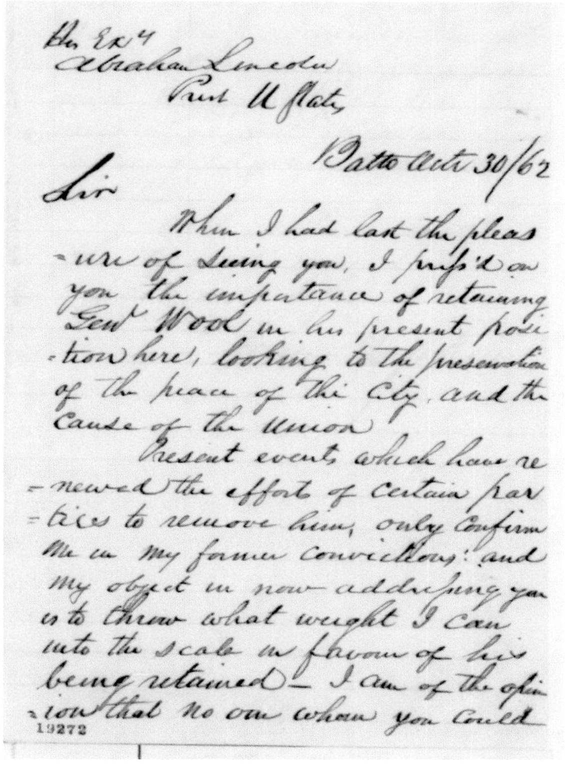
His Ex'y
Abraham Lincoln
Prest U States

Balto Octr 30/62

Sir
When I had last the pleas
-ure of seeing you, I press'd on
you the importance of retaining
Genl Wool in his present posi
-tion here, looking to the preservation
of the peace of the City, and the
Cause of the Union
Present events which have re
-newed the efforts of certain par
-ties to remove him, only confirm
me in my former convictions; and
my object in now addressing you
is to throw what weight I can
into the scale in favour of his
being retained— I am of the opin
-ion that no one whom you could
19272

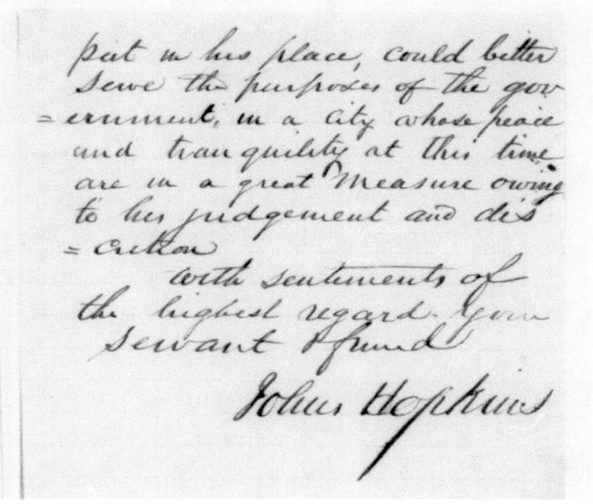
put in his place, could better
serve the purposes of the gov
-ernment, in a city whose peace
and tranquility at this time
are in a great measure owing
to his judgement and dis
-cretion
With sentiments of
the highest regard—your
Servant & friend
Johns Hopkins

Johns Hopkins, the Baltimore and Ohio Railroad's largest stockholder, and the B&O's president, John Work Garrett, overcame the objections of Southern sympathizers on the railroad's board and put the B&O's trains and tracks at the Union's service during the Civil War. Hopkins knew Abraham Lincoln well enough to send him a note offering advice about keeping the peace in deeply divided Baltimore; Garrett accompanied the president on his visit to the Antietam Battlefield on Oct. 3, 1862 and was photographed with him and then-Union commander George McClellan. Garrett is on the right in this photo, wearing a broad-brimmed, light-colored hat. (Thirty years later, the fundraising efforts and personal generosity of Garrett's daughter, Mary, ensured the opening of the Johns Hopkins School of Medicine.)

More federal troops were dispatched to place Baltimore under military oversight. Conflict between Southern sympathizers and Union supporters taxed the abilities of several commanders who were appointed as stewards of what the Union Army called the "Middle District," meaning the area of the mid-Atlantic states, headquartered in Baltimore.

The actions of one of these commanders, General John E. Wool, a native of Newburg, New York, and 78-year-old veteran of the War of 1812, fueled the insistence of some Baltimoreans that Lincoln find a replacement for him.[21] Johns Hopkins thought otherwise. An abolitionist, a strong supporter of the Union and an admiring acquaintance of Lincoln's, Hopkins believed that Wool was preventing additional upheaval in Baltimore. He took a piece of lined note paper and quickly wrote a letter to Lincoln—addressing it simply "His Exlcy [Excellency] Abraham Lincoln Prest U States," and dating it, "Balto Octr 30/62."

> Sir
>
> When I had last the pleasure of seeing you, I press'd on you the importance of retaining Genl Wool in his present position here, looking to the preservation of the peace of the city, and the cause of the Union
>
> Present events which have renewed the efforts of certain parties to remove him, only confirm me in my former convictions; and my object in now addressing you is to throw what weight I can into the scale in favour of his being retained—I am of the opinion that no one whom you could put in his place, could better serve the purposes of the government, in a city whose peace and tranquility at this time are in a great measure owing to his judgement and discretion
>
> With sentiments of the highest regard—your Servant & friend
>
> Johns Hopkins

■ When The Johns Hopkins Hospital opened in 1889, the maximum daily charge for a ward patient was $2. By 1927, daily costs of $6 forced a boost in the charge from $2.50 to $3—essentially a $1 price hike over 38 years in response to a doubling in costs. By 1937, a private room in the exclusive Marburg Pavilion cost $12 a day.

■ Mr. Johns Hopkins lived in downtown Baltimore at 81 West Saratoga Street (on a site where a parking garage later was built), but he also had a "country" home in a then-rural area northeast of the downtown. In 1838, Hopkins purchased a mansion in what now is the city's Clifton Park Golf Course, at 2107 St. Lo Drive. His former home, named Clifton, once was a showplace where Hopkins hosted such visitors as the Prince of Wales (later King Edward VII). Clifton was sold to the city in 1895 and Hopkins' mansion long served as the golf course's club house. Clifton, seen above, now is leased to Civic Works, a nonprofit youth training corps.

1866
October 25: The Peabody Institute of Music—funded by Hopkins' friend, former Baltimore merchant and multimillionaire George Peabody—dedicates its first building (completed in 1861 but not opened until after the Civil War). Hopkins is impressed by Peabody's establishment of an institute comprising the music academy, a library, art gallery and lecture series, and is known to have discussed these beneficences with him.

1867
August 24: Johns Hopkins incorporates both his university and his hospital.

Lincoln seemed to listen to Hopkins—at least for a while. Wool remained at his post in Baltimore until January 1863, when the president had an equally demanding job for him as Commander of the Eastern Department, headquartered in New York, a city with a larger, similarly violence-prone population.[22]

Hopkins' letter to Lincoln is one of his few surviving private documents of historical interest. He never gave a public speech and left behind a paucity of personal papers of consequence—only business records. Most of what is known of his private history and personality largely is anecdotal, contained in a slim, 125-page profile written in 1929 by his grand niece Helen Hopkins Thom. Its title is appropriate: *Johns Hopkins: A Silhouette*, since it provides only an outline of what the man must have been like, based entirely on recollections of those by-then elderly nieces and nephews who had known him, including Thom's father, Joseph Hopkins.[23]

Although hagiographic in tone, the book has charm and some candor. It is clear that Hopkins could be hard-nosed, heavy-handed and tightfisted, but he also had foresight and was shrewd, civic-minded and generous.[24] In addition to the colonial-style mansion downtown in which he died, Hopkins had a substantial Italianate villa, known as Clifton, located on 330 acres in what then were the rural outskirts of Baltimore. It became a Baltimore showplace, a stop on all sightseeing tours and a home in which he loved to entertain guests, among them, in 1860, the then-18-year-old Prince of Wales, later King Edward VII.[25]

Yet Hopkins also was renowned for dressing in old, drab suits and refusing to wear an overcoat even in the coldest, most inclement weather. Tall and thin, trudging along Baltimore's downtown streets in his somber, threadbare clothes, he probably appeared the living likeness of Ebenezer Scrooge.[26]

For all of his quixotic penny-pinching, however, Hopkins always had looked upon his wealth as a legacy to be guarded for future generations. That also was the philosophy of his friend, the Massachusetts-born George Peabody, who made his first fortune in Baltimore just as Hopkins was building his own. The music conservatory and eclectic library that Peabody, also a bachelor, had endowed in 1857—as well as conversations he is known to have had with Hopkins—may have given Hopkins the idea of leaving his money to found a university. What is certain is that only 10 months after the Peabody Institute's building was dedicated in October 1866 (its opening having been delayed by the Civil War), Hopkins secured the incorporation of The Johns Hopkins University and Hospital.[27]

It remains unclear why Johns Hopkins decided to endow a university, or even what the term "university" meant to him.[28] Why he wanted to found a "teaching hospital" is less mysterious. He had seen Baltimore ravaged by epidemics of yellow fever, small pox and cholera. He had contracted cholera himself and suffered its after-effects for the rest of his life. He served on the board of the State Asylum for the Insane (which stood on the site now occupied by The Johns Hopkins Hospital) and was a trustee of the Union Protestant Infirmary. He was well aware of the city's lack of medical facilities—and of the low quality of medical training and care.[29]

Hopkins knew he was a master of finance but readily recognized he understood little about education. As university and hospital trustees he chose 14 close friends

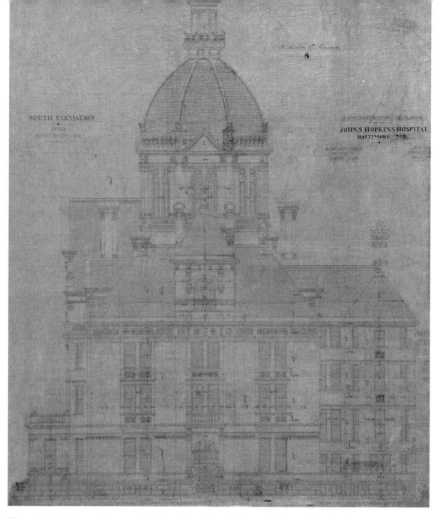

whose judgment he respected. Although the terms of his will may have been vague, a letter he wrote on March 10, 1873 to the hospital trustees about the implementation of its provisions, cited earlier, was specific and emphatic about what he wanted his hospital to be. Not only should it be associated with the university, but he insisted that it be open to all, regardless of race, sex or creed—free of charge, if necessary—and "compare favorably with any other institution of like character in this country or in Europe." In other words, he wanted it to be the best in the world. Hopkins told his trustees:[30]

The original Queen Anne-style Hopkins Hospital was designed by the Boston architectural firm of Cabot and Chandler, which completed the initial plans prepared by architect John Niernsee, who had followed John Shaw Billings' meticulous instructions for what he wanted the Hospital to feature. It was the first hospital in the nation to be equipped with central heating; most inside corners were curved to avoid the buildup of dust and dirt; and precisely calculated ventilation systems sought to guard patients from contaminated air. The Hospital was wired for electricity and telephones—even though no electrical service was yet available in that part of Baltimore.

"It will, therefore, be your duty to obtain the advice and assistance of those, at home or abroad, who have achieved the greatest success in the construction and management of Hospitals.

"I cannot press this injunction too strongly upon you, because the usefulness of this charity will greatly depend upon the plan which you may adopt for the construction and arrangement of the buildings....

"The indigent sick of this city and its environs, without regard to sex, age, or color, who may require surgical or medical treatment, and who can be received into the Hospital without peril to the other inmates, and the poor of this city and State, of all races, who are strickened down by any casualty, shall be received into the Hospital without charge, for such periods of time and under such regulations as you may prescribe....

"You will also provide for the reception of a limited number of patients who are able to make compensation for the room and attention they may require. The money received from such persons will enable you to appropriate a larger sum for the relief of the sufferings of that class which I direct you to admit free of charge....

"It will be your especial duty to secure for the service of the Hospital, surgeons and physicians of the highest character and greatest skill....

"I have felt it to be my duty to bring these subjects to your particular attention, knowing that you will conform to the wishes which I thus definitely express.

"In other particulars I leave your Board to the exercise of its discretion, believing that your good judgment and experience in life will enable you to make this charity a substantial benefit to the community."[31]

For the past 120 years, trustees have followed Hopkins' instructions with dedication and foresight. What now is Johns Hopkins Medicine has proven to be a "substantial benefit" not just to its community but to the world.

1870
June 13: The first meetings of the trustees for the university and hospital are held under the charters of 1867.

1873
March 10: In a letter to the trustees of The Johns Hopkins University and Hospital, settling forth the principles they are to follow, Hopkins states that the hospital must provide for "the indigent of the sick of this city and its environs, without regard to sex, age, or color, who may require surgical or medical treatment, and who can be received into the Hospital without peril to other inmates." The letter also directs that the hospital should accommodate four hundred patients and that a school of nursing and a school of medicine should be established in conjunction with the hospital.

September: During another national stock market and banking panic, Hopkins once more pledges his fortune to back the B&O and prevents financial ruin for the railroad and Baltimore.

December 24: Johns Hopkins dies at his residence at 81 West Saratoga Street in Baltimore at the age of 78.

Daniel Coit Gilman (1831–1908), first president of The Johns Hopkins University and first director of The Johns Hopkins Hospital, was an innovative, forward-looking and much-admired educator. A graduate of Yale who later became its librarian and then a professor of physical and political geography, Gilman was a favorite of the younger Yale faculty to be named the university's next president. Instead, passed over for that post, he was chosen in 1872 to become the young University of California's chief. Three years later, impressed by Gilman's accomplishments and vision, Hopkins' trustees selected him to be the guiding hand behind the creation of what would be the United States' first research and graduate studies-oriented university.

1875
Following Hopkins' instructions, the trustees learn all they can about higher education, confer with some of the nation's top educational leaders, and choose Daniel Coit Gilman, president of the University of California, to become the first president of The Johns Hopkins University. He begins preparing for the university's official opening.

1876
February 22: Daniel Coit Gilman is inaugurated as the first president of The Johns Hopkins University during ceremonies at the Peabody Institute at which the Peabody Orchestra performs. Gilman is determined that Johns Hopkins will become the first research-oriented graduate university in the U.S., based on German and other European institutions of higher learning.

The Founders

Two months after Hopkins died, the trustees met to adopt bylaws, appoint committees, and decide how to follow their friend's instruction to obtain good advice on appointing someone to head the university and undertake construction of the hospital.[32]

They were thorough. They gathered books and other publications on higher education to give themselves a complete grounding in its precepts and latest practices. They learned that instead of another college like the 400 or so already in the United States, what modern American education leaders almost universally called for was creation of a great *graduate* university, resembling the renowned institutions then in Germany. What the United States lacked, educators were saying, was a place where research would flourish and knowledge was expanded.[33]

Before selecting a president to implement this vision, the trustees conferred with Harvard's President Eliot, Cornell's President Andrew D. White, and Michigan's President James B. Angell. All three—independently—recommended Daniel Coit Gilman as the man for the job. Then president of the University of California, which, its name to the contrary, actually was an alliance of trade schools and colleges, Gilman proved an ideal choice.

Daniel Coit Gilman (1831–1908)
A graduate of Yale, where he had been a classmate and good friend of Cornell's Andrew White, Gilman had spent a postgraduate winter studying at the University of Berlin and absorbed the German graduate education philosophy. Returning to Yale in 1855, he became its librarian and later, as a professor of physical and political geography, was instrumental in establishing the research-oriented Sheffield Scientific School. During his tenure there, Sheffield developed its pioneering biology course for prospective medical students.[34]

Gilman was chosen in 1872 to become the young University of California's chief. In 1874, after receiving the glowing recommendations about him, the Hopkins trustees invited Gilman to visit Baltimore to discuss becoming president of their brand-new university. Gilman, sporting a thick mustache and expansive mutton-chop whiskers, was a man of magnetic personality. He impressed the trustees with his eloquently expressed views on the importance of recruiting faculty with proven abilities as researchers; on ensuring that teaching responsibilities would not interfere with the faculty's research; and on encouraging the faculty to publish the results of their inquiries. He got the job.[35]

By the time Gilman returned to Baltimore in 1875 to assume the Hopkins presidency, he had formalized his views on medical education. He foresaw small classes, rigid examinations and practical work in the hospital's departments. In his inaugural address at the university's formal opening on February 22, 1876, he criticized the lamentable standards of existing American medical schools but concluded with an expression of hope for the future: "We need not fear that the day is distant…which will see endowments for medical science as munificent as those now provided for any brand of learning, in schools as good as those now provided in any other land."[36]

Gilman believed that the "glory of the University should rest upon the character of the teachers and scholars here brought together, and not…upon the buildings constructed for their use." The Hospital trustees, however, knew that they could not offer a superior clinical program without appropriate facilities, so building the hospital became their next important task.[37]

John Shaw Billings (1838–1913)

In 1875, the Hospital board's president and head of its building committee, Francis T. King, invited five consultants to submit plans for buildings to fit the 13-acre site that Hopkins had purchased and given to the trustees. The most elaborate, indeed, exhaustive reply came from John Shaw Billings, an Army major, former Union battlefield surgeon and member of the U.S. Surgeon General's office. An organizational genius who established what would become the National Library of Medicine and the *Index Medicus* (which eventually became today's MedLine), Billings had published an extensively researched report on hospital design and management in 1870, was a nationally known expert in the field, and already met once with the hospital building committee.[38] His 46-page response to King convinced the trustees that they had found their man.[39]

Many previous chroniclers of Johns Hopkins Medicine's history consider Billings' contribution to the creation of the Hospital and new university's medical school to be second to none. He helped select and recruit the medical school's first faculty, advising Gilman and the trustees during the organization of the Hospital and the medical school; and his ideas on the importance of research were instrumental in the successful launching of both institutions.[40]

Billings' official tenure began in July 1876 and would last for the 13 years it took to build the Hospital. In the spirit of Johns Hopkins' instructions, the hospital trustees sent him on a three-month tour of Europe to consult with top medical educators and visit hospitals and medical schools. He was not particularly impressed. He wrote to the trustees: "It cannot be said that the general principles of hospital construction are as yet settled on any scientific basis… nearly as many opinions as persons." He also thought the purported hospital experts in Europe were ivory tower dwellers, "unbewildered by the slightest experience."[41]

One expert whose opinion he did value, however, was Florence Nightingale, the legendary nurse. He mailed her copies of the earliest sketched plans for Hopkins Hospital, as well as for a training school for nurses, a convalescent hospital and an

■ After shepherding the building of Hopkins Hospital from the drawing board to its completion in 1889, the multi-talented, former Union Army battlefield surgeon John Shaw Billings (1839–1913) spent his last years overseeing creation of the New York Public Library, supervising the merger of the Tilden Trust with the Astor and Lenox Libraries to form a single, magnificent municipal resource. He also oversaw the design of the library's main building, still a landmark on New York's Fifth Avenue; developed its classification and cataloging systems; and served as its first director.

A multi-talented, indefatigable former Union Army battlefield surgeon, John Shaw Billings (1838–1913) was an organizational genius recognized as one of the country's foremost experts on hospital design and management when Hopkins' trustees chose him in 1876 to spearhead construction of The Johns Hopkins Hospital—a job that would take 13 years to complete.

Florence Nightingale (1820–1910), dubbed the Lady with the Lamp for her legendary service as a battlefield nurse during Great Britain's Crimean War, was one of the few European medical authorities whose judgment John Shaw Billings valued. In 1876, he sent her preliminary sketches for Hopkins Hospital and its nursing school and asked for advice—which she provided less than two months later in 12 pages of notes. Billings, replying that he did not "think it probable that I should do otherwise than agree" with her recommendations, said that he would show her notes to the Hopkins trustees.

1876
July: John Shaw Billings, a former Union Army battlefield surgeon and renowned expert on hospital design and management who had originally been hired by the trustees in 1875, officially begins work as the designer of The Johns Hopkins Hospital. He also will have a profound impact on creating the School of Medicine, including working with Gilman to select its original faculty.

1878
June: Ground is broken for The Johns Hopkins Hospital. Because of financial constraints due to the weakness of the B&O Railroad stock, on which Hopkins depends for funding, construction of the Hospital will take 13 years.

1884
Pathologist William Henry Welch, recruited by John Shaw Billings, is named a professor on the faculty of the yet-unopened School of Medicine. By 1886, he opens a pathology laboratory and launches the first graduate training program for physicians in the country.

asylum for African American orphans, all of which Johns Hopkins also had wanted built with his bequest. "Knowing as I do the great interest you take in such subjects I shall consider it as a great favour if before my return you will, if your health permits, examine these plans … and let me know what you think of them," Billings wrote to the fabled Lady with the Lamp, then 56 but already largely a bedridden invalid (perhaps due to brucellosis, a chronic bacterial disease contracted during the Crimean War).[42]

Nightingale responded six weeks later with 12 pages of notes on the sketches and other materials Billings had sent her. "Your remarks shall be laid before the Trustees as soon as I return to America and I feel sure that they will be greatly interested in and influenced by your criticisms," Billings wrote. He offered his "sincere personal thanks" for her comments—adding that he did not "think it probable that I should do otherwise than agree with them."[43]

After two years of planning, ground was broken in June 1878 for what would become 17 hospital buildings, situated in a pavilion-like arrangement around an open courtyard. Construction proceeded by fits and starts, however, because only the income from Johns Hopkins' bequest could be used to fund it and the B&O stock on which it largely depended was slumping badly. The trustees could only let construction proceed to a certain point each year, then halt it and calculate how much they could afford the following year. (Similar financial strictures would delay opening of the medical school.)[44]

Johns Hopkins himself originally had no intention of putting his hospital on the site where it now sits. He had wanted it located on his peaceful Clifton estate. State lawmakers, however, already aware of his impending bequest, approached him about purchasing the 13-acre property at the crown of what then was called Loudenslager's Hill. It was home to the State Asylum for the Insane that Hopkins had helped manage for two decades and soon would be replaced by a new Spring Grove asylum in the countryside near Catonsville, Maryland. (The hill also had once been the site of a general hospital, built in 1807 following a devastating epidemic of yellow fever.)[45]

Because Hopkins' handpicked hospital trustees persuaded him that if he wanted to help the poor, it made better sense to build his hospital closer to where they lived—and perhaps because he also decided that placing the buildings on high ground would provide fresher air and safety from floods—Hopkins changed his mind and paid $150,000 for the site. Although much of the old insane asylum had been demolished by late 1873, Hopkins did not live long enough to see so much as a preliminary blueprint for what would go there.[46]

Even before the hospital's lumbering construction project got under way, the workaholic, walrus-mustached and often brusque Billings began scouting prospects for the medical school's first faculty. He was a key figure in securing the appointments of two of the four physicians who would shape the future of the Johns Hopkins School of Medicine and medical education worldwide: pathologist William Henry Welch and William Osler, the premiere internist of his era.

The Big Four

William Henry Welch (1850–1934)

Billings first met Welch in 1876, when the then-26-year-old researcher, a graduate of Yale and New York's College of Physicians and Surgeons, was training in a Leipzig pathology laboratory and Billings was on his European tour of hospitals and medical schools. Welch sensed that Billings was impressed with his understanding of science and the German philosophy about it. He was correct. As soon as Billings got back to his hotel, he told hospital trustee chair Francis King, who had traveled with him: "The young man should be, in my opinion, one of the first men to be secured when the time comes to begin the medical school." When Billings returned to the United States, he repeated that recommendation to Gilman.[48]

With progress on opening the hospital and medical school moving slowly, Gilman's offer to Welch did not come through until 1884, by which time Welch was on the faculty of New York's Bellevue Hospital.[49] Although some histories say Welch was the first member of the Hopkins School of Medicine's faculty to be appointed, Gilman already had named Billings, biologist Henry Newell Martin and chemist Ira Remsen to the university's "department of medicine" when The Johns Hopkins University opened in 1876. None of them, however, joined the medical school's faculty when the school opened in 1893, while Welch did—and also became the school's first dean.[50]

Best known to subsequent generations of physicians trained at Hopkins—indeed anywhere—as the quartet of robe-clad practitioners memorably depicted in John Singer Sargent's now-iconic 1906 painting, "The Four Doctors," the founders of the Johns Hopkins School of Medicine were far too colorful as individuals to be confined to the four corners of that artwork's gilded frame. The quirks of their personalities—as much as their remarkably important research and clinical achievements—became the source of a seemingly endless stream of admiring or amusing anecdotes that keep them as vibrant a presence at Hopkins now as they were when they walked its corridors. Echoing entirely accurate aspects of their characters, a writer once dubbed the first heads of Hopkins' medical departments "pathologist William Henry Welch (at left), a stout bachelor whose favorite pastime was a week of swimming, carnival rides and five-dessert dinners in Atlantic City; surgeon William Stewart Halsted (standing), whose severity with students masked an almost debilitating shyness; internist William Osler (seated center), king of pranks; and gynecologist Howard Kelly (at right), snake collector and evangelical saver of souls." They continue to fascinate historians and inspire physicians decades after their deaths.[47]

William Henry Welch (1850–1934), first professor of pathology and first dean of the Johns Hopkins School of Medicine, was almost single-handedly responsible for getting it up and running in less than a year, once the money for it became available. A popular lecturer and valued mentor nicknamed "Popsy" by his residents and fellows, Welch had extensive influence elsewhere, becoming president or chairman of some 19 major scientific organizations, including the American Medical Association and the American Association for the Advancement of Science. His commendation could launch a career. He had, one colleague said, "the power to transform men's lives almost by the flick of a wrist."

1886
Welch invites surgeon William Stewart Halsted, a friend from their days at New York's Bellevue Hospital, to become a fellow in his pathology laboratory. Halsted spends three years developing the painstaking methods of what will become known as the "Halsted School of Surgery." He is appointed the first surgeon in chief of Hopkins Hospital in 1890 and first professor of surgery in the School of Medicine in 1892, a year before the school actually opens.

1888
Internist William Osler, considered the nation's premier clinical physician and a superb teacher, is named Hopkins' first professor of medicine and physician in chief of Hopkins Hospital, which is nearing completion. At Hopkins, Osler later will create America's first medical residency program, perhaps his most profound contribution to the training of American physicians.

Before setting up his Hopkins laboratory, Welch returned to Europe for more grounding in the scientific method. By 1886, he had opened his pathology laboratory and was training 16 graduate students. All of them were committed research scientists with medical degrees and were attracted by Welch's well-established reputation for serious scientific study, particularly in bacteriology. His was the first graduate training program for physicians in the country.[51]

Hiring Welch set the Hopkins precedent of importing top-flight faculty from around the nation and the world at the risk of antagonizing the then-especially parochial Baltimore medical community. Welch's immediate popularity prevented such resistance. Blue-eyed, bearded, cherubic and charismatic ("a good mixer," one Hopkins colleague observed), Welch was no threat to the local physicians' clinical practices.[52] A warm and brilliant conversationalist who was charming and affable, he swiftly became a much-admired professor who let his students pursue their own interests. This was a deviation from the German practice of establishing a common theme for a laboratory's research but produced impressive results. Although Welch rarely suggested topics of study for his pupils, he was quick to recognize potential in what they were doing and sometimes made recommendations that launched careers.[53]

As a researcher himself, Welch was best known for his 1891 discovery of a micro-organism that was named after him, the *Bacilus welchii* or *Clostridium welchii*, a gas-producing bacterium later found to be connected to the gangrene suffered by wounded soldiers during World War I, as well as perforative peritonitis, urinary tract infections and lung hemorrhages.[54]

In his role as head of his laboratory and a teacher, Welch became mentor to an astonishing roster of future medical luminaries even before the Hospital opened. Among them were Walter Reed, James Carroll and Jesse Lazear, whose discovery that mosquitoes transmit yellow fever is considered one of the greatest achievements of modern medical science and led to the virtual eradication of the disease. Other Welch protégés included Halsted; Franklin P. Mall, who became Hopkins' first professor of anatomy (and worked with Halsted in Welch's lab to develop the distinctive intestinal suture that bears Halsted's name); William T. Councilman, later head of pathology at Harvard; Thomas Cullen, who succeeded Kelly as Hopkins' professor of gynecology; J. Whitridge Williams, an obstetrician who later became dean of the School of Medicine; and Simon Flexner, later head of the Rockefeller Institute of Medical Research (now Rockefeller University) and Welch's biographer.[55]

Welch also played a key role with Gilman and Billings in assembling the original medical school faculty, including Osler as well as Halsted. It was Welch, more than anyone, who created an invigorating atmosphere at Hopkins that focused attention on the revolutionary changes in medicine taking place in Baltimore. After the Hospital opened, it was through Welch's efforts that it quickly was recognized as an institution dedicated to the advancement of scientific medicine, and it became a powerful base for the immediate success of the School of Medicine.[56]

In 1916, Welch also founded the nation's first school of public health at Hopkins (now the Bloomberg School of Public Health), becoming its first dean. He then created and headed Hopkins' Institute of the History of Medicine, housed in the medical library that bears his name. He was instrumental in founding the Rockefeller Institute for Medical Research, serving as its first chairman of the board; and founding editor of the *Journal of Experimental Medicine*, the nation's first journal devoted exclusively to medical research.[57]

A lifelong bachelor, Welch never owned a home and was content with modest rented quarters—living in a St. Paul Street boarding house. When the landlady and her daughter moved from one rowhouse to another, so did he. In reality, he lived most of his life in his clubs.[58]

Despite being a magnetic, avuncular person with an almost parental interest in his students—so much so that they jovially called him "Popsy" (behind his back)— Welch also was strangely aloof and mysterious. A popular piece of student doggerel went: "Nobody knows where Popsy eats, nobody knows where Popsy sleeps, nobody knows whom Popsy keeps, but Popsy."[59]

Welch actually ate most of his meals at the Baltimore's University Club near his St. Paul Street rooms.[60] Since it was only a few blocks from his residence, he often arrived there when the dining room opened for breakfast, then tucked into "showers of eggs, griddle cakes, [and] sausages," recalled Simon Flexner. Portly all his life, he also hosted elaborate dinners either there or in the wood-paneled dining room of the nearby, fancier Maryland Club. Simon Flexner remembered plates laden with "diamond-backed terrapin, wild duck, pearly soft-shelled crabs, and of course the prodigious oysters" of the Chesapeake.[61]

■ An early innovation of the School of Medicine was experimental surgery, introduced by surgeon in chief William Halsted in 1895 and later supervised by future brain surgeon Harvey Cushing. Surgical students learned by practicing on anesthetized animals in what then and still is called the "Hunterian" building, a name that puzzles many. Cushing wanted to name the building after French physiologist François Magendie (1783–1855), but his name was well-known to Baltimore's vocal anti-vivisectionists, who opposed animal experimentation. So the building was named for Scottish physician John Hunter (1728–1793), the father of scientific surgery.

William Osler, Hopkins Hospital's first physician in chief (seated in a chair, right center), and Henry Hurd, the Hospital's longtime superintendent (seated in a chair, left center), pose in 1893 with the first residents and fellows who lived with them in the main, domed building of the Hospital itself. Lewellys Barker (seated second from the left) lived in the room next to Osler and recalled that he "could with relative safety set my watch at 10 p.m. each night when I heard [Osler] place his boots on the floor outside his bedroom door," leaving them to be shined by an orderly. Barker would succeed Osler as physician in chief in 1905. Seated on the far right in a chair is William Sydney Thayer (1864–1932), who also served as physician in chief from 1919 to 1921. Between Thayer and Osler is seated Simon Flexner (1863–1946), who would ensure that Hopkins received significant financial support when he later became head of the Rockefeller Institute for Medical Research.

William Henry Welch (1850–1934), the first dean of the School of Medicine, liked to spend his summer vacations at the beach, particularly near such seaside amusement parks as New York's Coney Island and Atlantic City, N.J., seen here. Of Atlantic City, he wrote in 1901, "Doubtless this is a vulgar place, but I am contented." He especially enjoyed repeatedly riding a roller coaster with a 360-degree loop. He described it as "the most terrifying, miraculous, blood-curdling affair called the 'Flip-flap railroad'."

Welch's conversation—frequently a low-modulated monologue—was fascinating. He awed his students with an apparent grasp of any subject. "A spell fell over the room as the quiet voice talked on, and the young men, some of them already a little round-shouldered from too much peering into the microscope… resolved to go to art galleries, to hear music, to read the masterpieces of literature about which Welch discoursed so excitingly," Flexner wrote.[62]

Something of a practical joker (although not in Osler's league), Welch so enjoyed his reputation for omniscience that sometimes he would bone up on subjects by reading the encyclopedia immediately before dinner, then steer conversations to the topic he just had studied, dazzling all with his knowledge. Even after the subterfuge was discovered, he continued the practice as an amusing game. He could be so persuasive in his discourses that once he actually convinced friends and family members that he had been taught how to fly an airplane by one of the Wright brothers—despite the fact that everyone knew he had never even learned how to drive a car. He also once hoodwinked Halsted by hiding a small rubber bulb under his shirt, connecting it to another in his pocket, and complaining to the unsuspecting Halsted of severe chest pains. When Halsted frantically pressed his ear to Welch's chest, he was dumbfounded to hear a monstrous pseudo-heartbeat that Welch produced by rhythmically squeezing the bulb in his pocket.[63]

By the time of his 80th birthday in 1930, Welch was universally acclaimed "the dean of American medicine." At a large, Washington, D.C. celebration of his milestone birthday, with Welch in attendance, President Herbert Hoover proclaimed to an audience of 1,700 guests: "Dr. Welch is our greatest statesman in the field of public health." Hoover's accolade was broadcast nationwide by all of America's major radio networks and across the sea by short-wave, where a London station retransmitted it over the British Isles and to Europe. Simultaneous ceremonies were held not only in London but New York, Paris, Geneva, Tokyo and Beijing.[64]

By then, with Welch's portrait on the April 14, 1930, cover of *Time* magazine over a caption saying, "… they call him 'Popsy'," everyone knew Welch's nickname, and he didn't seem to mind.[65]

William Stewart Halsted (1852–1922)

Halsted was the next of the Big Four to come to Hopkins, largely at Welch's instigation. Welch had become friendly with him in New York, where Halsted also had been on the staff of Bellevue Hospital. A graduate of Yale and the Columbia University College of Physicians and Surgeons, Halsted had studied abroad, worked simultaneously at five other New York hospitals besides Bellevue, and developed a sterling reputation as a surgeon, diagnostician and advocate of techniques aimed at avoiding infections.[66]

Welch recalled that he had known Halsted "in those days of his highest physical and mental vigor" in New York, but he scrupulously hid what he also knew: Halsted had lost much of those high spirits—though none of his research or clinical acumen—due to his unremitting battle against drug addiction. Using himself as a guinea pig, Halsted had experimented on the anesthetic properties of cocaine, and in 1884 he became the first to describe how an injection of the drug into a major nerve trunk could numb an entire limb or block the spinal cord. Although his experiments advanced the course of early anesthesia, they cost him dearly. He became addicted to cocaine and underwent treatment in a sanitarium for a year.[67]

When Halsted got out of the sanitarium in 1886, his New York medical career stymied, Welch invited him to come to Baltimore to join Hopkins' newly formed pathology laboratory. It was during Halsted's three years of concentrated research in Welch's laboratory that the "Halsted School of Surgery" evolved. It consisted of strict adherence to antiseptic methods, gentle handling of tissues, use of the finest silk suture material, small stitches and low tension on the tissue, and complete closure of wounds whenever possible. These procedures, now considered basic, had a far-reaching effect on the practice of surgery, making it safer and more effective than it ever had been before.[68]

Halsted often is credited with "inventing" the use of rubber gloves during surgery to prevent the spread of infection. Because of his immense renown as a surgeon, the fact that he began using rubber gloves around 1890 gave them an unqualified cachet and undoubtedly popularized their use worldwide, but a number of surgeons elsewhere actually had used gloves prior to Halsted.[69]

Ironically, this shift in Halsted's surgical procedures initially occurred only because the chief nurse in his operating room—Caroline Hampton, whom he later married—complained that the mercuric chloride with which she was supposed to wash gave her a rash. Halsted asked the Goodyear Rubber Co. to use a design of his to make thin rubber gloves to protect her hands. His surgical assistants also began wearing them during operations, saying the gloves made them more dexterous. The idea that the gloves also might help in germ control actually didn't occur to any of them for years, which Halsted admitted with some bemusement long afterward. "Operating in gloves was an evolution rather than an inspiration or happy thought," he said, "and it is remarkable that during the four or five years when as an operator, I wore them only occasionally, we could have been so blind as not to have perceived the necessity for wearing them invariably at the operating table."[70]

Halsted revolutionized surgery by insisting on subtle skill and technique, not swiftness and brute force. Using an experimental approach, he developed new

William Stewart Halsted (1852–1922), an oil portrait based on his last formal photographs, taken shortly before his death in 1922. Recognized as the father of modern surgery, he battled addiction to cocaine, with which he had experimented on himself in 1884 to test its potential as a local anesthetic. Although his self-experiments advanced the course of early anesthesia, they affected his health for the rest of his life.

■ Famed novelist and short story writer F. Scott Fitzgerald (1896–1940) once described the large "Christus Consolator" statue in Hopkins' Billings Building as gesturing "in marble pity over" the Hospital's entrance. He lived in Baltimore in the early 1930s, while his wife Zelda was being treated at Hopkins' Phipps Psychiatric Clinic. Fitzgerald himself suffered from tuberculosis and alcoholism and was admitted to Hopkins' Marburg Pavilion at least nine times. He wrote a 1932 short story entitled "One Interne," set in a Hopkins-like hospital and medical school.

Dr. William S. Halsted with present and former members of his staff at the twenty-fifth anniversary of the opening of Johns Hopkins Hospital, October 1914.

From left to right, standing: Roy D. McClure, Hugh H. Young, Harvey Cushing, James F. Mitchell, Richard H. Follis, Robert T. Miller, Jr., John W. Churchman, George J. Heuer; sitting: John M. T. Finney, William S. Halsted, Joseph C. Bloodgood.

"Soap and water and common sense are the best disinfectants."

—William Osler, Founding Physician in Chief, The Johns Hopkins Hospital; Director of the Department of Medicine, Johns Hopkins School of Medicine, 1889–1905

■ Actress Katharine Hepburn had a fond spot in her heart for the Johns Hopkins School of Medicine. Her father, Thomas Hepburn, graduated from Hopkins in 1905— two years after he had met the future star's mother, Kit Houghton, in Baltimore. Her sister, Edith Houghton, was a classmate of Thomas Hepburn, who became the first urologist in Connecticut. Katharine Hepburn always said her parents were the most remarkable people she ever knew.

operations for intestinal and stomach surgery, gallstone removal, hernia repair, disorders of the thyroid gland and, perhaps best known, the "Halsted radical," the lifesaving mastectomy procedure for breast cancer.[71]

In his own day, however, the "Halsted" meant something else to his surgical students and colleagues. They used the term to mean any drawn-out procedure that seemed to take longer than absolutely necessary. Halsted became so slow and intensely focused during surgery that some of his colleagues were driven to distraction.[72]

Osler believed that Halsted's formerly dazzling personality and early, remarkable speed as a surgeon had been changed forever by his struggle with drug addiction. He became shy and reticent with his students, went out of his way to avoid meeting people in the hospital's hallways, and had few close friends. Although he was appointed the first surgeon in chief of Hopkins Hospital in 1890, he was not named the first professor of surgery in the medical school until 1892, when the trustees finally were persuaded that he had kicked his drug habit. Prior to then, he had been given the title "acting professor."[73]

About six months after Halsted was promoted, however, Osler was surprised one day to find him shivering unnaturally. Halsted then confided to Osler that he still was taking morphine. "Although he had never been able to reduce the amount to less than three grains daily," Osler wrote, "on this he could do his work comfortably and maintain his excellent physical vigor (for he was a very muscular fellow). I do not think that anyone suspected him—not even Welch."[74]

Halsted may have remained addicted for the rest of his life. To Osler, Halsted's "proneness to seclusion, the slight peculiarities, amounting to eccentricities at times (which to his old friends in New York seem more strange than to us) were the only outward traces of the daily battle through which this brave fellow lived for years."[75]

Halsted's "slight peculiarities" became legendary. His obsession with perfection led to his insistence that he personally select the leather from which his European-manufactured shoes were made; he shipped dozens of his dress shirts to Paris or London to be washed, considering Baltimore's laundries unsuitable; his fireplaces at home were fueled only by white oak or hickory wood, specially cut and aged to Halsted's exacting specifications and shipped from North Carolina;[76] and he was as painstakingly meticulous about his meals as he was about his surgeries.[77] He carefully compiled the menus, often went to Lexington Market himself to buy the food (a duty then usually reserved for servants or wives), selected the flowers that graced the table, even insisted that the tablecloth be ironed in place to remove any trace of wrinkles. He personally chose special Turkish coffee beans and oversaw their grinding, producing a powerful, postprandial black coffee that his guests later said would keep them up all night.[78]

His internal, lifelong struggle notwithstanding, Halsted had a profound impact on both the practice and instruction of surgery, training residents to be not only surgeons but teachers of the skill. He succeeded well: Seven of his seventeen resident surgeons became professors of surgery at such schools as Harvard, Yale and Stanford, and another six went on to other positions at medical schools or teaching hospitals. Eleven of Halsted's residents set up residency training programs in their new posts, thereby spreading the system even wider. Halsted selected four of his assistant residents to organize and direct the surgical subdivisions of orthopedics, otolaryngology, radiology, and urology. He left an indelible mark on an entire generation of American surgeons and on the practice of surgery. The 1931 building housing Hopkins' surgical services was named in his honor and Osler's.

William Osler (1849–1919)

Welch would claim that he overcame Billings' initial resistance to the appointment of Osler. A widely respected internist, born and trained in Canada, Osler was named Hopkins' first professor of medicine and physician in chief of the Hospital in 1888, shortly before it opened.

Osler, previously a professor of medicine at Montreal's McGill University and then at the University of Pennsylvania, already was considered America's premier physician and a superb clinical teacher, so Billings eventually relented. Welch was delighted, writing to his sister that Osler "is the best man to be found in the country and it is a great acquisition for us to secure him. I know him well and have the highest opinion of him as a scientist and as a man."[79]

Billings may have been persuaded that Osler was a great choice, but he did not waste much time on pleasantries when extending the job offer. As Osler recalled in 1913, the career- and life-altering "interview" he had with Billings "illustrates the man and his methods." Billings, he wrote, showed up one day at his Philadelphia apartment. "Without sitting down, he asked me abruptly, 'Will you take charge of the Medical Department of the Johns Hopkins Hospital?' Without a moment's hesitation I answered: 'Yes.' 'See Welch about the details; we are to open very soon. I am very busy to-day, good morning,' and he was off, having been in my room not more than a couple of minutes."[80]

Even Osler's protégé and biographer, pioneering neurosurgeon Harvey Cushing, questioned whether his mentor's account of his hiring was entirely accurate. Osler, a short, wiry man with an extravagant handlebar mustache, had an almost uncontrollable urge to embellish his tales and pull practical jokes that often got him in

"It is much more important to know what sort of a patient has a disease than what sort of disease a patient has."

—William Osler

Sir William Osler (1849–1919), perhaps the most influential internist of the late 19th and early 20th centuries. His 1892 textbook, *The Principles and Practice of Medicine*, written in a small study inside The Johns Hopkins Hospital, remained in print for a century. A charismatic, warm person, he gave favorite fellows and residents keys to his home so they could use his private medical library. The group became known as the "latchkeyers." After he moved to Oxford University in 1905, he and his wife, Grace Revere Gross (1854–1928), dubbed their home "The Open Arms," and it always was filled with guests.

"Variability is the law of life, and as no two faces are the same, so no two bodies are alike, and no two individuals react alike and behave alike under the abnormal conditions which we know as disease."

—William Osler

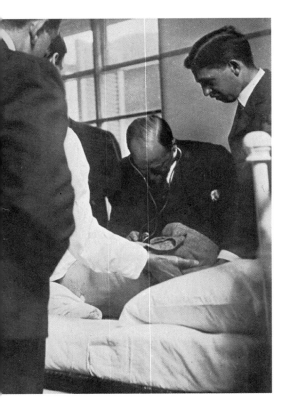

William Osler going on rounds in the Hopkins Hospital, instructing students in the process. He famously observed, "He who studies medicine without books sails an uncharted sea, but he who studies medicine without patients does not go to sea at all."

trouble.[81] Two years before he arrived at Hopkins, he bamboozled a respected medical journal by submitting outlandish research papers under a pseudonym, "Egerton Yorrick Davis," a supposed retired Army surgeon in Canada. When the editor discovered the chicanery, he refused ever to publish Osler's legitimate work—or believe any of it.[82]

Despite his mischievous streak, Osler was revered as a physician for a reason. The breadth of his knowledge was staggering, his warmth and empathy were captivating, and students adored him. Although he had studied overseas, he never formed close ties with the scientific community in Germany and instead represented the best of the English hospital schools as they had been transplanted to Canada. The consummate clinician and teacher, his appointment strengthened the influence of the English tradition of clinical medicine at Hopkins. His landmark 1892 book, *The Principles and Practice of Medicine*, written in the Hopkins Hospital, became the standard internal medicine text for decades. Translated into many languages, it was continually updated by Osler's Hopkins successors for more than a century. (The 23rd edition came out in 1996.)[83] Among Osler's key precepts—presciently appropriate for today's emphasis on "individualized medicine," with its concentration on the genetic and cultural backgrounds of patients—was his oft-quoted dictum: "It is much more important to know what sort of a patient has a disease than what sort of a disease a patient has."[84]

Perhaps Osler's greatest contribution to American medicine was the establishment of the medical residency program. Under his direction, young physicians-in-training actually lived in Hopkins Hospital—hence the now-universal term "residents"—ensuring their constant attention to patients. It was a concept that spread across the country and remains in place today in most training hospitals, although residents usually live in their own homes now. Through Osler's system, young physicians made up much of the Hospital's medical staff. The success of Osler's residency system depended largely on its pyramidal structure, with interns, a few assistant residents and a chief resident.[85]

Osler also instituted another first by making sure his medical students were brought to patients' bedsides early in their training. He insisted that they should be taking patient histories, performing physicals and doing laboratory tests—examining secretions, blood and the like—rather than sitting in a lecture hall, dutifully taking notes. An extraordinary coiner of aphorisms still quoted by physicians today, he famously observed, "He who studies medicine without books sails an uncharted sea, but he who studies medicine without patients does not go to sea at all."[86]

So powerful has Osler's legacy been, and so influential was it on the generations who later went through Hopkins' grueling "Osler Service," which generally is recognized as the finest training program in the medical profession, that throughout the nation and the world, its alumni still honor him every Friday by donning "The Tie." It is a plain blue necktie adorned with shields that bear one of Osler's favorite expressions, *Aequanimitas*, which is Latin for imperturbability. It symbolizes not only his approach to medicine but their membership in a select club.[87]

Howard A. Kelly (1858–1943)

The final member of the "Big Four," Kelly was not invited to join the faculty until after the Hospital opened in 1889. The personal choice of Osler, who had known him at the University of Pennsylvania, the then-baby faced Kelly was only 31 but already renowned for his skill as a gynecological surgeon. Kelly was American-born and trained. A graduate of the University of Pennsylvania's medical school and subsequently a gynecologist at the Kensington Hospital in Philadelphia, he had been recruited for Pennsylvania's faculty by Osler. At Hopkins, Osler backed Kelly for the chair of gynecology and obstetrics just as Welch had backed Halsted for the surgery chair.[88]

Kelly is credited with establishing gynecology as a true specialty. He concentrated mainly on developing new surgical approaches to women's diseases and to understanding the underlying pathology of them. He invented numerous medical devices, including a urinary cystoscope and what is known as the "Kelly clamp," a scissor-like forceps with a locking mechanism and no teeth, used for holding tissues during gynecological surgeries. He also introduced the use of absorbable sutures at Hopkins. When radium was discovered, Kelly was among the first to try it for cancer treatment (some sources say he got a sample directly from Marie Curie), and he founded his own Kelly Clinic in Baltimore, which once was the nation's leading center for radiation therapy.[89]

Kelly also was partly responsible for bringing the German artist Max Broedel, the father of medical illustration, to Hopkins. Broedel originally had come to Baltimore in 1894 to work for Franklin P. Mall, the first anatomy professor, but Mall proved to be too busy to use his services, so "loaned" him to Kelly—who kept him very busy. Broedel (or Brödel, as it written in the German manner) later became head of Hopkins' Department of Art as Applied to Medicine—the first of its kind in the country and progenitor of the finest practitioners in the field ever since. Famed gynecologist Howard W. Jones Jr., a 1935 graduate of the School of Medicine (and 100 years old at the time of this writing) worked in Kelly's private clinic during the mid-1930s and had tea with him occasionally. "I once asked him what he thought his greatest contribution to medicine was and I was surprised at his answer. His answer was that he was responsible for bringing Max Broedel here from Germany to illustrate his operative gynecology and other papers that he wrote," Jones recalls. In the days before high-speed, multicolored and three-dimensional photographic images of internal organs and surgical procedures, Broedel's illustrations of the work performed by Kelly, Cushing and other Hopkins pioneers were pivotal to the advancement of medical education—and Hopkins' renown.[90]

As for being "a snake collector and evangelical saver of souls," Kelly was all of that and more. Recognized as an expert among amateur herpetologists, he kept a substantial collection of snakes in his five-story, eight-bedroom Bolton

Howard A. Kelly (1858–1943) seated in his "sanctuary," the 100,000-volume library and curio-filled personal museum where he spent most of his leisure hours and was happiest. Among its treasures were centuries-old medical texts and Bibles, volumes on astronomy, botany, conchology, etymology, geology, oceanography and theology. He also had a mind-boggling assortment of coins, mineral samples, Mexican pottery, oil lamps, snake skins—even a saber-toothed tiger's skull, several shrunken heads, and a stuffed monkey, perched on a small tree. Kelly would point to it and say jovially, "Oh, yes, there is my family tree!"

■ Howard A. Kelly (1858–1943), one of the founding "Four Doctors" of Hopkins Medicine, was a fervent fundamentalist Christian and often wore a small blue button on his left coat lapel that bore a question mark. Whenever asked the meaning of the question-mark button, Kelly would reply, "That asks what is the most important thing in life?" He then would answer that the most important thing was a person's relationship to Christ—and, if allowed, proceed to talk about his Christian faith.

Daniel Coit Gilman knew that Johns Hopkins—an avowed foe of sectarian bigotry—had wanted his University and Hospital to have no religious affiliation. Nevertheless, Gilman also remembered that when the University was dedicated in 1876 without so much as a benediction, many in Baltimore considered it blasphemous.[94] In his speech at the Hopkins Hospital's dedication, Gilman sought to bridge that gap by linking the results of medical research to the advancement of Christian charity and the relief of individual suffering. He hoped that "some friend of this hospital place beneath this dome a copy of Thorwaldsen's *Christus Consolator* [statue], with outstretched hands of mercy," as a reminder to physicians, nurses and students of "the constant spirit of 'good will to man.'"[95] Gilman knew that this then well-known 1820 work by a Danish sculptor, Bertel Thorwaldsen, located in Copenhagen's cathedral, *Vor Frue Kirke* (Church of Our Lady), was admired by William Wallace Spence, a wealthy Baltimore businessman who had been a close friend of Johns Hopkins—and a generous contributor to both the University and the Hospital. Spence subsequently paid the $5,360 needed for an exact replica of the 10½-foot statue—cut from a single block of Carrara marble. Unveiled in the rotunda in 1896, it has become less a specific religious icon than a symbol of comfort, hope and healing to those of many faiths—its toes polished by the frequent, brief touches of individuals seeking reassurance and consolation (or, in the case of medical students burdened by studies, good luck).[96]

Hill brownstone, where the snakes periodically escaped from their glass tanks and slithered through the house, much to the amusement of his nine children.[91]

A devout fundamentalist Christian who read the Bible in the original Greek and Hebrew, Kelly would call a prayer meeting before every operation. At home, his family's lavish Sunday dinners inevitably were followed by his lengthy sermons. Putting his beliefs into practice, Kelly fought political corruption and organized prostitution in Baltimore, seeking out prostitutes and inviting them to his home in an effort to convert them to a righteous life.[92]

Baltimore's iconoclastic newspaperman H.L. Mencken frequently took issue with Kelly's crusades and fundamentalist preaching, writing in 1921: "Before cock-crow in the morning he has got out of bed, held a song and praise service, read two or three chapters in his Greek Old Testament, sung a couple of hymns, cut off six or eight legs, pulled out a pint of tonsils and eyeballs, relieved a dozen patients of their appendices, filled the gallstone keg in the corner, pronounced the benediction, washed up, filled his pockets with tracts, got into a high-speed automobile … and started off at 50 miles an hour to raid a gambling house and close the red-light district in Emory Grove, Maryland."

Yet Mencken, referring affectionately to Kelly as his "old sparring partner," also wrote, "put a knife in his hand, and he is at once master of the situation, and if surgery can help the patient, the patient will be helped."[93]

The Hospital Opens

On a sunny and mild May 7, 1889, after a dozen years of periodically interrupted construction, The Johns Hopkins Hospital at last opened officially—"magnificent in her beauty and her glory," according to the *Baltimore American*. It was believed to be the nation's largest hospital at that time, with 17 buildings, 330 beds, 25 physicians, 200 employees, and an annual budget of some $85,000.[97]

In the soaring rotunda of the domed main building, which would be named for Billings in 1976, he and Gilman were the chief speakers at the ceremony. They addressed a large crowd of medical, political and society luminaries, packed together on the marble floor of its central area, along the rails of its octagonal galleries, and down the adjoining hallways, radiating from it like the spokes of a wheel.

Billings' speech focused on what he believed the Hospital was poised to accomplish. He stressed the importance of linking it with the medical school, allying clinical care with research and teaching, since the patient would be the ultimate beneficiary. He concluded with a memorable—and prophetic—vision of the Hospital's future:[98]

"Let us hope that before the last sands have run out from beneath the feet of the years of the nineteenth century it will have become a model of its kind, and that upon the centennial of its anniversary it will be a hospital which shall still compare favorably, not only in structure and arrangement, but also in results achieved, with any other institution of like character in existence."[99]

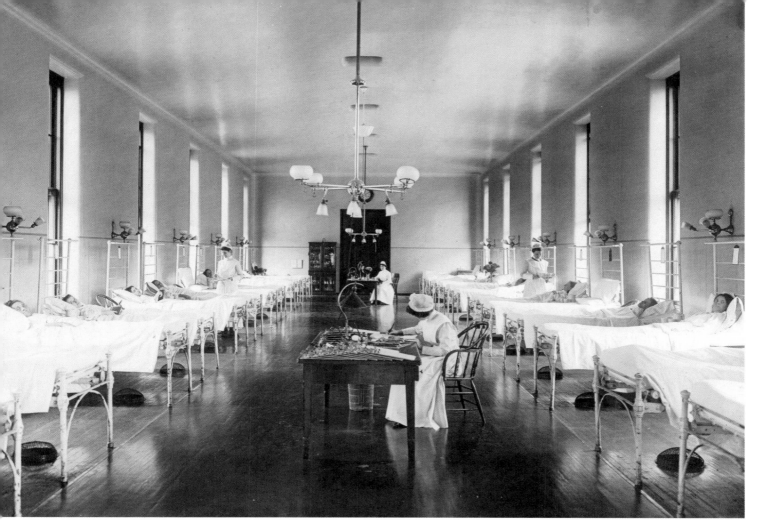

When Hopkins Hospital opened in 1889, the Common Wards contained twelve beds on each side, with individually adjustable, dome-shaped ventilators under every bed to supply fresh air. Large windows provided natural illumination, which was augmented by gas lighting fixtures that were designed especially for the hospital. These were simple and had round glass globes that were easy to clean. Billings wisely had electrical wiring included in the original construction, recognizing that the age of electric lighting was on its way. The wards had no running water, however, and no provision for privacy.

For the first week following its dedication, the Hospital mainly served as a tourist attraction. Elaborate arrangements were made so everyone who wished to do so could visit and examine all its rooms and appurtenances. Seven days after the inauguration ceremonies, the first official patient—suffering an aortic aneurysm—was admitted.

Hopkins Hospital was an immediate success. Within two weeks, Osler sent a card to a friend in Philadelphia: "'Spital booming—very busy." During its first fractional fiscal year of seven and a half months—May 15 to Dec. 31, 1889—nearly 800 patients were under treatment in the several wards available. (Ironically, despite the continuing flood of patients, for its first eight decades, Hopkins Hospital never covered its expenses. Because of its adherence to Mr. Hopkins' direction to treat patients who could not pay, it operated in the red from the day it opened. The deficits were covered by wealthy trustees. Robert Heyssel, head of the Hospital from 1972 to 1992, instituted the changes that finally began to generate a surplus on operations in 1977.)[100]

As impressive as Hopkins Hospital's success was the youth of the early Hopkins leaders—something that Gilman and his successors attempted to maintain. Gilman wanted vibrant young men, untainted by traditions he considered obsolete, heading Hopkins' medical enterprise. Welch had been just 34 when he was appointed; Osler barely 40; Halsted was 37; Kelly, 31, as was the first professor of anatomy, Franklin P. Mall. Others who would be instrumental in Hopkins' swift rise to prominence included pharmacologist John Jacob Abel, only 36 when he was appointed, and physiologist William Henry Howell, another 31-year-old.[101]

"There is nothing really difficult if you only begin—some people contemplate a task until it looms so big, it seems impossible, but I just begin and it gets done somehow. There would be no coral islands if the first bug sat down and began to wonder how the job was to be done."

—John Shaw Billings, designer of the original Johns Hopkins Hospital and key developer of the first curriculum of the Johns Hopkins School of Medicine, 1877–1889

1889

May 7: The Johns Hopkins Hospital is opened officially at ceremonies attended by about six hundred people. Daniel Coit Gilman's wife writes to their daughters: "I think all the speeches struck a very high note. There was no self-glorification, no mere spread eagle and empty oratory but a tone of earnest responsibility in the presence of a great trust. I think everyone must have been struck with it." Billings, in his speech, expresses the hope that Hopkins Hospital will prove to be "a model of its kind" that will come to "compare favorably, not only in structure and arrangement, but also in results achieved, with any other institution of like character in existence."

Gynecologist and surgeon Howard A. Kelly becomes the last of "The Big Four" founding physicians of Hopkins Medicine to join the faculty as head of gynecology and obstetrics. He is credited with establishing gynecology as a true specialty and being among the first physicians to use radiation as a cancer treatment.

September: The failure of Baltimore & Ohio Railroad stock, on which much of the university's finances depends, creates a crisis. Professors accept reduced salaries and fees are increased. Gilman says the 1889–1890 school year will begin as usual—but the opening of the School of Medicine continues to be delayed.

October 9: The Johns Hopkins Hospital's School of Nursing formally opens with Isabel Hampton as the first superintendant of nurses.

■ Sir William Osler (1849–1919), Hopkins Hospital's first physician in chief, was one of the most influential medical educators in history—but he also was a renowned practical joker. For example, as a youth in Canada, he unscrewed all the desks in his one-room schoolhouse and hid them in the attic. Later, as an acclaimed physician, he used a phony name, Egerton Yorrick Davis, to send detailed articles about fictitious patients to medical journals.

Early Leaders

As noted earlier, long before the Hospital and medical school opened, Gilman had begun appointing faculty to the university's "department of medicine." Several were of immense importance to the history of the University and Hopkins Medicine, including the first heads of its preclinical departments.

Henry Newell Martin (1848–1896)

A key and ultimately tragic individual, Martin was the first professor named to the Hopkins medical faculty when the university opened in 1876. A brilliant Irish-born biologist and physiologist—indeed, one of the first researchers ever to receive a degree in physiology at Trinity College in Cambridge, England—he was a protégé of Thomas Huxley (1825–1895), the British biologist known as "Darwin's Bulldog" for his tenacious advocacy of Charles Darwin's theory of evolution. (Huxley gave the keynote address at the opening day ceremonies for The Johns Hopkins University.) [102]

Henry Newell Martin, a brilliant scientist whose life was tragically short.

Although only 28, Martin already was embarked on a distinguished scientific career at Cambridge when Gilman offered him the Hopkins job. His appointment reflected Gilman's desire to choose "young men who would be heard from" and Billings' belief that the biological sciences were of primary importance.[103]

At Hopkins, Martin established the first American biological laboratory and developed a well-organized premedical and graduate training course, the so-called Chemical-Biological Program. As a researcher studying the effects of drugs on the mammalian heart, he devised methods of inquiry that have been acclaimed as among the most important contributions ever made in an American physiology laboratory. His work at Hopkins has been credited with laying the foundation for medical science in this country.[104]

Attesting to the success of Martin's program was the excellence of his students. Two eventually were appointed to the first faculty of the School of Medicine: John Jacob Abel, who would become the School of Medicine's first professor of pharmacology, and William Henry Howell, who would become the first head of the Department of Physiology. That post originally was expected to be filled by Martin himself, but deteriorating health due to alcoholism compelled him to resign from Hopkins in 1893, just before the medical school opened. He returned to England and died there three years later, at the age of 47.[105]

John Jacob Abel (1857–1938) (in white cap), was America's first full-time pharmacologist. Among his landmark achievements was invention of the first "artificial kidney" dialysis device, seen below. He called the machine's process "vividiffusion," meaning the removal of blood from living animals, cleansing it of impurities by diffusion, and returning the cleaned blood to the animal. Although invented between 1912 and 1913 and successfully tested on animals, the device wasn't used on humans with kidney failure until decades later. A tireless worker, Abel ignored accidents and ailments. In 1900, a laboratory explosion sprayed glass shards that destroyed his right eye. Although he never discussed the accident, most photos of Abel after that date were taken from his left side.

John Jacob Abel (1857–1938)

Abel became America's first full-time pharmacologist—and essentially the father of modern pharmacology—when he was named the head of the Department of Pharmacology in 1893. A kind of Cal Ripken Jr. of the faculty, he held the post for an astonishing 39 years—even after losing an eye in a 1900 laboratory explosion, suffering debilitating headaches, and being hit by a car when he was nearly 70, fracturing his leg.[106]

A graduate of the University of Michigan, where he studied physiology with a former student of Martin's, he moved to Baltimore in 1884 for a year of biological study at Hopkins under Martin himself. He then went to Germany for seven more years of study. Returning to Michigan as a professor of materia medica (as pharmacology then was known), he stayed for only a year before joining the Hopkins faculty.[107]

Above all a researcher, Abel believed that chemistry held the clues to solving problems in medical science. The early focus of his work was on isolating and characterizing the hormones of the endocrine system, especially adrenaline and insulin.[108] He succeeded in isolating what he called epinephrine, now also known as adrenalin, a natural substance produced in the adrenal gland that quickens the heart beat, strengthens the force of the heart's contraction, opens up the lung's airways and has numerous other effects. It is a key stimulant used to counteract severe allergic reactions and heart attacks.[109] Among other critical discoveries were his invention of a prototype for an artificial kidney for renal dialysis, which he called "vividiffusion," and his crystallization of insulin, making it safe to use for treating diabetes.[110]

As a teacher of medical students, the thin, aesthetic-looking and goateed Abel replaced lecture-room demonstrations with laboratory sessions that required the students' participation—a practice that many of his colleagues would adopt. As a mentor to a generation of postdoctoral fellows, he was warm, enthusiastic and supportive.[111]

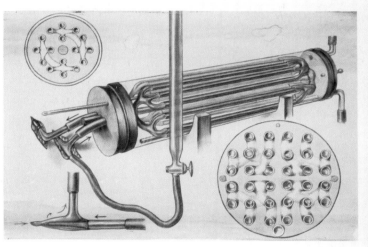

■ Hopkins Hospital's first chief radiologist, Frederick Henry Baetjer (1874–1933), a 1901 graduate of the School of Medicine, was so devoted to exploring radiology's potential that he over-exposed himself to radiation before its dangers were understood fully. It cost him all of his fingers and one eye—but *The New York Times* reported in 1926 that he planned to "continue his work as long as he lives, fingers or no fingers."

Abel also participated in the founding of some of the nation's first scientific journals, including the *Journal of Experimental Medicine* (with Welch), the *Journal of Biological Chemistry* and the *Journal of Pharmacology and Experimental Therapeutics*. His eminence in his field led his colleagues to elect him the first president of the American Society for Pharmacology and Experimental Therapeutics.[112]

William Henry Howell (1860–1945)

Of the first heads of Hopkins' original four preclinical laboratories, Howell was the only one not to have studied abroad when such training was deemed essential. He nonetheless was a prolific and accomplished researcher, an influential mentor, and eventually a dean of both the School of Medicine and what then was the School of Hygiene and Public Health (now the Bloomberg School of Public Health).[113]

Born and educated in Baltimore, Howell intended to become a physician when he entered Martin's new Chemical-Biological Program to prepare to go to medical school. A lack of funds prevented him from pursuing the medical degree after graduation, however, so instead he obtained an undergraduate degree in physiology. He left Baltimore in 1889 to become head of the physiology department at the University of Michigan; in 1892, he joined the faculty of Harvard. He was there only a year before Gilman recruited him to return to Hopkins and its brand-new medical school.[114]

Howell's career at Hopkins spanned 38 years, a quarter century of them in the School of Medicine and 13 more in the School of Public Health. He was best known to two generations of medical students for his *Textbook of Physiology*, which first appeared in 1905 and went through 14 editions. His own research accomplishments were many. He is honored eponymously by the Howell-Jolly bodies, particles in red corpuscles, and he also was among the first researchers to suggest that the two lobes of the pituitary gland are functionally different and to provide evidence for the chemical nature of the nervous influences that control the heart rate. After 1909, his research was devoted almost exclusively to the study of blood coagulation. In 1915, along with second-year medical student Jay McLean, he discovered the critical anticoagulant heparin.[115]

Franklin P. Mall (1862–1917)

Franklin P. Mall, the first professor of anatomy, gave the field a sound scientific basis and had a profound influence on medical education nationwide, training fellows who became the leaders of anatomy departments across the country.[116]

Mall left lecturing duties to his subordinates and focused his own course on independent laboratory work. He scoffed at pedantic instruction that tested only a student's memory, not intelligence. Each entering student was assigned to a

William Henry Howell (1860–1945), a Ph.D. in physiology, was that rare medical school dean who wasn't a physician. His protégé Joseph Erlanger, an 1899 graduate of the School of Medicine, won the 1944 Nobel Prize in physiology or medicine.

■ The first use of an X-ray for neurological purposes occurred at Hopkins in September 1896. The patient was a female shooting victim who had a bullet lodged in her spine. To find its location, two residents used the Hospital's newly arrived, hand-cranked X-ray machine. Samuel Crowe, future chief of otolaryngology–head and neck surgery, had to crank the handle for 45 minutes, while future pioneering brain surgeon Harvey Cushing exposed the X-ray plate.

part of a cadaver and told to go to work. Mall expected them to thrash about until they discovered for themselves the principles of anatomy. (This prompted the tale that when Mall's young wife sought advice on how to bathe their infant daughter, he purportedly replied, "Put her in a tub and let her work it out for herself.")[117]

The results of Mall's emphasis on this inductive method of education varied. Some students foundered, while others flourished. Many would later admit that difficult as it had been, they owed to Mall's method their intellectual awakening and the first stirrings of their interest in becoming independent scientists. Mall also was eager for Hopkins to adopt a liberal elective system so that students could follow their own interests.[118]

Mall's own areas of research spanned three areas: embryology, the structural adaptation of adult organs to their function, and anthropology. He is best known for work he did on the heart, liver and spleen, and he brought to its fullest development the idea that for each organ there is a structural unit that is a unit of function. Like his other early Hopkins colleagues, he helped to found several scholarly journals in his field and led scientific societies. He also established a research institute for embryology.[119]

Henry M. Hurd (1843–1927)

Henry M. Hurd became the first director of the Hopkins Hospital because John Shaw Billings turned down the job—the first instance, perhaps, of a number of fortuitous rejections in Hopkins' history. Billings did agree to be a "consulting director," but in 1895, he became director of the Department of Hygiene at the University of Pennsylvania and director of its University Hospital in Philadephia. Soon thereafter, he became the first director of the New York Public Library, a merger of the collections of the Astor, Lenox and Tilden Foundation libraries. He oversaw construction of its landmark building at Fifth Avenue and 42nd Street, where the entrance is flanked by two haughty but noble stone lions.[120]

In the winter of 1888–1889, just prior to the Hospital's opening, the trustees turned to Gilman himself to take over its leadership as the first superintendent. (The title for head of the Hospital later would change to "director" and, in 1963, became "president.") Gilman agreed and began to divide his day between the University and the Hospital, plunging into an exhaustive study of hospital organization.[121]

As soon as he could find an administrator equal to the task, however, Gilman was happy to relinquish the Hospital's leadership. In August 1889, with the Hospital humming along, he appointed Hurd, a psychiatrist and the first head of the Eastern Michigan Asylum for the Insane, as superintendent. Hurd also was named the Hospital's first psychiatry chief.[122]

A small, thin, birdlike man with wispy mutton-chop whiskers, Hurd was a strong organizer, a prolific writer and avid historian. Over the next 22 years, he would oversee massive construction campaigns. Among other important structures

Franklin P. Mall (1862–1917), the first professor of anatomy, turned his lecturing duties over to subordinates and scoffed at teaching that tested only a student's memory, not intelligence. Each student entering his class was assigned a part of a cadaver's body and expected to figure out the principles of anatomy independently.

■ In a 1937 *Baltimore Sun* series on The Johns Hopkins Hospital, H.L. Mencken (1880–1856) reported that the Hospital purchased a half-million quarts of raw milk annually from a private dairy herd in Maryland's Worthington Valley. It was pasteurized by the Hospital itself for the use of patients, physicians and staff. The Hospital even made its own ice cream, of which 145 quarts were consumed each day. Mencken also wrote that Hopkins used 75,000 quarts of cream and purchased "enough eggs to keep 2,000 hens busy the year round."

Henry M. Hurd (1843–1927), superintendent (a position now called president) of Hopkins Hospital from 1889 until 1911, as well as a professor of psychiatry. Not only a fine physician and administrator, Hurd also was an exceptional writer and editor. His founding editorship of *The Johns Hopkins Hospital Bulletin* and the *Johns Hopkins Reports*—as well as his unrelenting insistence that Hopkins' top physicians write about their work for these journals—proved essential to creating and advancing the reputation and influence of both the Hospital and its practitioners and researchers.

begun under his supervision were the Phipps Psychiatric Clinic and the Harriet Lane Home for Invalid Children, both the first of their kind associated with a teaching hospital and subsequent models in patient care and research. As a psychiatrist, Hurd enthusiastically embraced the many new ideas then developing in the field, including the abolition of physical restraints on patients, employment of the mentally ill, and the introduction of homelike comforts into institutional life. [123]

In addition to his great skill as an administrator and psychiatrist, Hurd brought writing and editing abilities of a high order to Hopkins. It was his founding editorship of *The Johns Hopkins Hospital Bulletin* and the *Johns Hopkins Reports* that proved crucial to the early establishment of Hopkins' world-wide reputation.

Gynecologist Thomas Cullen (1868–1953), who began his 62-year Hopkins career in 1891 as a pathology resident under Welch, later wrote that he "always felt that Hurd should have been included in Sargent's portrait of the 'big four.' It should have been five. In his life Dr. Hurd did as much for Johns Hopkins as any of the four and it might easily be shown that he did as much for them as he did for Hopkins. Welch and Osler and Halsted and Kelly would never have had the reputations they made, except for Hurd."[124]

"The four others recognized in theory that they should record what they were doing and finding out but it was Hurd who made them do it. He kept after them all the time to write up their experimental work and their interesting cases, their clinical observations and laboratory findings, for the *Bulletin* first and then for the *Reports*; never let them rest until they had done it.

"And that more than anything else, as they all recognized, was what made Hopkins' name, and theirs. Especially in Europe. *Hopkins Bulletins* were known in the European clinics before the hospital had been going three years. So, thanks to Hurd, were the names of the big four. ..."[125]

Hopkins Hospital's venerable 300-seat amphitheater, dedicated in 1932, was named the Henry M. Hurd Memorial Hall and is located, appropriately, between the clinical towers named after Osler and Halsted.[126]

The School of Medicine Opens

The financial constraints that had delayed the Hospital's opening continued to bedevil plans for launching the School of Medicine. It was unclear when—or whether—it would open at all.

Mary Elizabeth Garrett (1854–1915) & Friends

Into the monetary breach stepped a group of women, four of whom were trustees' daughters: Mary Elizabeth Garrett, Mary (Mamie) Gwinn, Elizabeth King and Martha Carey Thomas. Led by Thomas, dean and later president of Bryn Mawr College, they formed a Women's Fund Committee in 1890 to raise the $100,000 that Gilman said was needed to open the medical school. They offered to provide the money on the condition that women would be accepted as students in the medical school "on the same terms as men." An astonishingly wide range of prominent women contributed to their fundraising drive, including First Lady Caroline Harrison (wife of President Benjamin Harrison), Elizabeth Agassiz, wife of the president of Radcliffe College, Jane Stanford, wife of the founder of Stanford University, and Julia Ward Howe, erstwhile abolitionist and author of "The Battle Hymn of the Republic."[127]

Even before the women's committee completed its fundraising—with Mary Elizabeth Garrett providing $47,787.50 of her personal fortune to the $100,000 total—the trustees upped the ante, saying that $500,000 now was needed to open the medical school. Mary Elizabeth Garrett again unclipped her considerable purse and offered in April 1891 to add another $100,000 to the women's gift.[128]

After another year passed, Garrett grew frustrated with the slow pace of fundraising. She decided to contribute the balance herself—the precise sum of $306,977. Accompanying her munificent December 1892 offer was a rigorous, unprecedented set of academic conditions for admission to the medical school she would underwrite. Her terms, influenced in part by her close friend Thomas, set a profound standard for medical education in this country.[129]

Garrett insisted that entering medical students have a bachelor's degree from a first-class college and a reading knowledge of French and German (languages in which many key medical texts of the time were written). She also decreed that the medical school be an integral part of the university, have a four-year course leading to a doctorate in medicine, and that the students be required to pass examinations in preliminary medical courses as well as their medical school studies before receiving their degrees. Under Garrett's terms, the Hopkins medical school would become the first in the United States with such stringent admission requirements—prompting Osler to say, "Welch, we are lucky to get in as professors, for I am sure that neither you nor I could ever get in as students."[130]

When the Johns Hopkins School of Medicine opened in 1893, it was the first major medical school in the United States to admit women on an equal basis with men, due to the insistence of a group of women benefactors led by **Mary Elizabeth Garrett**, daughter of founding Hopkins trustee **John Work Garrett**. Within the group, three others also were daughters of founding trustees. Garrett herself contributed the bulk of the funds—$455,000.

Bessie King (upper right), Julia Rogers (upper left), Mary Garrett (lower right), M. Carey Thomas (lower left), and Mamie Gwinn (seated).

1890

May 2: Five Baltimore women—four of them daughters of university trustees—organize the Women's Fund Committee. Mary Elizabeth Garrett, M. Carey Thomas, Mary Gwinn, Elizabeth King, and Julia Rogers intend to raise the money needed to establish the School of Medicine, provided that the school will accept women with the same high qualifications required of male applicants. Women in Baltimore, Boston, Philadelphia, Washington, New York, Chicago, San Francisco, and other cities contribute to the fund.

1892

William Osler publishes *The Principles and Practice of Medicine*, written in a study at Hopkins Hospital. It becomes the standard internal medicine text for decades and is updated regularly for the next century.

March 15: Dr. Henry Hurd, superintendent of Hopkins Hospital, recommends that a separate ward for African American patients be built, ensuring continuation of the Hospital's unwritten segregation policy, which reflects contemporary Baltimore practices, but also improved the facilities for black patients—which many other city hospitals will not accept. The *Baltimore American* reports that the proposed building will accommodate 56 patients. "Special care will be taken to see that the heating and ventilation apparatus is as perfect as possible. A sun balcony will be erected on each floor... for convalescents.... The building will be fireproof throughout."

December 22: Mary Elizabeth Garrett announces that she will give more than $300,000 so that the Women's Fund Committee can meet its goal to open the medical school. Garrett stipulates that the school must be a four-year graduate school for men and women, who are to be admitted without distinction between them. University trustees accept Garrett's gift and her conditions on December 24. In an editorial, *The Sun* says, "The women of the country are to be congratulated upon finding so generous a champion as Miss Garrett has proved to be, and the university has no less cause to be congratulated. All will wish for the success of this important step toward the education of women in a profession where equal opportunities with men have been so long denied."

William Henry Welch sits proudly surrounded by the first graduates of the Johns Hopkins School of Medicine, the Class of 1897. Well, not all of them. Absent from this photo for unrecorded reasons was Mary S. Packard, one of the three women admitted in 1893 to the first class. She was the only one of the women to graduate and then became one of the first Hopkins Hospital interns. Her classmates did, however, have a dog—presumably the class mascot—in the picture.

Indeed, as tough as Garrett's terms were, they actually reflected concepts originally proposed privately to the trustees years earlier by Gilman and Welch, who to some extent had just been hoist by their own petard. Thirty years later, Welch wrote that Garrett and Thomas "naturally supposed that this was exactly what we wanted," adding ruefully, "It is one thing to build an educational castle in the air at your library table, and another to face its actual appearance under the existing circumstances."[131]

The foundations of Hopkins' medical educational castle now had been built, however, thanks to Mary Elizabeth Garrett and her friends. Because the faculty initially feared that the admission requirements might scare away students and none would show up, no formal opening ceremonies were held for the School of Medicine in October 1893.[132]

1893–1903

The founding faculty need not have worried: The Johns Hopkins School of Medicine was a success from the start. On the first day, 16 students enrolled. "Considered a good beginning," Welch wrote, given the school's "extremely lofty and severe requirements for admission...." A few days later, two more enrolled, bringing the first class to 18 students, three of whom were women.[133]

Of the first three women students—Mabel S. Glover, Cornelia O. Church, and Mary S. Packard—only Packard graduated with her classmates in 1897 and then became one of the first group of interns at Hopkins Hospital. Glover dropped out after a year to marry Franklin Mall; Church dropped out two years later, having become a Christian Scientist, a conversion she attributed to Osler, whose methods of treatment, she said, reminded her of Mary Baker Eddy's precepts. Osler would

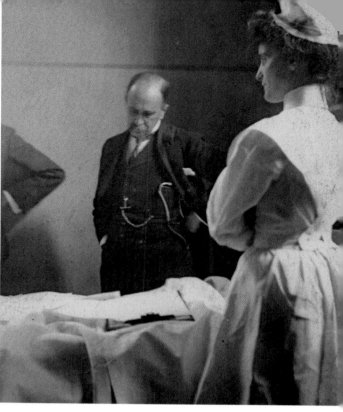

William Osler, stethoscope in vest pocket, ponders a case at a Hopkins patient's bedside. Renowned not only for his exceptional diagnostic skill but his remarkable ability to coin memorable aphorisms, he observed that in difficult cases, the physician's best option was: "Look wise, say nothing, and grunt. Speech was given to conceal thought."

find this and the attrition among the first three women students amusing. Something of a chauvinist, he would quip that the success of admitting women to Hopkins was proven by the fact that thirty-three and a third percent of the first female enrollees "were engaged to their professors at the end of the first year."[134]

The jest perhaps also reflected the relief he and the rest of the medical school faculty felt at its instant popularity. Its second-year enrollment was 40 (eight of whom were women), and soon classes were growing almost too fast for the limited teaching space.[135]

The essential element in the success of the hospital and medical school was the idea that the hospital should form part of the medical school and thus part of the university. As an instrument of medical education, the hospital was linked to the medical school in a unique way: a single individual headed both the hospital department and the corresponding department in the medical school. Instead of the responsibility for control of hospital beds being dispersed among a host of local practitioners, each hospital—and medical school department—had a single chief.[136]

Following Johns Hopkins' instruction to attract a staff of "the highest character and skill," the trustees turned their backs on the provincialism that then marked most hospital and medical school appointments. They searched worldwide for the most qualified individuals. These teachers had a quality new to American medicine: their proven ability as researchers. Most had studied in Germany and brought back one of the most important influences of German medicine upon America: the university ideal. Hopkins' medical school created the conditions that allowed such an ideal to flourish.[137]

The Hopkins faculty was capable not only of teaching what was known but also developing new knowledge and training others to do so as well. Well-equipped laboratories, rare in other medical schools, were provided in the preclinical sciences. The journals created at Hopkins for disseminating the findings of the school's researchers also were among the first such publications in the nation.[138]

Another unusual feature of the new hospital and medical school was the preclinical faculty's freedom to earn a living by concentrating on their research without having to practice medicine. The medical school and university paid them salaries, placing them on the so-called full-time basis. Clinical professors also received a salary but were permitted to have some private patients—a policy that ultimately would be changed following heated debate.[139]

Students were not seen primarily as a source of revenue, as at proprietary medical schools, but as the recipients of existing knowledge and as partners with the faculty in developing new knowledge. The hospital and medical school also included two advanced groups previously unknown in American medicine: the residents in clinical medicine and the research fellows in the preclinical medical sciences.[140]

Like the pieces of a jigsaw puzzle, all of these novel features fitted together to form a model for American medical education and medical care, making The Johns Hopkins Hospital and the university's School of Medicine the most advanced institutions of the day.[141]

■ Sir William Osler (1849–1919), founding physician in chief of The Johns Hopkins Hospital and first director of the Department of Medicine, was born in Canada and knighted by Great Britain's King George V, but had a close connection to an American Founding Father. In 1892, Osler married Boston-born widow Grace Revere Gross (1854–1928). She was the great-granddaughter of Paul Revere, the Boston silversmith best remembered for his famous April 1775 ride to warn patriots of the approach of British troops. Osler's son, Edward Revere Osler (1895–1917), born in Baltimore but raised in Oxford, England, joined the Royal Artillery during World War I and died at 21 of battlefield wounds despite the heroic efforts of American surgeons who were at nearby American Army bases, knew Osler, and rushed to his son's aid. Among them was Hopkins-trained Harvey Cushing (1869–1939), a protégé of Osler who had known Revere since he was a child.

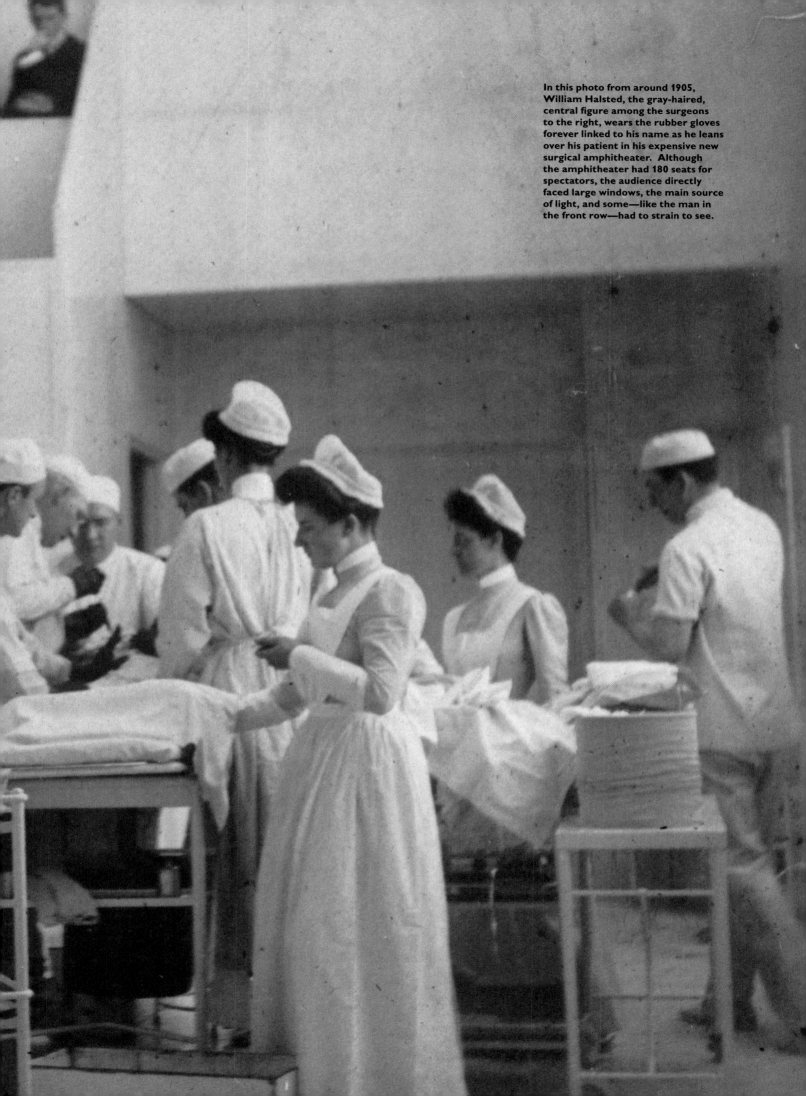

In this photo from around 1905, William Halsted, the gray-haired, central figure among the surgeons to the right, wears the rubber gloves forever linked to his name as he leans over his patient in his expensive new surgical amphitheater. Although the amphitheater had 180 seats for spectators, the audience directly faced large windows, the main source of light, and some—like the man in the front row—had to strain to see.

Three of the Founding Four left to right, William Halsted, William Osler and Howard Kelly pose on the lawn in front of the Hopkins Hospital around 1897.
    Halsted, always meticulous about his dress, sports an impressive top hat. Kelly, just 31 when he was appointed gynecologist in chief the year before, was so youthful-looking that some patients refused to believe he was a physician.

■ Hopkins' pioneering urologist Hugh D. Young (1870–1945) was the moving force behind the effort to have a statue placed in the U.S. Capitol's Statuary Hall honoring Crawford W. Long (1815–1878), the long-forgotten Georgia country doctor now acknowledged to have been the first physician to use ether during surgery—in 1842. Others later claimed that distinction, but an 1897 research paper by Young substantiated Long's claim. Young officiated at the 1926 ceremony in the Capitol when Georgia unveiled a marble statue of Long.

"Learning By Doing"

Gilman, Billings and Osler all were enthusiastic proponents of teaching students how to learn, rather than force-feeding information to them. "An important part of the higher education of modern times is the teaching how to increase knowledge," Billings said, "and the best way of teaching this, as of many other things, is by doing it, and by causing the pupils to do it." In both the basic and clinical sciences, "doing" required the use of all the senses; to "learn by doing," medical students left the lecture halls for the laboratories and wards.[142]

In advocating such a method of instruction, Gilman, Billings and Osler were in the vanguard of a slowly evolving revolution in American higher education. Although bedside teaching for medical students already was being discussed and probably implemented by bits and pieces in other institutions prior to the opening of Hopkins' medical school, Hopkins was the first to have control of a teaching hospital, which enabled it to make the principle of learning by doing a matter of policy.

The First Curriculum

To begin their medical education, Hopkins' first students were immersed in rigorous, if informal, preclinical courses taught by the professors of anatomy, physiology, pathology and physiological chemistry. The students found themselves listening to researchers who were conducting pathbreaking research; and in the laboratories, they enjoyed a close professional relationship with these pioneering scientists. For example, the course in pathology taught to second-year students by Welch included actual work with bacteria in the laboratory, as well as observation of postmortem examinations in the Hospital. Specimens and microscopic sections were studied in the laboratory.[143]

Although other schools relied on lectures as the primary teaching method in the final two years of clinical instruction, at Hopkins, the clinical courses continued to emphasize student participation. The highlight of the third year probably was Osler's medical clinic. Fourth-year students were divided into four groups, each spending two months in medicine, surgery, obstetrics, and gynecology—with students in Osler's medicine department becoming heavily involved in the Hospital's daily routine. In what Osler called his "natural method of teaching," the student "began with the patient, continued with the patient, ended his studies with the patient, using books and lectures as tools, as means to an end."[144]

Over Hopkins' first three decades, the curriculum gradually changed. As medical specialties developed, the curriculum was stretched to accommodate them. Formal revisions of the curriculum occurred in 1927 and again in 1936, but "learning by doing" stayed a cornerstone of Hopkins education, as it does to this day. Advances in medical knowledge, technology and teaching philosophy led to subsequent curriculum changes in 1975, 1992 and 2010, yet that basic principle remains.

Early Faculty Stars

Halsted had a remarkable knack for deciding what sort of specialty would suit early residents. In a somewhat imperious way, he determined the careers of a number of future medical superstars. In three particular instances, he essentially planted the seeds for Hopkins' role as the birthplace of both urology and neurosurgery.

Hugh Hampton Young (1870–1945)

When Hugh Hampton Young, a graduate of the University of Virginia, arrived at Hopkins in 1896 as an assistant resident surgeon, his interest was general surgery. Halsted, however, identified him as the man to develop the specialty of urology. In his 1940 memoir, Young recounted the 1897 conversation that changed his career. Halsted abruptly asked him to take charge of genitourinary surgery. "I thanked him and said: 'This is a great surprise. I know nothing about genitourinary surgery.' Whereupon Dr. Halsted replied, 'Welch and I said you didn't know anything about it, but we believe you could learn.'"[145]

Learn Young did. He performed the first perineal prostatectomy in 1902 and invented the radical perineal prostatectomy for removing a cancerous prostate gland, performing the first such operation at Hopkins on April 7, 1904, with Halsted actually serving as his assistant. He invented numerous operating instruments such as the Young punch, used to excise the prostate gland, and devised several surgical procedures for treating genitourinary diseases. In 1915, as will be described later, he founded the Brady Urological Institute, which still is the top urology center in the nation. During World War I, he was named director of the Division of Urology for the American Expeditionary Force sent to France and wrote a 300-page book, the *Manual of Military Urology,* which was the AEF's first text on any medical subject. In 1926, he published *Young's Practice of Urology,* a two-volume treatise on the diagnosis and treatment of many urologic conditions based on the experiences gained in the care of 12,500 patients at the Brady Institute. Widely sold and immensely influential, it was among the first texts to provide a comprehensive view of urology. As director of the institute until his retirement in 1942, he had a profound influence on the practice of urology for decades, personally training dozens of residents who would become leaders in the field.[146]

Hugh Hampton Young (1870–1945) was a 27-year-old resident in surgery at Hopkins when his boss, surgeon in chief William Halsted, pre-emptorily told him that he'd been chosen to specialize in urologic surgery—a field about which he then knew nothing. Promptly named head of genitourinary surgery, he learned quickly. His invention of pioneering methods and surgical implements for treating enlarged, inflamed prostates and removing cancerous ones earned him international renown. A New York surgeon quipped: "The prostate makes most men old, but it made Hugh Young."

Harvey Cushing (1869–1939), above, at his Hopkins laboratory desk, around 1907. Cushing spent his 16 years at Hopkins establishing the foundation for neurosurgery. Pioneering neurosurgeon Walter Dandy (1886–1946), at right, a 1910 School of Medicine graduate, was an equally brilliant medical visionary who spent his entire career at Hopkins.

■ Hopkins-trained physicians played a central role in launching the era of antibiotics in American medicine. Alexander Ashley Weech (1895–1977), a 1921 graduate of the School of Medicine, and Francis Howell Wright (1908–1992), a 1933 graduate, were the first to use the German-discovered antibiotic sulfachrysoidine (Prontosil) in the United States. They administered it in 1935 to a 10-year-old girl critically ill with leptomeningitis at New York's Babies Hospital of the Columbia Medical Center. The condition of the child improved briefly but she died. Her autopsy, crucial to determining what caused her death and how future such cases might be handled, was performed by Dorothy Hansine Andersen (1901–1963), also a Hopkins-trained physician, School of Medicine Class of 1926. Perrin H. Long (1899–1965), head of Hopkins' then-Department of Preventive Medicine, became so renowned for his pioneering use of the so-called sulfa drugs in the mid-1930s that he received a personal request for help from Eleanor Roosevelt, whose son, Franklin, Jr., was seriously ill. He provided the medication that cured the president's son.

Harvey Cushing (1869–1939)

Halsted also chose Harvey Cushing to develop the specialty of neurosurgery. Cushing, a Harvard medical school graduate, arrived at Hopkins in 1896 as a resident. His next 16 years at Hopkins laid the foundation for a distinguished career in which he became one of the most influential physicians in the history of medicine.

When Cushing came to Hopkins, he brought with him a hand-cranked X-ray machine, a device that had been invented only the year before. He introduced the use of X-rays in preparation for surgery, as well as the practice of monitoring blood pressure during operations. Although Cushing specialized in neurosurgery, his wide-ranging interests helped him—and Hopkins—to gain an international reputation. He discovered the hormonal function of the pituitary gland, founded the medical specialty of endocrinology, and, as will be described later, established the concept of the clinician-scientist by becoming director of the nation's first experimental surgery laboratory, the Hunterian, named for the famous English surgeon John Hunter (1728–1793). Cushing wrote the first definitive text on neurosurgery and, more than anyone else, developed it as a specialty.[147]

Cushing remained at Hopkins until 1912, when he returned to Harvard as a professor of surgery at the newly opened Peter Bent Brigham Hospital in Boston. His career exemplifies one of Hopkins' most important accomplishments: the training of clinicians and investigators who go on to assume key posts at other university medical centers. Cushing's wide-ranging interests and abilities included exceptional talent as a medical illustrator and as a biographer. His two-volume *The Life of Sir William Osler* won the 1926 Pulitzer Prize for biography.[148]

Walter Dandy (1886–1946)

Halsted's sharp eye for exceptional talent also was drawn to Walter Dandy, a 1910 graduate of the School of Medicine who would become one of Hopkins' most famous medical visionaries.

Serving as a surgical resident under Halsted, Dandy met Cushing in the Hunterian Laboratory and became his assistant there. They did not get along. When Cushing left Hopkins to go back to Harvard in 1912, he pointedly advised Dandy that he would not be accompanying the other Cushing trainees headed for Boston. Dandy—whose subsequent professional conflicts with Cushing became part of neurosurgical lore—completed his residency under Halsted, who ultimately made him head of neurosurgery.

Dandy would spend his entire professional career at Hopkins, serving as head of neurosurgery until his death in 1946. Considered by many to have been as great a pioneer in the field as Cushing, Dandy did work in both basic and clinical neurosurgery and neurology research, as well as critical care neuroradiology. A surgical genius—albeit a volatile, emotional person in the operating room—he devised procedures still in use today. He developed new methods for operating on brain tumors, cranial nerve lesions and injuries, Ménière disease (an inner-ear disorder causing dizziness) and vascular lesions in the brain, as well as surgery for spinal cord tumors and ruptured intervertebral discs. The magnitude of his contributions to neurosurgery is recognized every day at Hopkins—and around the world.[149]

Early Dropout

Not every early medical student or resident at Hopkins was a resounding success. Perhaps the most celebrated School of Medicine dropout of this period (or any other) was avant-garde writer, poet, art patron, librettist, literary mentor and self-proclaimed genius Gertrude Stein (1874–1946).[150]

Stein entered the medical school in 1897 as one of 12 women in the 55-member Class of 1901. Although she successfully completed her first two years of basic science, she found the two subsequent clinical years "boring," and by her final year she was failing four of her nine classes—laryngology and otology, ophthalmology, dermatology, and obstetrics.[151]

By then, Stein was experiencing enormous personal turmoil. Recognizing and accepting her lesbianism and falling in love for the first time, she became embroiled in a difficult triangle involving one of her classmates, Mabel Haynes, and a Bryn Mawr College graduate, May Bookstaver. It ended unhappily during her fourth year of medical school, which also was when her brother Leo decided to abandon his own graduate studies at Hopkins and leave Baltimore.[152]

On the surface, Stein seemed committed to obtaining her degree, but her general performance was deemed unsuitable by some key faculty members. One former Hopkins instructor who attempted to salvage her medical education was Lewellys F. Barker, a pathologist at Hopkins from 1892 to 1899 and by 1902 an anatomy professor at the University of Chicago. That year, he tried to get the *The American Journal of Anatomy* to publish a paper Stein had written. When the journal's secretary raised questions about the paper, Barker (who in a few years would succeed Osler as Hopkins' head of medicine) urged Stein to revise it. She flatly refused in a hastily typed letter, replete with a flurry of word-separating slashes and x-ed out typos:

> "My xx dear Dr Barker,
>
> "I have just received your note and hasten to answerxxx it. I will not be able to do any further work/on the paper that I have xsent/to you as I am going/abroad for an indefinite period.Furthermore I xx feel that I have done all that x I can with it so I xxxxwillingly leave it in your hands to do with as you like....
>
> "In biddingyou good by I want to thankyou for yourmanykindnessesand to hope that my workwill be of some service toyou you.
>
> "Sincerely yours,
>
> Gertrude Stein" [153]

Stein's "indefinite period" abroad lasted for the rest of her life. She became famous for her unconventional writing and the celebrated literary and artistic salon that she and her eventual lifelong companion, Alice B. Toklas, maintained in Paris as a meeting place for everyone from F. Scott Fitzgerald and Ernest Hemingway to Pablo Picasso and Henri Matisse.

Avant-garde poet, writer, art patron, literary mentor and self-proclaimed genius Gertrude Stein (1874–1946) was one of 12 women who entered the Hopkins medical school in 1897. She grew bored with medicine, however, and was failing four of her classes in her final year when she dropped out to go to Europe "indefinitely"—a period that would last the rest of her life. Her Paris apartment became a meeting place for everyone from Ernest Hemingway to Henri Matisse.

■ After changing the course of medical education as the first physician in chief of the Hopkins Hospital, William Osler (1849–1919) became Regius Professor of Medicine at Oxford University in England. He died there—but saw to it that part of him would always be in the U.S. As he lay dying from pneumonia, Osler directed that his brain be sent to the Wistar Institute in Philadelphia, where anatomical studies were made of the brains of prominent individuals to see what might have distinguished them from others. An examination in 1927 didn't reveal much—but his brain's still there.

1893
William Welch, who is named the first dean of the School of Medicine, reviews the credentials of applicants. Eighteen students are accepted—three of them women. (The following year, the class contains 40 students, eight of them women.)

1896
October 14: The 10½ -foot, Carrara marble statue, *Christus Consolator*, the gift of William Wallace Spence of Baltimore, is unveiled in the rotunda of the Hospital.

1897
June 15: The first class graduates from The Johns Hopkins University School of Medicine. The 15 graduates include Mary Packard, one of the three women who had entered with the class. (Of the other two, one dropped out of school to marry Franklin Mall, the first professor of anatomy; the other left school after deciding to convert to Christian Science.) Twelve of the 15 graduates—including Packard—become the first Hopkins Hospital interns.

Winford Smith, a 1903 graduate of the School of Medicine, succeeded Henry Hurd as Hopkins Hospital's superintendent/director in 1911. He would remain in the job for next 35 years, shepherding the Hospital through the turmoil of World War I, the Great Depression and World War II. Careful of expenditures in good times and lean, he was not reluctant to criticize the costly accessories in the elegant Phipps Clinic or upbraid faculty about unauthorized purchases of equipment.

The Residency System

From the 15 members of the first graduating class of 1897, 11 men and one woman became the Hospital's first interns. Known as "house medical officers," these young physicians spent four months each on the medical, surgical and gynecological services—with one month of the gynecology rotation spent on the obstetrical service. Supervising them directly would be a resident physician, a surgeon, and a gynecologist, who were supported by assistant residents. After 1900, the interns spent the entire year on a single service. Most of these internships and residencies went to Hopkins graduates. Not until 1914 did graduates of other medical schools begin to fill a substantial proportion of the Hopkins internships.[154]

Early Hopkins medical residents remained for so long that during his 16 years at Hopkins, Osler trained only five resident physicians—one of whom, William S. Thayer (1864–1932), was resident for seven years, nearly half of Osler's tenure. Thayer's lengthy service revealed the wisdom of the residency system. He would be in charge of the Hospital's medical service in Osler's absence and acquired the experience and skill to take over much of the teaching of third- and fourth-year students, and he'd be extremely well prepared to care for patients when he finally entered private practice. He organized the first course in clinical microscopy for the medical school, served with distinction as chief medical consultant to the American Expeditionary Forces in France during World War I, and ultimately became one of Osler's successors as professor and head of the Department of Medicine from 1919 to 1921.[155]

First Research Grant, Endowed Professorship, Lectureship, Overseas Research and New Specialties

During this period, Osler obtained Hopkins' first gift for medical research—an 1898 pledge from an anonymous donor of $750 a year for five years for the study of tuberculosis. Two years later, the University also was able to create its first endowed professorship with a bequest of $24,000 from Henry Willis Baxley, a well-known Baltimore physician who had died in 1876, the year the University opened. The interest had been allowed to accumulate, so that by the time the Baxley Professorship in Pathology was established in 1900, the fund contained more than $50,000. Three years later, Christian A. Herter (1865–1910), an independently wealthy physician and biomedical researcher who had trained under Welch, donated the funds to create the Herter Lectures, the first of many endowed lectureships at Hopkins. It remains the best-endowed and most prestigious of these events, its principal purpose being to disseminate the research of overseas investigators in medical science.[156]

The inaugural Herter Lecture was delivered by the renowned German scientist Paul Ehrlich, a future Nobel Prize winner. At the time, he already was in the process of developing his "magic bullet" remedy for syphilis (an arsenic compound he called Salvarsan). He also had developed his method for staining blood corpuscles, conceived of his "side chain theory" of immunity to diseases caused by bacteria, and coined the term "chemotherapy" for his theories concerning chemistry-based medications designed to treat specific infectious diseases.[157]

New specialties were emerging during this period, as surgical pathology, obstetrics and orthopedics were recognized as independent fields of study. Surgical pathology had become the first, albeit unofficial, specialty division of the Department of Surgery in 1895, when Halsted appointed his fourth resident, Joseph C. Bloodgood (1867–1935), to develop this field. By 1913, Bloodgood was among the founders of the American Society for the Control of Cancer, forerunner of the American Cancer Society; and in 1927, he became one of the first physicians to link tobacco use to mouth cancer.[158] Fulfilling Kelly's long-standing wish, a separate Department of Obstetrics was established in the Hospital in 1899; and a clinic devoted to orthopedic surgery was created under William S. Baer (1872–1931).

Baer would become an innovative spine and hip surgeon, mentor to some of the finest orthopedic surgeons of the 20th century, and president of the American Orthopaedic Association in 1924.[159]

By 1899, the medical school and hospital were starting to reach outside their walls. Postgraduate courses, which had begun before the medical school opened, were organized more systematically, prompting a corresponding increase in enrollment by local physicians. The university also sponsored its first overseas medical research effort, with two faculty and two medical students joining an 1899 commission that traveled to the Philippines to study tropical diseases.[160]

As The Johns Hopkins Hospital and School of Medicine entered the 20th century, they offered the best of American medicine at the time. The innovations that seemed excellent in the early 1900s retain their luster today, and so embedded in the traditions of medical education, practice and research have Hopkins' contributions become that they now are accepted as ordinary. More than any other factor, the advanced clinical training developed at Hopkins has been identified as the reason for the preeminence of American medicine.[161]

In 1886—seven years before the School of Medicine opened—William H. Welch began instruction for 16 post-graduate students in his brand-new pathology laboratory, launching the first post-graduate training program for physicians in the country. In this 1899 photo, a class of post-graduate students pose in the Hopkins Hospital's amphitheater with (left to right) instructor in surgery and future brain surgery pioneer **Harvey Cushing**, gynecology and obstetrics professor **Howard Kelly**, physician in chief and medicine professor **William Osler**, and **William S. Thayer**, associate professor of medicine and later a successor to Osler as physician in chief, all seated in the foreground.

1898
William Osler obtains Hopkins' first grant for medical research when an anonymous donor gives $750 a year for five years for the study of tuberculosis.

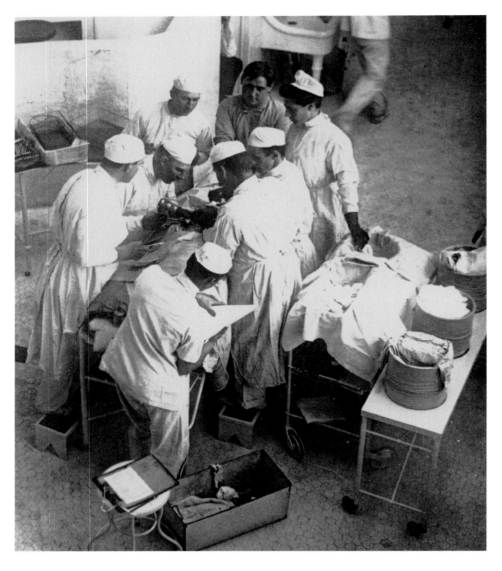

1903–1913

Hopkins Medicine's second decade was marked by an extraordinary expansion of its influence and scope. The landmark 1910 Flexner Report proclaimed the School of Medicine and Hospital the models that all others should emulate, and medical education began changing throughout the United States. The founding of the Phipps Psychiatric Clinic created the first such institution associated with a teaching hospital. The Harriet Lane Clinic for Invalid Children, the birthplace of American pediatrics, was founded. The Department of Art as Applied to Medicine became the first of its kind in the country.

All of these and Hopkins' other impressive achievements of this period, however, were preceded by a disaster—a municipal calamity that seriously threatened the institution's financial underpinnings.

By 1901, Hopkins' rapid development of new fields of surgery and surgical procedures attracted so many patients that the Department of Surgery was cramped. Surgeon in chief William Halsted requested that a new building be erected with a surgical amphitheater just for his department (something the equally busy Department of Gynecology had gotten in 1892). Halsted's new building—costing $159,573.53 and paid for entirely by the Hospital—opened in 1904. Senior members of the surgical staff held their own dedication "ceremony," seen above, an operation to remove the upper part of a patient's femur (thighbone) afflicted by a chronic infection (osteomyelitis). Instead of having ordinary residents work beside him, Halsted assembled his senior assistants—James Mitchell, J.M.T. Finney, Harvey Cushing, Joseph Bloodgood, Hugh Young, Richard H. Follis and F.H. Baetjer—to participate in what Cushing later called the "All-Star Performance," a widely publicized and photographed surgery. Halsted is in the center of the group, on the left, facing the camera and leaning over the patient.[162]

The Great Baltimore Fire of 1904

The progress of the hospital and medical school almost came to a sudden halt when a mammoth fire swept through Baltimore's downtown and waterfront between February 7 and February 9, 1904, demolishing much of the city's business district. More than 70 blocks were destroyed. Although the fire did not reach Hopkins Hospital, it consumed more than 60 buildings that Johns Hopkins had bequeathed to the university—and when they went up in flames, so did the income from them.[164]

Rescue came from an unexpected source. A few years earlier, Osler had received an unsolicited letter from Frederick T. Gates, a close advisor to oil magnate John D. Rockefeller, telling him how important Osler's textbook on medicine had been in the founding of the Rockefeller Institute for Medical Research. Fortunately, at the time of the fire, Welch was head of that institute's Board of Scientific Advisors. After the fire, Rockefeller donated $500,000 to Hopkins—the first of what ultimately would be many millions in gifts that Hopkins Medicine would receive over ensuing decades from the Rockefeller-endowed General Education Fund and the Rockefeller Foundation.[165]

The First of the Big Four Departs

Six months after the Great Baltimore Fire, Osler accepted the Regius Professorship of Medicine at Oxford University in England. In an August 6, 1904 letter to William Thayer, he explained that he simply was worn out.

"I am on the downgrade, the pace of the last three winters has been such that I knew I was riding for a fall. Better to get out decently in time, & leave while there is still a little elasticity in the rubber … ." In another letter, he wrote that the Regius Professorship was too good a job to turn down—"a dignified easy position, almost a sinecure."

"I know how much I give up—the very best clinical position in the English-speaking world, a very large income, a host of warm friends, an enviable position in the heart of the profession in America—I know it and have considered the cost. It will be a terrible wrench but it is I believe for the best—best perhaps for the School, best certainly for me."[166]

Osler, much beloved, left Baltimore in May 1905. A group of the residents to whom he had given keys to his home on Franklin Street—the self-dubbed "latch-keyers" (among whom was Cushing)—parceled out among themselves some pieces of furniture and other items that Osler and his wife had decided not to take to Oxford. Precisely a century later, Osler's library chair returned to Hopkins Hospital. Pulmonologist Henry Thomas III, a 1961 graduate of the School of Medicine, son of Hopkins cardiologist/epidemiologist Caroline Bedell Thomas, and grandson of Osler latch-keyer Henry Thomas, Hopkins' first neurologist, fulfilled a provision in his mother's will and gave the library chair to the Hopkins Medical Archives. It now stands in the Osler Textbook Room, the small hideaway in the Billings administration building where Osler wrote his *Principles and Practice of Medicine*.[167]

Knighted in 1911, Sir William Osler relaxes in this 1913 photo, below. Right before leaving Hopkins in 1905 to become regius professor of Medicine at Oxford, he gave a final lecture that dogged him for the rest of his days. A noted practical joker, Osler, then 55, jocularly suggested that men are most creative prior to their 40th birthdays and become progressively "useless" thereafter. He cited an Anthony Trollope novel in which was proposed "the admirable scheme of a college into which men at sixty retired for a year of contemplation before a peaceful departure by chloroform." The uproar caused by the speech—which elicited headlines nationwide proclaiming "Osler Recommends Chloroform at 60"—never completely subsided. Although he repeatedly insisted he had been joking, "Oslerization" became

a synonym for euthanasia, and even *The New York Times* headline on Osler's 1919 obituary said, "Noted Physician Stirred World With Theory That Man's Great Work Is Done Under 40. 'CHLOROFORM AT 60' A JEST."

Lewellys F. Barker (1867–1943), a Canadian-born physician who was among the first fellows at Hopkins, then a professor of anatomy at the University of Chicago, succeeded William Osler as director of medicine and physician in chief of Hospital Hospital in 1905. He joked in his autobiography that his unconventional first name probably should have been "Lewellyn," but that he had been "christened by a misprint."[163]

■ Hopkins' pioneering psychobiologist Curt Richter (1894–1988), who spent 60 years on the faculty, introduced the term "biological clock" in a 1927 paper describing his observations of the cyclical internal mechanisms that drive the eating, drinking and sexual behavior of animals.

1899
March 22: A group of Hopkins physicians embarks on a 'round-the-world expedition that will take them to Hong Kong, Japan, and the Philippines, where they set up headquarters in Manila to study tropical diseases.

1900
Three protégés of William Welch—Walter Reed, James Carroll and Jesse Lazear—are members of the U.S. Army Yellow Fever Commission in Cuba, which discovers that mosquitoes transmit the disease. Reed, an Army major, is the head of the commission, and Lazear dies of yellow fever contracted during its experiments. The commission's work is considered one of the greatest achievements of modern medical science and virtually eradicates yellow fever.

Lewellys F. Barker and the Creation of Clinical Medicine Laboratories

Osler was a master clinician and diagnostician, but he believed that experimental work ought to be confined to the preclinical departments. Although he was sympathetic to the idea that experimental studies should supplement patient care on the wards, he did not understand or vigorously promote inquiries into the fundamental nature of disease. His successor as director of medicine and physician in chief of the hospital, Lewellys F. Barker (1867–1943), was a new type of clinician. He was trained in the basic sciences and was instrumental in bringing laboratory investigation into clinical medicine, first at Hopkins and later, by extension, throughout American medicine.[168]

A native of Canada as Osler had been, Barker was an 1890 graduate of the University of Toronto medical school who had come to Baltimore in 1892 as an assistant resident to Osler—and lived in a room in the Hopkins administration building right next to him. Years later, Barker wrote that he "could with relative safety set my watch at 10 p.m. each night when I heard [Osler] place his boots on the floor outside his bedroom door," leaving them to be shined by an orderly.[169] After a year as an Osler resident, Barker obtained a fellowship and served a residency in pathology under Welch. After the medical school opened in 1893, he joined Mall in the anatomical laboratory and for the first time appreciated the excitement of laboratory work. He was appointed as an associate professor of anatomy in 1897, then recruited away by the University of Chicago to become a professor of anatomy there.[170]

The 1905 selection of Barker as Osler's successor represented a move from one era to the next. Soon he had organized three full-time research divisions within the Department of Medicine to provide his clinical staff with opportunities for investigation. His laboratories were unique because they were established for this specific purpose. They began a movement that changed the character of university clinics in the United States and helped to create the scientific base of modern medical practice.[171]

The first of Barker's three clinical research laboratories was the Physiological Division, which opened in 1905. Its purpose was to study individual cases of disease from the standpoint of the body's functions. Next to open was the Biological Division, which began operation in 1906. Its scientist/physicians applied bacteriologic and immunologic methods to the study of patients. The Chemical Division also opened in 1906.[172]

The individuals whom Barker chose to head these laboratories and those who followed them in the medicine department created a new academic breed, the clinical scientist, who bridged the gap between the preclinical researcher and the practicing physician. Kelly and his associates and successors in the gynecology department followed suit, developing a pathologic laboratory that became a model of its kind; and Halsted advanced the practice of surgery by breaking with the age-old tradition that it simply was an empirical skill, instead relating it to physiologic research.[173]

The Flexner Report

In 1908, Abraham Flexner (1866–1959), an 1886 graduate of Hopkins and protégé of Gilman, wrote a book, *The American College*, which was highly critical of the lecture-heavy methods of instruction in most U.S. colleges and universities of that time. A native of Lexington, Kentucky, he had established a private academy there, the Flexner School, which was so successful that its graduates even caught the eye of Harvard's President Charles Eliot. He sent Flexner a note observing that boys from the Flexner School were entering Harvard at a younger age than other freshmen and graduating more quickly. Eliot wanted to know, "What are you doing?"[174]

Flexner believed that what he was doing was implementing Gilman's overall educational philosophy, with its emphasis on small classes and close, personal instruction by accomplished teachers. His book on American colleges intrigued Henry Pritchett, president of the Carnegie Foundation for the Advancement of Teaching, founded by steel baron philanthropist Andrew Carnegie. Pritchett asked Flexner to examine the state of American medical schools. Taking 18 months to conduct his study, Flexner visited 155 medical schools in the United States and Canada, examining their admission standards, faculty size and training, endowment and tuition, laboratory quality, and access to a teaching hospital. He produced a 337-page study titled "Medical Education in the United States and Canada," published in 1910, which quickly became known simply as "The Flexner Report."[175]

Although Flexner had graduated from Hopkins before either the Hospital or School of Medicine existed, he knew both institutions well. He not only had been close to Gilman but also provided financial support to his older brother, Simon Flexner (1863–1946), while he was studying under Welch at the Hospital between 1890 and 1892, before the School of Medicine opened, and later when he was serving on its early faculty. The quality of the medical education at Hopkins stood in stark contrast to what Abraham Flexner found elsewhere, with most medical schools being mainly profit-oriented businesses. His report was a scathing indictment of American medical education. Of all 155 medical schools he had visited, only five received lukewarm praise from him—but of Hopkins he wrote: "This institution, fortunate in its freedom from all entanglements, in its possession of an excellent endowed hospital, and, above all, in wise and devoted leadership, set a new and stimulating example precisely when a demonstration of the right type was most urgently needed."[176]

The Flexner Report called upon other medical schools to adopt the policies Hopkins maintained—academically qualified students, exposed to professors who were full-time teachers and researchers, which would ensure the best available training to every physician to bridge the gap between what was known to medical science and what was known to the average practitioner. It launched a national movement to reform medical education that made Hopkins the standard by which all others would be measured.[177]

Abraham Flexner (1866–1959) knew a lot about Johns Hopkins long before he wrote his landmark 1910 study of U.S. medical schools that named Hopkins the best. Flexner was an 1886 graduate of The Johns Hopkins University (before it had a hospital or medical school) and a protégé of its first president, Daniel Coit Gilman (1831–1908). He also was the younger brother of Simon Flexner (1863–1946), whose expenses he helped cover when Simon was a fellow studying pathology from 1890 to 1892 under William H. Welch (1850–1934). Simon later became a professor on the Hopkins medical faculty.

1901
June 11: Daniel Coit Gilman gives his farewell address as president of Johns Hopkins. He congratulates the university on the selection of acclaimed chemist Ira Remsen as his successor. Remsen, a member of the first faculty, is the founder of the *American Chemical Journal*, acknowledged as the first really scientific journal of research chemistry in America.

Max Broedel (1870–1941), founder of the Department of Art as Applied to Medicine, the first of its kind, in his studio. Broedel's drawings were so realistic that it is said a Hopkins nurse once tried to pick up a gallstone that she thought had been glued to one of his pictures.

■ During Prohibition, Max Broedel (1870–1941), founding director of the Department of Art as Applied to Medicine, developed a much-appreciated sideline: beer baron. He had been given beer yeast obtained in New York by his great friend, famed newspaperman H.L. Mencken (1880–1956). Broedel grew and perfected it in the Hopkins bacteriological laboratory, creating a pure strain that he then used to provide Mencken and other friends with good beer throughout the Prohibition drought.

Entitled "The Saint—Johns Hopkins Hospital," this 1896 caricature by Max Broedel shows the apotheosis of William Osler, lifted aloft by a whirlwind and driving a multitude of germs, bacteria and other pestilence from Hopkins Hospital, seen far in the distance.

■ In the 1920s, The Johns Hopkins Hospital's carpentry shop used cherry wood to make tongue depressors, because the physicians thought they would taste better.

The Department of Art as Applied to Medicine

By 1906, Max Broedel's mastery of medical illustration, as well as his skill at its instruction, had earned him a worldwide following. William J. Mayo himself tried to persuade Broedel to leave Hopkins and come to his new clinic in Rochester, Minnesota, offering him the then-substantial annual salary of $5,000.[178]

Kelly countered by matching Mayo's offer and successfully arguing that the work Broedel was doing at Hopkins was more important. In 1910, Mayo sweetened his offer by saying Broedel could devote most of his time to teaching. Broedel was tempted. He wanted to establish a sort of "School of Broedel," where a steady stream of students could be taught medical illustration.[179]

Thomas S. Cullen, who would succeed Kelly as head of gynecology at Hopkins, was dismayed at the thought of losing Broedel. He had been among those to have their research and surgeries illustrated by the intensely dedicated yet amiable, curly-haired artist. To keep him here, Cullen sought funding from millionaire railroad man and art connoisseur Henry Walters (1848–1931), founder of Baltimore's Walters Art Museum. Walters was blunt in reply: "I am not interested in medical illustration. I took twenty lectures in medicine at Harvard and nearly vomited my boots up." He nevertheless admitted that he understood the value of medical illustration and agreed to give Hopkins $5,000 a year for three years to found the Department of Art as Applied to Medicine in 1911. The nation's first such department, its launch ensured that Broedel would get his "School of Broedel."[180] Walters ultimately endowed the department with a donation of $110,000. All of Walters' gifts were anonymous and his role in funding the department was not known publicly for years.[181]

Broedel aimed to draw pictures that showed more than any photograph could. "The camera copies as well, and often better, than the eye and the hand," he said, but in a medical drawing, "full comprehension must precede execution." This required an exquisite understanding of anatomy that could be gained only by dissection or watching surgery, as well as an extraordinary artistic technique.[182]

Broedel's determination to understand completely what he was drawing led him to become a research investigator—and even to devise some new surgical approaches. For instance, he recommended that surgeons start fishing for kidney stones from the avascular part of the kidney, in order to limit damage to the organ's filtering mechanisms, which are in the vascular areas. This insight, and a sturdy, triangular stitch still known as Broedel's suture, developed from his in-depth study of a kidney in the autopsy room. Kelly was so impressed by Broedel's scholarship that he had him write the anatomical chapters in his book, *The Vermiform Appendix and Its Diseases*.[183]

Broedel had a profound impact, training some of the world's leading medical illustrators. When other schools began launching their own medical illustration departments, seven of the eight earliest ones were headed by artists who had trained under Broedel—and three of those seven were women, who found ready acceptance in the field.[184]

Broedel's students also made important contributions to the development of medicine. Dorcas Padget, for example, helped Dandy with his studies of the brain; Annette Burgess worked with the Wilmer Eye Institute, illuminating details of the eye not visible through photographs; William F. Didusch worked with Hugh Young in the Brady Institute; and Leon Schlossberg collaborated brilliantly for many years with members of the surgical staff. [185]

The Henry Phipps Psychiatric Clinic

As Osler was about to depart Hopkins and Baltimore for Oxford in 1905, he voiced only one regret about the School of Medicine's enormous accomplishments. "You need a department of psychiatry," he told Welch.[186]

Psychiatry was coming into its own as a medical specialty by the early 20th century. As Sigmund Freud's and Carl Jung's theories on the subconscious were catching fire in Europe, American researchers were rushing to investigate the causes of insanity and learn more about the brain's anatomy. In 1908, a shocking book called *A Mind That Found Itself*, by Clifford Beers, a recovered psychiatric patient, described the horrifying conditions in America's turn-of-the-century insane asylums.[187]

Henry Phipps, a Philadelphia steel magnate and one-time partner of Andrew Carnegie, had been a major benefactor to Hopkins, establishing the Phipps Tuberculosis Dispensary in 1905. On a May 1908 visit to the Hospital to see how his TB clinic was operating, Phipps asked Welch if any other projects needed funding. Welch promptly handed him a copy of *A Mind That Found Itself*. He pointed out that it had been published with the help of Adolf Meyer, a Swiss-born and -trained pathologist who then was a professor of psychiatry at Cornell, as well as the world's first psychobiologist, intent on determining whether biological factors and mental problems were inseparable. Welch liked Meyer's thinking and told Phipps that Hopkins needed to become a leader in this new field of psychiatry, too. Within a month, Phipps agreed to donate $1.5 million to fund a psychiatric department and clinic.[188]

Adolf Meyer (1866–1950), founding director of the Phipps Psychiatric Clinic, transformed psychiatry from the almost exclusive concern of "asylum" superintendents into the province of university-affiliated physicians. Swiss-born and heavily accented, Meyer is credited with coining the term "mental hygiene." He was convinced that much mental disease was preventable. His program linked the fields of psychiatry with neurology. His penetrating gaze was offset by a droll sense of humor. Some thought his heavy mustache and beard were maintained to hide his grin.

1903
Christian A. Herter of New York, an independently wealthy physician and biomedical researcher who is a protégé of Welch's, joins his wife in giving $25,000 to found a lectureship in medicine "designed to promote a more intimate knowledge of the researches of foreign investigators in the realm of medical science." The first Herter lectures in 1904 are delivered by German scientist Paul Ehrlich, a future Nobel Prize-winner acclaimed for his discovery of a "magic bullet" treatment for syphilis. The Herter becomes the oldest and among the most prestigious of Hopkins Medicine's endowed lectureships.

The Phipps Psychiatric Clinic had light-filled wards separated by an outdoor courtyard with a pond and cloistered walk, seen here in a circa 1913–1915 photo.

■ The Phipps Psychiatric Clinic was named for Henry Phipps, a Philadelphia-based trustee and benefactor of the Hospital. At the formal dedication of the original Phipps building in 1913, psychiatry co-existed with superstition: The plaque above the building's entrance says "1912"—because a year ending in "13" was considered unlucky.

1904
February 7: The Great Baltimore Fire destroys 70 investment properties, most of them warehouses, belonging to the Hospital. The buildings had provided one-half of the Hospital's annual endowment income, which helps fund its free dispensary. In May, the Hospital receives a letter from John D. Rockefeller Jr., stating: "In view of the high character of work which the hospital and medical school are doing in medical instruction and research, including the training of nurses, which work he understands will otherwise be materially curtailed because of the losses, my father will give five hundred thousand dollars to Johns Hopkins Hospital." It is the first of many subsequent donations from the Rockefellers that Hopkins Medicine will receive.

October 5: William S. Halsted, director of surgery and surgeon in chief of the Hospital, performs the first operation in his new surgical amphitheater. A team of residents handpicked by Halsted to assist him includes Joseph Bloodgood, Richard H. Follis, James Mitchell, Harvey W. Cushing, Hugh H. Young, and J.M.T. Finney. Given the future prominence of those assisting Halsted, the operation later is dubbed by Cushing as "The All-Star Performance."

Welch swiftly recruited Meyer to become the new department's director in 1908. Meyer became psychiatrist in chief of Hopkins Hospital in 1909 and oversaw the building and development of the Phipps Psychiatric Clinic, of which he also became the first director. By 1910, ground was broken for the elegant building, which would have marble floors, gardens, porches, fireplaces, and even a pipe organ in a spacious auditorium. Its formal dedication occurred on April 16, 1913—although the modern psychiatric concepts it represented coexisted with superstition: The date on a plaque above its main entrance says "1912," because a year ending in "13" was considered bad luck.[189]

Adolf Meyer (1866–1950) and the Phipps Clinic proved a perfect fit. With his bushy Van Dyke beard, heavily accented English, and penetrating gaze, Meyer seemed the stereotype of the practicing psychiatrist, yet like the clinic he headed, he broke new ground. Credited with coining the term "mental hygiene," Meyer was convinced that much mental disease was preventable. He established a brilliant program that linked the fields of psychiatry and neurology.

He stressed direct observation of the patient and insisted that residents and students collect all their information by this means before formulating their concepts of a patient's case and plan of treatment. He taught his students to take the broadest possible view of psychiatric problems in order to prevent the premature creation of theories that eventually would only hinder progress in the field. He showed that manifestations of mental disease could be studied with precision and advocated flexibility in diagnosis. If one method of treatment failed, the psychiatrist should be prepared to try another, he contended. The goal was to figure out how best to help individual patients and continue to learn about the psychiatric problems they had.[190]

Within a decade, Meyer's program was so renowned that international dignitaries were traveling to Hopkins for psychiatric treatment. That led to a famously unanticipated encounter between Meyer and Queen Marie of Romania, who was visiting the Hospital in 1919. She purportedly knocked on Meyer's office door, explaining that she had wanted to meet him for some time. She introduced herself as the Queen of Romania. Meyer reportedly replied with his usual courtesy and appearance of genuine concern, asking: "And how long have you felt this way?"[191]

Among Meyer's actual well-known patients was Zelda Fitzgerald, wife of the novelist F. Scott Fitzgerald. She wrote her autobiographical novel, *Save Me the Waltz*, during several months at Phipps in 1932—while her husband was working on his own novel, *Tender is the Night*, in an apartment across the street from Hopkins' Homewood campus. His book was about a psychiatrist married to a wealthy mental patient.[192]

A residency in the Phipps Clinic became a cherished prize for psychiatrists from all over the world, who then returned home to establish facilities for the extension of Meyer's ideas. The high standards he set in teaching, practice, and research soon spread to other cities as his disciples joined the faculties of other schools. By the time he retired in 1941, he had trained approximately 10 percent of the teachers of psychiatry in the country.[193]

The Harriet Lane Clinic

When Hopkins Hospital opened, care for children was undertaken by the Department of Medicine, not a separate pediatric facility. A change in that procedure was made possible in 1912 by the generosity—prompted by personal bereavement—of Harriet Lane Johnston (1830–1903).[194]

Johnston was the niece of former President James Buchanan, the nation's only bachelor chief executive. During Buchanan's dismal term in office, 1857 to 1861, Johnston, then simply Harriet Lane, was perhaps the one bright spot in Washington as the Civil War approached. Tall, blonde, lively and charming, she was her uncle's official White House hostess. Presiding over glittering parties, she became the first to regularly be referred to as the "First Lady" in that capacity.[195]

In 1866, she married Baltimore banker Henry Johnston. They had two sons, but both only lived into their teens, dying of rheumatic fever within a year of each other. Devastated, the Johnstons—aware that the newly opened Johns Hopkins University would have a medical school and that the Hopkins Hospital was under construction—decided to donate the bulk of their substantial estate to a home for invalid children to be built near the Hospital. Henry Johnston died in 1884, and when Harriet Lane died in 1903, her $400,000 bequest was forthcoming.[196]

Executors of Harriet Lane's estate realized that if an adequate site near Hopkins was to be purchased, buildings erected, and a sum set aside for a permanent endowment, her bequest would only be able to provide a hospital capable of caring for about 25 children at a time. An alliance with Hopkins Hospital therefore seemed advantageous. An agreement was reached in 1906 under which Hopkins would provide the site and the medical care for the patients, while the Harriet Lane Home, for its part, would contribute a modern, 100-bed hospital building specifically designed for treating children's diseases.[197]

Lewellys Barker wanted to hire a director of the new Department of Pediatrics immediately to supervise construction of the home and organize its facilities. He appointed Clemens von Pirquet, a 34-year-old native of Vienna who already was well-known for discovering the phenomenon of acquired hypersensitivity and coining the term "allergy" to describe it. He also had developed the skin test for tuberculosis that bears his name.[198]

Von Pirquet arrived at Hopkins in 1909, but within a year he was offered the professorship of pediatrics at the University of Breslau. He said he'd remain at Hopkins if his salary was raised from $7,500 to $10,000 a year. The trustees were unwilling to meet that demand, and Pirquet, by then also being offered a professorship of pediatrics in Vienna, took it. Having anticipated this turn of events, the trustees immediately authorized Welch to offer the Hopkins' top pediatric job to John Howland (1873–1926), then professor of pediatrics at Washington University in St. Louis.[199]

Under Howland's direction, Hopkins' Harriet Lane Clinic became the birthplace of American pediatrics, the first children's clinic in the country associated with a medical

Because Baltimore, located below the Mason-Dixon line, practiced segregation at the time the original Hopkins Hospital opened, Hopkins followed that policy—but not in every department or division. The Harriet Lane Home for Children never was segregated, as this circa 1913–1915 photo shows. Below, children clamber over the car used by the Department of Social Services, founded in 1907, to transport patients from home to the hospital.

■ In her will, Harriet Lane Johnston (1830–1903) the benefactress whose generosity funded Hopkins' Harriet Lane Clinic, predecessor of the Johns Hopkins Children's Center and the first children's hospital associated with a university school of medicine, requested that each year on her birthday (May 9), the director of pediatric nursing should put flowers on her grave in Baltimore's Green Mount Cemetery.

John Howland (1873–1926), head of the Harriet Lane Clinic from 1912 to 1926, is considered the father of American pediatrics. The American Pediatrics Society named its lifetime achievement award, the Howland Medal, after him.

Edwards A. Park (1877–1969), a warm, gentle giant of a man, was a protégé of Howland's and succeeded him as head of Hopkins pediatrics from 1926 to 1946.

school. He established the nation's preeminent research laboratory in the field, and for the first time, the techniques of biochemistry were applied systematically to the study of childhood diseases. At Hopkins, Howland began the research that resulted in new modes of treatment for and prevention of rickets, and along with his laboratory associates, he discovered a new fat-soluble vitamin they dubbed Vitamin D. So influential was Howland in his field that he now is recognized universally as the father of modern pediatrics, with the American Pediatrics Society naming its most prestigious, lifetime achievement award in his honor.[200]

Having had no preexisting pediatric staff, Howland was able to select his own group of four associates—every one of whom later became head of a pediatrics department, including Howland's successor at Hopkins, Edwards A. Park (1877–1969). Working under Howland from 1912 to 1921, Park left Hopkins to found the pediatric department at Yale. After Howland's death in 1926, he returned to take over his old mentor's position as professor of pediatrics and pediatrician in chief of the Harriet Lane Home.[201]

Park, a descendant of 18th century American theologian Jonathan Edwards, was a gentle giant of a man. Affectionately known as "Uncle Ned," his philosophy concerning residents was that they learned more quickly by assuming greater responsibility. As head of pediatrics until 1946, he created divisions of cardiology, endocrinology, neurology and psychiatry within the department, while focusing his own research on studies of rickets, scurvy and lead poisoning.[202]

Park had a gift for spotting promising young candidates for leadership. Among his protégés was Helen Taussig, whom he chose in 1930 to head his new pediatric cardiac clinic. She went on to be one of the developers of the landmark "blue baby" operation. Another was Leo Kanner, whom Park and Meyer chose in 1930 to head the first child psychiatric clinic in the United States. Kanner later was the first to identify autism. Yet another notable young physician Park mentored was Lawson Wilkins, selected in 1935 to start a pediatric endocrinology clinic. He would conduct landmark studies of thyroid deficiencies while also training an entire generation of pediatric endocrinologists. They—like the young physicians subsequently trained by Taussig, Kanner and other Park picks—would spread the Hopkins influence in pediatrics everywhere they went.[203]

The Full-Time Controversy

When it opened in 1893, Hopkins' medical school became the first in the country to mandate that all of the research scientists for its preclinical faculty would be hired on a "full-time" basis. This meant that they received a guaranteed, set salary and in return would concentrate exclusively on their research and teaching, without doing outside work—such as treating private patients—to supplement their income.[204]

This innovative policy was considered one of the keys to Hopkins' early research successes. It set new standards for medical teachers across the country and dramatically demonstrated the increasing applicability of the preclinical sciences to medical care. In turn, this prompted a movement to make the clinical departments full-time, as well, so clinicians could adopt a similar research orientation.[205]

IT DOES GO A BIT HARD SOMETIMES

Welch, Mall and Barker all were outspoken advocates for establishing a full-time system in Hopkins' clinical departments, but knew that doing so would prove difficult. A key problem was that the University lacked the funds to give clinicians salaries comparable to those provided for research scientists. The clinicians' pay was so low that many on the staff had to support themselves by private practice and give only a limited part of their time to the medical school.[206]

Osler initially opposed adoption of a full-time system for clinicians. He believed those working under such restrictions would become "clinical prigs," laboratory workers isolated from the surrounding medical community of practicing physicians and out of touch with the real-life world of clinical practice in which young residents eventually would have to live. Other clinicians, accustomed to augmenting their medical school salaries with fees from private patients, were loath to give up that income.[207]

When Osler left for Oxford and was succeeded by Barker, the full-time concept (also dubbed the "university plan") gained a staunch proponent—at least for a while. Welch, while still favoring the concept, also wavered in his enthusiasm for it, being deterred periodically by the seemingly insurmountable challenge of obtaining a sufficient endowment to fund clinician salaries.[208]

After the Flexner Report was published in 1910, the application of the full-time concept to clinical departments gained momentum. The Rockefeller Foundation's General Education Board began to promote the idea and considered Hopkins a natural place to institute the system, given Flexner's enthusiasm for the medical school and Hospital. It took Welch several years of quiet, behind-the-scenes persuasion to get the University's board of trustees and faculty to agree to adoption of a full-time system for clinicians. His lobbying likely was enhanced in 1913, when Hopkins' then-president, Ira Remsen, resigned due to poor health and the trustees prevailed upon Welch to serve as chairman of an administrative committee of the faculty to run the university until a replacement could be found. By October 1913, he could write to the Rockefeller Foundation that the full-time negotiations had been completed successfully. Two days later, the General Education Board gave Hopkins $1.5 million to establish "The William H. Welch Endowment for Clinical Education and Research" and implement the full-time system for the clinical faculty in 1914.[209]

Welch—much relieved to be released from his university leadership duties in the spring of 1914—believed that flexibility in adopting the full-time system was crucial. Fixed rules would be harassing, improper and unnecessary, he said. Salaried clinicians could be allowed to act as consultants at the Hospital and to admit a limited number of patients to the private wards, with the professional fees derived from such patients going to the Hospital or the University.[210]

Ironically, by 1914, Barker's personal view on the full-time system had changed. As adopted, it offered clinical faculty a $10,000 annual salary (nearly $200,000

The controversy within Hopkins over whether to adopt a "full-time" system for its clinicians—requiring them to give up their private practices and focus exclusively on teaching medical students, conducting research and treating Hospital patients—had become public by the time this cartoon appeared in the *Baltimore American* of Oct. 30, 1913. One might think Hopkins physicians opposed to the full-time system would find this drawing offensive, but physician in chief Lewellys Barker, a one-time supporter but ultimately foe of the full-time system, thought it was funny enough to save—and reprint in his 1942 autobiography.

■ Anatomist Lewellys F. Barker (1867–1943), the successor to William Osler as director of the Department of Medicine and physician in chief of Hopkins Hospital, recalled in his 1942 autobiography that as a young instructor at Hopkins, he taught the histology of the central nervous system—and one of his students was Gertrude Stein (1874–1946), the avant-garde writer. "I have often wondered whether my attempts to teach her the intricacies of the medulla oblongata had anything to do with the development of the strange literary forms with which she was later to perplex the world," he wrote.

When Florence Sabin, a 1900 graduate of the School of Medicine, was promoted to associate professor of anatomy in 1905, the *Baltimore News* commented that "the appointment of a woman to a position of this rank in an institution of such distinction as The Johns Hopkins Medical School, and other than a college for women, is without parallel among American universities, in the Eastern section of the country at least." In 1917, Sabin became the first woman at Hopkins to receive appointment to a full professorship. She is honored by a statue in the U. S. Capitol's Statuary Hall.

1905
June: Pathologist William H. Welch, surgeon William S. Halsted, and gynecologist Howard A. Kelly of the original medical faculty travel to London to meet their former colleague William Osler and sit for a portrait by John Singer Sargent. It will become one of the great icons of American medicine and art entitled *The Four Doctors*. Mary Elizabeth Garrett pays for the mammoth new painting, which is 10 feet, 9 inches in height and 9 feet, 1 inch in width.

in today's money) in return for their agreement to forgo private practice. By then, however, Barker was the father of three, including a disabled son, and privately scoffed that Welch, a bachelor living in a boarding house, "was almost as naïve about personal finances as the average debutante." He resigned from the faculty and, by 1919, was making five times his Hopkins salary and living in a 38-room mansion. He nevertheless maintained his association with the School of Medicine as a visiting professor—even becoming Welch's successor as chairman of the medical board in 1926—and routinely persuaded wealthy patients to donate substantial sums to Hopkins.[211]

Given its controversial nature, implementation of the full-time system was gradual. The departments of medicine, surgery and pediatrics were the first to be enrolled in 1914. Obstetrics joined the list in 1920, psychiatry in 1924, and ophthalmology in 1926.[212]

1914–1924

When Barker balked at adoption of the full-time system and stepped down as head of the Department of Medicine, he was succeeded by Theodore C. Janeway (1872–1917), who became Hopkins' first full-time professor of medicine. Janeway was one of the first modern internists to think in terms of the natural history of a disease and its dynamics—and also the first to measure blood pressure routinely in his office. He defined treatment as "a therapeutic experiment based on a theory. This theory we call diagnosis." His belief that the experimental approach was "the only sure basis for the successful practice of the art of medicine" was unique for that time.[213]

Janeway's greatest contribution to clinical science was his work on blood pressure. He was among the first in America to study hypertension and its experimental and clinical relation to the heart and kidneys. He also was one of the first to advocate the long-term use of digitalis for minor traces of cardiac insufficiency. In 1914, at least five years before anyone else mentioned the subject, he prescribed digitalis for patients who had normal cardiac rhythm but cardiac insufficiency due to chronic high blood pressure.[214]

The Brady Urological Institute

In 1912, multimillionaire James Buchanan Brady, famed as "Diamond Jim" for his extravagant jewelry, came to see Hopkins' urologic surgeon Hugh Hampton Young. The flashily dressed and rotund railroad equipment magnate, equally renowned for his gargantuan appetite, was worried.[215]

Surgeons in Boston and New York had refused to operate on Brady for an inflamed and enlarged prostate, saying his case was too dangerous due to multiple complicating factors: his diabetes, high blood pressure, angina and Bright's disease, or kidney failure. He saw Young as his last chance for a treatment that would ease his severe prostate problems.[216]

The urbane, *pince nez*-wearing Young found Brady "a rough diamond, but a fine fellow" who "looked his nickname," with a huge diamond stickpin punctuating his necktie and additional diamonds sparkling on his vest, his watch chain, cuff links and the head of his walking stick. Young eased his fears. Contrary to what the New York and Boston surgeons had said, an operation indeed was possible.[217]

"I explained that I had recently invented an instrument [the Young Punch] with which the operation could be done entirely through the urethra, and without making any external cut, and that he would not have to have general anesthesia, which was considered dangerous in his case," Young wrote in his 1940 memoir.[218]

Soon thereafter, Brady was on Young's operating table. Young injected cocaine into his urethra as a local anesthetic, quickly and deftly wielded the Young punch to excise the troublesome prostate obstruction, and "in a few minutes Brady was off the table, pleased that the operation had been carried out so quickly and without pain," Young recalled.[219]

The jewelry-bespangled Brady soon considered Young among his best buddies—and Young, knowing that Brady had "spent hundreds of thousands of dollars" on lavish parties for Broadway actresses and other friends, thought there were better purposes to which his fortune could be put. He began to think that Brady "might be persuaded to build a hospital as a monument to himself." No slouch as a salesman, Young proposed to Brady that he consider funding a urologic institute.[220]

"Such a hospital would carry Mr. Brady's name forever," Young told his prospective benefactor. "From it would come a great series of clinical and scientific papers that would reach all quarters of the earth and carry on each publication the name of James Buchanan Brady," Young recalled.[221]

Brady was sold. On January 21, 1915, the eight-floor James Buchanan Brady Urological Institute opened, and at its formal dedication the following May, Brady was called upon to speak. He "arose and said, with evident emotion, 'The sky was never so blue and the grass never so green as they are this day for me,' and sat down," Young wrote.[222]

When Brady died of heart failure in 1917, he bequeathed another $300,000 to maintain the institute bearing his name. It has fulfilled every promise Young used to pitch its potential and today is recognized as the finest urologic center in the nation.

1906

May: John Singer Sargent's *The Four Doctors* is the centerpiece at a private showing before the opening of the 138th exhibition of Great Britain's Royal Academy of Art. The New York *Sun* reports that the huge painting is considered "the noblest work of art that has been hung in the Royal Academy since Sir Joshua Reynolds was president of it…." The painting will be unveiled at Hopkins' McCoy Hall on Jan. 19, 1907. It later is moved to the Welch Medical Library and remains there as perhaps the most valuable piece of art in the University's collection.

1908

June: Another generous gift from industrialist-turned-philanthropist Henry Phipps enables Hopkins to establish America's first clinic for the treatment of mental illness and endows a professorship of psychiatry at the University, thereby giving birth to the Phipps Psychiatric Clinic and the Department of Psychiatry. Heading both is the renowned Adolf Meyer—credited with coining the term "mental hygiene" and creator of a brilliant program that links psychiatry with neurology. The Phipps Building is dedicated in 1913.

Hugh H. Young (above) not long before he first met and treated James Buchanan "Diamond Jim" Brady (1856–1917), (above, left). The rotund multimillionaire known for his extravagant jewelry and mammoth meals had been told by surgeons elsewhere that they could not operate on him for an inflamed, enlarged prostate, saying such surgery was dangerous given his girth and other physical problems. Young knew differently. Using an implement he had invented, the Young Punch, he excised Brady's troublesome obstruction deftly and safely.

Although 67 years old when the U.S. entered World War I in 1917, William Henry Welch would not be denied an opportunity to serve. He wangled a commission as a major in the medical section of the Officers' Reserve Corps. Thomas B. Turner, future School of Medicine dean, who knew the short, portly Welch, wrote, "A less likely combination—Welch and the Army—can scarcely be imagined."

■ Johns Hopkins played a key role in conquering yellow fever. Three of the four physicians who led the U.S. Army's 1900 Yellow Fever Commission, which proved that yellow fever is transmitted by mosquitoes, were Hopkins-trained: Walter Reed, James Carroll and Jesse Lazear. All were graduate students under William Welch, Hopkins' first pathologist and medical school dean. Allowing themselves to be bitten by infected mosquitoes, Carroll contracted yellow fever and survived; Lazear, only 34, died. A 31"x 37" bronze plaque honoring him is in the Hopkins Hospital hallway connecting the Billings Building and the Wilmer Eye Institute. The Army's former medical center in Washington was named for Reed.

World War I

Department of Medicine head Theodore Janeway, having a family to support, grew disenchanted with the full-time system and announced his intention to resign from Hopkins in 1917. Before officially doing so, however, he responded to the United States' entry into World War I by becoming director of heart disease research in the Office of the Surgeon General. Within a few months, he was dead of pneumonia, having been worn out by commuting back and forth from Baltimore to Washington to carry on his work both at Hopkins and in the Surgeon General's Office.[223]

William S. Thayer would succeed Janeway as head of medicine, but before he could assume the position in 1919, he had to complete his own service during World War I as a chief medical consultant to the American Expeditionary Force, headed to France. He joined many of Hopkins' other leaders and medical students in making significant contributions to the war effort.[224]

Welch, a close friend of the Army's surgeon general, William C. Gorgas, was commissioned as a major in the medical section of the Officers' Reserve Corps. When the Spanish Influenza pandemic erupted in the summer of 1918, soldiers at Camp Devens, northwest of Boston, were among those devastated by the outbreak. Welch visited there and was the first to recognize that the disease was not just another form of influenza but "some new kind of infection or plague." He initiated the effort to find its cause and develop a medical response to it—and also came down with the virus in the course of his work, but survived the attack. (Others at Hopkins Hospital, swamped with cases, were not so lucky. Four nurses, three physicians and three medical students were among some 1,400 victims of the flu in Baltimore.)[225]

Winford Smith, superintendent of Hopkins Hospital, became chief of the hospital division of the Surgeon General's Office and oversaw the organization of many military hospital units, among them Hopkins' own Base Hospital Unit 18, the first university-center medical unit to go to France. A 1,000-bed, barrack-style hospital located at Bazoille-sur-Meuse in the Vosges district, it was headed by

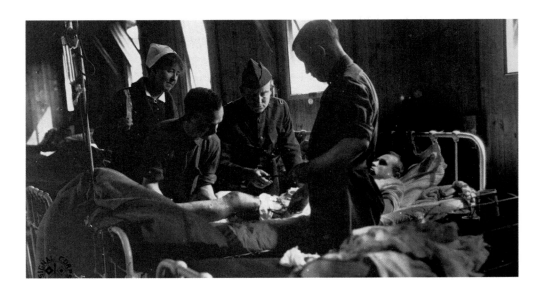

Physicians and a nurse attend a soldier in Hopkins' Base Hospital Unit 18 in Bazoille-sur-Meuse, France, during World War I. The 1,000-bed, barrack-style hospital was the first university-center medical unit to go to France.

J.M.T. Finney, successor to Halsted as director of surgery at Hopkins, and staffed by some 34 Hopkins physicians and medical students. As an evacuation hospital during the 1918 Battle of the Marne, providing primary surgical care to the wounded, it maintained nearly round-the-clock operating rooms amid the carnage.[226]

For their work overseas, Finney, Thayer and Young each received the Distinguished Service Medal. Welch, who also had gone to Europe during the war, received the Distinguished Service Medal and was promoted to brigadier general in the Officers' Reserve Corps in 1921.[227]

The First Faculty Begins Fading, a Second Faculty Begins Working, a Flood Tide of Exports Begins Expanding Hopkins' Influence

The First World War may also have claimed an indirect casualty in William Osler—by then "Sir William," thanks to a baronetcy he received in the coronation honors bestowed by Great Britain's newly crowned King George V.[228]

Osler's son, Revere, born in Baltimore in 1895, had joined the British Army and been killed in the war in 1916, shortly after turning 21. Osler, understandably, was heartbroken. He died of pneumonia in 1919. Frederick P. Mall had died in 1917, and Halsted would die in 1922. Of the original senior faculty of the School of Medicine, the only full-time members still active by the early 1920s were pharmacology director, John Jacob Abel; Walter Jones, head of the Department of Physiological Chemistry; and obstetrician J. Whitridge Williams, by then dean of the medical school. Welch had founded the School of Public Health in 1916 and focused his attention there. Kelly had become an emeritus professor. Finney, Barker, Cullen and Thayer were part-time faculty.[229]

The need to replace many senior faculty, coupled with abundant financial resources flowing from the Rockefellers' General Education Board and a 1918 bequest from Joseph R. DeLamar, a mining millionaire, led to a time of rapid change and growth at both the medical school and Hospital.[230]

Between 1916 and 1927, new department heads were appointed for anatomy, medicine, gynecology, pathology, pediatrics, physiological chemistry, physiology, and surgery—and the faculty grew as these new directors brought in junior associates. The creation of the Wilmer Eye Institute in 1925 and the Institute of the History of Medicine in 1929 further increased the number of faculty—and were key elements of the building boom of the '20s that would see not only Wilmer and the Welch Medical Library built but other landmark construction projects undertaken.[231]

Hopkins Clones

A flood tide of Hopkins exports began in the early 1920s. Many faculty and alumni left to oversee the founding, restructuring or upgrading of medical schools elsewhere, modeling them all on Hopkins. Among these offspring were the medical schools of Washington University in St. Louis; Vanderbilt University, the University of Rochester and Duke University. Hopkins alumni also became influential deans of the medical schools at the University of Wisconsin and Yale.[232]

1910
April 10: The Carnegie Foundation of New York issues *Medical Education in the United States and Canada*, a comprehensive report by Abraham Flexner that singles out Johns Hopkins as a leader in medical education. Flexner notes that Hopkins, "fortunate in its freedom from all entanglements, in its possession of an excellent endowed hospital, and, above all, in wise and devoted leadership, set a new and stimulating example precisely when a demonstration of the right type was most urgently needed." The Flexner Report, as the study becomes known, ensures that the Hopkins School of Medicine becomes the model against which all other American medical schools are measured.

1911
The Department of Art as Applied to Medicine—the first of its kind in the nation—is created. Its director is Max Broedel, the masterful medical illustrator whose work at Hopkins since 1894 had made him internationally famous.

1912
September: Announcement is made that a new hospital building named for Charles L. Marburg (whose heirs gave $100,000 to Hopkins in 1907) will be built for the use of private patients.

November 20: The Harriet Lane Home for Invalid Children opens, becoming the first children's clinic in the United States that is associated with a medical school. It is named for Mrs. Harriet Lane Johnston, widow of Baltimore banker Henry Johnston. She left a sum of more than $400,000 at her death in 1903 to establish the Harriet Lane Home for Invalid Children as a memorial to two sons who had died in childhood. She was the niece of President James Buchanan and served as First Lady of the White House during her uncle's term of office (1857–1861).

Trustees of Peking Union Medical College. William Welch, center, in white suit, with hands folded, stands next to John D. Rockefeller, Jr., PUMC's largest donor (holding hat).

1913
January: Ira Remsen steps down as president of The Johns Hopkins University due to ill health. The university's board of trustees asks William H. Welch, former dean of the School of Medicine, to serve as chairman of an Administrative Committee of the Faculty to run the university until a successor to Remsen can be found.

1914
The "full-time system" (also called the "university system") is adopted for clinicians on the medical school faculty, as it had long been implemented for the research faculty. Under the full-time system, the clinicians are given a guaranteed, set salary and in return concentrate exclusively on their research and teaching, without doing outside work—such as treating private patients—to supplement their income. Given its controversial nature, implementation of the full-time system is gradual. The departments of medicine, surgery and pediatrics were the first to be enrolled in 1914. Obstetrics joined the list in 1920, psychiatry in 1924, and ophthalmology in 1926. Some prominent clinicians on the faculty, such as Howard Kelly, decline to join the full-time system and instead retain their private practices. They become part-time faculty, however; still teach, and usually put their patients who require hospitalization in Hopkins.

October 5-8: The 25th anniversary of the Hospital is observed.

The Peking Union Medical College and Hospital

Even before Hopkins spawned its clones in the United States, it was instrumental in establishing one in China, the Peking Union Medical College (PUMC), which became the most prestigious medical school in that nation, and Peking Union Medical College Hospital (PUMCH). Missionary groups from the United States and Great Britain had joined forces with the Chinese government to establish the PUMC in 1906 and maintained it until 1913. When the Rockefeller Foundation was created that year, it established a commission to study the status of medical education in China and provide the funds to improve it. Among that commission's members were both William Welch and his protégé, Simon Flexner. They were leaders of a Rockefeller-sponsored group that went on an Asian tour in 1915, spending three weeks in Peking (now Beijing) to confer with officials of the college and hospital. Under the guidance of Welch, Flexner and the others, the PUMC and PUMCH were reorganized along Hopkins' lines. Welch even insisted that women be admitted to the school, noting that female physicians already were important in China. When an entirely new, multimillion-dollar, 14-building PUMC was ready to be dedicated in 1921, John D. Rockefeller Jr. insisted that Welch travel with him to Peking to attend the ceremonies. "It is of the utmost importance that you should be there," Rockefeller wrote. "It will mean more to the medical world than the presence of any other man."[233]

The Building Boom of the 1920s Begins

In addition to the Rockefellers' General Education Fund, other philanthropies answered Hopkins' call for assistance to build the additional facilities it required following World War I. Technological advances in patient care, a surge in medical research on many fronts, and the needs of a modern medical education mandated major upgrading of existing facilities and construction of new ones if Hopkins wanted to remain at the forefront of medical progress. So desperate was the need to modernize Hopkins Hospital and the School of Medicine that administrators considered reducing the size of medical school classes until new buildings could go up.[234]

A $3 million grant from the General Education Board for physical expansion made class cutbacks unnecessary. With this and gifts from others, such as the Carnegie Foundation and the Harkness family's Commonwealth Fund, Hopkins embarked on a 10-year building boom that added 12 major facilities to the medical institutions, providing more than 800,000 square feet of space, at a cost of more than $10 million.[235]

1925–1935

The roster of structures begun during the 1920s building boom include the five-story Women's Clinic Building for obstetrics and gynecology and the Welch Medical Library, completed in 1923 and 1928, respectively, and both paid for largely by the General Education Fund; the Carnegie Dispensary, an eight-story outpatient building completed in 1927 with a $2 million grant from the Carnegie Corporation; and the seven-story Osler and Halsted clinical buildings, begun in

1929 with a $3 million gift from Edward S. Harkness, head of the Commonwealth Fund, later supplemented by the General Education Fund's last great gift to Hopkins. Dedicated in 1932, the two symmetrical wings of the Osler and Halsted buildings had located between them the Henry M. Hurd Memorial Hall, a three-hundred-seat amphitheater. With its curved, ascending rows of padded, pew-like benches, wood paneling and curtained stage, it was regarded as the epitome of a modern auditorium appropriate for both lectures and semipublic meetings.[236]

The Wilmer Eye Institute

While the foundations bearing the names of John D. Rockefeller, Andrew Carnegie and the Harkness family were established by remarkable and fascinating individuals, perhaps no benefactor of Hopkins Medicine was as colorful as the lady whose determination to honor ophthalmologist William Wilmer led to creation of the Wilmer Eye Institute.

Aida de Acosta Root Breckinridge (1884–1962) may have been the first woman ever to pilot a dirigible—and decades later, as a sufferer of glaucoma, she made it her mission to save the sight of others.[237]

The vivacious daughter of wealthy Cuban-born parents, young Aida de Acosta was a student at the Sacred Heart Convent in Paris in 1903 when she was entranced by the exploits of balloonist Alberto Santos-Dumont, who then was testing his dirigible, the *Santos-Dumont IX*, by flying it around the Eiffel Tower. She purportedly persuaded him to give her lessons in the aircraft's operation, and on June 29, 1903—six months before the Wright Brothers went aloft in their heavier-than-air flying machine in Kitty Hawk—the 19-year-old Aida went on an impromptu, five-mile flight in Santos-Dumont's dirigible, with the balloonist shouting instructions to her as he sped along on his bicycle down below.[238]

Five years later, she married Oren Root, a nephew of Secretary of State Elihu Root, and became well known for staging impressive charitable drives by employing such tactics as having Enrico Caruso sell government bonds to Wall Street passersby by singing to them from the steps of the New York Treasury Building during World War I.[239]

Afflicted with glaucoma in the early 1920s, she lost the sight in one eye but was treated by Wilmer, then the most prominent ophthalmologist in Washington, D.C., who saved the sight in her other eye. By then she had divorced Root and married Henry Breckinridge, a former assistant secretary of war, and was a well-known society hostess.[240]

1914
October 9: James Buchanan "Diamond Jim" Brady makes a surprise visit to East Baltimore to inspect the new urologic building for which he is paying. He is so pleased with it that he draws up a new will to provide money for its maintenance and improvements. The *American* reports that, "as usual, Mr. Brady wore a princely array of diamonds." The Brady Urological Institute is dedicated on Jan. 21, 1915.

October: William Welch happily relinquishes leadership of Johns Hopkins when the university's board of trustees elects Frank Goodnow, a political scientist with a long record of public service to both U.S. and foreign governments, as the third president of Hopkins.

1916
June 13: William H. Welch announces that The Johns Hopkins University has received a grant from the Rockefeller Foundation to establish a school of hygiene. Welch will become its first director.

Ophthalmologist **William H. Wilmer (1863–1936)**, who counted among his patients eight **Presidents of the United States (from William McKinley to Franklin D. Roosevelt)**, became the first director of the Department of Ophthalmology in 1925.

Socialite and medical benefactor Aida de Acosta Root Breckinridge (1884–1962), seen around 1910 in this photo, was the vivacious former patient of William H. Wilmer—who had saved her sight following a battle with glaucoma. Believed to have been the first woman ever to pilot a dirigible, she decided in 1922 to organize fundraising to establish the Wilmer Eye Institute.

1917

Thirty-two medical students are among those who serve at the Johns Hopkins Base Hospital 18 in Bazoilles-sur-Meuse, France during World War I. They spend their senior year attending lectures and rounds with their professors and tending to the wounded in a 1,000-bed facility built early in the war by the French. By late 1917, the base hospital is filled with American victims of gassing, as well as battle and bomb casualties. Two medical students and two Hopkins nurses die from diseases contracted in the hospital. The remaining 30 students receive their medical degrees in April 1918 while still in France.

1921

Millionaire railroad executive and art collector Henry Walters, founder of Baltimore's Walters Art Museum, anonymously donates $110,000 to endow the Department of Art as Applied to Medicine, the first such department in an American medical school. Walters had been supporting the department with anonymous yearly donations since its founding in 1911.

1924

January 9: The Women's Clinic is dedicated at Hopkins Hospital. Gynecologist Thomas S. Cullen and obstetrician J. Whitridge Williams, a former dean of the School of Medicine, are the clinic's co-directors.

Acosta Root Breckinridge decided in 1922 that she would create a Wilmer Foundation to endow and build an independent Wilmer laboratory and hospital in Washington. Seeking advice about fundraising from the General Education Board's Abraham Flexner, the ever-faithful advocate of Hopkins, she was steered toward Baltimore instead of Washington as the site for a Wilmer Institute. Although Wilmer shunned the idea of soliciting his former patients to raise money, Acosta Root Breckinridge had no such scruples. She contacted numerous former patients who had been treated—and charmed—by the tall, charismatic Wilmer and got the money she needed to launch the foundation bearing his name.[241]

Fortunately, among Wilmer's other patients had been Wallace Buttrick, then-head of the Rockefeller Foundation's General Education Board. He told another one of his eye doctors, Hopkins ophthalmologist Alan C. Woods, that a large institute of ophthalmology was something that would be good for Hopkins—and Woods concurred enthusiastically. Another former Wilmer patient arranged for Woods and Wilmer to meet, and by 1924, Wilmer agreed to postpone acceptance of a Georgetown University offer to locate his institute there if Hopkins could raise the $3 million necessary to open it in Baltimore. He promptly was recommended for appointment as a professor at Hopkins and named first director of the yet-unbuilt eye institute to be associated with the School of Medicine.[242]

The General Education Board offered Hopkins $1.5 million to match a like sum by the Wilmer Foundation to create the new institute. By late 1924, Hopkins' fundraising effort had slowed to a crawl—but then the 75-year-old warhorse Welch stepped into the breach. Mrs. Breckinridge gave him a list of Wilmer's wealthy patients and he proceeded to work his magic on them. In one morning alone, he got $600,000 in contributions, according to a letter the astounded Wilmer wrote to a friend. One of the men Welch had visited told Wilmer "that he was much relieved when Dr. Welch left, because Dr. Welch was so charming and beguiling that he could have gotten all of his possessions had he remained a little longer," Wilmer wrote. By January 1925, all but $100,000 of the necessary $1.5 million had been raised, and The Johns Hopkins University decided to make up the difference.[243]

Thus was the Wilmer Institute born. When its five-story building was completed in 1928, it had ward, semiprivate, isolation and nursery beds, as well as an emergency operating room, social services and outpatient departments, research laboratories and rooms for teaching, among other facilities. Like Hopkins itself, the Wilmer Institute became a model of its kind and an outstanding, integral part of the institution.[244]

Popsy Bows Out

With his last successful forays into fundraising completed, Welch could gaze on a Johns Hopkins campus that would remain largely untouched for the next quarter-century. Except for minor additions and renovations, no further major building projects or reconstructions were undertaken until the 1950s. The structures for teaching, research, and patient care built in the 1920s would form the basic physical plant of Johns Hopkins for more than 25 years.[245]

In October 1929—less than two weeks before the stock market crash that plunged the nation and world into the Great Depression—the Welch Library and the Department of the History of Medicine were dedicated with ceremonies during which Welch said his only duty was to "sit and blush" at all the accolades heaped upon him.[246]

His repeated request to retire having finally been accepted in 1930, Welch still made occasional appearances at the Hospital and School of Medicine. Howard Jones recalls seeing him attend the famous Feb. 29, 1932, lecture in the newly dedicated Hurd Hall in which Harvey Cushing, visiting from Harvard, presented his landmark paper to the Johns Hopkins Medical Society on the pituitary tumor-related hormonal disorder that became known as "Cushing Syndrome." Jones remembers Popsy tweaking the apparent pride of the one-time Hopkins junior faculty member on his great discovery.[247]

Hurd Hall was packed, Jones recalls. "The professors were there—they occupied the center of the rows of seats…; the chairmen of the departments were there, regardless of what the subject was. And the medical students would occupy the two lateral rows of seats, and I was sitting down low, on the left side."

"Popsy Welch always occupied the first pew, I'll call it, on the right side of the auditorium, facing the front. And no one else sat in that pew but him. He was kind of a roly-poly guy and he'd put his arms out on the back of the bench and he kind of occupied the whole front seat."

Cushing presented nine cases of a tumor of the pituitary gland that caused a group of symptoms he had dubbed "Cushing's Syndrome," Jones recalls, pronouncing "syndrome" as Cushing had done: "syn-dro-mee," not "syn-drome." After Cushing completed his talk, the Medical Society's then-chairman, renowned clinician Louis Hamman, asked Welch if he wished to comment on Cushing's presentation.

"So Popsy got up," says Jones, "and sat on the table that's down in front of Hurd Hall and swung one of his legs as he usually did and said, 'Well, I don't really know anything about what Dr. Cushing has said, but I was awfully interested in how he pronounced the word 's-y-n-d-r-o-m-e'…. Popsy said, 'You know, if I want to go to the movies, I go down on Eutaw Place to the Hippodrome Theatre. I never heard anybody call it the 'Hippo-dro-mee,' and I think if Dr. Cushing is going to use this word again, he better consider how he pronounces the word 's-y-n-d-r-o-m-e.' And that was Popsy Welch's entire discussion of Harvey Cushing's classic paper…. I think Popsy wasn't right on that one, but that was his discussion."[248]

Like many elderly, accomplished individuals, Welch felt he had earned the right to say and do as he pleased—which he did. At the Maryland Club, he co-opted a small room as his private office; at the University Club, he spent hours in the second floor library, ensconced beside a bay window in a special chair in which no one else dared to sit. There, constantly puffing on five-cent cigars, he would read until 2 or 3 a.m., then toddle off to his boarding house. He didn't worry about being robbed at such an hour, he said, because "I never go home until after the bandits have gone to bed."[249]

Simon Flexner (left), first director of the Rockefeller Institute for Medical Research and a protégé of William Welch, watches approvingly as Welch (center) receives yet another beneficence from John D. Rockefeller, Jr. (right). The Rockefeller General Education Board was an immensely generous benefactor to Hopkins Medicine. By the 1920s, it had given more than a quarter of its early multi-million dollar education grants to Welch and his Hopkins colleagues.

1926
February 22: The University celebrates its 50th anniversary. A building boom at the Medical School and Hospital is in full swing.

1927
March 19: Plans are announced for construction of a library for medical and public health students and the Hospital. When built, the library is named for William H. Welch and houses the Department of the History of Medicine (later called the Institute of the History of Medicine), of which Welch becomes the first director. Hopkins' alumni magazine notes that the library has "a room for a secretary" of the director, but no secretary. "Dr. Welch's associates are wondering what he is going to do about that. He has not, throughout his life, engaged a secretary. He prefers to conduct all correspondence himself."

1929
May 7: The Johns Hopkins Hospital marks its 40th anniversary. The Hospital that opened with 330 beds now has 743. During the Hospital's first year, 1,825 patients were treated; during 1928, 11,697 were treated.

October 15: The William H. Wilmer Ophthalmological Institute is dedicated. It is the first in the United States associated with a university. It is under the direction of William Holland Wilmer of Washington, D.C.

1933
As the Great Depression worsens, the medical faculty joins other Hopkins faculties and administrators in voluntarily contributing a portion of their salaries to help avoid a university deficit. *The Sun* observes: "The money itself is extremely important to the university at this juncture; but we doubt that even the money will be as important, in the long run, as this demonstration that the members of its staff believe in it so strongly that they are willing to back their belief with hard cash."

Esther Richards, a 1915 graduate of the School of Medicine and psychiatrist-in-charge of the outpatient department of the Phipps Clinic from 1920 to 1951, lectures to Hospital nursing students in Hurd Hall in 1939. Most likely, she spoke on her specialty, the behavioral and mental problems of children and adolescents.

In February 1933, Welch ate his last dinner at the University Club, then sat at a library table, writing. Later that night, he entered Hopkins Hospital for cancer surgery. He never left it. He spent the last 14 months of his life in the Hospital, mostly cheerful, talkative and just as mysteriously reticent about personal feelings as ever, dying on April 30, 1934.[250]

1935–1939

The Last of the Big Four
With Welch's death, Howard Kelly became the last of the Big Four. Although long since retired from the Hopkins faculty, he still continued to perform and observe operations at the Hospital and on rare instances instructed clinical students into the late 1930s.

"I saw Howard Kelly operate once, but I was not part of the team," Howard Jones remembers. "While I was an intern in GYN at the Hopkins [between 1935 and 1937], Howard Kelly came over and operated on a missionary. He was great on foreign missions. But this person, a woman, was admitted and as I recall it she had fibroids or something like that, and so he scrubbed up…along with Tom Cullen, who was then the chairman of the department—T.S. as we called him. T.S. really did the operation, but Howard Kelly was there and went through the scrubbing up process, so I can say that I saw Howard Kelly operate."[251]

Kelly's last operation at Hopkins was in January 1938, when he was nearly 80. He removed an ovarian cyst and some abdominal adhesions, performing the surgery "with the same ease, dexterity and exactness as he had always done," recalled Cullen in 1943.[252]

Kelly also retained his interest in the School of Medicine—giving a clinical lecture now and then for Cullen and even occasionally helping out financially strapped students.

Dealing with the Great Depression
Thomas Turner (1902–2002), whose association with Hopkins Medicine spanned 75 years, including a term as dean of the School of Medicine from 1957 to 1968, recalled that as the vibrant 1920s gave way to the distressing 1930s, "depression psychology set in—notices went up to turn out unused lights, surgical dressings were used less lavishly, the annual report was reduced in size, and patients were pressed more vigorously for higher payments." [253] Care for those who could not pay was maintained, however, with 66 percent of all the patients in the public wards being treated for free during the Depression's depths.[254]

Growth of Hopkins' clinical facilities was achieved by developing affiliations with other local institutions, such as Sinai Hospital, then located directly across Monument Street from Hopkins Hospital; the Syndeman Hospital for Infectious Diseases, the Children's Hospital School, the Baltimore City Hospitals, and, for

psychiatric rotations, Spring Grove State Hospital. The preclinical departments were allowed to remain small, but for the first time, it was conceded that research must be funded in general from outside sources. It was during this period that Hopkins received its first federal grant for research, when the U.S. Public Health Service provided money for a study of syphilis in 1937.[255]

Dean Alan M. Chesney, who had taken over the deanship from Lewis Weed in 1929, right before the Depression began, succeeded in managing the medical school's finances prudently throughout the 1930s. Despite budget and salary cuts, the faculty was not pressured to produce income for the School of Medicine; and even under severe financial stress, it maintained a nucleus of senior teachers who did not have to engage in private practice to earn an income for themselves or their institution.[256]

The School of Medicine's deficits increased in the early '30s, peaking at $124,000 for the fiscal year of 1934-35. By 1936, however, its budget was in the black and would remain so for the next three years.[257]

Baltimore's Greatest Glory

The Great Depression's economic pressures mandated self-sufficiency, and in that Hopkins had become a master. In a series of articles about Hopkins Hospital written for the *Baltimore Sun* during the summer of 1937, H.L. Mencken marveled at its size and the scope of the in-house facilities it maintained to meet patient needs.[258]

Reporting not just on Hopkins' infrastructure but its medical care, research, teaching and social services, Mencken—not given to dispensing superlatives casually—concluded, "The Johns Hopkins does an invaluable and magnificent work, and is unquestionably the greatest of all the glories of Baltimore."[259]

Thomas Turner believed that perhaps the Great Depression saved Hopkins' full-time system. The basic salary provided inevitably was low, simply because sufficient funds were never available to afford both an increased staff and higher salaries. Yet with young and middle-aged executives in business and industry regularly losing their jobs to Depression-caused layoffs, "the low full-time salary at Hopkins looked relatively better," mused Turner—who was one of those receiving that low full-time salary at the time.[260]

Despite the Depression, by the end of the 1930s, the directors of Hopkins' basic science departments all enjoyed international reputations and each mentored a generation of department heads for other medical schools. With World War II looming, it was to these professors, among many others, that national leaders turned for advice and support on medical and scientific matters, as well as for a significant segment of the research directed to the medical problems encountered by troops during the war. All of this would lay the groundwork for the federal government's mammoth participation in research during the 1950s and beyond.[261]

Alan Mason Chesney (1888–1964), at left, was a 1912 graduate of the School of Medicine who became its longest-serving dean, overseeing it from 1929 to 1953. A fierce advocate of researchers' rights, he opposed efforts by anti-vivisectionists to restrict animal research in Baltimore. When a supplier of laboratory animals was arrested in 1948 on charges brought by the Society for the Prevention of Cruelty to Animals, Chesney and his wife put up their home as bail security. In 1950, Chesey successfully led the fight to defeat a proposed anti-vivisection law in Baltimore.

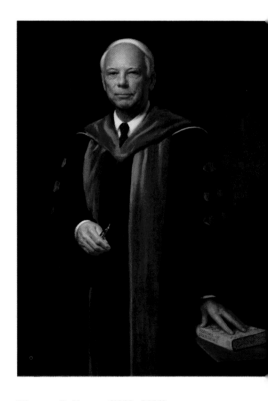

Thomas B. Turner (1902–2002) spent nearly 75 years of his career at Hopkins, including nine years as dean of the School of Medicine (1957–1968). The last dean to have known all of his predecessors except Osler, he regretted the expansion of Hopkins that he helped to foster but which reduced, he believed, its intimate atmosphere. A witty observer, Turner offered a series of tips for longevity when he turned 90, including: "Almost all soups can be improved by a dash of sherry," "If given a book, thank the giver within 48 hours; otherwise you really will have to read it," and "Hold onto the bannister when going downstairs." He lived to be 100.

Hopkins Medicine Discoveries and Breakthroughs, 1889–1939

1890 — Surgeon in chief William Halsted and other Hopkins colleagues begin using rubber gloves during surgery, popularizing a practice that ultimately will become universal.

1896 — Neurosurgery resident Harvey Cushing, later to become the father of modern neurosurgery, performs the first neurologic application of X-rays, using Hopkins Hospital's hand-cranked, first X-ray machine to locate a bullet lodged in the sixth cervical vertebra of a Baltimore woman who had been shot.[262]

1897 — John Jacob Abel, first professor of pharmacology, discovers what he calls "epinephrine," also now known as adrenalin, a natural substance produced by the adrenal glands. It becomes a crucial, first-line treatment for heart attacks and severe allergic reactions, among other conditions.[263]

1906 — Anatomist Ross Granville Harrison takes spinal cord cells from a frog embryo and a drop of lymph sac fluid to create the first tissue culture. Harrison's method forms the basis of modern tissue culture technique, a cornerstone of today's medical research, enabling scientists to study isolated living cells in a controlled environment. Harrison's student, John Enders, used this method to devise a test for mumps and to grow polio virus cultures, leading to development of the Salk polio vaccine. Harrison's technique still is the most efficient way of studying living cells and tissue and is used both in *in vitro* fertilization and the cultivation of viruses for modern vaccines.[264]

1909 — Urologic surgeon Hugh H. Young invents his punch resectoscope for treating enlarged prostates without resorting to invasive surgery. Young had performed the first perineal prostatectomy in 1902 and the first radical perineal prostatectomy to remove a cancerous prostate gland in 1904.[265]

1913 — John Jacob Abel invents the first "artificial kidney" dialysis device.[266]

1913 — Pioneering neurosurgeon Walter Dandy publishes the work that first will make him famous, a series of 10 reports on hydrocephalus, an accumulation of fluid within the cranium.

1914 — Internist Theodore Janeway becomes the first to prescribe digitalis for patients who have normal cardiac rhythm but cardiac insufficiency due to prolonged high blood pressure.

1915 — Second-year medical student Jay McLean discovers the critical anticoagulant heparin, vital for preventing dangerous blood clots. Since the substance is derived from the liver, McLean's professor, William Henry Howell, bases its name on the ancient Greek word for liver, *hepar*.[267]

1915 — Biologists Warren and Margaret Lewis refine Ross Granville Harrison's tissue culturing technique, developing a clear culture medium to study individual cells and provide the first detailed description of the living cell.[268]

1916 — William Henry Welch founds Johns Hopkins School of Hygiene and Public Health, now the Bloomberg School of Public Health, the first such institution in the world and still the largest and best of its kind.[269]

1917 — Neurosurgeon Walter Dandy revolutionizes the diagnosis of brain tumors and the field of neurosurgery by inventing ventriculography, the injection of air into the ventricles of the brain, thereby for the first time enabling X-rays to reveal the precise location of tumors.[270]

1917 — Florence Sabin, a 1900 graduate of the School of Medicine and authority on blood cells and the body's defenses against infection, becomes the first woman appointed a full professor at Hopkins. In 1925, she becomes the first woman elected to the National Academy of Sciences. The impact of her pioneering research will be recognized by placement of her statue in the U.S. Capitol's Statuary Hall in 1959.[271]

1923 — Neurosurgeon Walter Dandy creates a 24-hour, specialized nursing unit for critically ill neurosurgical patients, establishing what is considered the forerunner of today's intensive care units.

1924–1927 — Pharmacologist John Jacob Abel becomes the first to purify and crystallize insulin, a key to developing a safe treatment for diabetes.[272]

1926 — Urologic surgeon Hugh H. Young reports on the Brady Urological Institute's first use of a mercury-based medication he and colleagues have dubbed "mercurochrome" to combat local and general infections. This is the first important effort to apply German scientist Paul Ehrlich's principles of chemotherapy.[273]

1929 — Walter Dandy is the first to recognize that protruding, ruptured or "herniated" spinal discs were one of the most frequent causes of back pain and sciatica, and operates to remove the loose cartilage.[274]

1935 — The era of antibiotics in American medicine is launched by Perrin H. Long, head of Hopkins' then-Department of Preventive Medicine, and three Hopkins School of Medicine alumni, Alexander Ashley Weech, Class of 1921, Francis Howell Wright, Class of 1933, and Dorothy Hansine Andersen, Class of 1926. Working at New York's Babies Hospital of the Columbia Medical Center, Weech and Wright are the first to use the German-discovered antibiotic sulfachrysoidine (Prontosil) in the United States, administering it to a 10-year-old girl critically ill with leptomeningitis. The child's condition improves briefly but she is so ill by the time the drug is used that she dies. Her autopsy, crucial to determining how future such cases might be handled, is performed by Andersen. Long is so renowned for his pioneering use of so-called sulfa drugs in the mid-1930s that he is contacted by the White House in 1939 to provide such medication for a son of President Franklin D. Roosevelt.

1936 — Gynecologic endocrinologist Georgeanna Seegar Jones first demonstrates that the pregnancy hormone (now called chorionic gonadotropin) originates in the placenta rather than in the pituitary gland, as previously thought, establishing the foundation for the development of pregnancy tests that are used today.[275] In 1939, she becomes the first chief of Hopkins' division of endocrinology and infertility—the first of its kind in the world. She remains head of it until 1978.

1937 — Walter Dandy establishes the field of vascular neurosurgery by performing the first surgical clipping of an intracranial aneurysm—one of many firsts he would achieve in this specialty.[276]

1937 — Urologist Hugh H. Young conducts experiments that show the effectiveness of sulfanilamide in treating gonorrhea and later uses it and other sulfa drugs to treat genitourinary infections.[277]

School of Medicine Deans, 1889–1939

William Henry Welch (1893–1898)

Welch carried almost total responsibility for getting the School of Medicine ready for classes in less than a year, once the money became available to launch it. This monumental task included writing all of his correspondence in longhand—since he never used a secretary and could not type. Welch disliked administrative work, however, and began submitting his resignation as dean after only two years—but it was rejected repeatedly until, finally, after two classes had graduated, he at last was released from the job. He remained as head of pathology and in 1916 became the first dean of the School of Hygiene and Public Health.[278]

William Osler (1898–1899)

As anticipated, Osler succeeded Welch as dean of the School of Medicine—but didn't like the administrative burden any more than Welch did. He requested that a replacement be found promptly, and one was.

William H. Howell (1899–1911)

A physiologist with a Hopkins Ph.D., Howell was the rare medical school dean who wasn't a physician. His career at Hopkins spanned 38 years, 12 of them as dean of the medical school and 13 as dean of the School of Public Health. During his tenure as dean, he wrote a textbook that for 40 years was the physiology bible for medical students. He also maintained most of his teaching load and continued his research. After he relinquished his duties as the medical school dean, he devoted all his attention to a problem that fascinated him: blood clotting. He and one of his students, Jay McLean, discovered the anticoagulant heparin. He eventually became dean of the School of Public Health before his so-called retirement in 1931. He actually kept up his laboratory research until his death in 1945, two weeks shy of his 85th birthday.[279]

J. Whitridge Williams (1911–1923)

An obstetrician descended from a long line of physicians who began practicing medicine in the 18th century, Williams was a protégé of Hopkins' first gynecology professor, Howard Kelly, and served as obstetrician in chief at Hopkins Hospital for 12 years. A popular teacher, nicknamed "The Bull" because of his massive size and rumbling voice, Williams was known for his punctuality and love of routine. Every day at 9:04 a.m., he greeted his resident with his version of good morning: "Well, whom did you kill last night?" Williams' droopy walrus mustache was tobacco-stained from incessant pipe-smoking. He kept seven pipes in constant rotation, each identical except for a Roman numeral he carved on the side of each bowl, corresponding to the days of the week. As dean of Hopkins' Medical School, he helped reorganize Baltimore City Hospitals, the municipal facility that now is known as Johns Hopkins Bayview Medical Center, setting up full-time staffs in pathology, medicine and surgery, and opening up its clinics so students from all of Baltimore's medical schools could train there. During his career, he studied the abnormal pelvis and the normal placenta, and is credited with being largely responsible for the establishment of adequate prenatal care. In 1931, the year he died, Williams worked for the repeal of a federal law forbidding the sending of birth-control information through the mail.[280]

Lewis H. Weed (1923–1929)

A 1912 graduate of the School of Medicine, Weed was an anatomist who specialized in experimental neurology, making exhaustive studies of the relations between cerebrospinal fluid and blood plasma. In his 39 years at Hopkins, he led the Department of Anatomy, adding anthropology, comparative anatomy and histochemistry to its fields. He saw the medical school through a massive rebuilding phase, and gained national distinction by serving as chairman of the medical sciences division for the National Research Council and on the Council of National Defense.[281]

Alan M. Chesney (1929–1953)

After serving as Weed's assistant for two years, Chesney took over as dean. His 24 years in the post was the longest tenure of any Hopkins medical school dean, seeing the school not only through the Great Depression but also World War II. Like Weed, also a 1912 graduate of the School of Medicine, Chesney spent his earlier years in research and patient care. He directed Hopkins' laboratory for the experimental study of syphilis, established in 1921, and most of his research revolved around the basic mechanism of immunity in venereal disease. While dean, he began a three-volume history of Hopkins Hospital and the School of Medicine. It covers the founding of both institutions and their development until 1914. His efforts to preserve the early records that he located during his historic research led to creation in 1978 of the now much-expanded Alan Mason Chesney Medical Archives, considered among the finest such repositories in the country.[282]

1940–1988:
Hopkins Medicine Moves into the Modern Era

Among the best-selling books of 1939–1940 was *Miss Susie Slagle's*, a sentimental novel about an unmarried, motherly landlady who looked after the Johns Hopkins medical students living in her Biddle Street boarding house. Set in the years just prior to World War I, it was written by Augusta Tucker Townsend (1904–1999), a slim, energetic and prolific freelance writer who published under the name Augusta Tucker. When her New York editor suggested that she write a medical novel, she unexpectedly produced a major hit. The tale of Miss Susie and her frequently harried Hopkins students, all struggling to become great doctors, vaulted to the *New York Times* best-seller list and went through 23 printings in four languages. It even was made into a Hollywood movie.[1]

"She became a town personage after the sudden and startling success of this book," recalled James Bready, a longtime book columnist for the *Baltimore Sun* and its *Evening Sun*. "It was a book that persuaded students to apply to Hopkins. It was the best possible propaganda for this institution. Hopkins gave her her own white coat with her name sewn above the breast pocket, doctor-style."[2]

The world portrayed in *Miss Susie Slagle's* was familiar to the Hopkins students who had gone through the School of Medicine from its founding until the end of the 1930s, a period when certain traditions, rules and mores prevailed despite the passage of decades. Much of that world would vanish in the next 10 years. The upheavals of World War II, as experienced both on the home front and overseas, swept away a lot of it—with some aspects long overdue for discarding. What remained soon would be battered and dispatched by the boom-time of the 1950s, tumult of the 1960s, and the enormous changes of the 1970s and 1980s. In its place arose a far larger, more complex, increasingly diverse—and better—Johns Hopkins.

Ranice W. Crosby (1916–2007) was a protégé of Max Broedel, founder of the Department of Art as Applied to Medicine, and was just 29 years old when she became head of the department in 1943. She held that post for an astounding 40 years, longer than any other clinical or basic science department head in the School of Medicine's history. Despite retiring from the directorship, she continued teaching until shortly before her death.

1940s

During the early 1940s, Hopkins lost the last of its founding fathers. Max Broedel, director of the Department of Art as Applied to Medicine since it opened in 1911, retired in 1940 at the age of 70 and died the following year. His immediate successor was James Didusch, who previously had concentrated his artistic work in the Department of Embryology. After two years, Didusch was succeeded by Broedel's protégé Ranice Crosby, then just 29 years old. Not only was she the first woman to direct a department, she held the directorship for an incredible 40 years before stepping down in 1983—although she continued teaching. Her successor, Gary Lees, a veteran of the faculty since 1970, has been head of the department since then.

Howard Kelly, the last of the Big Four and the man who considered his greatest accomplishment the employment of Max Broedel, died of pneumonia in 1943, six weeks shy of his 85th birthday and about six hours before his wife of 53 years, Laetitia, died in the hospital room next to his. They were unaware that they had been hospitalized simultaneously and did not know the gravity of each other's condition. A double funeral was held. Mourned by the entire medical community, Kelly had maintained his multiple interests even as he grew frail in his later years. Richard W. TeLinde, one of Kelly's successors as head of gynecology at Hopkins, recalled visiting his mentor's jam-packed library about a year before Kelly's death and finding the venerable surgeon happily holding a large tarantula in the palm of his hand. "See," he told TeLinde, "it won't bite you if you don't annoy it."[3]

In February 1945, as World War II was drawing to a close, the last of the School of Medicine's original faculty—pioneering physiologist William Henry Howell—died precisely two weeks before his 85th birthday, and just two days after he'd last visited his laboratory. Howell, who entered The Johns Hopkins University with its third undergraduate class in 1878, obtained his physiology Ph.D. in 1884 and became the School of Medicine's first professor of physiology when it opened in 1893. In addition to his groundbreaking work on blood coagulation and landmark physiology textbook,

■ In 1940, neurosurgeon Walter Dandy created the prototype of the modern baseball batter's helmet (far right). Asked by Larry MacPhail, then president of the Brooklyn Dodgers, to create batting headgear that would look like a normal cap yet protect players from being beaned, Dandy and orthopedic surgeon George Bennett came up with a cap that had zippered pockets on each side into which curved, plastic protective shields could be inserted. It first was used in 1941. Among those who pioneered use of the new cap was Dodger left fielder Joe Medwick, showing the innovative plastic shield.

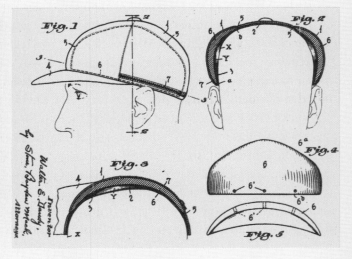

the soft-spoken, Southern-accented and Van Dyke-bearded Howell was revered for his superb teaching and warm personality. His protégé Joseph Erlanger, an 1899 graduate of the School of Medicine who won the Nobel Prize in physiology or medicine in 1944, wrote of him: "Howell was one of the best loved of American physiologists. A kindly disposition and unpretentiousness of manner endeared him to all who knew him well."[4]

Blalock Becomes Head of Surgery

In 1941, an unexpected but fortuitous decision to select Alfred Blalock to become director of the Department of Surgery launched an extraordinary era in the history of Hopkins.

When Halsted died in 1922, his immediate successor as acting director of surgery had been one of his protégés, J.M.T. Finney. He was an 1888 graduate of the Harvard school of medicine about whom Halsted had heard good things. He invited Finney to the opening of Hopkins Hospital in 1889, and literally hired him on the spot that day. Finney would earn great renown as a superlative surgeon. He was elected the first president of the American College of Surgeons when it was founded in 1913 and once even turned down the presidency of Princeton. Yet he considered himself too old at 59 to succeed Halsted when his old mentor died.[5]

Instead, Finney headed surgery on an acting basis for three years until a permanent successor to Halsted could be found. In 1925, Dean DeWitt Lewis, a professor of surgery at the University of Illinois in Chicago, was chosen to head surgery and be surgeon in chief at Hopkins Hospital. In 1938, he had a massive stroke but remained as head of the department until resigning in 1939.[6]

It took the School of Medicine's search committee two years to sift through possible surgery directors before finally settling on Blalock, a 1922 graduate of the School of Medicine who by 1940 was a well-known professor of surgery at Vanderbilt.[7]

Blalock (1899–1964) was a native of Culloden, Georgia, who never lost his pronounced Southern drawl and easy-going charm—at least outside of the operating room. After graduating from the School of Medicine, he spent three years on the surgical house staff but was not chosen to complete his residency at Hopkins. Bitterly disappointed, he moved to Vanderbilt as a surgical resident, then joined its faculty.[8]

First at Hopkins and later at Vanderbilt, Blalock conducted landmark research on surgical shock, which many had thought was caused by an unidentified toxin, but which Blalock proved was due to a drop in blood volume and loss of fluid.[9] By the mid-1930s, he was a recognized authority in the field. It is certain that his studies on the crucial importance of administering blood transfusions led to superior results in the treatment of surgical and traumatic shock among servicemen during World War II. [10]

Alfred Blalock (1899–1964), director of the Department of Surgery from 1941 to 1964, was best known for launching modern cardiac surgery with the famous blue baby operation in 1944. However, he believed that equally important was his landmark research on surgical shock, revealing it to be due to a drop in blood pressure and loss of fluid. His studies led to superior treatment of surgical and traumatic shock among wounded servicemen during World War II.

■ Britain's King Edward VIII (1894–1972), better known by his post-1936 abdication title, the Duke of Windsor, made use of Hopkins Hospital's services. In 1941, the Duke came to Hopkins with his Baltimorean wife, Wallis Warfield Simpson (above), who was consulting a Hopkins gastroenterologist. The Duke suddenly decided that he'd like to be examined, too. Renowned internist Benjamin Baker (1902–2003) obliged the ex-king, who paid for the service with a $50 check drawn on the Bank of England and signed simply "David" (the name by which he was known in the royal family).

Vivien Thomas (1910–1985), for decades the laboratory technician and surgical assistant to Alfred Blalock, was long the unheralded contributor to development of the procedures used in the groundbreaking blue baby operation. Acclaim for the 1944 surgery was universal—for everyone but Thomas, whose role was left in the shadows due to the racism of that era. By the 1960s, however, he received full recognition, including the portrait above, commissioned by the surgical residents he had trained, and an honorary degree from the University.

"There was quite a bit of tension around the surgical department, and around the Hospital, when it was evident that they were going to proceed with this new and revolutionary approach to the treatment of a cyanotic infant."

— Denton A. Cooley

1942

June: As an emergency wartime measure, an accelerated program of courses is instituted, made possible by eliminating the summer vacation but not reducing class hours. Its adoption reflects the federal War Manpower Commission's requests for more physicians. In effect until March 1946, the program enables two classes to graduate each year, one in February and the other in November. Some of the rigid admission requirements, such as the need for a bachelor's degree and competence in German and French, also are lifted.

Johns Hopkins Medicine personnel staff two 500-bed military hospital units in the Pacific during World War II. General Hospital 18 spends two years in Fiji, then moves on to the India-Burma Theater. General Hospital Unit 118 is based successively in Australia, Papua New Guinea, and the Philippines and treats more than 40,000 patients.

The Blue Baby Operation and the Birth of Cardiac Surgery

When Blalock returned to Hopkins in July 1941, he brought with him from Vanderbilt his longtime technical assistant, Vivien Thomas (1910–1985), an African American carpenter's son from Louisiana whose dreams of becoming a surgeon had been dashed by the Great Depression. Forced to leave college in 1930, he obtained a job in Blalock's Vanderbilt laboratory, where he soon became indispensable as a technician and surgical assistant. By the time he accompanied Blalock to Baltimore, Thomas was a skillful surgeon able to perform single-handedly many intricate procedures for which other surgeons would have required assistance, having independently refined and perfected his abilities while practicing on animals.[11]

Although the Hopkins of 1941 followed the customs of a Southern city of that day and was strictly segregated, Thomas Turner believed that Blalock "undoubtedly informed Hopkins administrative authorities" that his closest aide was a black man and must be hired, too. Hopkins "found no objection" to Vivien Thomas' employment in Blalock's lab, "for it was known that Blalock had previously declined a major surgical post at another institution principally because of its restrictive policy on the employment" of African Americans, according to Turner's 1974 history of Hopkins, *A Heritage of Excellence*.[12] That may be so, but the sight of Vivien Thomas walking the hallways of Hopkins in a white lab coat still caused heads to turn.[13]

Determined to maintain his research despite his ongoing surgical, teaching and now-heavy administrative burdens, Blalock was back in his old haunts in Hopkins' Hunterian laboratory within days of his return to Baltimore. He began to shift his focus from shock to problems of cardiac surgery, still a largely uncharted realm into which many surgeons were reluctant to venture. Most still held the opinion voiced in 1896 by British physician Stephen Paget that surgery on the heart had "probably reached the limit set by nature to all surgery. No new method and no new discovery can overcome the natural difficulties that attend a wound of the heart."[14]

Blalock didn't believe that. He and Thomas already had conducted experiments on dogs at Vanderbilt that offered possibilities for undertaking prolonged operations to correct heart defects. He was determined to push that boundary and open the way to new surgical procedures.

Knowing of Blalock's interest in heart surgery, Edwards Park, head of pediatrics and the Harriet Lane Clinic, asked him in 1943 to examine a child with a congenital narrowing of the heart's main vessel (a condition known as coarctation of the aorta). Park thought an operation to bypass such constrictions might be developed. Blalock's work on this problem brought him into closer contact with Hopkins' pediatricians, especially Helen Taussig (1898–1986), then head of pediatric cardiology.[15]

Taussig was concerned not only with aortic stenosis (the inability of the aortic valve to open fully) but an even thornier condition that afflicted and inevitably killed young patients. It was a quartet of congenital defects that prevented the children's small hearts from functioning properly and sending

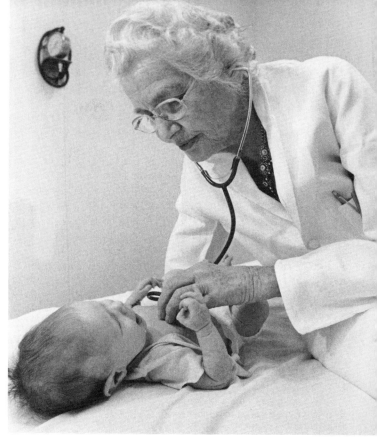

enough blood to the lungs so that when it returned to the heart, sufficient oxygen-enriched blood could then be pumped throughout their bodies. Its clinical name was tetralogy of Fallot, a combination of the word for congenital abnormalities (*tetralogy*) and the name of French physician, Étienne-Louis Arthur Fallot (1850–1911), who first described the combination of anatomical problems that led to the condition. Its more common name was "the blue baby syndrome," since the skin of the children who struggled to survive on oxygen-low blood frequently turned blue (or cyanotic, in medical terms).[16]

Taussig was a 1927 graduate of the School of Medicine who had been denied admission to Harvard because she was a woman. A protégé of Park, who had named her head of the Harriet Lane cardiac clinic in 1930, she suggested to Blalock that procedures might be devised to "reconnect the pipes" in the defective hearts of the blue babies, thereby ensuring an increased blood flow between the heart and lungs and curing the condition. Blalock and Thomas, building on the work they already had done on dogs to address other heart problems, spent months experimenting on several hundred dogs to re-create the blue baby condition in them and then fix it. They finally devised methods that they believed could be tried safely on humans.[17]

On November 29, 1944, with Taussig observing from the head of the operating table and Vivien Thomas standing on a step stool right behind Blalock, the surgeon performed the first blue baby operation on a frail, 18-month-old girl named Eileen Saxon. Thomas, who had performed far more of the experimental procedure on dogs than had Blalock and also devised the adapted surgical clamps and needles for it, provided guidance as the delicate operation proceeded on the faintly beating heart—something surgeons for centuries had warned against attempting.[18]

Colleagues of Blalock's and medical students jammed the two-level observation gallery overlooking the Halsted clinic operating room. Anesthesiologist Merel Harmel, Blalock's chief surgical resident William Longmire, and nurse Charlotte Mitchell also were in standing beside the operating table—as was a lanky class of 1944 Hopkins Medical School graduate, a six-foot, four-inch 24-year-old intern named Denton Cooley. He was holding the chest retractor and standing ready to administer fluids to the tiny patient. Cooley would become one of the world's premier heart surgeons. He performed the first repair of an aortic aneurysm, the first removal of a pulmonary embolism, and co-developed the first heart-lung machine in 1955 (with Michael DeBakey of Houston's Methodist Hospital), which ever since has enabled surgeons to keeps patients' blood circulating during open-heart surgery. Cooley also implanted the first human artificial heart in 1969 (along with Domingo Liotta at the Texas Heart Institute, which Cooley founded), performed dozens of heart transplants, and set the international record for the number of heart bypass operations. Ninety years old at the time of this writing, Cooley remembers the details of the first blue baby operation vividly.[19]

Helen Taussig (1898–1986), co-developer of the blue baby operation, or what became known in medical circles as the "Blalock-Taussig Shunt," became a pioneering pediatric cardiologist despite dyslexia, hearing impairment, and prejudice against women physicians. She would share with Blalock the 1954 Albert Lasker Award for outstanding contributions to medicine; receive the Presidential Medal of Freedom in 1964; and become the first woman elected president of the American Heart Association in 1965.

Hopkins' blue baby operation captured the public's imagination and garnered immense publicity, such as the story above in the Feb. 17, 1947 issue of *American Weekly*, a widely distributed Sunday newspaper feature section.

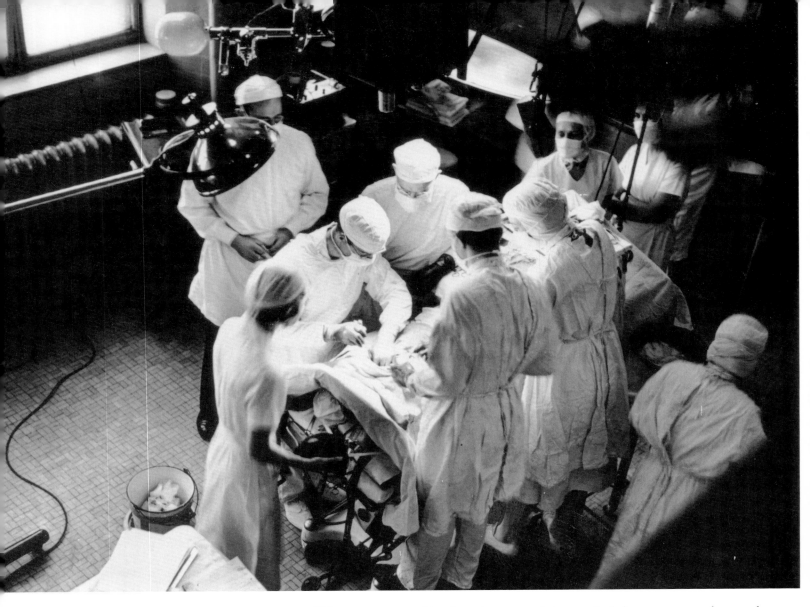

On Feb. 28, 1947, one of the early blue baby operations was performed on a then-revolutionary closed-circuit television broadcast. Alfred Blalock can be seen in the center of the group on the left side of the operating table, leaning over the patient. Standing behind him, hands folded and partly obscured by one of the TV lights, is his long-time surgical technician and assistant, Vivien Thomas. Also shown are nurse anesthetist Olive Berger and first assistant resident surgeon Henry Bahnson.

■ Although dialysis first became a vital treatment for patients with defective kidneys in 1946, Hopkins pharmacology chief John Jacob Abel (1857–1938) devised the first machine for cleansing impurities from the blood 33 years earlier. In 1913, Abel hooked up his device to a rabbit and successfully eliminated salicylic acid from the bunny's bloodstream, mimicking the normal operation of a kidney.

"There was quite a bit of tension around the surgical department, and around the Hospital, when it was evident that they were going to proceed with this new and revolutionary approach to the treatment of a cyanotic infant," Cooley recalled in 2008. "Some people even thought it was sort of a human sacrifice even to undertake such an operation."[20]

Blalock, humane and warm outside of the operating room, could become volatile and abusive to associates during surgeries. He also frequently asked his assistants, "What should I do now?" or "How should I do this?"—actually as a way of teaching his young associates but prompting some observers to think he was insecure or inept. Far from it. Although he often asked Thomas for suggestions during the first blue baby operation, Cooley says that "Dr. Blalock knew what he was going to do: He just liked to have everybody feel like they were part of the team."[21]

As recreated dramatically 60 years later in the 2004 Home Box Office movie, *Something the Lord Made*, little Eileen Saxon's blue color changed almost instantly when Blalock completed the rearrangement of her heart's physiological plumbing.

"I remember when Dr. Blalock took the clamp off that subclavian artery [a vessel that carries blood away from the heart] and the blood began to flow back into the pulmonary artery [which carries deoxygenated blood to the lungs] and was exposed to the lung. Dr. Blalock looked over the ether screen, as we called it, and Dr. Taussig said, 'Al, the baby is a glorious pink color.'

"And from that day forward, every time we did the same thing, took the clamp off and watched the baby turn from blue to pink, it was very inspirational to all of us," Cooley says.[22]

It inspired more than just the surgeons in the Halsted clinic's operating rooms. Cardiac surgery literally was "born that day," according the 2008 history, *Pioneers of Cardiac Surgery*, written by William S. Stoney, an emeritus professor of cardiac and thoracic surgery at Vanderbilt University. "It really opened up the field and revealed that we could start making more advances in…congenital heart disease," Cooley says.[23]

Word of the operation's success spread rapidly. Although Eileen Saxon died shortly after her surgery due to unrelated complications, subsequent operations had resoundingly happier outcomes. When Blalock and Taussig published their landmark Johns Hopkins Medical Society paper on the procedure in the May 14, 1945, issue of *Journal of the American Medical Association*, the acclaim was universal—for everyone except Vivien Thomas. His contributions to development of it, known but unacknowledged at Hopkins, were left in the shadows because of the racism that generally prevailed in society during that era. The depth of his role was not revealed to the general public until decades later. In 1968, the surgeons who Thomas had done so much to train, including Cooley, commissioned a portrait of him to be hung in the Alfred Blalock Clinical Sciences Building; and in 1976, Hopkins gave Thomas an honorary doctorate. At the 1971 portrait presentation ceremony, Thomas displayed an extraordinary lack of bitterness and an enviable sense of humor. "You have to be nice to your students," he said, "because you never know if one will come back as your boss."[24]

The relationship between Blalock and Taussig, while mutually respectful outwardly, was strained considerably by an intense competitiveness over who deserved the lion's share of credit for what became widely known as the blue baby operation or the Blalock-Taussig shunt procedure.[25] Blalock gave Taussig credit for her contributions in the original report on the first operation, but she would not mention his name in her subsequent articles about the procedure, which she suggested had been entirely her idea. Blalock would contend that Taussig's description of the problem was not equivalent to devising a solution for it.[26]

For her part, Taussig, who was dyslexic and partially deaf, later expressed disappointment over what she saw as an effort by the male-dominated medical profession to deny appropriate credit for her accomplishments. "It hurt when Dr. Blalock was elected to the National Academy of Sciences and I didn't even get promoted from assistant to associate professor," she said decades later. She did not become a full professor until 1959.[27]

Regardless of the tension between the parents of the blue baby operation, its success was a triumph for Hopkins. Having gone through a difficult period of restricted growth and recognition imposed by the Depression and World War II, the dawn of a new epoch in cardiovascular surgery at Hopkins brought with it a steady flow of patients and professional visitors. The blue baby operation put Hopkins back on the map as the pre-eminent exponent of medical research and clinical advancements.[28]

1943
February 28: The SS *Johns Hopkins* Liberty ship is christened. It is one of more than 2,700 Liberty Ships launched during World War II to transport cargo to military outlets overseas. It was damaged by a mine in Marseilles but was repaired and returned to action.

1944
November 29: Cardiac surgery is born as surgeon Alfred Blalock performs the first "blue baby" operation after collaborating with pediatric cardiologist Helen Taussig and laboratory technician Vivien Thomas to devise the procedure for correcting congenital heart defects that result in poor blood circulation. One grateful mother wrote of her son's subsequent surgery: "All of a sudden the rosy complexion of his face struck me. His lips were bright red…. The following day he was sitting up in his bed. The blood circulation in his lungs was fully restored by a miracle of intelligence and skill, intense work, and exacting attention which Professor Blalock and his assistants accomplish every day."

Although mutually respectful, **Alfred Blalock**, left, and **Helen Taussig**, competed intensely for credit as developers of the blue baby operation. Taussig would contend that the idea for the innovative surgery to "reconnect the pipes" in the hearts of children with the blue baby syndrome was hers, while Blalock would insist that proposing development of a procedure to repair the physical problem of the blue babies' congenital heart condition was not the equivalent to devising how to do it. Taussig also believed that prejudice against women prevented her from receiving full recognition for her achievements in the mid-1940s. Hopkins did not promote her to a full professorship until 1959.

First Lady Eleanor Roosevelt visited Hampton House, the residence of Hopkins' nurses and nursing students, early in 1945 and had tea with them in the lounge area. Mrs. Roosevelt earlier had visited Army Medical Corps' 118th General Hospital unit, staffed by Hopkins physicians and nurses in Sydney, Australia.

Members of Hopkins' 118th General Hospital Unit during World War II (lower right) sit in front of one of their Sydney, Australia barracks in 1943. Above the door they've nailed a small sign saying "The Maryland Club," a mordantly amusing allusion to the venerable (and exclusive) private dining and social institution in Baltimore. During the course of the war, the physicians and nurses of the 500-bed 118th treated more than 40,000 casualties.

■ In 1947, Triple Crown champion Assault, winner of the 1946 Kentucky Derby, Preakness and Belmont Stakes, funded a neurosurgery fellowship at Hopkins. The thoroughbred was owned by Texan Robert Kleberg, proprietor of the King Ranch, the world's largest. A close relative of Kleberg's had been a patient at Hopkins, and when Assault won Pimlico's 1947 Dixie Stakes (often called the Dixie Handicap), Kleberg donated the winnings to Hopkins for creation of a neurosurgery fellowship.

Hopkins Hospital and School of Medicine During World War II

As the meticulous research that produced the blue baby operation was being pursued, elsewhere within Hopkins the impact of World War II was felt every day. After Pearl Harbor, the Hospital made plans for handling a sudden influx of civilian casualties (not unlike the planning that took place 60 years later following the September 11, 2001 terrorist attacks). To guard against air raids, Hospital windows were blacked out at considerable cost. Some 88 faculty joined the Army, 19 signed up for the Navy and five entered the U.S. Public Health Service, while house officers also went directly into the military.[29]

With the ranks of medical school applicants depleted by the military enlistment and drafting of college graduates, the school needed to bend its rigid requirements and admit a few students without bachelor's degrees. This could be done only after getting specific permission from Bryn Mawr College. Among the provisions of Mary Elizabeth Garrett's gift had been the warning that if the School of Medicine ever violated any aspect of the high admission guidelines it had pledged to maintain—such as the requirement that all applicants admitted to the school have a bachelor's degree—the money would be revoked and transferred to Bryn Mawr. With Bryn Mawr's OK to overlook the bachelor's degree requirement during wartime, among those who got into Hopkins without a B.A. were future genetic medicine legend Victor McKusick, admitted in 1942; and Richard Johns, admitted in 1944 and later the founder and longtime director of the Department of Biomedical Engineering. The medical school also lifted the requirement that applicants have competence in German and French.[30]

The nation's need for more physicians brought changes in medical education, requested by the federal War Manpower Commission. An accelerated program of courses, made possible by eliminating the summer vacation so that students could graduate early without a reduction in class hours, was in operation from June 1942 until March 1946. One new class would graduate in February of each year, the

other in November, a production of physicians that certainly was a contribution to addressing the professional personnel shortage during the conflict.[31]

In the first six months of the war, activation of two Hopkins Hospital-affiliated military medical units removed 38 members of the faculty at one stroke, and 25 others were given leaves of absence for military service. Ninety-two members of the faculty were in uniform by June 1942. Back in Baltimore, researchers at Hopkins made medical discoveries that greatly reduced the death toll from disease in the Pacific theater of operations.[32]

Hopkins' 18th General Hospital Unit—the same one that had operated in France during World War I—was reactivated. It was based first in New Zealand, then in Fiji and finally in the India-Burma theater of operations. (After the 18th General arrived in New Zealand, the Army shipped it boxes of blankets that turned out to have been packed in 1917 for its France-based predecessor.)[33] Another unit, the 118th General Hospital, first was based in Sydney, Australia, and later moved to New Guinea. Hopkins physicians and nurses staffed each of these hospital units, which could care for up to 500 patients at a time. During the course of the war, the 118th unit treated more than 40,000 casualties. Other Hopkins-trained physicians served throughout the military and on battlefields on every front, including on transport ships for the troops that landed on Omaha Beach on D-Day.[34]

On the home front, Lewis Weed, head of the anatomy department, served as chairman of the National Research Council in Washington. Weed also was vice chair of the federal Committee on Medical Research, which offered insights on wartime medicine.[35]

In the Hospital and medical school, infectious disease expert Frederick Bang oversaw an intensive program among several clinical divisions to study new anti-malarial drugs—an effort that earned him the Legion of Merit. In another breakthrough, Department of Pharmacology chief Eli Kennerly Marshall Jr. developed sulfaguanidine, a drug that helped American and Australian forces to recover speedily from a severe outbreak of dysentery, enabling these troops to successfully repulse the 1942 Japanese attack on New Guinea's vital Port Moresby. Marshall also worked on developing better treatments for malaria, and his pioneering work made him a principal architect of the scientific age of chemotherapy.[36]

With wartime demands on faculty imposing severe strains on the Hospital and School of Medicine, leaders in both who had been planning to retire instead remained in harness for the duration. Winford Smith, director of the Hospital since 1911, kept at his post until 1946, as did the Hospital's physician in chief and director of the Department of Medicine, Warfield T. Longcope, who was named to those jobs in 1922 and had hoped to retire in 1942. When the war ended, these and other aging faculty and administrators finally stepped down. They were replaced by younger men with new attitudes and values that would dramatically change Hopkins.[37]

By 1944, Hopkins' 118th General Hospital Unit had been transferred from Australia to the island of Leyte in the Philippines. During the hurry-up-and-wait periods pending the suddenly swift arrival of casualties to treat, the physicians and nurses could spend some R&R time on the beach.

■ Howard G. Bruenn (1905–1995), a 1929 graduate of the School of Medicine, became chief of the Vanderbilt cardiac clinic and a renowned author of studies on internal medicine and cardiology. Yet Bruenn was best known for a single patient he treated for just 13 months: President Franklin D. Roosevelt. Bruenn, a Lieutenant Commander in the U.S. Navy and chief of cardiology at Bethesda Naval Hospital during World War II, became FDR's personal physician in March 1944 and served him continuously until Roosevelt's death from a cerebral hemorrhage on April 12, 1945. Ordered by his Navy superiors to keep the facts about FDR's medical condition confidential, Bruenn did not write about Roosevelt's chronic hypertension and congestive heart failure until 1970— when urged by FDR's family to do so.

The last photo of President Franklin D. Roosevelt, taken the day before he died.

A. McGehee (Mac) Harvey (1911–1998), a 1934 graduate of the School of Medicine, was only 34 when he was named the youngest director of the Department of Medicine and physician in chief in the history of The Johns Hopkins Hospital in 1946. Seen to the left in this photo, Harvey was a brilliant diagnostician and a superb teacher at patients' bedsides, drilling into his students the importance of taking a careful history, performing a thorough physical examination, and exhaustively analyzing the data to attain an accurate diagnosis. He believed that "the more intelligent the question you put to nature, the more intelligible will be the reply." Harvey also was the quiet instigator behind the desegregation of Hopkins Hospital and ending the anti-Semitism that had prevented any Jewish faculty from being promoted to head departments. At Harvey's insistence, master pathologist Arnold Rich, seen to the right in this photo, was in 1947 the first Jew appointed a department director.

1946
To sustain the World War II-inspired camaraderie among physicians from Hopkins and Guy's Hospital medical school in London, the two institutions arrange an annual exchange of doctors. Alfred Blalock becomes one of the first Hopkins physicians to take advantage of the opportunity and introduces the blue baby operation to England. In subsequent years, hundreds of physicians and administrators participate in the program.

The Immediate Post-World War II Period

Longcope had been a respected clinician and researcher, named to the National Academy of Sciences (an honor rarely given to clinicians) and recipient of numerous awards and honorary degrees. The committee formed to pick his successor followed the time-honored Hopkins penchant for individuals poised on the threshold of great careers and chose the youngest candidate, A. McGehee Harvey, a 1934 graduate of the School of Medicine and former member of the house staff who was only 34 in 1946. He had been on active duty during the war with Hopkins 18th General Hospital Unit in the Pacific. Now the youngest physician in chief in Hopkins' history, he would have a profound impact on the Hospital and School of Medicine until his retirement in 1973.

Ending Segregation and Anti-Semitism

Some of Harvey's most important actions as the Hospital physician in chief and head of medicine had nothing directly to do with curing patients, teaching students or conducting research—but everything to do with making Hopkins a better place. Without fanfare or confrontation, he quietly ensured that the Hospital would be desegregated and ended the longstanding but unwritten policy of denying School of Medicine departmental directorships to Jews.

Mr. Johns Hopkins' instructions concerning his hospital's treatment of African Americans were explicit: They were to be admitted and cared for by the best physicians, as were all others, regardless of race, sex or religion. When the Hospital opened in 1889, its second patient was African American—the first of many. By the end of the Hospital's first full year of operation, 13.6 percent of the patients were listed as "colored." By 1900, it was 20.7 percent—all at a time when many other hospitals in Baltimore did not admit blacks.[38]

According to a detailed history of segregation and desegregation at Hopkins, written in 1992 by former Hospital administrator Louise Cavagnaro, no provision appears to have been made for segregated wards when the original Hopkins Hospital was built, despite the fact that the city, located below the Mason-Dixon line, the traditional demarcation between north and south, practiced segregation. By December 1894, however, a separate ward for black patients had been built. It represented just one example of the Hospital's racial policies in a city where everything else—schools, neighborhoods, stores and public facilities—was segregated. [39]

The Hospital had separate rest rooms and dining facilities, waiting rooms, and even separate shelves in the blood bank. Water fountains were marked "colored" or "white." Curiously, however, these practices were not observed throughout the Hospital. Some public facilities, such as the emergency room and the Harriet Lane Home for children, never were segregated. Neither were the communicable disease unit on the fifth floor of the Osler clinical building nor the pediatric surgery unit on the third floor of the Halsted building.[40]

From the beginning, physicians at Hopkins accorded all patients, regardless of color, the same quality of care and respect. "Mac" Harvey, as he was universally known, clearly thought that in a postwar Baltimore, the time had

In this 1959 photo, Hopkins nurses, all wearing the traditional white caps now rarely seen, listen intently and take notes while reviewing cases with a physician.

come to extend that practice to the private as well as public wards. In 1946, a black patient arrived at one of Hopkins' private clinics, having made an appointment by a letter that gave no indication of his race. Internist James Bordley called Harvey, then new in his job as director of medicine, for advice. Harvey told him to see the patient as he would any other.[41]

Slowly but surely, beginning really in the early 1950s, integration took place, recalled Cavagnaro, who was involved closely with the changes. In 1956, the administrator of the private services instructed the admitting office to implement the change gradually. Three years later, Blalock approved the full integration of the surgical ward services. The Osler nursing units were integrated by around 1960. By 1968, psychiatry became the last inpatient service to desegregate. The psychiatric outpatient service always had provided care to blacks, but those requiring inpatient care had been referred to a hospital operated by the state—which itself had separate hospitals for black and white psychiatric patients.[42]

No general announcements or proclamations were made to implement these changes. In an interview in 2009, a year before her death at 90, Cavagnaro was asked why she thought Harvey had launched his quiet but steady desegregation initiative, which began before the U.S. Supreme Court rulings that fostered desegregation nationwide. She replied simply: "Because it was the right thing to do!"[43]

Admission of African American medical students took longer to accomplish. In this, Hopkins' policy reflected that of many top medical schools of the time—and society in general. The National Medical Association, a professional organization for African American physicians, was founded in 1895 and developed as a parallel group to the American Medical Association and its state and county affiliates precisely because the AMA refused to accept blacks for decades. (Although a few AMA state branches in the north accepted the occasional African American member, it was not until 1968 that the AMA's constitution was amended to permit expulsion of state societies for racial discrimination in membership policies.)[44]

As Hopkins internist Peter Dans has chronicled, following the medical school reforms prompted by the Flexner Report, only some 70 medical schools remained open in the U.S., and only two of them, Howard University and Meharry Medical College, were primarily African American. The few other medical schools that admitted blacks did so under a rigid quota system. From 1920 to 1964, only 2 to 3 percent of medical students were black. In the 1940s,

■ Renowned Hopkins urological surgeon Hugh Hampton Young (1870–1945), founding director of the Brady Urological Institute, and press baron William Randolph Hearst (1863–1951) helped Maryland obtain a half-dozen artistic treasures. In 1933, Young learned that full-length portraits of the six Lords Baltimore, painted between 1580 and 1771, had been auctioned to several purchasers at Sotheby's in London. He succeeded in acquiring the portraits of the first, third, fourth, fifth and sixth Lords and brought them to Baltimore to be displayed at a 1933 celebration of the 300th anniversary of Maryland's founding. Young later learned that the missing portrait of the second Lord Baltimore had been bought by Hearst, an omnivorous collector of artwork. In 1940, Young told the Enoch Pratt Free Library that if it bought the second Lord Baltimore's portrait from Hearst, he'd donate the other five. The paintings still remain on display in the main room of the Central Library on Cathedral Street.

Levi Watkins Jr. (right), shown here conferring with anesthesiologist Jackie Martin, came to Hopkins in 1970 as an intern in surgery, and ultimately became the first black chief resident in cardiac surgery. A pioneering surgeon renowned for performing the world's first implantation of an automatic heart defibrillator in a patient, he joined with another African American colleague, ophthalmologist Earl Kidwell, to begin a successful nationwide drive to recruit talented minority students to Hopkins.

1948–1949
Hopkins researchers led by David Bodian, Isabel Morgan and Howard A. Howe verify three strains of the poliomyelitis virus. This verification is an essential step in the development of the Salk polio vaccine. In 1957, Bodian, Morgan and Howe are honored for their work by election to the Polio Hall of Fame.

A two-decade postwar building boom, unlike anything since the 1920s, gets under way and lasts until 1968. More than 500,000 square feet of research and teaching space are added, much of it partially funded with federal money. Projects include the Clinical Science Building (it is renamed the Blalock Building in 1964); the Basic Science Building (opened in 1959, later named for W. Barry Wood); the Biophysics Building (completed in 1963); the Children's Medical and Surgical Center (opened in 1964); the Wilmer Eye Institute research wing (opened in 1964).

1949
January: Detlev Bronk, credited with formulating the modern theory of the science of biophysics, becomes president of the university a month after the retirement of Isaiah Bowman. He picks up where Bowman had left off reshaping the postwar university.

just 10 to 20 African Americans graduated annually from schools other than Howard and Meharry. Twenty-six of the 78 medical schools in the nation during the 1940s were in southern and border states, and all of them—including Hopkins—were closed to black applicants. It was not until the early 1960s that more than half of all southern medical schools—14 of the 26—began to admit black students, and Hopkins was among them.[45]

The first black student to enroll in the School of Medicine was not an American but a Nigerian, James Nabwangu, who was admitted in 1962 and graduated in 1967. Robert Gamble became the first African American student at Hopkins in 1963 and also graduated in 1967. Within a decade, the number of African American students in the School of Medicine had grown to 32, and by the beginning of the 1989-90 academic year, the number had reached 50, or 11.2 percent of the 447 students enrolled.[46]

The efforts to increase the enrollment of minority students proceeded slowly. When Alabama native and Vanderbilt medical school graduate Levi Watkins Jr. came to Hopkins in 1970 as an intern in surgery, he was astonished to find Baltimore not much different than the segregated south he had known. He would become a major player in changing the Hopkins he had entered.[47]

In 1975, at the request of the newly appointed dean, Richard Ross, Watkins and fellow African American faculty colleague Earl Kidwell, an ophthalmologist and 1973 graduate of the School of Medicine, began a concerted nationwide drive to recruit talented minority students who were interested in studying medicine. Within a few years, Hopkins was attracting black students from all over the country who were convinced—correctly—that Hopkins wanted them. They not only came to Hopkins, they later joined the faculty, in some cases. (The success of the Hopkins minority recruitment campaign soon made it a model imitated by other medical schools.) By the mid-1990s, the annual reception for black students, house staff and faculty that Watkins hosted had grown from 10 attendees to more than 100.[48]

From their beginnings, the Hospital and School of Medicine had admitted Jews, not only in accordance with Mr. Hopkins' instructions but also because several of the institution's founders were foes of anti-Semitism. As a graduate medical student in Berlin, William Osler was appalled by the attacks he witnessed on Jews in Germany's medical profession, despite their considerable accomplishments. In 1884, Osler wrote from Berlin to a friend, "One cannot but notice here, in any assembly of doctors, the strong Semitic element; at the local societies and at the German Congress of Physicians it was particularly noticeable, and the same holds good in any collection of students. All honour to them!"[49]

William Welch similarly appeared without prejudice toward Jews. Among his earliest protégés was pathology fellow Simon Flexner, whose appointment to the faculty and professional advancement he advocated eagerly. When Flexner was offered the chair in pathology at Cornell in 1898, Welch and Osler saw to it that he was promoted to professor of pathological anatomy at Hopkins to keep him here. The following year, when the University of Pennsylvania offered Flexner a pathology professorship at a much higher salary, Osler tried to persuade him to stay at Hopkins by pointing out that Penn was an anti-Semitic institution with only one Jew on its faculty. "Tell them to go to hell!" Osler advised. Yet Flexner had to consider his career development—and family's finances—so he could not afford to follow Osler's advice and took the job at Penn. It wasn't long, however, before Welch secured an even better position for him: appointment in 1901 as the first director of the Rockefeller Institute, the board of which Welch chaired. In time, Flexner would more than repay Hopkins' devotion by steering to it millions in Rockefeller grants.[50]

Others at Hopkins then or later did not share the openness toward Jews demonstrated by Osler and Welch. As was common at most major medical schools in the first five decades of the 20th century, Hopkins admitted Jewish students based on a quota limiting the number allowed in each entering class. Jews also were appointed to the faculty, but their advancement—and personal acceptance—was restricted. Hematology pioneer Maxwell Wintrobe (1901–1986), who was on the faculty from 1930 to 1943, found Hopkins to be everything he desired clinically and scientifically. However, he never advanced beyond the rank of associate physician, below the rank of associate professor, and was deeply saddened by the fact that in his 13 years as an admired Hopkins teacher and researcher, not one of his colleagues ever invited him or his wife to their home for dinner. He left Hopkins to become head of the department of medicine at the new medical school of the University of Utah—immediately jumping from the lowest of academic ranks to the top.[51]

Obstetrician Whitridge Williams, School of Medicine dean from 1911 to 1923, probably considered himself enlightened when he told Alan Guttmacher, a 1919 graduate who was on the faculty from 1926 to 1952, that "if a Jew and a gentile have equal merit, I prefer the gentile, but if the Jew is the better man, I'll take him." After 26 years on the faculty, Guttmacher (1898–1974) remained an associate professor of obstetrics at Hopkins, despite becoming chief of obstetrics and gynecology at Baltimore's Sinai Hospital and winning a 1947 Lasker award for distinguished contributions to planned parenthood. After leaving Hopkins and Baltimore, he became director of obstetrics and gynecology at New York's Mt. Sinai Hospital, national president of Planned Parenthood, and founder of the renowned Guttmacher Institute for advancing sexual and reproductive health worldwide.[52]

No formal proscriptions ever were written, but the Hopkins policy was clear: no Jew could become head of a clinical department—even if he was the "better man."[53]

Pathologist Simon Flexner (1863–1946), above, a protégé of William Welch and William Osler, advanced through academic ranks to a professorship in pathological anatomy, despite the then-unspoken but general refusal to promote Jewish faculty. Two years after he left Hopkins in 1899 to become professor of experimental pathology at the University of Pennsylvania, he was named the first director of the Rockefeller Institute for Medical Research, thanks to Welch, head of the institute's board.

Pioneering hematologist Maxwell Wintrobe (1901–1986) established the modern methods for measuring a blood sample's volume of red and white blood cells and platelets; invented the Wintrobe hematocrit, the most commonly used instrument for performing these measurements; and single-handedly wrote *Clinical Hematology*, the most authoritative work in the field. Yet in his 13 years at Hopkins, due to its subtle but pervasive anti-Semitism, he never advanced beyond the rank of associate physician. He left to become head of the department of medicine at the University of Utah – immediately jumping from the lowest academic rank to the top.

■ In 1943, Perrin H. Long, an acclaimed Hopkins virologist and infectious disease expert who helped launch the antibiotic era in modern medicine, was serving as a U.S. Army colonel and consulting physician to the Allied Force Headquarters in Italy when he witnessed the infamous "slapping incidents" in which fabled – and volatile – Lt. Gen. George S. Patton Jr. hit and berated soldiers being treated for shell shock. Bypassing the chain of command, Long sent his Aug. 16, 1943 report, entitled "Mistreatment of Patients in Receiving Tents of the 15th and 93rd Evacuation Hospitals," directly to Gen. Dwight D. Eisenhower, who reprimanded Patton and made him apologize to the soldiers.

"The prejudice was very real in terms of senior faculty," recalled neuroscience pioneer Vernon Mountcastle, a 1942 graduate of the School of Medicine and director of physiology from 1964 to 1980. "Some of the best brains that ever graced our scene were in Jewish heads," Mountcastle said, but professional advancement for these "brains" was blocked—at the direction, some said, of the University's top leadership. Once again, Mac Harvey was the one who saw to it that Hopkins changed.[54]

The man whose cause Harvey championed was Arnold Rich (1893–1968), a 1919 graduate of the School of Medicine who was named an associate professor in pathology in 1923 and soon became one of the most acclaimed researchers and teachers in the field. When William McCallum, head of the pathology department, suffered a stroke late in 1941, Rich became acting director of the department. The Hospital's Medical Board wanted the appointment made permanent, and a petition to the medical school's Advisory Board, signed by virtually all associate and assistant professors in the School of Medicine, expressed their admiration for Rich and urged his selection. The Advisory Board balked. It chose to postpone a final choice until after the war. The stalemate continued until Harvey was chosen to become head of medicine and physician in chief in 1946.[55]

"I was told that when Mac Harvey was offered the chair of medicine, he made as a condition of his acceptance the promotion of and full appointment of Arnold Rich as Baxley Professor and director of the Department of Pathology," Mountcastle recalled. "Thus we owe to him—and to those who agreed—the end of a rather sorry story in the Hopkins history.... Looking back on this history, it is amazing to me that Hopkins could have retained its eminent position in American medicine in spite of this prejudice—for it was widely known."[56]

Today, the president of The Johns Hopkins University is Jewish, and many deanships, department and division leadership posts, and top faculty positions are held by Jews, African Americans, Asians and other minorities throughout the institution.

In 1947, pathologist Arnold Rich (1893–1968) became the first Jew to head a department at Hopkins, thanks to the quiet but firm insistence of A. McGehee Harvey, the then-new director of the Department of Medicine. A 1919 graduate of the School of Medicine, Rich spent his entire career at Hopkins, conducting research and inspiring medical students to think for themselves. His book, *Pathogenesis of Tuberculosis*, became the standard text in the field. A man of encyclopedic interests and an accomplished violist, he had an excellent sense of humor and could be as charming in private as he was intimidating in scientific debate. Once, when visiting the modest apartment of a member of the house staff, the small-statured Rich sat in the tiny rocking chair of the young physician's five-year-old daughter. The child protested that he was sitting in *her* chair. Rich arose and said graciously, "My dear young lady, I may occupy the Baxley Chair of Pathology across the street, but here, I agree, the chair is yours."

Association with Guy's Hospital in London

One postwar development that marked the beginning of a happy tradition was the creation of a lasting bond between Hopkins and the renowned Guy's Hospital in London.[57]

The link between Hopkins and Guy's was forged during World War II, when Brigadier General Edward "Bo" Boland, a Guy's physician in charge of the British Medical Services and consulting physician to the Allied Forces Headquarters in North Africa, met his American counterpart, Colonel Perrin Long, an acclaimed Hopkins virologist and infectious disease expert who was among those pioneering the antibiotic era in modern medicine.[58] (Earlier in the war, Long had witnessed the infamous "slapping incidents" in which Lieutenant General George S. Patton Jr. hit and berated American soldiers being treated for shell shock in Sicily. Bypassing the chain of command, he sent his August 16, 1943, report, titled "Mistreatment of Patients in Receiving Tents of the 15th and 93rd Evacuation Hospitals," directly to General Dwight D. Eisenhower, who reprimanded Patton and made him apologize.)[59]

Boland, also known as a tough-as-nails physician, became such a close friend of Long's that they decided it would be good for Hopkins and Guy's to form a permanent bond.[60] When Boland was named dean of Guy's Hospital Medical School after the war, he promptly sent a letter to his Hopkins counterpart, Alan Mason Chesney, proposing a new exchange program.

"The object," Boland wrote in his August. 28, 1946, letter, "would be to maintain the friendship, co-operation and exchange of ideas which has been one of the better things which have come out of this War. Many Guy's men have derived great advantages and have happy personal memories of former associations with your great Hospital… and we should value a closer association with you."[61]

With the enthusiastic endorsement of the medical faculty's advisory board and the university president, Chesney happily replied, "We welcome it whole-heartedly."

In the fall of 1947, the Johns Hopkins and Guy's Hospital exchange began—in grand style—with a visit to London by Alfred Blalock and Helen Taussig. Blalock performed the blue baby operation on eight desperately ill British children, and he and Taussig gave a combined lecture on the procedure at the Great Hall of the British Medical Association, which was packed for the presentation. It was a memorable occasion, as Lord Russell Brock (1903–1980), Blalock's host at Guy's, recalled years later:

> "The hall was quite dark for projection of [Blalock's] slides which had been illustrating patients before and after operation, when suddenly a long searchlight beam unerringly picked out on the platform a Guy's nursing sister, dressed in her attractive blue uniform, sitting on a chair and holding a cherub-like girl of two and a half years with a halo of blonde candy hair and looking pink and well; she had been operated on at Guy's a week earlier. The effect was dramatic and theatrical and the applause from the audience was tumultuous—a perfect climax—[which] left nothing more to be said by the lecturer—no one there could possibly forget it."[62]

It was the beginning of a beautiful friendship. Ever since, for more than six decades, two physicians a year from both schools have crossed the ocean to spend a month working at the other institution. Hundreds of medical students from each school have also traded places to do their electives.[63]

In the fall of 1947, the six-decade-old exchange program between The Johns Hopkins Hospital and Guy's Hospital in London began auspiciously with a visit by surgeon Alfred Blalock and his blue baby operation co-developer, pediatric cardiologist Helen Taussig. Blalock, seen standing in the center of this photo with colleagues from Guy's, wowed his British counterparts, performing eight blue baby operations on desperately ill English children and lecturing on the procedure.

■ Leslie N. Gay, a 1917 graduate of the School of Medicine, became famous for discovering that Dramamine was a cure for motion sickness—but motion sickness wasn't his specialty. Gay (1891–1978) was an allergist. He founded Hopkins' allergy clinic and directed it for 36 years. In 1948, a woman with hives came to him for treatment. He gave her an antihistamine known as compound 1694. She later mentioned that while she had always suffered terrible motion sickness when taking the trolley to Hopkins, once she had begun taking the compound, the motion sickness ceased. Intrigued, Gay and colleague Paul Carliner conducted subsequent studies—including one approved by Gen. Omar Bradley that involved 1,500 soldiers on a troop ship sailing from the U.S. to Germany— and proved that compound 1694, later called Dramamine, not only prevented seasickness but also helped relieve seasickness after it developed.

When the half-century ban on marriage for Hopkins residents and interns was ended after World War II, spouses and offspring became familiar figures at the Broadway Garden Apartments—better known as "The Compound"—an unpretentious, slightly ramshackle square of 120 residences built by Hopkins and opened across from the Hospital in 1957.

Biostatistician Lowell Reed (1886–1966) had been dean of what now is the Bloomberg School of Public Health (1937-46), vice president of the University (1946-49), and vice president of the University and Hospital (1949-53), when he was asked to become president of the University in 1953. He held the post until 1956 – the year that the nine-story dormitory then being built for Hopkins medical students was named for him.

The Marriage Barrier Falls for Interns and Residents

Osler believed that young physicians in training should not marry. He wanted his protégés to spend uncounted hours in the wards and at the laboratory table, and to write research papers to deliver at professional gatherings and submit for publication.[64]

The Medical Board of the Hospital took Osler's dictum seriously. For more than a half-century, Hopkins interns and residents were not allowed to marry except under extraordinary circumstances—such as if a prospective married resident had "special merit" or an intern's wife was going to be living elsewhere other than Baltimore during his internship.[65]

World War II changed all that. Many already-married young physicians and medical students returned to Hopkins following their military service. Hopkins adjusted its rules, slightly at first, to accommodate them. In time, Hopkins built residences for married residents across from the Hospital in an effort to keep them close to it.

1950s

The boarding houses in which the successors to Miss Susie Slagle provided accommodations to six decades' worth of Hopkins students began to disappear in the mid-1950s. The residences along a nine-block stretch of the west side of Broadway were purchased and demolished to make way for a large postwar expansion of the campus, and Hopkins built its own housing for medical students, interns and residents.[66]

Reed Hall and "The Compound"

Reed Hall, a nine-story dormitory for medical students, was built right across Broadway from the Hospital and opened in 1957 with rooms for 220 men and 30 women, along with dining and recreational facilities. A 13-story addition opened in 1966, providing what now is a combined 300 units. It is named for biostatistician Lowell J. Reed (1886–1966), a 38-year Hopkins veteran who had been dean of the School of Public Health from 1937 to 1946, vice president of the University from 1946 to 1949, vice president of the University and Hopkins Hospital from 1949 to 1953, and president of the University from 1953 to 1956.[67]

Reed Hall's reputation has had its ups and downs. Even when it was new, student occupants often found it uncongenial and tried to move out within a year or so. At one point, tenants had T-shirts printed up saying, "I Survived a Year in Reed Hall." In recent years, however, its desirability as a residence improved dramatically. Students like its convenience to the Hospital (to which they even have access via an underground tunnel), its location beside the Cooley Recreation Center, its central air-conditioning, 24-hour security, and relatively low-cost monthly rent. The popularity of Reed Hall increased to such an extent that by the eve of its 50th anniversary in 2006, so many students wished to continue living in it after their first year of medical school that a lottery had to be set up for rooms.[68] By 2010, construction had begun on a major graduate student residential tower just north of the campus to replace the aging Reed Hall.

More consistently beloved as a housing complex was an unpretentious square of 120 apartments that also opened in 1957, behind Reed Hall, as residences for hundreds of married Hopkins interns and residents. It quickly became known simply as "The Compound." It was an urban oasis, with a swimming pool and a big park in the center, usually full of moms, babies and toddlers, where young physicians (mostly men back then) and their families could live pretty well on next to nothing.[69]

"Fundamentally, it was a commune where we enjoyed good friends, good times and safety," said Wilbur Mattison, a 1952 graduate of the School of Medicine, in a 2004 interview with *Hopkins Medicine* magazine. Mattison has been credited with giving the place—officially, the Broadway Garden Apartments—its enduring nickname. The families who lived there forged remarkable bonds that have lasted a lifetime. They often hold reunions and love recounting the Compound's quirks, such as the fact that because it was located on a hill sloping from Bond Street to Caroline Street, all of their kids' tricycles and wagons would end up on the doorstep of the unit at the bottom of the street.[70]

Despite its stolid nickname, the structure of the Compound was not that sturdy. The walls were thin enough to enable neighbors hear each other singing in the bathtub or to inadvertently punch holes that opened up an instant avenue of communication between tenants.[71]

To Mattison, "the most astounding thing of all was how you could live there with no income." Interns in the late 1950s were paid just $25 a month; residents only $166. Between what he made as a house officer and later as a hematology fellow with a slightly higher salary, plus "selling blood at the blood bank, that was income," he remembered.[72]

The Compound was torn down in 1986 to clear space for Hopkins Hospital's new outpatient center, opened in 1992.

Two residents of the Compound have an excellent view of the Billings Building's iconic dome from a window in apartments built by Hopkins for residents and interns. Despite its stolid nickname, construction of the Compound was less than sturdy. Neighbors could hear each other singing in the bathtub or accidently knock holes in the thin walls. The Compound was torn down in 1986, making way for Hopkins' new outpatient center, opened in 1992.

1950
December 5: Russell Morgan, chief of radiology, appears on *The Johns Hopkins Science Review* and conducts the first inter-city diagnosis and consultation ever seen on television. The *Science Review* is a nationally broadcast show that is the first weekly network program produced by a university. On this particular evening, Morgan demonstrates his revolutionary X-ray fluoroscope, which provides clear, moving images of the body's beating heart and inflating lungs. Using the machine he created, Morgan and physicians in New York and Chicago examine a Baltimore machinist who had been in an industrial accident. They recommend relatively minor surgery to remove metal shards in the man's back. *The Johns Hopkins Science Review* had a successful 12-year run, from 1948 to 1960. It wins two George Foster Peabody Awards, television's top accolade.

1951
George Gey, director of the Department of Surgery's tissue culture laboratory, establishes the world's fist continuously multiplying human cell culture, HeLa, with cervical cancer cells obtained from Henrietta Lacks. HeLa cells are instrumental in the research that produces multiple medical advances.

■ In the 1950s, Anna Baetjer (1899–1984), a Hopkins pioneer in occupational medicine, was among the first to link workplace chemicals to cancer. A 1924 graduate of what now is the Bloomberg School of Public Health, Baetjer is renowned for linking exposure to chromium in the workplace to cancer. Baetjer also warned of air pollution perils as early as the 1950s and founded one of the country's first research and training programs in environmental toxicology at Hopkins in 1963.

Federal Involvement in Postwar Building and Research

Federal funds had helped to pay for the Compound's construction—and virtually every other building project undertaken between 1945 and 1968. In the two decades after the war, Hopkins experienced a building boom unlike anything since the 1920s, with more than 500,000 square feet of research and teaching space added.[73]

Its changing physical plant reflected the growth of medical specialization, the advance of technology, and the requirements of the medical school and Hospital. New departments and divisions needed new facilities for both research and teaching.[74]

The federal government's influence was pervasive. Its money enabled the construction, and its requirements—especially its desire to foster medical advancements—led Hopkins to develop methods for achieving both its ends and the government's aims. The federal funds may have been meant mostly for encouraging research, but Hopkins also managed to replace or remodel the site of every one of its original inpatient beds. All three of Hopkins' missions—research, teaching and patient care—were growing, and all three needed additional space. The buildings erected during this time were designed to serve multiple purposes.[75]

Among these projects was the Clinical Science Building, which took from 1951 to 1963 to complete and was renamed the Blalock Building in 1964. The National Heart Institute gave $485,000 to support construction of its cardiovascular research facilities, and the National Cancer Institute gave $750,000 for oncology research labs.[76] Another was a new 10-story Basic Science Building, opened in 1959 thanks in part to a U.S. Public Health Service contribution of $1.5 million to its $5 million price tag. It later was named for W. Barry Wood (1910–1971), a 1936 graduate of the School of Medicine. Wood was a pioneer in the use of penicillin and served as a vice president of the University and the Hospital before becoming a much-admired director of the Department of Microbiology in 1959. He held that post until his death. Yet another project was the Children's Medical and Surgical Center, which was opened in 1964.[77]

To the left of the Billings Building, the original Hopkins Hospital, arises the superstructure of the new Children's Medical and Surgical Center, opened in 1964. A grant from the National Institutes of Health provided the money for construction of four floors for surgical research, including research in orthopedics, as well as research on mental retardation. The rest of the building comprised facilities for treating children.

Federal Research Funding Surge

In 1945, the National Institute of Health sent a letter to the deans of all American medical schools, modestly advising them: "We have limited funds available for research purposes. If you have investigators who need these funds, let us hear from you by return mail." It was a muted toot on a trumpet that would sound a huge, unprecedented nationwide charge toward Washington coffers by academic medical centers seeking federal funding for an ensuing explosion in scientific achievements.[78]

Americans had been immensely impressed by the practical, immediately useful contributions of medical research and attendant technological advances during World War II. For example, the death rate for wounded soldiers was half that of World War I, partly because of the availability of blood and plasma for the treatment of hemorrhage and shock, plus the use of sulfonamides and penicillin—in all of which research by Hopkins physicians had played a significant role.[79]

Once creation of the atomic bomb for military use demonstrated the potential of scientific research to the government and people, it was only a short step to the belief that the nation's future would depend on development of its intellectual capital.[80]

In the years immediately following World War II, the greatest influence on Hopkins Hospital and the School of Medicine was the overwhelming financial dominance of the federal government, which greatly affected Hopkins' direction—and indeed, that of most academic medical centers in the United States. The numbers tell the story: At Hopkins, the medical school's total grant receipts increased from a little more than $2.6 million in 1947 to about $7 million a decade later. They increased fourfold between 1957 and 1968, to almost $28 million. Research and training grants from the U.S. Public Health Service to the Department of Medicine alone increased from $620,102 to nearly $2.8 million between 1957 and 1967.[81]

The influx of federal dollars for medical research supported the hiring of additional faculty and fellows to carry it out. The number of full-time assistant, associate and full professors more than doubled between 1959 and 1968, from 131 to 304; the number of fellows grew from 393 to 616; the number of house staff also grew from 235 to 328. Patient volume, growing because of the new clinical buildings the federal government helped fund, soared as well. Hospital receipts increased threefold between 1957 and 1968, from $10.5 million to $30.7 million.[82]

Thomas Turner was the School of Medicine dean between 1957 and 1968, the period when Hopkins began receiving more federal money than ever before and burgeoning in size accordingly. Although Turner recognized the immense value to Hopkins of the federally fueled expansion and the importance of the research and clinical advances this brought, he was not particularly happy about losing aspects of the old Johns Hopkins that he had known.

White-haired, witty, patrician and pipe-smoking, Turner always dressed in a three-piece suit, with a watch nestling in one pocket of his vest and its gold chain stretching across it. He also always wore a hat, a homburg in the winter, a straw hat in the summer. His executive assistant, Claudia Ewell, called one method she devised for keeping him updated "communication by hatband." She'd prepare notes

■ In 1951, Hopkins biochemist Elmer V. McCollum (1879–1967) was dubbed "Dr. Vitamin" by *TIME* magazine, which said he "had done more than any other man to put vitamins back in the nation's bread and milk, to put fruit on American breakfast tables, fresh vegetables and salad greens in the daily diet." McCollum discovered Vitamins A and D and their link to improving health.

1953
August: Detlev Bronk resigns as president of Hopkins in order to become the head of the Rockefeller Institute for Medical Science, which was renamed Rockefeller University under his leadership as its first president.

September: Lowell Reed, former head of Hopkins' School of Hygiene and Public Health and the university's vice president in charge of medical affairs, was 67 and had just retired from a distinguished 35-year career as a research scientist in biostatistics and public health administration at Hopkins when he was asked to serve as university president. He will serve three years as the university's president.

1956
July: Milton S. Eisenhower, president of Penn State, a former career official in the U.S. Department of Agriculture—and the youngest brother of then-U.S. President Dwight Eisenhower—is elected president of Hopkins following Lowell Reed's retirement. Eisenhower will be the only Hopkins president to serve two non-consecutive terms. During his first term, the university income tripled, the endowment doubled and new construction included the athletic center and the library at Homewood that would bear his name. Upon his retirement in 1967, he was given the title president emeritus in recognition of his devoted service. He was coaxed out of retirement for 10 months in 1971-72 after his immediate successor, Lincoln Gordon, resigned.

Thomas B. Turner (1902–2002), shown here during his tenure as dean of the School of Medicine from 1957 to 1968. He oversaw an immense growth in the medical institutions but lamented losing the intimate aspects he had relished in the pre-World War II Hopkins, where everyone seemed to know everyone else.

1957

The Hopkins-developed electronic AC defibrillator device, created by electrical engineer William B. Kouwenhoven, cardiologist Willian Milnor and biophysicist Samuel Talbot, is first used on a human subject. Anesthesiologist Peter Safar, a member of the Hopkins faculty, develops new methods for mouth-to-mouth resuscitation while working at the old Baltimore City Hospitals (now Johns Hopkins Bayview). Both Kouwenhoven and Safar are dubbed the "father of CPR" for their work.

Physiologist and pioneering neuroscientist Vernon Mountcastle discovers the column-like organization of cells in the brain's cortex, forever changing the study of the body's most mysterious organ.

Cardiologist and pioneering geneticist Victor McKusick founds Hopkins' Division of Medical Genetics and begins concentrating on the work that will earn him the recognition as "the father of medical genetics."

for him on important matters he had to address and slip them in his hatband so he'd have them to review on the way home. In a memoir written in 1981, when he was 80, he reflected on the dramatic changes that had occurred to his view of the dome atop Hopkins' Billings Building as he approached it on his daily two-mile walk to the Hospital from his Bolton Hill townhouse. As the pace of Hopkins' growth accelerated during his deanship, the once-grandly isolated Queen Anne's-style landmark seemed to diminish in prominence as it was surrounded by modern structures he considered sterile and ugly. The intimate atmosphere of the prewar Hopkins, in which everyone seemed to know everyone else, also was disappearing as the faculty grew. Turner wrote that his mind rebelled at the rapid changes. "*No! It cannot be allowed to happen!*" he recalled himself thinking.

Yet he knew it must. Those nondescript buildings were full of vitally important research laboratories and up-to-date patient beds; the new faculty was doing extraordinary work and burnishing Hopkins' name. Although progress had its price, "[n]ot to have embraced the opportunity for such enrichment would have left Johns Hopkins in an educational backwash and pragmatically non-competitive with the better medical schools in America," Turner wrote.[83]

Medical Breakthroughs

The Birth of CPR

Two Hopkins scientists—one of whom was not even a physician—each have been called the "father of CPR" for groundbreaking work in cardiopulmonary resuscitation: electrical engineer William B. Kouwenhoven (1886–1975) and anesthesiologist Peter Safar (1924–2003). Dealing with different aspects of the critical need to restore both a patient's steady heartbeat and breathing, Kouwenhoven and Safar each earned that honorific in the 1950s for their lifesaving innovations, all of which made Hopkins the birthplace of CPR.

Kouwenhoven, dean of Hopkins' Whiting School of Engineering, had begun his work on cardiopulmonary resuscitation in the 1920s, when electric utility companies, concerned about the increasing number of linemen killed by electrical shock, sought his assistance in studying the physical damage inflicted by high voltage. Conducting studies with colleagues on animals and humans, Kouwenhoven determined that ventricular fibrillation occurred mainly from low-voltage shocks, while respiratory paralysis resulted from higher voltages.[84]

In 1947, surgeon Claude Beck, a 1921 graduate of the School of Medicine, working at Case Western Reserve University in Cleveland, reported the first open-chest defibrillation (restoration of a patient's regular heartbeat) by the placement of electrodes directly to the heart's surface. Beck also found that squeezing the exposed heart by hand could provide circulation of the blood.[85]

Kouwenhoven and Hopkins cardiologist William Milnor (1921–2008) began work on developing a closed-chest defibrillator in 1950. They soon discovered that a brief AC current of 20 amperes would defibrillate a dog's heart successfully. They perfected their procedures while working with biophysicist Samuel Talbot (1903–1964), who also was studying heart arrhythmias (irregular heartbeats), conducting hundreds of tests in the laboratory. The Hopkins AC defibrillator device they developed first was used on a human subject in March 1957.[86]

When Kouwenhoven retired from the deanship of the Whiting School in 1956, he transferred to the Department of Surgery as a lecturer. Tests with the defibrillator device had shown a rise in blood pressure when a countershock was applied to the heart. Electrical engineer G. Guy Knickerbocker (1932-), a Kouwenhoven protégé, noticed that even before the countershock was applied, a slight rise in blood pressure occurred when the electrodes were pressed on a dog's chest. This, according to Hopkins cardiac surgeon James Jude (1928-), was a wonderfully "chance observation to a fertile young mind." Jude immediately recognized that Knickerbocker's discovery could be developed into a method for external cardiac massage, a means not only for restarting a stopped heart but keeping it beating long enough to sustain survival. Surgeon Henry T. Bahnson (1920–2003) first used this technique in February 1958 to successfully resuscitate a two-year-old patient whose heart was in ventricular fibrillation. The same year, Jude successfully applied external cardiac massage to a woman in her 40s who had developed ventricular fibrillation while undergoing anesthesia.[87]

During the same period, anesthesiologist Peter Safar was perfecting the mouth-to-mouth method of lung ventilation that earned *him* the title of "father of CPR," a sobriquet he always declined to accept but which *The New York Times* used to describe him in his obituary. Safar devised what became the standard procedures for oxygenating the blood, although it does not ensure its circulation. Safar had developed his cardiopulmonary resuscitation methods in part by hiring Hopkins medical students at $150 apiece to undergo a potentially harrowing experiment. Among the willing guinea pigs was third-year medical student Chester Schmidt Jr., later a longtime director of psychiatry at Johns Hopkins Bayview, who felt Safar's offer was too good to turn down—even if it meant being injected with a drug that briefly paralyzed his diaphragm. "That was a lot of money then," he recalls with a chuckle. While the students' diaphragms were momentarily incapacitated, Safar place a metal device he'd created, later dubbed the Safar Airway, in their mouths to ensure their continued ability to breathe.[88]

Safar's work helped define CPR's essential ABCs: maintaining a patient's airway, breathing and circulation. He later collaborated with a Norwegian company to create the first CPR training mannequin, Resusci Anne, well-known to all who have trained in lifesaving courses. In 1958, Safar also founded the nation's first medical-surgical, multidisciplinary intensive care unit (ICU), with 24-hour coverage by anesthesiologists, at Baltimore City Hospitals (now Johns Hopkins Bayview). In addition, he formulated the basic principles of emergency medicine at Hopkins, as well as contributed to the design of modern ambulances while on the Hopkins faculty. He later went to the University of Pittsburgh, where he founded the International Resuscitation Research Center, now known as the Safar Center for Resuscitation Research, in 1979.[89]

■ In the early 1950s, a Hopkins professor with no medical background whatsoever created a vital, life-saving medical device. William Kouwenhoven (1886–1975), above, a professor of electrical engineering and dean of what now is the Whiting School of Engineering, invented the closed-chest heart defibrillator in collaboration with cardiovascular physiologist William Milnor (1921–2008) and biopysicist Samuel Talbot (1903–1964). Kouwenhoven's protégé, electrical engineer G. Guy Knickerbocker (1932–) and cardiac surgeon James Jude (1928–) developed a method for external cardiac massage. Kouwenhoven received the first-ever honorary Doctor of Medicine degree from Hopkins in 1969 and in 1973, he received the Lasker Award for Clinical Medical Research.

1958
Peter Safar creates the first modern medical-surgical, interdisciplinary intensive care unit (ICU) at Baltimore City Hospitals. He will formulate the basic principles of emergency medicine and even contribute to the design of modern ambulances.

1960
July: William Kouwenhoven, professor emeritus and former dean of the School of Engineering; James Jude, a surgical resident; and surgical instructor Guy Knickerbocker, an electrical engineer, report in the *Journal of the American Medical Association* that they have developed a closed chest massage procedure to keep the inert or fibrillating heart pumping blood. Kouwenhoven's electrical defibrillation machine has been in use since 1957, but this new cardiopulmonary resuscitation (CPR) technique enables trained persons to provide life-saving therapy without machines or surgery.

In 1969, Kouwenhoven received the first honorary degree of doctor of medicine ever awarded by Hopkins, and in 1973, he received the Lasker Foundation Award for his contributions in proving that electric shock could reverse ventricular fibrillation, developing the devices for both open- and closed-chest defibrillation, as well as the technique of external cardiac massage.[90]

Hopkins' leadership in the CPR field has continued ever since its birth here. Cardiologist Myron "Mike" Weisfeldt, now director of the Department of Medicine, has been instrumental in efforts to place CPR devices at airports, stadia and other public places; and in 2010, he wrote an editorial for the *New England Journal of Medicine*, noting recent studies which indicate that "less may be better" in CPR. Chest compression alone, he wrote, is critical if bystanders who respond to a person stricken with cardiac arrest are untrained in or uncomfortable with mouth-to-mouth resuscitation. Weisfeldt nevertheless believes that CPR courses still should teach rescue breathing, "since it is important in cases of cardiac arrest due to obvious respiratory failure, which include most cardiac arrests in children and some adults." For most children who suffer cardiac arrest, such as drowning victims, "we must do rescue breathing," he wrote.[91]

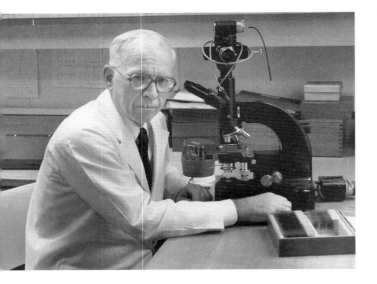

The "father of neuroscience," Vernon Mountcastle, director of physiology from 1964 to 1980, had many of what he calls "Ah-ha! moments" in his decades of research on the function of the brain. These include his 1957 discovery that the cells in the cortex are arranged in vertical columns, completely revolutionizing studies of the brain; and his refinement of "the waking monkey technique." By training monkeys to perform certain tasks and then using electrodes to pinpoint the specific neurons carrying out those tasks, he and colleagues have been able to identify the groups of brain cells directly involved in sensory perception. "If you do an experiment—and you discover something—that's the biggest thrill in the world," he says.

The Jacques Cousteau of the Cortex Unlocks Deep Mysteries of the Brain

One day in 1957, Vernon Mountcastle, a 1942 graduate of the School of Medicine, World War II battlefield surgeon in Europe, and protégé of Philip Bard, Hopkins' physiology director and medical school dean, had what he calls an *Ah-ha!* moment. It was a research breakthrough that would reveal the previously unknown structure of the brain's cortex and forever change the study of the most mysterious organ in the human body.[92]

Sitting at his desk in the office next door to his lab in Hopkins' physiology department, Mountcastle was working on the results of tests made on the brains of cats, recording the character of each cell from successive penetration layers. More than a half-century later, Mountcastle, now 93, recalls that "fortunately, I was writing them down vertically on a yellow piece of paper."[93]

"Suddenly I realized that they were all the same, all the way down. And in such a penetration you might identify the properties of 10 to 20 cells, from top to bottom."

What he had found was that the brain, unlike any other part of the human body, is divided into little sub-units—or columns—each with its own specific role: skin cells lay atop skin cells, joint cells atop joint cells and so on, extending in columns from the brain's surface all the way down through six layers of the cortex. "That was my 'Ah-ha' experience," Mountcastle says. He quickly shifted to examining the brains of monkeys, with the same result. [94]

Mountcastle's discovery of the "columnar" structure of the brain forever changed studies of it. Before his breakthrough, researchers had believed that brain cells were organized randomly, with each layer of the cortex having a specific function. His findings "hit neuro-anatomists where they lived," he recalls, and revolutionized the concept of how the brain is built.[95]

Two decades later, another *Ah-ha!* moment prompted Mountcastle to launch even more important research on the parietal lobe of the cortex, the region involved in such higher functions as the perception of sensory information and physical reaction to it. These experiments involved "the waking monkey technique," a method for inserting microeletrodes into the brain of a conscious test animal, pioneered by Canadian Herbert Jasper at Montreal's McGill University. "What I did was to combine monkey psychophysics and observation of cells in the brain—and that took five years to develop," Mountcastle explained decades later. He figured out how to record the activity of a fully alert monkey's brain, not one dulled by anesthesia, as the carefully trained animal reacted by pressing a key when seeing such stimuli as moving light beams.[96]

Mountcastle's discoveries and extraordinary record as a teacher have earned him the honorific "father of neuroscience" and practically every major scientific award, including the National Medal of Science, the nation's highest scientific honor, the National Academy of Sciences Award in Neurosciences, and the 1983 Albert Lasker Award for Basic Medical Research, often referred to as the "American Nobel." The Lasker citation called him "the intellectual progenitor of his field," the first researcher to ask, "How does the brain process, perceive and respond to the information gathered by the senses?"

He was the first president of the Society for Neuroscience, editor of the *Journal of Neurophysiology*, director of Hopkins' Department of Physiology from 1964 to 1980, head of the Philip Bard Laboratories of Neurophysiology, among the founders of Hopkins' Krieger Mind/Brain Institute, and author of landmark textbooks. A masterful teacher, he directed the training and research of more than 49 postdoctoral fellows, approximately 30 of whom since have become heads of their own departments and laboratories throughout the nation and overseas.[97]

His protégés invariably cite the phenomenal knowledge that Mountcastle maintains, his fiercely focused work ethic, the clarity of his speech and writing, his patience and civility, and his astounding capacity to handle multiple outside duties while somehow being ever present in the lab.

Aspostolos Georgopoulous, a physiology fellow from 1972 to 1974, later a professor of neuroscience, and now a professor at the University of Minnesota, says that working alongside Mountcastle as he put delicate microelectrodes into the brains of research animals was akin to "getting to go in a little submarine with him—to be like the Jacques Costeau of the cortex," voyaging into the unknown depths of the brain. Hopkins neuroscientist Solomon Snyder, whose discovery of the brain's opiate receptors (described later in the chapter) earned him a Lasker Award and National Medal of Science of his own, says that no matter how far the studies of brain cells and molecules take researchers in the future, neuroscientists ultimately will be drawn back to Mountcastle's work.

"The more and more we know of individual genes that regulate brain function, the more and more it becomes clear that molecular biology is just the beginning—and we need to return to the lessons of Vernon Mountcastle to put it all together."[98]

■ The Turtle Derby, a fixture on the School of Medicine campus since 1931, is an elaborate, fun-filled event held each spring by the entire Hopkins Hospital community, raising money for worthy causes by "racing" a pack of lumbering turtles. In 1940, Ogden Nash (1902–1971), so renowned for his light verse that the U.S. Postal Service issued a 2002 commemorative stamp in his honor, was a patient in Hopkins Hospital. Resident Daniel Labby asked Nash, a master of pun-filled rhymes, if he could help publicize the Turtle Derby. Nash promptly and graciously wrote a Turtle Derby poem in longhand, even making a reference in it to then-famous Spanish pianist José Iturbi (1895–1980):

Oh, come to the Turtle Derby
Or is it the Turtle Darby?
Where the terrapins leap
as swiftly
As the fingers of Señor Iturbi.

Come hippety-hop to Hopkins
Fill up on soda-bicarby
Is it the Turtle Derby
or is it the Turtle Darby?

David Bodian
Baltimore, Md.

Special Student

Remis. Fees

S.B.   Chicago   1931
"   "   1934
PhD. "   1937
M.D.

Position: Assoc. in Epidemiology

The "student card" for David Bodian (1910–1992), who arrived at Hopkins in 1938 as a research fellow in the Department of Anatomy laboratory of Howard Howe. There, collaborating with Howe, Robert Ecke, Talmage Peel, Kenneth Maxcy and Isabel Morgan, Bodian's research laid the foundation for development of the Salk and Sabin polio vaccines. Bodian's team demonstrated that the polio virus was transmitted through the mouth and digestive tract, and that there were, in fact, three distinct types of the virus. They demonstrated that for a vaccine against polio to be effective, it must include antibodies recognizing all three virus types—and succeeded in developing early polio vaccines that immunized rhesus monkeys against the disease. Bodian also developed a technique for staining nerve fibers and nerve endings—called the "Bodian stain"— and made major contributions to the knowledge of the basic structure of nerve cells.

Ensuring the Conquest of Polio

Hopkins' pivotal role in research on the cause and prevention of poliomyelitis began in 1936, when Howard A. Howe and his associates in the Department of Anatomy, Robert Ecke and Talmadge Peel, demonstrated in rhesus monkeys that the polio virus could spread along known nerve fiber pathways in the brain. In 1938, David Bodian joined Howe's group as a research fellow—a momentous event for the future of polio inquiries.[99]

By 1940, Bodian and his co-workers had established themselves as one of the nation's outstanding polio research teams. They were the first to demonstrate that chimpanzees could be infected with the polio virus orally, not just through the respiratory system, which previously was thought to be the major route by which the virus entered humans.[100]

Concurrent with the polio research being done in the School of Medicine, epidemiologist Kenneth Maxcy at the Hopkins School of Hygiene and Public Health (now the Bloomberg School) was researching how the polio virus spread throughout a community, along with its spread inside the human body. Maxcy, a 1915 graduate of the School of Medicine, wanted to establish a multidisciplinary research center at the school to study polio and other viruses. In 1941, he applied to the National Foundation for Infantile Paralysis for a long-term grant—an unusual request at the time, since research grants then usually were awarded yearly and for modest amounts. The Foundation, organizer of the fundraising "March of Dimes" campaign, had been created in 1938 by President Franklin D. Roosevelt, a polio victim, and was headed by his former law partner, Basil O'Connor. After several months of negotiations, it agreed to a five-year grant of $300,000 for The Johns Hopkins University to establish a center at the school of public health solely for the study of polio.[101]

Bodian and Howe became members of Maxcy's research team and were joined in 1942 by immunologist Isabel Morgan, daughter of Hopkins-trained, Nobel Prize-winning zoologist, Thomas Hunt Morgan. They succeeded in inducing resistance to the polio virus in monkeys by giving the animals injections of both the live virus and a strain of it that had been inactivated. In 1949, they also made the crucial discovery that there were only three major types of polio virus and that, in contrast to flu virus, they were genetically stable.[102]

By 1951, Morgan had convincingly demonstrated that rhesus monkeys could be immunized to the disease with inactivated viruses of all three basic polio types—and, indeed, the monkeys were resistant to the polio virus no matter which route it took to enter the body, orally or through respiration. Bodian, in turn, discovered the multiple effectiveness of the polio antibody in human gamma globulin, a protein in the blood that contains many antibodies that guard against bacterial and viral infections.[103]

These and other key discoveries by Hopkins researchers created a sound foundation for development of polio vaccines, beginning with Jonas Salk's 1954 vaccine using deactivated or "killed" viruses and Albert Sabin's 1957 oral vaccine using "live" viruses. [104]

The Birth of Medical Genetics

Skeptical colleagues of Victor McKusick (1921–2008), a 1946 graduate of the School of Medicine, thought he was foolish to abandon a promising career in cardiology in the mid-1950s to pursue the study of genetics, which some considered little better than the medical equivalent of stamp collecting—and possibly not even a science.[105]

His doubting peers "thought I was committing professional suicide because I had a reputation in cardiology and was shifting over to focus for the most part on rare, unimportant conditions," McKusick told *The Baltimore Sun* only a few months before his death in July 2008 at the age of 86. "But it didn't bother me. I felt certain it was going somewhere."[106]

Where genetics led, thanks to McKusick's relentless inquiries, meticulous record-keeping and indefatigable cataloging, was to a whole new discipline that many scientists and clinicians believe is destined to be the driving force behind medical breakthroughs in the 21st century.

McKusick became a towering international figure in genetics research, diagnosis and treatment by founding Hopkins' Division of Medical Genetics in 1957, forerunner of the more than 100 clinical genetics units that now exist in North America, training thousands of students. His creation of the annual, two-week Short Course in Medical and Experimental Mammalian Genetics at the Jackson Laboratory in Bar Harbor, Maine, first given in 1959, is widely credited with the post-graduate training of generations of genetic medicine practitioners and scholars, as is his launching of his classic book, *Mendelian Inheritance in Man*, in 1966. Now published only online, with its database updated daily, it remains the world's most comprehensive catalogue of genetic data and is considered an essential tool of the medical geneticist.[107] McKusick also was the first scientist to forcefully advocate the sequencing of the human genome. "I doubt that anyone would have conceived of the Human Genome Project if he had not shined the light on the value of genetics in so many human conditions," says Mike Weisfeldt, a successor to McKusick as William Osler Professor of Medicine.[108]

At the top of the Hop: Genetics pioneer Victor McKusick (1921–2008), to the left (above) in a white coat and pointing, relished the history of Johns Hopkins Medicine—much of which he had witnessed. Arriving as a medical student in 1943, he never left, establishing a record for the longest uninterrupted service of any faculty member since the medical school's founding in 1893. He enjoyed taking new medical students, residents, interns and others on a history-laden tour of the 1889 Billings Building that concluded with a challenging climb up the narrow, winding stairs inside its dome to the top. There, with a spectacular view of Baltimore beneath him, McKusick would point out municipal landmarks and locations with links to Hopkins.

Victor McKusick poses with the printed editions of his *Mendelian Inheritance in Man*, **beginning on the left with the first, relatively slim volume published in 1966, up to the far fatter, 1,545-page ninth edition, published in 1990. Now published only online, with its database updated daily, it remains the world's most comprehensive catalogue of genetic data.**

1961

November: The Johns Hopkins University Press "publishes" a work that will become its most financially successful item for years: "Mr. Bones," an 18-inch plastic replica of a human skeleton, designed by Leon Schlossberg, a renowned teacher and illustrator in the Department of Art as Applied to Medicine. Schlossberg devoted three years of painstaking work to creation of the skeleton model.

Hopkins opens its first cancer chemotherapy unit at Baltimore City Hospitals (now Johns Hopkins Bayview), one of the first university-based centers of its kind in the country. In 1977, the Hopkins Cancer Center moves to The Johns Hopkins Hospital, where by 1978 it is housed in a new $17 million facility.

1962

Nigerian James Nabwangu becomes the first black student to enroll in the School of Medicine. The first African American student to enroll is Robert Gamble, who enters the school in 1963 and graduates in 1967, as does Nabwangu.

Born in Parkman, Maine (estimated population, 550), McKusick was an identical twin whose parents, both former teachers, made education a priority for the five children they raised on a dairy farm. McKusick went to a one-room schoolhouse, studying under the same teacher for seven of the eight years of grammar school, then attended a high school that had no courses in science.[109]

He originally thought he might become a minister, but a major illness changed his mind. In 1937, at the age of 16, he contracted a severe streptococcus infection from cutting hay . Hospitalized for 11 weeks before finally being cured with the antibiotic sulfanilamide, which had been introduced just the year before, McKusick later said, "I decided I liked what doctors did; I decided I wanted to join them."

In 1940, McKusick enrolled in Tufts University, located in nearby Medford, Mass., because it had a medical school. He had read about Hopkins, however, in a 1939 *TIME* magazine and wanted to go there. Finding that he was able to take advantage of the Hopkins medical school's wartime relaxation of its requirement that applicants have a B.A, he dropped out of Tufts without completing his bachelor's degree and came to Hopkins in 1943. He never left, establishing a record for the longest uninterrupted service of any faculty member since the medical school's founding in 1893.[110]

Although he earned a medical degree from Hopkins and 22 honorary doctorates, McKusick joked that he did so as a college dropout, having never officially earned a bachelor's degree.

Initially concentrating his research on heart problems, McKusick found that cardiology led him to medical genetics. He became the first to describe the cluster of characteristics of Marfan syndrome, an inherited connective tissue disease characterized by heart defects, unusually tall height, and several other abnormalities. Studying Marfan patients and those afflicted with other familial diseases triggered his interest in learning about single genes that result in multiple deformities.

After founding the Medical Genetics Division, he began compiling extensive medical histories of members of the Old Order Amish of Pennsylvania to identify genes responsible for their inherited disorders, of which a kind of dwarfism, now known as McKusick Type Metaphyseal Chondrodysplasia, is unusually common. His name also is attached to McKusick-Kaufman syndrome, a developmental disorder marked by congenital heart disease, buildup of fluid in the female reproductive tract, and extra fingers and toes.

McKusick was named William Osler Professor of Medicine, director of the Department of Medicine and physician in chief of The Johns Hopkins Hospital in 1973. He held those positions until he was named University Professor of Medical Genetics in 1985.

At a birth defects meeting in 1969 in The Hague, McKusick proposed mapping the human genome, saying it would be the key to unraveling the mysteries of many congenital abnormalities and genetic diseases. Years later, McKusick said that his call for mapping the human genome "took a little nerve" because the main technological tools for accomplishing such a task expeditiously did not exist then. Many scientists dismissed it as wildly impractical, expensive and of little value.

Not a laboratory scientist himself, McKusick relied on other researchers for the actual gene-mapping, which originally took years. With the aid of today's automated sequencers, a gene can be mapped in a matter of weeks. The human genome sequence, containing more than three billion units of DNA, was published in February 2001.

In a 2000 interview, McKusick said: "I like to say the arrangement of genes on chromosomes is part of the microanatomy. Just as the gross anatomy in the Middle Ages was important to medicine, every [medical] specialty now uses mapping genes for diseases."[111]

Today, the importance of the links between various genes and disease is universally recognized, even among non-scientists. Finding a link between a particular gene and a disease rarely makes headlines now, reflecting the widespread acceptance of McKusick's fundamental approach to studying disease.

Lean and laconic, McKusick leavened his New England reserve with a sly sense of humor. Never without his small notepad and pocket camera, he constantly scribbled observations and snapped pictures of events, places and people. Perhaps his most memorable extracurricular exploit was a tour of the original Hopkins Hospital, now the Billings administration building, which he conducted for medical students and residents while giving a lecture on Hopkins history. For decades, he concluded the tour with a physically challenging (for others) climb up the narrow, winding stairs that lead to the top of the building's iconic dome, with its spectacular view of Baltimore.[112]

McKusick's honors were abundant. Among the most notable were the 1997 Albert Lasker Award for Special Achievement in Medical Science, the 2001 National Medal of Science, and the 2008 Japan Prize in Medical Genomics and Genetics. Three months before he died, he went to Tokyo April to receive that medal—and a 50 million yen ($470,000) award—at a formal presentation attended by the Japanese Emperor and Empress. He called the Japan Prize, considered that country's equivalent of the Nobel, an honor that also lauded "the contributions and support of Johns Hopkins, and of my colleagues and students over many decades."[113]

"Look wise, say nothing, and grunt. Speech was given to conceal thought."

—William Osler

Victor McKusick and his long-time associate, Kathryn "Kay" Smith, meet with leaders of Little People of America, a philanthropic organization for people no taller than 4 feet, 6 inches, prior to its national convention in 1968. The group was founded by 3-foot 9-inch actor Billy Barty, shown shaking hands with McKusick. Others pictured, left to right: Stanley Wermes; George Moore; Tina Baehm; and George Baehm III. The 6-foot, 2-inch McKusick was an honorary member of the organization.

■ The Johns Hopkins Precursors Study, founded in 1948 by Caroline Bedell Thomas (1904–1997), still is under way. For six decades, it has been keeping a running medical chart of more than 1,300 members of the School of Medicine's classes from 1948 to 1964—making it the longest continuous study of its kind in the country. Using detailed annual survey results provided by these alumni, the study has produced landmark findings linking high cholesterol to later heart disease and mental depression with heart attacks.

1964
Alfred Blalock retires as director of the Department of Surgery and is succeeded by George Zuidema, a 1953 graduate of the School of Medicine who then was 36 years old. Zuidema will undertake a significant restructuring and expansion of the department and remain head of surgery until 1984, when he is succeeded by John Cameron.

1967
July: Lincoln Gordon, a former ambassador to Brazil and assistant secretary of state for inter-American affairs, is chosen to succeed Milton Eisenhower as president of the university. He serves during four tumultuous years, not only for Hopkins but for the country. Students and faculty critical of the country's involvement in Vietnam for a brief time occupied the university's executive offices. Citing increasing criticism from faculty, Gordon resigned in March 1971.

Henrietta Lacks and HeLa

Certainly among the most far-reaching—and ultimately controversial—Hopkins Medicine achievements of the 1950s was cancer researcher George O. Gey's development of the first human cell line, known worldwide as the HeLa cell culture. Gey named the culture by taking the first two letters from the first and last names of the 31-year-old, terminally ill cervical cancer patient from whose tumor the original cells had come, Henrietta Lacks.[114]

The importance of HeLa cells to molecular and cell biology studies cannot be overestimated. Since 1951, scientists throughout the United States and overseas have used these "immortal," still-multiplying cells as the basis for research studies that have led to some of most crucial medical advances of the past 60 years. These included the polio vaccine, chemotherapy breakthroughs, cloning, gene mapping and *in vitro* fertilization. This line of research also led to Nobel Prize-winning work on the cancer-causing human papillomavirus (HPV) that resulted in vaccines to guard against it, and landmark research into HIV and tuberculosis. HeLa even were the first human cells sent into space to see if they could survive zero gravity.[115]

Gey (1899–1970), a 1933 graduate of the School of Medicine and director of the tissue culture laboratory in the Department of Surgery since 1928, had tried for decades to grow human cancer cells in test tubes. He and others wanted to study living cells under standardized conditions and subject them to all sorts of tests involving toxins, drugs, hormones and viruses that would not be possible to use on a patient. Such tests, he believed, would reveal important biological questions about how cancer develops and possibly lead to a cure. To establish such a cell line, Gey had arranged years earlier for his Hopkins clinical colleagues to give him cervical cancer cells from their patients. Obtaining tissue samples from patients for research was a common practice not just at Hopkins but throughout medicine in the 1950s, which was long before any regulations had been written, or even contemplated, for obtaining permission from patients for the use of their tissues in research. Such "informed consent" simply was not on medicine's radar screen in 1951.[116]

Every previous cell sample used by Gey to try establishing a constantly reproducing line of cells had died after a short time in the lab—until he received the cells from Henrietta Lacks. A poor African American from rural Virginia who moved to Baltimore County in 1943 with her husband, a steelworker, and their five children, Lacks went to Hopkins Hospital on Feb. 1, 1951. She told physicians about a painful "knot on my womb." The doctors' examination revealed advanced cervical cancer and syphilis. She was given radiation treatment for the cancer, but before it was administered, two dime-size samples of Lacks' tissue—one cancerous and one not—were taken and sent to Gey. The record is not clear on whether Lacks, who died shortly thereafter, was told that samples of her tissue were taken to be studied, since it was a routine procedure then to do so without mentioning it to patients or their relatives.[117]

Lacks' cells proved to be a biomedical marvel, multiplying and surviving in an unprecedented way, soon doubling in number every 20 to 24 hours. Why this is so still is unclear—although the importance of the development was apparent immediately. Gey predicted that they held the key to eradicating cancer and began distributing them for free to scientific colleagues anywhere in the world who wanted them. (Ironically, a clue to HeLa's remarkable ability to reproduce outside of the body may be linked to a discovery made decades later by Carol Greider, now a renowned Hopkins researcher. She discovered that an enzyme she named telomerase protects the tips of a cell's chromosomes and ensures the continued reproduction of cells, a finding that won her the 2009 Nobel Prize in medicine. Today it is known that the HeLa cells have an abundance of telomerase.)[118]

Hopkins never patented the HeLa cells, never sold them commercially, and neither it nor Gey ever made any money directly from them. Over the years, however, pharmaceutical firms and others that obtained free samples of the cells have made substantial sums by selling supplies of them and developing medicines based upon research using them. An estimated 60,000 studies have been conducted using HeLa cells, which still are employed regularly in scientific inquiries.[119]

Controversy over Hopkins' role in developing the HeLa cell culture emerged after Gey's death in 1970. He had adhered to the well-established tenets of patient confidentiality and never revealed Henrietta Lacks as the source of HeLa. Much to their astonishment, Lacks' surviving children did not learn about the existence of the cell culture until 1975, when a friend who was aware of HeLa asked them if they were related to Henrietta Lacks. Later, when a sample of the cell line somehow became contaminated, Hopkins researchers contacted her family and obtained samples of their cells to determine what had happened, but did not fully explain what they were seeking or follow up completely on what they found. The family felt hurt and unappreciated.[120]

In 2010, freelance science writer Rebecca Skloot, a past contributor to Hopkins Medicine periodicals, published *The Immortal Life of Henrietta Lacks*, a riveting 396-page, best-selling book detailing the HeLa story and focusing on the bioethical questions it raised. Hopkins, now home to the nationally known Berman Bioethical Institute, is a staunch advocate for making biomedical ethics an integral part of teaching, research and clinical care. It readily acknowledges not only the contribution to advances in biomedical research made possible by Henrietta Lacks and the HeLa cells but how insensitive the biomedical research profession was more than a half-century ago—and how far American medicine has come since then.[121]

Henrietta Lacks (1920–1951) with her husband, David, circa 1945-50. A young mother of five, she came to Hopkins in 1951 to be treated for cervical cancer. A sample of her cells (below), named "HeLa" by using the first two letters of her first and last names, were obtained by cancer researcher George Gey (left) in a tissue sample taken from the virulent tumor that killed her. Gey (1899–1970) had been trying to further his efforts to combat cancer by finding malignant cells that could be grown outside the human body. HeLa proved to be what he was seeking—"immortal" cells that continue to multiply in the laboratory six decades after Henrietta Lacks' death. HeLa cells have been instrumental in tens of thousands of experiments, aiding in the development of the polio vaccine, in studies of the effects of radiation and poisons, in analyses of how viruses work, and inquiries into the human genome.

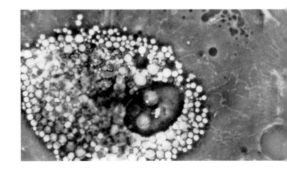

Jeanne Dougherty, then head nurse in the pediatric research unit, and a young patient positioned themselves on the third-floor porch of the Halsted Building in January 1962 to watch the early stages of the construction of the new Children's Medical and Surgical Center and a new Hospital cafeteria.

Robert Cooke, director of the Department of Pediatrics from 1956 to 1973, oversaw construction of the new Children's Medical and Surgical Center, successor to the beloved but dilapidated Harriet Lane Home. Seen here with young patients in the new center's recreation room, Cooke also was instrumental in creation of the John F. Kennedy Institute for the Rehabilitation of the Mentally and Physically Handicapped Child, now known as the Kennedy-Krieger Institute, located next to Hopkins Hospital.

1960s

The Building Boom Continues

Construction emphasizing the burgeoning postwar interest in research continued into the 1960s. The entire Hunterian II Building, completed in 1913, was renovated from top to bottom for modern research in 1961. (It was replaced by Hunterian III in 1987.) Additional space was provided for the Departments of Anatomy, Pediatrics, and Gynecology and Obstetrics. The Biophysics Building was completed in 1963 and a year later, the Wilmer Eye Institute opened a six-story research wing—described as the largest unified center for eye research in the world. Federal funds enabled the 1964 Children's Medical and Surgical Center to add four additional floors for surgical research, including research into orthopedics, and research on mental retardation. The building additionally contained new, much-needed facilities for the care of children.[122]

Robert Cooke, director of the Department of Pediatrics from 1956 to 1973, also was able to parlay his closeness to Eunice Kennedy Shriver and her husband, Sargent Shriver—the sister and brother-in-law of President John F. Kennedy—into federal money for creation of a major center for children with developmental problems, the John F. Kennedy Institute for the Rehabilitation of the Mentally and Physically Handicapped Child, now known simply as the Kennedy-Krieger Institute, which is located next to Hopkins Hospital.[123]

Early in 1963, Cooke was playing tennis with the Shrivers at their Maryland estate in the Washington suburbs when a call came from President Kennedy. He wanted advice from his sister on how to improve a speech about mental health and mental retardation, key concerns of the Joseph P. Kennedy Jr. Foundation, which Eunice Shriver headed.[124]

Cooke, the Kennedy Foundation's top medical advisor, had an idea. In the Shrivers' library, he drafted several paragraphs proposing academically affiliated clinical and teaching facilities for children with mental disabilities (among whom were two of Cooke's own children). "Much to my surprise," recalled Cooke in 2006, "some pieces of the paragraphs appeared in the president's message to Congress." The resulting landmark legislation established the federally funded program that would support such facilities—of which the Hopkins-affiliated Kennedy Institute was the first of more than 60 to be built nationwide. (Its name later was expanded to acknowledge a substantial gift by Hopkins trustee Zanvyl Krieger to enhance its programs and facilities.)[125]

Oncology Comes into its Own

In the 1960s, the word "oncology" was not part of the academic medicine vocabulary, few specific medical care facilities were set aside for it, no clearly recognized, academic discipline was concentrated on it—and "cancer" still was a word that patients and their families feared to utter. Yet clinical oncology was beginning to emerge as a specialty that could make a difference not only in the prospects for a patient's recovery, but in the quality of life that such patients would have while battling their disease. Albert Owens, a 1949 graduate of the School of Medicine who had joined the faculty in 1956, chose to devote his himself to fostering its advancement. He would play a leadership role in the development of oncology as a scientific discipline and clinical specialty.[126]

The year after Owens joined the Hopkins faculty, A. McGehee Harvey decided that with chemotherapy becoming a new weapon in the battle against cancer, along with radiotherapy and surgery, he could create a cancer research and treatment division within his Department of Medicine by connecting the clinical investigators in these areas. He tapped Owens to head the new oncology division, but because Hopkins Hospital at that time lacked the space for such a service, Owen moved a large part of his inpatient, clinical and research activities to Baltimore City Hospitals, where generations of Hopkins medical students and graduates were trained. In 1961, he opened Hopkins' first cancer chemotherapy unit there. It was one of the first university-based centers of its kind in the country.[127]

In 1973, Owens, a slightly bashful, bow-tie wearing researcher and clinician, was named the first director of the Johns Hopkins Oncology Center, which had won federal designation as one of the first of 18 National Cancer Centers in the country. In 1977, Owens was able to move the center to Hopkins Hospital, where by 1978 it was housed in a $17 million facility. By then, Owens also had become president of both the Maryland division of the American Cancer Society and the Association of American Cancer Institutes.[128]

"I'll pay a lot of money to anybody who can convert my eyes from 10 diopters, i.e., right up there with a…bat, to 20/20! Glasses and contact lenses don't count. I'm honored to be the recipient of your attention, more so to know that I've entertained some of the real heroes. The main reason we're put here is to make the world a better place. Hopkins and its people do just that. My only direct connection to Hopkins is that my father was once your mailman."

—Tom Clancy, author of *The Hunt for Red October* and other internationally best-selling novels, endower of the Tom Clancy Professorship in Ophthalmology in the Wilmer Eye Institute and the Kyle Haydock Professorship in the Division of Pediatric Oncology

A close teacher-student relationship long has been a hallmark of medical education at Hopkins, as show in this 1962 photo of a professor instructing a small group of students on how to read neurological X-rays.

■ A landmark medical achievement was accomplished first at Hopkins—but not recognized at the time by those who did it. In 1965, gynecology fellow Robert Edwards of Great Britain, working with Hopkins gynecologist and crytogenetics laboratory chief Howard W. Jones Jr. (above, right), conducted experiments that produced the first human egg fertilization in vitro. Because the criteria for in vitro fertilization then were different than what later was established, they didn't realize their achievement at the time and only reported their effort as an "attempted" in vitro fertilization. Along with gynecologist Patrick Steptoe, Edwards subsequently developed IVF therapy in Britain, which led to the world's first "test tube" baby in 1978, an accomplishment for which Edwards won the 2010 Nobel Prize in Medicine. Jones and his wife, gynecologist Georgeanna Seegar Jones (above, left), by then retired from Hopkins and heading their own Jones Institute of Reproductive Medicine at East Virginia Medical School, oversaw the birth of the first "test tube" baby in the U.S. in 1981.

Over the next decade, the number of cancer patients treated at Hopkins nearly tripled, rising from 6,000 in the 1970s to more than 17,000 by 1988, while funding for cancer research reached more than $19 million by 1988.[129]

Owens had announced plans in December 1986 to build a $110 million, 88-bed cancer center to replace the one opened at Hopkins Hospital in 1977. In January 1987, however, then-Hopkins Hospital president Robert Heyssel chose to step down in order to concentrate on being president and CEO of the newly incorporated Johns Hopkins Health System. It had been created in 1986 to oversee the finances of Hopkins Hospital, its affiliates and future acquisitions, but not the medical school. Owens agreed to assume Heyssel's old post and become the Hospital president while also remaining head of oncology.[130]

By July 1988, Owens decided instead to focus on plans to develop the new multi-million-dollar oncology center at Hopkins and relinquished the Hospital presidency, which Heyssel resumed. Before Owens did so, however, he made sure that in his role as a cancer specialist, he would leave a distinctive mark on the Hospital during his brief tenure in the top job: the Hospital became smoke-free the month he left its presidency.[131]

Launching the Genetic Engineering Revolution

Some revolutions begin with one big explosion. Others build up step by step before emerging full-blown. The modern revolution in genetic engineering was, perhaps appropriately, a bit of a hybrid—requiring a series of steps before bursting upon the scientific world with swift, immense force, generated significantly by the slow, quiet research done by two Hopkins scientists who'd spent their entire careers at the School of Medicine: microbiologists Daniel Nathans and Hamilton Smith.

In 1962, Swiss microbiologist Werner Arber of the University of Basel theorized that naturally produced chemicals, proteins he called restriction enzymes, existed—but he couldn't find them. Such enzymes could be used like a scalpel to dissect precise sequences of genes at specific—or restricted—points in an organism's DNA.[132]

In the fall of 1968, Ham Smith, a six-foot, five-inch, 1956 School of Medicine graduate who considered himself a "kitchen chemist," had included in his routine reading a paper by a pair of Harvard scientists describing work they were doing based on Arber's search for restriction enzymes. The Harvard duo had observed an enzyme that cut DNA, but only randomly. That made their chemical scalpel interesting, but not terribly useful to would-be genetic engineers, who required a more precise tool.[133]

At the time Smith read the paper, he and a Ph.D. student in his laboratory, Kent Wilcox, were exploring how DNA split and recombined. They did this by introducing foreign DNA into bacterial cells and manipulating various chemical processes in those cells to force them to accept the new DNA. As often happens in scientific experiments—as when Alexander Fleming discovered a "contaminated" lab growth that turned out to be penicillin—what Smith saw happen in his lab might have caused him to ignore or discount his investigation. He observed that the "donor" or foreign DNA was being destroyed, instead of being taken up by the host

cells. The experiment wasn't working—but instead of giving up, Smith suddenly remembered the concept that Arber had proposed for restriction enzymes. Yes, the foreign DNA was being destroyed, Smith could see, but in very specific ways at very specific sites, not randomly. Could the reason be that a restriction enzyme—a precise protein scalpel—was slicing the DNA at well-defined places? Shortly after this discovery, Smith's protégé Wilcox was drafted into the Army and a post-doctoral fellow, Thomas Kelly, assumed his place at the lab bench to assist Smith in continuing the study.[134]

Smith wrote to his longtime colleague and friend, Dan Nathans, who was on sabbatical in Israel, about what he had seen. Both agreed it was worth going after the enzymatic cutters involved in Smith's supposedly failed experiment.

In his own work, Nathans had been looking for better ways to analyze viruses known to cause cancer in animals. Nathans speculated that Smith's enzyme-cutters might be just the thing he needed to slice and isolate specific segments of DNA in the virus cell. The cutters did the job, and with them, Nathans and graduate student Kathleen Danna then were able to map the genes of the tumor virus. That, in turn, made it possible to identify specific genes required for the manufacture of the tumor-producing protein in the cell.

Nathans and his own group of researchers also used the site-specific restriction enzymes to create—in essence, engineer—mutations in the DNA of the cancer virus in order to demonstrate how the viral genes are regulated. He'd succeeded in developing the first useful application of restriction enzymes.[135]

Smith's and Nathan's discoveries would be immensely influential, launching an entirely new era in medical research—the era of recombinant (recombined) DNA—and earning them and Arber the 1978 Nobel Prize for Medicine. Cutting DNA into fragments of manageable size is the first step in cloning a gene, and cloning itself can be the first step in deciphering the changes that occur in disease. Cloning has permitted the preparation of DNA probes to determine which genes cause cancer, perform prenatal diagnosis, synthesize therapeutic agents in bacteria, and manufacture synthetic insulin. The application of DNA technology not only has enabled the sequencing of the entire human genome (the complete set of 23 chromosomes, which are made up of some 100,000 genes that comprise a person) but also has given medicine a powerful tool for determining the causes of disease—and lead to ways to prevent or cure it.[136]

Microbiologists Daniel Nathans (right) and Hamilton Smith shared the 1978 Nobel Prize for Medicine for their discovery and first use of restriction enzymes, which launched the genetic engineering revolution. It has led not only to the sequencing of the entire human genome but given medical researchers a powerful tool for determining the causes of disease—and lead to ways to prevent or cure it.

"Hopkins never left me. My body, mind and energies went to Vanderbilt, but my soul stayed on Broadway."

—Gottlieb C. (Bud) Friesinger, 1955 graduate of the School of Medicine, founding director of Hopkins' Coronary Care Unit, responsible for reducing the one-year mortality rate for myocardial infarction from 40 percent to 4 percent, later Vanderbilt University's first director of cardiology

All of the former residents of retiring surgeon in chief Alfred Blalock successfully lobbied behind the scenes to have the Clinical Science Building (above), completed in 1963, named in honor of Blalock in 1964. The announcement of the name change was made at a dinner celebrating the 75th anniversary of the opening of the Hopkins Hospital—much to Blalock's surprise and delight.

"Johns Hopkins must grow along with medicine and the nation, while preserving the essence of its greatness, to link the intellectual and cultural values of the Nineteenth and Twenty-first Centuries. The creative potential of the individual must be protected and cultivated, with machines, sophisticated instruments and social organization playing subordinate roles."

—Thomas B. Turner, microbiologist and Dean of the School of Medicine, 1957–1968

Zuidema Succeeds Blalock in Surgery

In 1964, Alfred Blalock retired after a momentous 23 years as director of the Department of Surgery. His impact on surgery's advances had been profound, not just at Hopkins but around the nation. Once wartime restrictions on expansion had ended, he was able to acquire the funds to put several of the Department of Surgery's divisions on a full-time basis, creating the Department of Radiology in 1946; the full-time Divisions of Urology, Neurosurgery and Anesthesiology in 1947; Otolaryngology in 1952; and Orthopedic Surgery in 1953. His special interest in pediatric surgery culminated with the opening of the Hopkins Children's Medical and Surgery Center, dedicated in 1964. Nationally, 38 of his residents went on to important academic appointments, including 10 who became department heads themselves.[137]

At a dinner celebrating the 75th anniversary of Hopkins Hospital, the Clinical Sciences Building was renamed the Blalock Building in his honor at the behest of all his former residents, which pleased him greatly. The subsequent appointment of his successor did not. The search committee shocked senior surgery professors all over the nation—and astounded an openly angry Blalock—by choosing George Zuidema, a boyish-looking, 36-year-old Hopkins medical school graduate who had received his degree barely a decade earlier and was just an associate professor of surgery at the University of Michigan.[138]

The selection of Zuidema was in the Hopkins tradition of picking young professors on the way up, as was his recruitment from another university. Blalock didn't accept that concept—at least when it came to his Department of Surgery. He considered Zuidema too young, inexperienced and unknown. Blalock insisted that one of his former residents should succeed him. Regardless of Zuidema's other attributes, what he had not been was a Blalock protégé.

In 1953, upon graduating from the School of Medicine, Zuidema had "jumped ship," as he observed drolly decades later, and gone to Massachusetts General Hospital for his internship and residency. Aspiring Hopkins surgeons then were supposed to fight to remain Blalock protégés—and many of those who did so subsequently attained positions of greater power elsewhere than Zuidema had at Michigan.[139]

Blalock continued his fierce opposition to Zuidema's appointment until it was announced officially. He then became reconciled—grudgingly—to Zuidema's accession to the surgery directorship and even agreed to remain on the faculty under him in an emeritus capacity. He would die only a few months later of cancer—but the controversy he had stirred would linger for an uncomfortably longer period.[140]

Zuidema was far from unaccomplished, despite his youth and associate professor's rank. By 1964, he already was legendarily prolific, having written more than 200 research articles, many of seminal importance; and he had received half a dozen prestigious scholarships and awards.[141]

Yet initially, Zuidema's appointment was met with resentment not just from Blalock but some disgruntled faculty who soon quit. That and other faculty problems made his first year "kind of crazy," Zuidema recalled.[142]

Once the dust had settled, Zuidema recognized three key things about the department he'd inherited: it needed to grow; to do that, it had to ensure its financial stability; and it required re-structuring in response to the intensifying pace of surgical specialization.[143]

To maintain the department's financial bottom line by expanding its growth, Zuidema developed a professional fee program and enlarged the full-time staff without antagonizing the part-time staff, a very important component of the practice. To update the department's structure, he sat down in 1972 and wrote out a reorganization chart that transformed the Department of Surgery into a Section of Surgical Sciences, with himself at its top.[144]

George Zuidema was only 36 years old and out of the Hopkins medical school just 11 years when he was named to succeed the legendary Alfred Blalock as head of surgery—much to the displeasure of Blalock and those who believed they were more qualified to succeed him. Richard Ross, former dean of the School of Medicine, recalls that stepping into Blalock's shoes was a tough assignment. "It was sort of like succeeding God. Blalock alumni… all felt that one of them should have gotten the job, and they made it pretty tough for George. But it's to his credit that he not only survived, but indeed, thrived on it."

Under Zuidema's plan, what now is called general surgery was elevated to full-time, departmental status, and neurosurgery, orthopedics, urology and otolaryngology also became departments. Formal status as services was given to transplantation and vascular surgery, cardiac surgery, plastic surgery, dental and oral surgery, and pediatric surgery. Overseeing this expanded organization were a director (Zuidema) and an associate director who would be elected from the heads of the surgical departments.[145]

In former dean Richard Ross' view, "not only was it the right thing to do, but it enabled him to recruit some excellent chairmen to run these various departments," among them Patrick Walsh in urology, soon to devise his renowned nerve-sparing prostatectomy surgery; cardiac surgeon Bruce Reitz, who would be a key figure in Hopkins' first multiple-donor transplant surgery; and otolaryngologist Michael Johns, a future School of Medicine dean.[146]

Zuidema's reputation as an indefatigable worker and skillful administrator inevitably led to many job offers. He turned them all down. "Hopkins is a place that gets in your blood, is unique in terms of collegiality, mutual respect, the ability to work with a lot of gifted people. It is not easy to leave that," he says.[147]

Then one day in 1984, he received a call from a recruiter for the University of Michigan. It was in the midst of a billion-dollar reconstruction and reorganization project and desperately needed someone to oversee it. Of particular concern was a problematic relationship between the university's medical school and hospital. The offer gave Zuidema, born in the small western Michigan town of Holland, a chance to go home to his native state and really make a difference. Within months, he was back in Ann Arbor, 20 years after leaving it, as vice provost for medical affairs and professor of surgery at the University of Michigan.[148] His successor as director of Hopkins' Department of Surgery was John Cameron, a 1962 graduate of the School of Medicine, who held the post for the next 19 years, building an international reputation as the world's foremost pancreatic cancer surgeon.

■ It was not until 1977 that The Johns Hopkins Hospital, which opened in 1889, at last got out of the red. Due to its commitment to Mr. Hopkins' mandate to care for those who couldn't pay, the Hospital operated at a loss throughout its first eight decades. Annual deficits were covered by gifts from trustees and dipping into the endowment. When Robert Heyssel became head of the Hospital in 1972, it was running a $1.2 million deficit. To help raise the $100 million needed to upgrade Hopkins, Heyssel persuaded the state's Health Services Cost Review Commission to allow higher fees for those who could pay, improved the Hospital's efficiency, and cut costs wherever he could. By 1977, the Hospital was making money.

Richard W. TeLinde (1894–1989), a 1920 graduate of the School of Medicine, maintained an extensive private gynecological practice even after being appointed director of the Department of Gynecology in 1939, but readily accepted a full-time appointment to the faculty in 1956, making gynecology the last department to adopt the full-time system—some 40 years after its creation. TeLinde established a division of what then was called GYN endocrinology and infertility in the department, making it the first division of reproductive medicine in any medical school in the world. His 1946 book, *Operative Gynecology*, still is considered the definitive work in the field.

■ One-time Soviet Union leader Nikita Khrushchev (1894–1971) and Russian revolutionary Leon Trotsky (1879–1940) each had a Hopkins Medicine connection. In 1969, Hopkins physician in chief A. McGehee Harvey (1911–1998) spent two weeks in Russia at the request of Khrushchev, deposed as First Secretary of the Soviet Community Party in 1964, to treat the former leader's lupus-afflicted daughter. Forty-four years earlier, when Trotsky was living in exile in Mexico, an assassin sent by rival Josef Stalin (1879–1953) struck him in the head with a pick-axe. Trotsky friends arranged for famed Hopkins brain surgeon Walter Dandy to fly to Mexico to treat Trotsky, but the case was hopeless and Trotsky died before Dandy got there.

Gynecology and Obstetrics Merge

When Richard W. TeLinde was named to succeed Thomas Cullen as head of the Department of Gynecology in 1939, it was a victory for those within Hopkins Medicine who favored independent departments of gynecology and obstetrics—a subject of internal debate since the separate Department of Obstetrics was created in 1899 at the behest of Howard Kelly, who originally had been head of both gynecology and obstetrics and preferred to concentrate on gynecology.

Over the next six decades, attempts to combine the specialties—the only ones dealing exclusively with the medical problems of women—foundered over such issues as adoption of full-time clinical personnel (which Kelly had opposed), gynecology's focus on surgery, and whether it should share facilities with obstetrics.[149]

TeLinde, unlike Kelly or Cullen, did not object to full-time faculty and would have preferred joining such a system, despite having a lucrative private practice. Ironically, he was not offered a full-time position until 1956, but when he accepted, gynecology became the last clinical department in the School of Medicine to adopt the full-time system—some 40 years after its creation and Kelly's opposition to it.[150]

When TeLinde retired in 1960, the same year that longtime obstetrics department director, Nicholson J. Eastman, also stepped down, the path toward unification of the two departments gradually began to clear. Most medical centers already had combined departments, and such an approach seemed appropriate. In 1964, a Department of Gynecology and Obstetrics was created—reuniting the specialties that Kelly had headed jointly from 1893 to 1899. Allan C. Barnes, who previously had headed GYN/OB departments at Ohio State and Case Western Reserve University, was named its director. He left Hopkins in 1970 to become a vice president of the Rockefeller Foundation. His successor, Theodore M. King, remained head of the GYN/OB until 1984, when he was succeeded by Edward E. Wallach.[151]

Neurology, Dermatology, Biophysics and Biomedical Engineering Departments Launched

Some of the "new" specialties that arose at Hopkins after World War II were not new areas of knowledge but segments of established departments that could not develop independently until federal and private funding became available. For example, a separate department of neurology had been under discussion since the 1920s, but money had not been available to cut neurology loose from its parent, the Department of Medicine.[152]

By 1966, however, funds finally became available—in part from the Joseph P. Kennedy Foundation and the Cerebral Palsy Foundation—to establish two professorships, and space for a separate department could be allocated in a new research building then being planned. Two more years elapsed before the selection of a director for the new department. In the spring of 1968, Guy McKhann, a Yale medical school graduate and former pediatric resident at Hopkins who then was on the faculty at Stanford, was chosen to head neurology.[153]

McKhann recruited outstanding, talented young neuroscience researchers, including Daniel Drachman and Richard Johnson, still central figures in the department, and John Freeman, who was picked to head the pediatric neurology division that now bears his name. McKhann also developed such a close relationship with the Department of Neurosurgery that by 1987, he characterized them as "virtually a single department, sharing clinical space (both inpatient and outpatient), administrative staffs, nursing staffs and teaching activities."[154]

Like neurology, dermatology had been one of the weaker clinical areas at Hopkins until it was given departmental status and a full-time director with the rank of professor. This was done in 1973, following a recommendation by medicine director Mac Harvey just prior to his retirement. Its first director was George W. Hambrick Jr., who had been a part-time member of the dermatology division in the Department of Medicine before being named a full professor in 1969, prior to the division's elevation to departmental status. Hambrick did such a fine job establishing the department that he was recruited away in 1976 by Cornell, which wanted him to establish its first dermatology department, too. (Hambrick later would also become the founder and president of the American Skin Association.) At Hopkins, Hambrick was succeeded by Irwin M. Freedberg, who himself was succeeded by Thomas T. Provost in 1981.[155]

The Department of Biophysics was founded in 1961 and the Department of Biomedical Engineering was created in 1970. As with neurology and dermatology, discussion about making these divisions into separate departments were ongoing as far back as 1946. Their ultimate, albeit sometimes glacial ascension to departmental status reflected not only the increasing medical importance of their disciplines, but Hopkins' judicious deliberateness in determining when the time was right to develop the resources necessary to ensure that whatever it did in these fields would be superior.

Guy McKhann, a one-time pediatric resident at Hopkins, was recruited from Stanford to become founding director of the new Department of Neurology in 1968. An acknowledged pioneer in the designing of clinical directives in the neurosciences, McKhann joined with John Freeman to create the Division of Pediatric Neurology in the department, establishing its postdoctoral training program as one of the finest in the nation. He also was founding director of the Hopkins Mind/Brain Institute and remains active on its faculty, as well holds an appointment in the Bloomberg School of Public Health's Center for Mind-Body Research. He has written books on subjects ranging from childhood neuropathies, Guillain-Barre syndrome, other diseases of the nervous system, the neurology of language, and on how to maintain physical and emotional health as one ages.

■ Solomon Snyder (1938-), founding director of the Department of Neuroscience, was a teenage classical guitar prodigy in Washington, D.C., when he gave a concert before legendary virtuoso Andrés Segovia (1893–1987), considered by many to have been the finest classical guitarist of the 20th century. Snyder, winner of the National Medal of Science and the Lasker Award, continues playing guitar. He created the Solomon H. Snyder Prize at Hopkins' Peabody Conservatory of Music to underwrite the New York debuts of talented Peabody guitar students.

1969
Appearing at a birth defects conference in The Hague, Victor McKusick proposes mapping the human genome, an idea initially scoffed at as impractical and of little value. When the human genome sequencing finally is completed in 2000, it is hailed as a signal development in medical history. McKusick attends the White House ceremony at which President Bill Clinton formally announces the breakthrough.

1970
Biomedical Engineering, a division of the Department of Medicine since 1946, becomes a separate department under Richard Johns. It soon is recognized as the finest in the nation. It also is among several new departments created during the '60s and '70s, including Biophysics (founded in 1961), Neurology (1968) and Dermatology (1973), which previously had been divisions.

The Maryland Health Services Cost Review Commission (HSCRC) is created by the General Assembly, the state legislature. The HSCRC is the nation's only state agency for regulating hospital rates. Its creation is helpful to Hopkins, although disputes periodically arise over the rates the commission sets for academic medical centers, given the costs of medical education in such institutions as Hopkins.

1970s

The House that Sol Built

Neuroscientist and psychiatrist Solomon Snyder became the youngest full professor in Hopkins Medicine's history when he rose to that rank in 1970 at the age of 32. He'd only been on the faculty for four years but already demonstrated the brilliance that would lead to his founding of the Department of Neuroscience that now bears his name.[156]

Growing up in Washington, D.C., Snyder was a classical guitar prodigy. He even performed once for Andrés Segovia, recognized by many as the finest classical guitarist of the 20th century. Although Snyder considered a career in music, he gravitated more strongly toward medicine—although not the laboratory. He didn't set out to become a neuroscientist. Initially he focused on psychiatry, a good choice, he has said, "for a nice Jewish boy who couldn't stand the sight of blood."[157]

After obtaining his undergraduate and medical degrees from Georgetown University in Washington, Snyder completed an internship at Kaiser Hospital in San Francisco in the early 1960s, then went to the National Institutes of Health and became a protégé of Julius Axelrod, a future Nobel Prize-winner for his discoveries of neurotransmissions, the exchange of electrical impulses between nerve cells.[158]

After two years with Axelrod, during which Snyder conducted and wrote scholarly papers about brain functions, he headed for Hopkins to begin his psychiatry residency. Instead, his impressive track record in brain studies at the NIH—plus the fact that he'd zoomed through college and medical school at Georgetown by the age of 23—caught the eye of Paul Talalay, the influential head of pharmacology. He wanted to establish psychopharmacology research at Hopkins and saw Snyder as the man who could do it. Snyder thought he ought to complete his psychiatry residency, but Talalay persuaded then-dean Thomas Turner to bend the rule that residents weren't allowed to hold faculty positions and arranged for Snyder to complete his residency while also being named to the faculty as an assistant professor of pharmacology in 1966.[159]

Within four years, Snyder had catapulted to full professor. His laboratory in the 1970s was leading research into the synapse, the space between nerve cells where complex molecular communication takes place, and on how opiate drugs such as heroin grip the nervous system. His group of investigators soon discovered that opiate molecules lock into specific receptors on the surfaces of nerve cells. To find the opiate receptors, they developed a technique called ligand-binding, which involved grinding brain cells and washing the mixture with radioactively tagged opiate molecules that would be easy to locate and study. In 1978, Synder's role in discovering the brain's opiate receptors with these radioactive tags earned him the Albert Lasker Basic Medical Research Award. His "grind and bind" technique became the basis for a highly successful biotech company he founded.[160]

Snyder's eclectic thinking spurred the pursuit of a slew of wide-ranging receptors during the 1970s: the GABA receptor where the drug Valium attaches,

the adenosine receptor, which caffeine blocks to cause mental stimulation, the bradykinin receptor connected to pain transmission, and the dopamine receptor where anti-psychotic drugs attach.

Snyder has always encouraged his graduate students and post-doctoral fellows not only to take chances in the lab but assume ownership of their labors. "He'd come up with a great idea, they'd work on it, and when they left his lab, they'd take it with them," recalls Gavril Pasternak, a Snyder protégé who now is professor of neurology at the Sloan-Kettering Institute and Memorial Hospital. "How many scientists would be willing to do that? He didn't care—he'd come up with another great idea!"[161]

After Snyder landed his Lasker Award, substantial overtures from other institutions arrived. In 1980, Rockefeller University made him so huge an offer to move to New York that he couldn't refuse it. So plentiful would be the money there that he no longer would have to apply for research grants. Snyder went to bid good-bye to then-dean Richard Ross, who made a last-minute counteroffer: Stay at Hopkins and form a new department with two other colleagues. "The idea was we would be a focal point to attract other neuroscience people, and I'd start building," Snyder recalls. He asked for three floors to house his department, got them, and agreed to stay.[162]

In 1970, neuroscientist and psychiatrist Solomon Snyder (above) became the youngest full professor in the history of the School of Medicine at the age of 32. He calls himself a "klutz" in the laboratory, prone to breaking things, and says his postdoctoral and graduate students figure out how to put his experiment ideas into practice. And what ideas he's had. His discovery of the opiate receptors in the brain and other groundbreaking research earned him the Albert Lasker Medical Research Award in 1978 and the National Medal of Science in 2005.

"We want the best of the best, but if they're a schmuck, they can go somewhere else."

—Solomon H. Snyder, professor and founding director (1980–2006) of what now is the Solomon H. Snyder Department of Neuroscience, describing the sort of graduate students and faculty he seeks to recruit

"I came to Hopkins 32 years ago as an assistant professor, eager to develop my own research program and to teach. What I found was an ideal setting for both. I discovered that teaching is a sacred trust signaled by example, that only our best efforts are worthy of us .... There's something special about this place, emanating from an inspiring history. It has to do with extraordinary freedom, collegiality, and unrelenting pursuit of excellence."

—Daniel Nathans (1928–1999), 1978 Nobel Prize-winner, 1993 recipient of the National Medal of Science, interim president of The Johns Hopkins University (1995–1996), director of the Department of Molecular Biology and Genetics, senior investigator of the Howard Hughes Medical Institute at Hopkins

■ Every Johns Hopkins Medical School class contains remarkable individuals, but the Class of **1964** included two physicians who gained renown in wildly different fields—genetics and jazz. Leroy Hood, an acclaimed genetic researcher and entrepreneur, created the DNA sequencer, which played a crucial role in mapping the human genome, and co-founded the influential Institute for Systems Biology in Seattle. His classmate Dennis Zeitlin, an associate professor of psychiatry at the University of California, San Francisco, is far better known as an improvisational jazz performer once dubbed the "thinking man's pianist" by a music critic for *The New York Times*. He has released more than **30** celebrated CDs and recorded more than **100** of his own compositions—including songs for *Sesame Street* and the film score for the **1978** version of *Invasion of the Body Snatchers*.

Educational Innovations

Evolving Student Interests, Needs and Attitudes Foster Curriculum Changes

The School of Medicine had a history of adjusting its curriculum as the times—and medical knowledge—changed. During the 1960s and 1970s, significant developments both in medicine and society prompted a steady review of policies that had been in place since the 1930s. The results were creation of a number of important innovations, from establishment of Hopkins' unique "Firm System" for the training of young physicians, to development of one of the nation's finest M.D./Ph.D. programs, the launching of a unique and much-admired cultural affairs office, and implementation of several new curricula.

The Curriculum Reviews of 1975 and 1987

Over the course of its first three decades, the School of Medicine added courses as new specialties developed, ultimately leading to so crowded a curriculum that students were spending eight hours a day in classes five days a week—with another seven hours of classes on Saturdays.[163]

When a group of noted British physicians visited Hopkins in 1920, they found much to criticize in the jam-packed curriculum then in place. Hopkins administrators took their critique to heart. A curriculum committee was established in 1921 and undertook a lengthy review that led to implementation of a far more flexible curriculum in 1927. Nine years later, another review resulted in changes that were implemented in 1936. That curriculum remained essentially the same for the next 30 years—although in 1959, an innovation called the Year V Program was adopted, nearly a quarter century after it had been proposed by pathology chief Arnold Rich. It enabled highly qualified undergraduate students who had decided on a career in medicine to begin their medical studies early and thus get a head start on obtaining their medical degrees.[164]

The Year V Program lasted until 1978, when it was replaced by the Human Biology Program as part of a significant curriculum revision initially implemented in 1975. The Human Biology Program was intended to eliminate one year from the conventional eight-year course of study for the bachelor's and medical degrees to increase all students' interdivisional access to the various courses related to the study of human biology. The Human Biology Program was succeeded in 1983 by the FlexMed Program. FlexMed encouraged students to take a more flexible approach to senior-year courses or to take time out between college and medical school to broaden their experiences and background.[165]

Indeed, the changing interests of medical students nationwide during the 1950s and early 1960s had contributed to the acceleration of curriculum reforms throughout the country.[166] Similarly, changes in the medical curricula nationwide during the early 1960s reflected the more "scientific" orientation of medical schools, with a corresponding emphasis on academic medicine. Full-time physician-scientists provided very visible, popular role models. At Hopkins, the medical students' traditional participation on the ward team

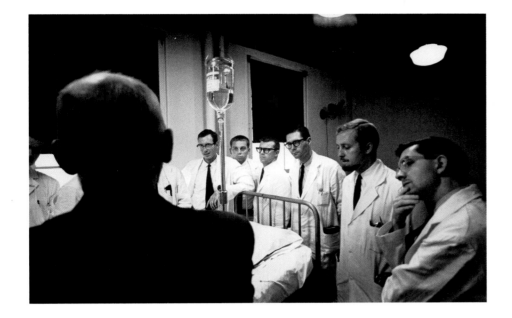

Bedside teaching, begun by **William Osler** in 1893, still remains a fixture at Hopkins, where this group of medical students in 1960 practice what Osler preached seven decades earlier: "The student begins with the patient, continues with the patient, and ends his studies with the patient, using books and lectures as tools, as means to an end."

yielded new enjoyment, as technological advances were put to use by a patient-care team that included a member of the full-time clinical faculty along with a mature, knowledgeable resident staff. The year 1965 was the peak of this "scientific era" of medical student interest, a period when the best students at all levels across the country looked upon full-time academic medicine as the most prestigious career.

Within only a few years, however, the entire atmosphere in medical schools had changed. Nationally, it was a time of ferment, with controversy over the Vietnam War and cutbacks in federal funding for medical research. Students insisted on "relevance" in the curriculum, and faculties were responsive to their demands. Many medical educators in the 1960s believed the basic curricula of the period again had become overloaded and advocated changes to accommodate the flood of new scientific information.[167]

By 1968, 80 percent of the medical schools belonging to the Association of Academic Medical Centers (AAMC) had instituted or were developing new curricula. The number of medical school graduates seeking postdoctoral research training peaked that year at 5,419; by 1978, the number had plummeted to 1,883. This striking change illustrated that the student activism of the period had begun to replace the commitment to scientific medicine that students had shown in the 1960s.[168]

Other medical schools felt these changes more than did Hopkins. When the 1975 curriculum revision was adopted, it was not in response to the prevailing social upheaval but rather to a particular scientific phenomenon: the rapid growth of molecular biology since the 1960s. This growth had blurred the boundaries between the traditional basic science departments of anatomy, biochemistry, physiology and pharmacology.

In 1987, another curriculum review committee was created in the wake of a 1985 national study, the *Report of the Panel on the General Professional Education of the Physician*, produced by a group headed by none other than Steven Muller, then president of Hopkins. The panel's findings were that medical faculties once again needed to reduce the number of scheduled lecture hours to curtail the passive form of medical education they reflected.

1972
February 1: Following Milton Eisenhower's 10-month, interim second term as Hopkins' president, the university trustees elect Steven Muller—who had held the post of provost and vice president for only those 10 months—to succeed Lincoln Gordon. Later in the year, he also becomes president of Hopkins Hospital, a distinction held only by Daniel Coit Gilman. Muller will relinquish the Hospital presidency in 1983 but remain president of the university until 1990.

Baltimore City Hospitals' physicians (many of whom are on the Hopkins faculty) create one of the nation's first successful faculty practice payment plans, Chesapeake Physicians. It becomes a prototype for other such faculty practice plans, including Hopkins' own Clinical Practice Association, created in 1976. The CPA soon garners praise from other medical schools as perhaps the best in the nation because it separates collection of fees from compensation.

1973
The last class of the Hospital's School of Nursing graduates. Nursing education is revived in 1983 with the opening of The Johns Hopkins University School of Nursing, a degree-granting division that fulfills the long-held dream of nursing alumnae for university affiliation. The School of Nursing opens its own building in January 1998.

1974
The beloved but dilapidated Harriet Lane Home is demolished to make way for the Russell Nelson Patient Tower and the Oncology Building, marking the beginning of the most extensive construction and remodeling program in the history of the medical institutions up until that time. The Nelson/Harvey Towers open in 1977, with Nelson devoted to patient care and Harvey to teaching. The designation of the front door to the Hospital, which had been located on Broadway since 1889, is changed in 1979, when the main entrance is moved to the first floor of the then-new Nelson Building, with its streamlined lobby facing Wolfe Street.

In 1975, molecular biologist and geneticist Robert Siliciano (above, standing) was one of the first School of Medicine students to enter its newly created, six- to seven-year M.D./Ph.D. program—which he now heads. Here he is seen in a 1995 lab meeting with students, including Nia Banks (right), then an M.D./Ph.D. candidate.

■ When The Johns Hopkins Hospital opened in 1889, one of its first employees was listed as "an apothecary, who purchases medicines and delivers prescriptions." That's still a pretty good job description—but the role of the "apothecary" now is handled by hundreds of employees in the Department of Pharmacy. Today, under senior director Daniel Ashby, it is responsible for dispensing medications around the clock for an average population of 700 inpatients and provides additional services at six outpatient locations, including the Johns Hopkins Bayview Medical Center and Howard County General Hospital. The department also has a specialty residency training program in pharmacy administration and leadership.

The M.D./Ph.D. Program

Amid all these studies, reconsiderations and revisions, Hopkins chose to institutionalize something it long had fostered: the ability of medical students to obtain both M.D. and Ph.D. degrees while in the medical school. Although some Hopkins medical students from the beginning had emulated such notable faculty as John Jacob Abel and Eli Kennerly Marshall Jr., both of whom held M.D. and Ph.D. degrees, the School of Medicine did not have a formal M.D./Ph.D. program until 1975. Its implementation did not come easily.[169]

As early as 1961, when federal support for basic science training was booming, then-director of the Department of Medicine A. McGehee Harvey used a proposal written by Kenneth Zierler, a professor of medicine and physiology, to obtain an interdepartmental training grant from the NIH that enabled residents and fellows in clinical departments who wanted to choose a career in medical science to pursue a Ph.D. Yet when the NIH launched its Medical Scientist Training Program (MDTP) in 1964, offering federal grants to medical schools that agreed to operate a formal M.D./Ph.D. program, Hopkins didn't apply for it. Even though the program would provide financial assistance to a highly motivated, handpicked cadre of top medical students, some Hopkins officials foresaw significant problems.[170]

At the heart of their reluctance were existential questions. What did the School of Medicine want to be? Where should it place its emphasis: An institute producing only world-class scientists? A school turning out only world-class clinicians? A blend of both?[171]

The rules for the NIH-sponsored program worried medical school leaders. During their first two years, students would focus on their M.D. courses, then break away for three or four research years before finally returning to the M.D. program to end with their clinical rotations. In other words, they would begin medical school with one set of classmates and graduate with another; and because the federal stipend would allow this small, elite group to graduate virtually debt-free, it was feared it would be flooded with applicants. Medical Scientist Training Program students would receive intense mentoring and could devote substantial time, unavailable to other students, to focusing on one problem. They would be treated differently than their classmates and that, argued those opposed to the program, would change the very nature of the School of Medicine itself. Professors in clinical fields also were telling Paul Talalay, a forceful advocate of the M.D./Ph.D. concept, that they did not want him to "to send over scientists to care for sick people."[172]

Those opposed to the M.D./Ph.D. program lost the battle. When the School of Medicine acquired its first federal grant for a Medical Scientist Training Program, Talalay became its architect and first director. He began recruiting students via an announcement sent only to those already in the School of Medicine—and immediately the number of them interested in earning both an M.D. and a Ph.D.

tripled (from three to nine in one year). Among those signing on for the new program were two of its future directors, Steve Desiderio, now head of Hopkins' Institute for Basic Biomedical Sciences, and Robert Siliciano, the program's current head.[173]

The philosophic and practical problems envisioned by the Medical Scientist Training Program opponents did not materialize significantly—although it did continue to stir tensions between clinical and basic science faculty.[174]

While the conflict over the program's focus continued, the M.D./Ph.D. students almost immediately became involved in some of the most significant research to come from Hopkins. They contributed to landmark discoveries in neuroscience and the genetics of cancer while publishing research findings of their own in prestigious scientific journals.[175]

Between 1975 and 1988, 95 students graduated with M.D./Ph.D. degrees, and as medicine grew increasingly sophisticated, with concerns such as genetics, translational research and public health becoming ever more relevant to patient care, the program has kept pace. Among the 40 U.S. medical schools with a Medical Scientist Training Program, Hopkins now has the most government-funded slots (50), and during the program's last formal federal review, Hopkins received the highest ranking ever achieved—along with the reward of one additional stipend. Extra points were awarded for recruiting the most minority students, as well as for their exceptional achievements, Siliciano told *Hopkins Medicine* magazine. "They finish on time [averaging 7.3 years], publish an average of 6.3 peer-reviewed papers while in the program, and 80 percent go on to academic careers. They are phenomenal."[176]

Kenneth Zierler (1918–2009), who conducted research that spanned many fields, from physiology to endocrinology, bioengineering and biomathematics, was an advocate of interdisciplinary collaborations. In 1961, he wrote a proposal that then-Department of Medicine director A. McGehee Harvey used to obtain an interdepartmental training grant from the National Institutes of Health, enabling residents and fellows in clinical departments to pursue a Ph.D. This initial, informal program later served as the model for the School of Medicine's formal M.D./Ph.D. Medical Scientist Training Program, originally headed by pharmacologist and molecular scientist Paul Talalay. The bow tie-wearing Zierler loved all levels of teaching medical students, graduate students and postdoctoral fellows. He estimated that he had taught more than 5,000 Hopkins medical students during his more than half-century career.

The outstanding quality of today's M.D./Ph.D. students at Hopkins—photographed here with Peter C. Maloney (center), associate dean for graduate student affairs—have earned Hopkins' M.D./Ph.D. program the highest ranking, with more government-funded slots (50) than any of the 40 such programs in the country. "They are phenomenal," says Robert Siliciano, head of the program.

When Victor McKusick (seated, center, amid medical students) became the William Osler Professor of Medicine, director of the Department of Medicine and physician in chief of The Johns Hopkins Hospital in 1973, he was concerned about the exponential growth in the size of the house staff brought about by the 1967-70 consolidation of the Osler (ward) and Marburg (private) medical residency services. He feared losing the intimate collegiality and compatibility that was unique to Hopkins' physician training. His solution to the potential problem was creation of "the Firm System," adapted from the residency training program at Guy's Hospital in London. There, the large residency group had been divided into smaller, equal units called "firms," giving each a separate identity and *esprit de corps*.

■ When renowned pediatric cardiologist Helen Taussig (1898–1986) retired in 1963, her former fellows commissioned a soon-to-be famous artist to paint her portrait—a picture Taussig so disliked she hid it in her attic. The portraitist was Jamie Wyeth, son of acclaimed artist Andrew Wyeth (1917–2009). Jamie Wyeth was only 17 when hired to paint Dr. Taussig's portrait—which actually was his first such commission. He produced a stark, realistic portrait of the then-65-year-old Taussig. It shocked her friends and did not please her. Wyeth thought she had destroyed it and was delighted to learn years later that it now is a prized possession of Hopkins' Alan Mason Chesney Medical Archives. Copies of it are on display in the Medical School and in the office of Janice Clements, vice dean for the faculty.

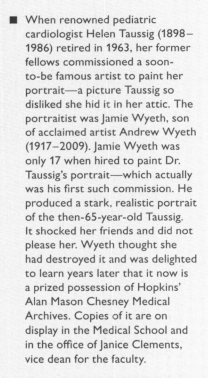

The Firm System

When the increasing cost of medical care in the late 1960s moved the practice of medicine from the hospital to the outpatient setting, Hopkins' traditional pattern of residency training in medicine had to change accordingly. It no longer was practical to maintain two separate medical residencies, one for private patients and one for those on the wards, because the number of patients who received treatment in the wards had declined as Medicare and Medicaid enabled them to have private physicians. The Osler (ward) and Marburg (private) medical residency services therefore were integrated gradually between 1967 and 1970.[177]

Medicare and Medicaid regulations also required every patient to have a physician of record other than a resident physician. In addition, patients who in earlier times would have been hospitalized were receiving outpatient treatment. The cost of hospital care and more rapid cures of many illnesses led to a reduction in the average length of stay in the hospital and an accompanying rise in the number of primary admissions. As a result, more residents were needed, and the size of the house staff gradually became unmanageable. It also was in danger of losing its unique Hopkins character.[178]

The house staff numbered nearly 80 following the consolidation of the Osler and Marburg services, recalled Victor McKusick, then head of the Department of Medicine and physician in chief of the Hospital. That was "almost four times the number there had been when I was chief resident…in 1951-52. We were very concerned that as a result we were going to lose the collegiality and compatibility that had always defined Hopkins training."[179]

The solution came from a concept picked up by the Hopkins faculty who had been visiting Guy's Hospital in London, where the residency training service was divided into smaller, equal units called "firms." The idea was simple: Take a large group and break it into pieces, giving each of the smaller groups a sense of identity, place and purpose. Writing in *The Johns Hopkins Medical Journal* in April 1975, McKusick explained, "Each firm will be headed by a junior faculty member whose experience and competence are comparable to those of our current chief residents…. The title will be *Assistant Chief of Service* (ACS)."[180]

Although initially McKusick proposed naming the services simply Osler A,B, C and D, the decision later was made to name each firm for one of the four seminal chiefs of medicine in line between Osler and Mac Harvey, who had been McKusick's immediate predecessor: Lewellys F. Barker, Theodore C. Janeway, William Sydney Thayer and Warfield T. Longcope. By giving each firm a name, a sense of personality was imbued as well, and with it, a friendly, but not infrequently intense, sense of rivalry.[181]

Initially, each firm had an inpatient nursing unit of 27 beds and an outpatient clinic. Patients assigned to a firm would return to that firm on subsequent hospitalizations and would receive follow-up care in the outpatient section of that firm—improving both continuity of care for the patients and the educational benefits to the residents of maintaining long-term contact with a patient. The ACS served as physician of record for all the patients of that firm, signing off on all treatment orders and thereby complying with the new government rules covering medical reimbursements. With the rise of Medicare and Medicare, McKusick envisioned even more changes in the organization of hospital-based medical practice and designed the firm system with an eye toward that future. In his 1975 article, he wrote optimistically that "one consideration prompting change is that National Health Insurance or its equivalent is expected to provide professional fee capability for all participants in the next few years."[182] (Well, not quite yet….)

Craig Smith, who later helped found Hopkins' Division of General Internal Medicine and then became a successful biotechnology entrepreneur, was one of the first four ACSs in the new firm system and recalled in 2007 that at first there was concern that the ACS role as the ultimate decision-maker somehow would diminish the responsibility of the residents, a hallmark of Hopkins medical training. "It was tough getting people to accept the new approach, but it didn't affect the independence of the house staff. I think what it did was improve the teaching. Now as ACS I had to go and see every patient. We were spending much more time at the patient bedside with the house staff instead of sitting in an office," Smith told *Hopkins Medicine* magazine. "It was incredibly grueling and intense, but there isn't a better way to be more intimately involved in medicine. It was an incredibly rich experience."[183]

Landon King, now vice dean for research and director of the Division of Pulmonary and Critical Care Medicine, believes there is something quintessentially Hopkins about the system's strong sense of individual firm identity, which may explain why few other American medical schools have adopted similar programs. "We have a particularly rich history, and we pay more attention to history here than at other schools," he says. "The idea that we organize teaching around the names of people who were prominent in the past is another way in which we respect that history. It's reflective of the fact that we have a lot of pride in the achievements and contributions of this place."[184]

McKusick (standing) meets with the first "Assistant Chiefs of Service"—the chief residents—he chose to lead the four new "firms" created in 1975 out of the buregoning residency training group. Seated, left to right, are Ernest Arnett, Jeffrey Gelfand, Kenneth Baughman, and Craig Smith. Arnett would join the Hopkins faculty and become an assistant professor of medicine, specializing in cardiology. Gelfand would become a professor of medicine at Harvard, specializing in infectious diseases. Baughman served as head of cardiology at Hopkins from 1992 to 2002 before becoming leader of Harvard's programs in advanced heart disease. He died tragically in 2009 when he was struck by a car while jogging. Smith later helped found Hopkins' Division of General Internal Medicine and then became a successful biotechnology entrepreneur.

A physician confers with a patient in the new $42 million Adolf Meyer Center, opened in 1982 for the departments of psychiatry, neurology and neurosurgery. With new molecular and biological approaches to medicine, it was determined that neuroscience studies could benefit from a close association of these disciplines, so each department was given three floors each in the nine-story center. Its construction was the single largest financial commitment by the University and Hospital to any group of departments in Hopkins Medicine's history up to that time.

1975
Heart surgeon Levi Watkins, Jr., joins with fellow African American faculty colleague, ophthalmologist Earl Kidwell, begin a concerted nationwide drive to recruit talented minority students to the School of Medicine. The success of the Hopkins minority recruitment campaign soon makes it a model imitated by other medical schools.

The changing interests of medical students and new disciplines, particularly the rapid growth of molecular biology, lead the School of Medicine to adopt a significantly revised curriculum, which is updated again in 1987.

A formal M.D./Ph.D. program is created. By 2010, Hopkins' program is ranked the finest in the nation by the National Institutes of Health's Medical Scientist Training Program (MSTP), which underwrites more M.D./Ph.D. students at Hopkins than at any other medical school.

Hopkins' residency training program adopts the "Firm System," following the gradual integration of the Osler (ward) and Marburg (private) medical residency services. The resulting large pool of residents is divided into smaller, equal units called "firms," named after four early chiefs of medicine: Lewellys F. Barker, Theodore C. Janeway, William Sydney Thayer and Warfield T. Long-cope. This arrangement preserves the esprit de corps of the smaller residents' groups that is a quintessential aspect of Hopkins training.

A New Building Boom Begins

The decline in federal support after 1968 required Hopkins' leaders to find new sources of construction funds. Building began again in the mid-1970s when the State of Maryland's new Health and Higher Education Act allowed the Hospital and University to issue bonds, enabling Hopkins to borrow money for construction projects at reasonable rates. Foundations and other private donors also became increasingly important in sponsoring facilities. The patterns of construction, communication, research, education, and patient care in the next 10 years would reflect the efforts of the medical school and Hospital to maintain their traditions in financially difficult times.[185]

The demolition in 1974 of the beloved but dilapidated 1912 Harriet Lane Home building to clear space for the Russell Nelson Patient Tower and the Oncology Building marked the beginning of the most extensive construction and remodeling program in the history of the medical institutions up until that time.[186]

Several other new buildings replaced older structures. The Adolf Meyer Center opened in 1982, with its $42 million construction representing the single largest financial commitment by the University and Hospital to any group of departments in Hopkins Medicine's history. With the advent of molecular and biological approaches to medicine, the study of the neurosciences could benefit from a close association between psychiatry, neurology, and neurosurgery. Consequently, each of those departments was given three floors in the nine-story Meyer Building—which itself was a physical embodiment of Adolf Meyer's belief that the mind and organic brain are integrated elements. The building incorporated 90 beds for neurology and neurosurgery and 88 beds for psychiatric inpatients, as well as operating suites, faculty offices, laboratories, clinics and other outpatient programs, including a day hospital for psychiatric patients. (Psychiatry's old home, the Phipps Building, was converted to administrative offices and teaching rooms.)[187]

With the 1977 opening of the Nelson/Harvey Towers (with Nelson devoted to patient care and Harvey to teaching), the proximity of Nelson to the 1931 Osler and Halsted buildings enabled a horizontal reorientation of obstetrics-gynecology, medicine and surgery, bringing them into a layered pattern, with each department occupying two or three floors of the three adjoining buildings. All of the Hospital's imaging resources—X-ray, computerized tomography, and magnetic resonance—were placed in the buildings' lowest levels and the intensive care units were located on the buildings' respective seventh floors, at the same level as the operating rooms.[188]

The Osler and Halsted buildings underwent renovation during this period, as did the Children's Medical and Surgical Center, the Physiology Building, the Biophysics Building and the Wood Basic Science Building. In 1982, the Preclinical Teaching Building opened, as did the Maumenee Building of the Wilmer Eye Institute, adjacent to the main corridor of the Hospital. Additional administrative offices and computer facilities for both the Hospital and School of Medicine were housed at 1830 East Monument Street, a nine-story building completed in 1988.[189]

Even the designation of the front door to the Hospital, which had been located on Broadway since 1889, was changed in 1979, when the main entrance was moved to the first floor of the then-new Nelson Building, with its streamlined lobby facing Wolfe Street.[190]

The A. McGehee Harvey and Russell Nelson Towers—with Harvey devoted to teaching and administration and Nelson to patient care—opened in 1977, facing Wolfe Street, previously the rear of the Hopkins Hospital but now its new front door. The 59-foot-long, 15-foot, 2-inch-high and 18-foot wide steel beam sculpture in front of the entrance was created by noted American-born artist Beverly Pepper and entitled "The Great Ascension." It was donated to the Hospital in 1985 by University trustee Robert H. Levi, a 1936 graduate of the Krieger School of Arts and Sciences, and his wife Ryda, both renowned civic leaders, collectors of modern art and longtime Hopkins benefactors. The sculpture later was moved to the entrance most frequently used by the growing number of patients who receive treatment in the Outpatient Center.

Addressing the Rise in Health Care Costs, Adjusting to Marketplace Changes

Maryland Creates its Unique Heath Services Cost Review Commission

The cost of hospital care in Maryland was rising so rapidly in the late 1960s that it was becoming one of the three or four most expensive states for hospital treatment. The Maryland General Assembly decided it had to stop the upward spiral. In 1971, the state legislature created Maryland's Health Services Cost Review Commission (HSCRC)—which would become the nation's only state agency for regulating hospital rates. Hopkins supported such an effort, given the skyrocketing expense of providing uncompensated care to poor patients with little or no insurance coverage.[191]

To establish a uniform method for determining what rates hospitals in Maryland could charge, the HSCRC created what was called an "all-payer" system, which required every medical care insurer, including Medicare and Medicaid, to pay the same rates for the same services rendered by any one hospital. Rates could vary among hospitals based on legitimate differences in cost structures, such as the special costs associated with the teaching mission in academic medical centers. Because Medicare and Medicaid were federal programs, the HSCRC had

1978

May 17: The Alan Mason Chesney Medical Archives is dedicated. It is named in honor of the former School of Medicine dean who wrote a three-volume history of the first 14 years of Johns Hopkins Medicine, entitled, *The Johns Hopkins Hospital and The Johns Hopkins University School of Medicine: A Chronicle.* The Chesney Medical Archives becomes one of the largest, most respected repositories of medical history in the nation.

October 12: Daniel Nathans and Hamilton Smith, both professors of microbiology in the School of Medicine, are awarded the Nobel Prize in Medicine or Physiology (with Swiss microbiologist Werner Arber) for their work on restriction enzymes. Their discoveries pave the way for the genetic engineering revolution.

1980

The Department of Neuroscience is founded under Solomon Snyder, discoverer in 1978 of the brain's opiate receptors. He will lead the department for the next 25 years, and it will be named for him upon his retirement in 2005.

1981

The Denton A. Cooley Recreation Center opens. Cooley, an acclaimed heart surgeon and 1944 graduate of the School of Medicine who had been a championship basketball player at the University of Texas, contributed $750,000 toward the complex's $2.6 million price tag. Returning to Hopkins to participate in the center's opening, the 6'4" surgeon inaugurates its full-size basketball court (which he had specifically instructed be built) by scoring the first basket with a vintage 20-foot, one-hand push shot.

■ Former Hopkins School of Medicine dean and professor of pharmacology William Henry Howell (1860–1945) often is cited for the 1916 discovery of the important blood thinner (anticoagulant), heparin—but Jay McLean (1890–1957), a second-year medical student, actually found it first. Working in Howell's laboratory on finding blood clotting agents (coagulants), McLean extracted from a dog's liver a substance that turned out to be an anticoagulant. Howell later named it heparin (hepar being the Greek word for liver). Howell mentioned Maclean's work in his initial paper on heparin, but Maclean's role in discovering it wasn't noted publicly until 1945.

to obtain a "waiver" from them to allow it to set the Maryland rates they would pay. A formula for this was devised, stipulating that Maryland's hospital rates could not be more than the average of hospital-rate increases nationwide. Hospital costs in Maryland began a gradual decline until they were more in keeping with what hospitals in other states charged. For Hopkins, creation of the HSCRC was helpful, although disputes periodically arose over the rates the commission set for academic medical centers, given the complexity of the cases treated in a place such as Hopkins, and the added costs it incurs because of its teaching and research commitments.[192]

Hopkins Launches Its Medical Practice Plan

By the 1970s, medical schools nationwide were hard-pressed to meet all their expenses. Few sources of income could be augmented sufficiently to cover the greater costs, so for most schools to survive in the new health care economic environment, they had to increase the income they received from the practice of medicine.[193]

At Hopkins, the key to solving the problem was to use its faculty's income fairly while preserving the fundamental principle of the full-time plan: No member of the faculty should receive direct payment from patients. One possible solution seemed to be establishment of a faculty practice plan, in which the income earned by each faculty physician would be put in a pool and divided among the plan's members.[194]

A faculty practice plan for the Johns Hopkins School of Medicine—the progenitor of Hopkins' Clinical Practice Association—was approved in January 1976. Faculty salaries were set by the departmental directors and approved by the dean. The salary of each faculty physician was determined by that individual's contribution to the institution in teaching, research, and patient care. Marketplace adjustments in salaries—the "marketplace" being other academic medical centers—were necessary, however, to maintain strength in the clinical departments, particularly in the surgical specialties. A distinguishing aspect of the Hopkins plan, however, was that compensation was based on overall productivity, and there was no mathematical relationship between salary and a department's total earnings.[195]

The combination of individual incentive and group reward worked well. In the practice plan's first 14 years, income grew 25-fold. The medical school's budget reflected this growth. In 1987, nearly one-third of the income came from practice, one-third from government grants and contracts, and one-third from all other sources.[196]

As often occurred throughout the history of Hopkins, what happened in East Baltimore proved to be a model for others to follow. Many deans in other medical schools praised Hopkins' practice plan as probably the best in the country because it separated collection of fees from compensation. Yet establishment of such a plan was peculiarly Hopkins-centric, since it relied on Hopkins' long adherence to the full-time system, in which fees for service did not divert directly to the physicians, and salaries were determined without regard to overall clinical earnings.[197]

"I'd Like to Get Confirmation...." and "Holy Cow!"

Around 8:30 on the morning of October 12, 1978, a wire service reporter telephoned the Northwest Baltimore home of Daniel Nathans, director of Hopkins' Department of Microbiology, to tell him that he and his colleague Hamilton Smith just had won the Nobel Prize for their discovery of restriction enzymes only 11 years earlier. Nathans, 49, short, balding, soft-spoken and reserved, was astounded—and certainly didn't begin celebrating. His initial response, according to his wife, Joanne, was, "Well, let's wait." He wanted to hear it directly from the Nobel Prize people in Sweden.[198]

Hopkins officialdom also was caught completely unaware and unprepared both for the honor that Nathans and Smith had received and for the tumultuous reaction to it that followed. Ever since Thomas Hunt Morgan, a 1915 Hopkins Ph.D., won the Nobel Prize in Medicine in 1933, a dozen Hopkins-trained physicians and Ph.D.s had received Nobel Prizes—but all after they had left Hopkins to go elsewhere. See Appendix B for the full list of Hopkins Medicine Nobel laureates. This was the first time scientists still on the Hopkins faculty had bagged the greatest honor in their field—and they had done so in the blink of an eye, by Nobel standards, so that no one had anticipated it.[199]

Ham Smith, then 47, similarly was nonplussed by word of his award. His initial reactions, he later said, were "Holy cow!" and, "I'm flabbergasted."[200]

Like the good, dedicated Hopkins faculty that they were, both Nathans and Smith simply proceeded to teach their morning classes. The students in Smith's lecture were subdued and made no fuss. Nathans' class, however, was giddy. When he showed up at 10 a.m. and began to lecture about cells and chromosomes, turning his back to draw illustrations on the blackboard, the microbiology students started laughing, produced bottles of sparkling wine they had purchased, and began handing it around.[201]

By 12:30, any pretense that this was an ordinary day was gone. Euphoria spread throughout the entire East Baltimore medical complex. An improvised

The stunned—but delighted—Daniel Nathans (right) and Hamilton Smith adapt to the unaccustomed role of media celebrities and answer questions at a press conference following announcement of their receipt of the 1978 Nobel Prize for Medicine for their discovery and first use of restriction enzymes.

The latest Hopkins Medicine-related Nobel laureate is Sir Robert Edwards (above, right), whose 1965 in-vitro fertilization experiments at Hopkins as a fellow under Howard W. Jones Jr. (above, left) proved crucial to Edwards' and British surgeon Patrick Steptoe's later success in helping produce the world's first IVF "test tube" baby in 1978—which won Edwards the 2010 Nobel for Medicine.

■ Johns Hopkins' renowned Pithotomy Club (1897–1992) was believed to be the oldest medical fraternity in the United States when it finally closed after 95 years. According to club lore, its name "pithotomy" was coined by William G. MacCallum, a member of the School of Medicine's first graduating class of 1897 by combining the Greek words pithos, meaning vessel, and otomos, meaning to open. To the "Pithotomists," that translated into "to tap a keg" of beer—something they did often, as depicted in the club's emblem (above), designed by Max Broedel. The Pithotomy Club began as a sort of honorary society for the top members in the School of Medicine's senior class and was meant to encourage informal student-faculty contact. Its most famous function was an annual show full of songs and skits that ridiculed leading faculty in an irreverent—and raunchy—way. For decades, all the leading faculty members attended, sometimes inviting distinguished guests, such as writers F. Scott Fitzgerald and H.L. Mencken. The club's 1982 show, however, was so over-the-top in its portrayal of one female faculty member that she threatened to sue if the club was not punished. Its reputation severely harmed, the club limped along for another 10 years but eventually faded away.

The raucous, irreverent and frequently salacious annual Pithotomy Club show often drew celebrity guests as well as top faculty. In this late 1940s photo, seen seated in the center right is famed Baltimore newspaperman H.L. Mencken, looking warily at the camera. On the left beside him is pathology director Arnold Rich, holding one of the Club's traditional mint juleps. Immediately in front of Rich, with his head partially bowed, is Department of Medicine head A. McGehee (Mac) Harvey, who has his hand on the shoulder of surgeon in chief, Alfred Blalock.

paper banner bearing the single word "CONGRATULATIONS" was stretched along one of the halls of the Wood Basic Science building on Wolfe Street, where both men had their laboratories, and a press conference was scheduled for a lecture hall there.[202]

After completing their morning classes, Nathans and Smith headed off for the press conference. Medical students and colleagues packed the hall, filled the foyer, and jammed into the rooms adjacent to it. Accompanied by Dean Richard Ross and university provost Richard Longacre, representing Hopkins' president Steve Muller, who was out of town, the two new Nobel laureates elbowed their way through an applauding and cheering crowd to the podium. They both still seemed stunned and awed by the honor they had received and the inevitable excitement it elicited.[203]

Responding to questions from the press, Nathan said his first reaction to receiving the news was disbelief. "I'd like to get confirmation before I invest so much emotional energy," he said, prompting a roar of laughter from the audience. He also was asked what he intended to do with the Nobel Prize money. He said he really didn't know how much it was. Someone in the crowd shouted "$165,000!"

"Thank you," Nathans replied, "but I leave all the finances to my wife."[204]

Both men praised the academic freedom and friendly interchange of ideas at Hopkins that had encouraged them to remain here. Nathans also was philosophical—and cautious—about the impact of the discoveries he and Smith had made, even though the importance of them soon would prove epochal. "I hope our work will lead to benefit for man, but as I look at the history of scientific development, it is clear that the course of that development begins with very basic investigations and that applications may not always be known by those who help clear the path."[205]

At a gathering of the medical school faculty six days later, Richard Ross reflected on the joy Hopkins felt by the recognition Nathans and Smith had achieved—and especially by how they responded to it at the press conference.

"I remember many of their comments, but one in particular sticks in my mind," Ross said. "A reporter asked Dr. Nathans if he had…the Nobel Prize on his mind for years and he replied without a moment's hesitation that he had never given it a thought. He told her that he did science because he enjoyed it and that was that.[206]

"People outside the institution who do not know scientists assume that our laureates are strange, reclusive individuals who only care for science. They are surprised when I tell them that I know them and that they also teach, they talk, they walk, they laugh, they labor on committees, and do all the things the rest of us do. Those of us who know them realize that it couldn't have happened to two nicer people."[207]

1980s

Although it would be a decade of continued expansion, achievement and celebration, the 1980s got off to a slightly rocky start with a challenging two-week strike by some 1,400 maintenance and food service workers who were members of Local 1199-E of the National Union of Hospital and Health Care Employees.[208]

The Hospital's blue-collar workers had joined the union in 1969, and in 1974 joined fellow Maryland members in a 10-day strike against Hopkins, Maryland General Hospital and the Greater Baltimore Medical Center.[209]

The disputes that led to the 1980 strike also were not confined to Hopkins. All told, the union's local had some 3,600 members employed at six Baltimore-area hospitals—Hopkins, Lutheran, Maryland General, the Greater Baltimore Medical Center, Sinai and Provident.[210]

The union conducted its negotiations with each hospital separately, but since the main issues were identical with each, it vowed to strike all of them if talks stalled. When the negotiations reached a stalemate on Dec. 1, 1980, however, the union struck Hopkins alone. As a reporter for the Baltimore newspaper, *The Evening Sun,* observed at the time, "The union's tactic was to take a hard line with the prestigious Hopkins, which it said was the key medical facility in the labor strife. As Hopkins went, it was felt, so would go the other hospitals."[211]

The union called another strike at Sinai Hospital on December 2, but never did strike the other hospitals.[212] The strike would last 16 days, imposing a substantial challenge on Hopkins' ability to operate smoothly without an estimated one-fourth of its workforce. Ron Peterson, then a young hospital administrator, was put in charge of coordinating the re-assignment of non-union staffers and outside volunteers. He deployed some 90 volunteers, about 30 more than ordinarily helped out daily, and shuffled the work assignments of some 350 non-striking accountants, administrators and other white-collar employees who usually worked in areas not directly involved in patient care. They found themselves performing maintenance chores, cooking, washing, drying and folding linens, and handling support assignments in the wards.[213]

The Hospital's occupancy rate remained at a steady 94 percent, no patients were turned away, and everything was kept running on an even keel—although sometimes on a bare-bones level and admittedly behind schedule periodically, with meals and linen changes arriving late.[214]

Finally, at 3 a.m. on December 16, both parties in the labor dispute reached an agreement to end the strike, following an all-night bargaining session involving the union, Hopkins and a veteran federal mediator, James Williams of the Federal Mediation and Conciliation Service.[215] The episode proved to be an educational experience for both sides. No employee strike has occurred at Hopkins since.

Pediatric surgeon J. Alex Haller Jr., named children's surgeon in charge in 1964, established the nation's first regional trauma center for children at Hopkins and created a training program for pediatric surgeons.

"Some words and phrases come to mind concerning that nasty Pithotomy Club show: obscene beyond belief, horribly demeaning to a fine faculty of famous professors, sexist and politically incorrect to the nth degree, utterly vulgar, promoting public drunkenness (students and faculty), absolutely no redeeming value. I participated in it; I loved it!"

—Henry V. P. Wilson, 1961 graduate of the School of Medicine

1982
The Adolf Meyer Center opens in 1982 on the site of the 1923 Women's Clinic, which was demolished in 1979. The $42 million price tag on the Meyer Center represents the single largest financial commitment by the University and Hospital to any group of departments in Hopkins Medicine's history. With the advent of molecular and biological approaches to medicine, the study of the neurosciences benefit from a close association between psychiatry, neurology, and neurosurgery. Each of those departments is given three floors in the nine-story Meyer Building—which itself is a physical embodiment of Adolf Meyer's belief that the mind and organic brain are integrated elements.

The Preclinical Teaching Building opens on the site of the 1899 Women's Fund Memorial Building (so named to commemorate the fundraising drive of Mary Elizabeth Garrett and her friends that led to the opening of the School of Medicine),which was demolished to permit construction of the teaching building.

The Maumenee Building of the Wilmer Eye Institute opens adjacent to the main corridor of the Hospital.

1983
The Heart and Heart-Lung Transplant Service is inaugurated. Comprising more than 100 members, from surgeons to nurses to anesthesiologists and support personnel, it is the only one of its kind in the Maryland-Washington, D.C. area.

New Clinical Services, New Centers—and a Gym—Open

The Denton A. Cooley Recreation Center

The demolition of the old Physicians Dining Room to make room for the 1964 Children's Medical and Surgical Center removed an important, comfortable location for communication that still is mourned by those who remember it fondly, but its role as a communal gathering place may have been eclipsed nearly 20 years later with the 1981 opening of the Denton A. Cooley Recreation Center.

Cooley, a championship basketball player at the University of Texas (later named to the National Collegiate Athletic Association's list of the 100 most influential student athletes), contributed $750,000 toward the complex's $2.6 million price tag. Returning to Hopkins to participate in the center's opening, he properly was accorded the privilege of inaugurating its full-size basketball court (which he had specifically instructed be built). A representative from *Sports Illustrated* happily reported that "the 6'4" surgeon ceremoniously scored the first basket in the new gym – a vintage 20-foot, one-hand push shot." [216]

In 2002, the center received a major $800,000 renovation and updating. This increased the number of exercise studios to six. The new studios provided space for such activities as spinning classes, yoga and Pilates, and a cardiovascular training area with satellite-fed televisions lining the walls. Free to students from the schools of medicine and public health, Cooley also is open to paid members. Just as the Physician's Dining Room was more than simply a place to eat, the Cooley Center has proven more than a place to work out. Its melting pot atmosphere, with members spanning every branch of Johns Hopkins Medicine huffing and puffing side-by-side, makes it a location where unexpected, potentially important contacts can be made. "You never know if the person riding the bike next to you is going to perform surgery on your sister," said John Johnson, an employee of Hopkins Medicine's information services, told *Dome*. [217]

New Orthopedics Center

The 1982 opening of the Meyer Building, described earlier, was followed in 1983 by several more centers designed to bring together specialists whose disciplines were related and whose closer association would benefit patients.

What orthopedics director Lee H. Riley called the most modern and complete orthopedics center then available in the country opened in 1983 on the first floor of the Blalock Building. With 10 examination rooms, two X-ray rooms, and a room for creating such important therapeutic items as body casts, the new 12,000-square-foot center was far larger, better equipped and more comfortable for patients than the old facility on the first floor of the Carnegie Building. [218]

Built entirely with donations from several anonymous philanthropists and patients from the local region as well as around the world, the new center was immensely successful. In its first full month of operation, 973 patients were treated there, whereas on average, just 800 patients were treated each month at the old center. Riley was pleased that the center not only had additional facilities for patient care, but also more space allocated for Hopkins orthopedists to maintain their research in bone injuries and disorders. [219]

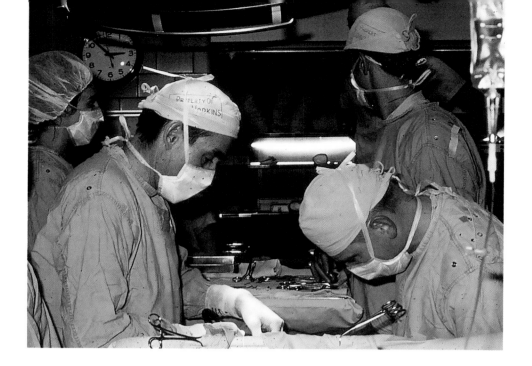

Vincent Gott (left), then-cardiac surgeon in chief at Hopkins, and Harvey Bender Jr. (right), then an assistant professor of surgery, were photographed while performing the first heart transplant at Hopkins Hospital on Nov. 25, 1968, less than a year after South African surgeon Christiaan Barnard performed the first such procedure. The six-hour operation at Hopkins, which provided a 56-year-old Baltimore real estate broker with the heart of a 28-year-old apparent suicide victim, involved a team of 10 physicians, led by Gott. During Gott's 55-year career at Hopkins, he was a primary researcher in the development of the pacemaker, co-inventor for the Gott-Daggett Valve and the Gott Shunt, and trained 50 of the cardiac surgery residency program's 93 graduates up to the time of his retirement. Bender would go on to become chairman of the Department of Cardiothoracic Surgery at Vanderbilt, renowned for his skill as a pediatric cardiac surgeon.

Expanding Cardiology, Cardiac and Transplant Services

By late 1983, the fifth floors of five adjacent buildings—Blalock, Halsted, Harvey, Nelson and the Children's Center—had been renovated to create what then was called the Clayton Heart Center.[220]

The center (now absorbed into the new Johns Hopkins Heart and Vascular Institute) was named for William S. Clayton, father-in-law of Hopkins' legendary professor of medicine, Benjamin Baker. Clayton had been the undersecretary of state who hand-wrote the first draft of what became known as the post-war Marshall Plan. By the time the Clayton Center was completed, 800 open heart surgeries were being performed annually at Hopkins and the new center's creation was designed to facilitate collaborations on patient care and research by adult and pediatric cardiologists, radiologists, cardiac surgeons and anesthesiologists, as well as make it easier to share the expensive medical equipment their specialties required.

Among the Clayton Center's major components was the E. Cowles Andrus Cardiac Clinic, named for an internationally known cardiologist who was on the Hopkins faculty for more than 50 years. Andrus (1896–1978) created Hopkins' first cardiovascular division and was especially well-known for his early and perceptive recognition of the relationship between the central nervous system's functions and the causes of cardiovascular disease. The Andrus Clinic brought together all of Hopkins' cardiac outpatient services.[221]

Heart and Heart-Lung Transplant Service

On July 16, 1983, Hopkins' new Heart and Heart-Lung Transplant Service was inaugurated when a 27-year-old man from Virginia got a new heart during a four-hour surgery conducted by Bruce Reitz, a one-time Hopkins intern who had gone on to perform the world's first successful heart and lung transplant at Stanford; William Baumgartner, whom Reitz had brought with him from Stanford to join Hopkins' expanded cardiac surgery team; and Vincent Gott, who had performed Hopkins' only previous heart transplant in 1968.[222]

The new heart and heart-lung transplant service was the only one of its kind in the Maryland-Washington, D.C. area. Comprised of a team of more than 100 members, from surgeons to nurses to anesthesiologists and support personnel, it completed 10 heart transplants by June, 1984—all successful but one.[223]

■ Today, "domino donor" transplant surgeries at Hopkins are most closely associated with Robert Montgomery's groundbreaking, multi-patient kidney transplants—but the organ exchanged in Hopkins' first "domino donor" transplant operation was the heart. In 1987, heart surgeons Bruce Reitz and Bill Baumgartner led a Hopkins medical team to successful completion of the nation's first "domino donor," three-way transplant operation. The marathon, 17-hour surgery involved Reitz's removal of a cystic fribrosis (CF) patient's healthy heart and diseased lungs; the harvesting of an accident victim's heart and lungs to give to the CF patient; and then Baumgartner's "piggyback" placement of the CF patient's healthy heart on top of the diseased heart of a second patient. Fortified with immunosuppressive drugs that Reitz also helped pioneer, the patients went home to new lives a few weeks later.

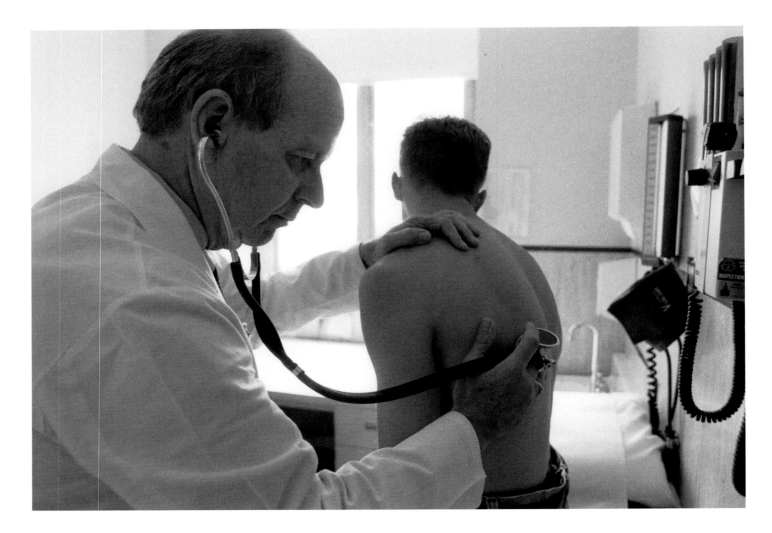

Infectious disease expert John Bartlett (above) attends to an **AIDS** patient early in the outbreak of what became known as the "Plague of the '80s." In the 26 years that Barlett led the Department of Medicine's Division of Infectious Diseases, it grew from just three full-time staff and a budget of $200,000 in 1980 to a roster of 55 faculty and a staff of 177 in 2006. It treats more than 5,100 patients annually and has a research budget of $40 million—one of Hopkins' largest. Prior to concentrating on HIV/AIDS, the rumpled, bald, soft-spoken but intensely focused Bartlett was best known for discovering what he calls "the bug," *Clostridium difficile* or *C. difficile*, which causes antibiotic-associated colitis, a chronic, debilitating diarrhea that sometimes afflicts patients on antibiotics. He was famous for working 100-hour weeks treating **HIV/AIDS** patients, administering trials of potential medications for them, answering questions from other physicians nationwide, teaching, lecturing, serving on national committees and organizations, and writing hundreds of scientific articles, book chapters and books. One colleague said that every piece of Bartlett's writing is done "intelligently, meticulously, comprehensively, and just *after* the deadline." Bradley Bender, a former post-doctoral fellow who worked with Bartlett, added, "Getting an article from John can only be described as 'the three too's': too long, too late and too good to be true!"

AIDS, the "Plague of the '80s," Offers Challenges, Inspires Renewed Commitment

John Stobo, named head of the Department of Medicine in 1985, knew something about AIDS, which first was reported in 1981. He had opened an AIDS outpatient clinic when he was chief of the Moffitt Hospital of the University of California San Francisco, considered the ground zero of the AIDs epidemic. Arriving at Hopkins to become not only director of medicine but physician in chief of Hopkins Hospital, one of the first things Stobo did was establish a dedicated AIDS ward, the third such unit in the nation and the only inpatient AIDS service between Washington and Philadelphia.[224]

By 1988, Hopkins scientists had the biggest AIDS research budget of any academic medical center in the nation, with $40 million in scientific study and education grants, far more than medical centers in either San Francisco or New York, which were considered the most high-volume AIDS cities. The Hopkins outpatient AIDS clinic, opened in 1984, even before Stobo's arrival, was treating 700 patients annually and expansion plans were in place to increase its size from the original nine-bed AIDS unit to 30 beds by the end of 1988.[225]

Hopkins' official "AIDS Team" modeled its multidisciplinary formation along the lines of its colleagues in oncology. The AIDS group consisted of 37 primary-care providers, including physicians from infectious diseases, infection control specialists, social workers, a nutritionist, physician assistants and Clinical Research Center nurses. Also brought into the picture were specialized services of neurology, psychiatry, oncology, pulmonary medicine, gastroenterology, surgery, ophthalmology and dermatology.[226]

Heading the Hopkins AIDS Care Service was John Bartlett, who would become world-renowned for his work as chief of the division of infectious diseases. In the beginning of the then-still mysterious AIDS epidemic, caregiver burnout—and sometimes fear—were a constant concern.[227]

"I guess if you're looking for heroes, it would be the people who accept the fact that you're going to take care of a group of patients who are critically ill and who always die, which is kind of morbid by itself," Bartlett said, reflecting the then-grim reality of AIDS in the mid-1980s.

"It's the worst death there is, a lingering death that's characterized by constant infections, debility, emaciation and dementia. To deal with it on a regular basis is painful for everybody in the care system. And at the end of it all, they get paid less. How's that for heroes?"[228]

While some members of the AIDS team only saw the patients on the unit and others were involved in global efforts to fight the disease, all of them shared the feeling that they were "part of a bigger mission," Bartlett said at the time, adding presciently, "I thinking preserving that team effort probably is very important."[229]

Bringing City Hospitals on Board

By the early 1980s, Baltimore City Hospitals, a venerable municipally-operated facility that could trace its lineage back to the Baltimore City and County Alms House of 1773, had fallen on hard times. It was staggering under annual losses topping $7 million, its physical plant had not undergone significant renovation since the 1930s, and then-Baltimore Mayor William Donald Schaefer wanted to get it off the hard-pressed city's ledgers.[230]

Despite its troubled situation, City Hospitals was a storied institution with an inspiring history of medical firsts—including invention of landmark CPR techniques and creation of the country's first intensive care unit. It also had an association with Hopkins Hospital that even pre-dated the opening of the School of Medicine in 1893, with William Welch himself obtaining pathology specimens for examination from it. Many of its physicians held Hopkins faculty appointments and Hopkins medical students long had done some training there.[231]

Hopkins Hospital's then-president, Robert Heyssel, was a savvy analyst of health care economics and believed Hopkins needed to expand to counteract potential competition. He did not want a for-profit firm to acquire City Hospitals and its significant attributes, which included a pioneering faculty practice group, the region's only burn unit and important federal research institutes. He believed that if Hopkins could manage City Hospitals for a trial period and reduce its mammoth deficits, perhaps acquiring it would make sense.[232]

After negotiating a contract with the Baltimore City to take over day-to-day management of the beleaguered facility, Heyssel turned to Ron Peterson, a young administrator to whom he'd taken a liking, and tapped him to run the debt-ridden, down-on-its-heels hospital and see if he could get it back on its feet

A unique educational development of this period was the establishment in 1977 of the Office of Cultural Affairs, the first such medical school program in the country. Its creators were neurosurgeon George Udvarhelyi (above) and the Homewood campus' legendary professor of humanities Richard Macksey, both polymaths *nonpareil*. Udvarhelyi believed that physicians and medical students needed to include the humanities in their educational curriculum. He and Macksey brought to the East Baltimore campus such prominent writers, actors, musicians, artists and authors as Isaac Bashevis Singer, Yehudi Menuhin, Aaron Copland, Isaac Stern, Richard Leakey—even Phyllis Diller—for lectures, performances and discussions.

1984
The Johns Hopkins Hospital acquires the Baltimore City Hospitals on Eastern Avenue. A venerable but financially shaky municipal facility that can trace its history back to 1773, City Hospitals long had provided space for clinical services, teaching and rounds for Hopkins Medical students. Re-named the Francis Scott Key Medical Center and headed by young Hopkins Hospital administrator Ronald R. Peterson, it undergoes a period of stunning revival, renovation and rebuilding, culminating in its re-naming as the Johns Hopkins Bayview Medical Center in 1994.

Hopkins opens an outpatient AIDS clinic, which treats 700 patients annually. In 1985, the Department of Medicine establishes a ward dedicated to the inpatient treatment of AIDS patients. It is only the third such unit in the nation and the sole inpatient AIDS service between Washington, D.C. and Philadelphia. By 1988, Hopkins scientists have the largest AIDS research budget of any academic medical center in the nation, with $40 million in scientific study and education grants. The Johns Hopkins AIDS Service is headed by John Bartlett, chief of the division of infectious diseases and a pioneer in the battle against the then still mysterious AIDS epidemic.

1985
The School of Medicine drops the Medical College Admissions Test as a requirement for applicants in an effort to encourage students to pursue broader educational opportunities before entering medical school.

Born in San Juan, Puerto Rico, where her father, an Army officer, was stationed, Judy Reitz became an Army nurse and later earned a doctorate in health finance and management from Hopkins while serving as Hopkins Hospital's director of nursing practices. In 1984, she joined Ron Peterson's management team at what now is Johns Hopkins Bayview as vice president for nursing. She subsequently became its executive vice president and chief operating officer. When Peterson became president of Hopkins Hospital, Reitz returned there as senior vice president for operations. In 1999, she became executive vice president of Hopkins Hospital—the first woman to hold that position. In 2003, she became vice president for quality improvement for Johns Hopkins Medicine.

An aerial view of the 130-acre campus of the old Baltimore City Hospitals, taken around the time Hopkins acquired it in 1984. Its physical plant had not been upgraded in decades and it was running a $7 million deficit. Beginning in 1982, two years prior to its acquisition, a young Hopkins management team of Ronald Peterson, Kenneth Grabill, and William Ward II, working under contract to Baltimore City, had cut the deficit significantly within a year. Upon Peterson's recommendation, a joint committee of University and Hopkins Hospital trustees agreed to acquire the facility, which first was re-named the Francis Scott Key Medical Center. It was re-christened the Johns Hopkins Bayview Medical Center in 1994, after a decade of tremendous advances and success.

and operating in the black. The challenges were daunting—but the potential, as Peterson saw it, was enormous.[233]

Beginning in 1982, Peterson and colleagues Kenneth Grabill and William Ward II cut City Hospital's losses by more than $7 million in one year. By 1984, following lengthy negotiations with the city, a joint committee of trustees from the University and Hopkins Hospital accepted Peterson's recommendation that Hopkins acquire the facility, appointing Peterson its president. On July 1, 1984, Hopkins took over City Hospitals, even though it still was mired in deficits and operating in decrepit buildings that hadn't been upgraded in decades.[234]

Adding Judy Reitz (now executive vice president and CEO of Hopkins Hospital) to his management team, Peterson and his colleagues launched more than two decades of extensive, financially successful expansion of buildings, staff and services on the 130-acre, Eastern Avenue campus. Renamed the Francis Scott Key Medical Center at the time of its acquisition, it was re-christened the Johns Hopkins Bayview Medical Center on the 10th anniversary of its Hopkins affiliation. As Peterson observed in 2009, "the acquisition of Baltimore City Hospitals has proven to be the win-win-win transaction we hoped it might be."[235]

"It has been a win for the City of Baltimore—no more red ink; a win for Johns Hopkins—130 acres of land on which to expand every dimension of our tripartite mission; and a win for the community, with dramatically improved facilities and services provided by Hopkins faculty and highly competent, caring staff."[236]

Those who work for The Johns Hopkins Hospital and the School of Medicine are well-versed in the institutions' rich history and protective of its heritage—but also not above poking gentle fun at it, too. Here (left to right) School of Medicine Dean Richard Ross, Department of Surgery Director George Zuidema, Department of Medicine director and physician in chief Victor McKusick, and Department of Pathology Director Robert Hepinstall sit for the camera around 1984, assuming the same poses as William Welch, William Halsted, William Osler and Howard Kelly in John Singer Sargent's iconic 1906 painting, "The Four Doctors." (McKusick even found a quill pen to hold, as Osler had in the original.) The sitters' reverence for Hopkins' founders is clear—as is their ability not to take themselves too seriously.

Change—and Continuity

The postwar rise of federal support changed not only the size of the Johns Hopkins Medical Institutions, but also what had been their intimate character. For its first 50 years, Hopkins had been relatively small and informal. Friendly interaction between faculty was a common occurrence. As the faculty spilled over into new buildings, federal support became a centrifugal force, posing a challenge to interactions between the Hospital and medical school that would continue for several decades.[237]

Thomas Turner believed that Hopkins had to retain its original ethos even as it entered a new age.

"Johns Hopkins must grow along with medicine and the nation, while preserving the essence of its greatness, to link the intellectual and cultural values of the Nineteenth and Twenty-first centuries," Turner wrote in 1981. "The creative potential of the individual must be protected and cultivated, with machines, sophisticated instruments and social organization playing subordinate roles."[238]

A centennial celebration of the Hopkins medical institutions was approaching, and in their home city, the importance of their contributions were a source of wonder—and pride.

In an editorial titled "Baltimore's Hopkins," the *Baltimore Sun* observed: "In fiscal 1938, the hospital served 16,492 bed patients and 312,480 outpatient visits. In fiscal 1988, the Johns Hopkins Health Service and its 10,000 employees treated 56,665 bed patients and 899,504 outpatients."

"In the Hopkins medical century, other great Baltimore institutions have withered. The Johns Hopkins Medical Institutes are still growing, still enlarging Baltimore."[239]

■ Thomas B. Turner (1902–2002), dean of the School of Medicine from 1957 to 1968, had a unique and highly unusual means of communication established to remain updated on important appointments and decisions. A gentleman of the old school, Turner wore a hat every day—a homburg in the winter, a straw hat in the summer. His executive assistant, Claudia Ewell, knew that his main concern at the end of each day was what was on his schedule that evening and what he had to do first thing the following morning. "We communicated by hatband," Ewell recalled in 1985. Before she left every evening, she would leave an index card in Turner's hatband, providing him with the information he needed on those matters.

Hopkins Medicine Discoveries and Breakthroughs, 1940–1988

1940–1942 — Eli Kennerly Marshall Jr., director of pharmacology and physiology, develops sulfaguanidine, a powerfully effective treatment for dysentery, and determines the proper dosage of quinolines to combat malaria. Both vital discoveries help maintain the health of U.S. troops during World War II combat in the Pacific. Infectious disease expert Frederick B. Bang oversees an intensive, coordinated program to study new anti-malarial drugs, research that earns him the Legion of Merit.[240]

1944 — Alfred Blalock, Helen Taussig and Vivien Thomas collaborate to develop the landmark "blue baby operation," a procedure that marks the beginning of cardiac surgery. It corrects deadly congenital defects in the hearts of children who otherwise would have died and inspires future advances in the field.

1946 — Physiologist Caroline Bedell Thomas launches the Johns Hopkins Precursors Study, the nation's first and now longest-running longitudinal health survey, to collect detailed medical information on a single group of individuals—the 1,337 members of the Hopkins School of Medicine classes of 1948 through 1964. Still continuing, its data have been the source for landmark studies correlating high cholesterol in early adulthood to later heart disease, as well as links between mental depression and heart attacks and youthful obesity to subsequent diabetes.[241]

1948 — Allergist Leslie N. Gay and colleague Paul Carliner discover that Dramamine could treat, even prevent, motion sickness.[242]

1951 — George Gey, director of the Department of Surgery's tissue culture laboratory, establishes the world's first continuously multiplying human cell culture, HeLa, with cervical cancer cells obtained from Henrietta Lacks. Gey distributes the HeLa cells for free to scientific researchers worldwide. Over the next 60 years, they prove instrumental in development of the polio vaccine, human papillomavirus (HPV) vaccines, chemotherapy breakthroughs, cloning, gene mapping, in vitro fertilization, and landmark research into HIV and tuberculosis.

1953 — Hematologist C. Lockard Conley and research fellow Ernest W. Smith use inexpensive homemade equipment to create a simple device for separating the components of hemoglobin on filter paper by using an electrical current charging method known as electrophoresis. Previously, a huge piece of machinery had been required for researchers elsewhere to perform only a few such hemoglobin analyses, but Conley and Smith's device enables them to do hundreds of analyses in a day. Researchers from all over the country come to Hopkins to see how their methods work.[243]

1954 — At Baltimore City Hospitals (now Johns Hopkins Bayview), Hopkins pediatricians Harold Harrison and Laurence Finberg publish a description of the oral rehydration therapy (ORT) that they have developed for preventing potentially deadly diarrhea in babies. The British medical journal *The Lancet* proclaims the therapy "potentially the most important medical advance this century," and it remains a process that still saves an estimated one million infants a year worldwide. In 1983, Harrison and his wife, fellow pediatrician and researcher Helen Harrison, become the first husband-and-wife team to be awarded the American Pediatric Society's top honor, the Howland Award, for their outstanding contributions to the health and welfare of children.[244]

1957 — The Hopkins AC closed-chest heart defibrillator, developed by William B. Kouwenhoven, Willliam Milnor and Samuel Talbot, is first used on a patient.[245]

1958 — Heart surgeons Henry T. Bahnson and James Jude, in separate cases, successfully apply the chest compression method of external heart massage, developed at Hopkins, to restart the hearts of their patients—one of whom is a two-year-old child, the other a 40-year-old woman.[246]

1963 — Henry Wagner, director of a joint effort between the departments of medicine and radiology known as the Division of Nuclear Medicine, oversees the first use of radioactive tracers for the rapid diagnosis of pulmonary embolism. Within a few years, both lung scans and pulmonary arteriography are being used widely.[247]

1965 — Gynecology fellow Robert Edwards of Great Britain, working with Hopkins gynecologist and crytogenetics laboratory chief Howard W. Jones Jr., conducts experiments that produce the first human egg fertilization in vitro. They don't realize their achievement at the time and only report their effort as an "attempted" in vitro fertilization, because the criteria for in vitro fertilization then were different than what later was established. Along with gynecologist Patrick Steptoe, Edwards subsequently develops IVF therapy in Britain, which led to the world's first "test tube" baby in 1978, an accomplishment for which Edwards will win the 2010 Nobel Prize in Medicine. Work by Jones and his wife, gynecologist Georgeanna Seegar Jones, by then retired from Hopkins and overseeing their Jones Institute of Reproductive Medicine at East Virginia Medical School, will achieve the first "test tube" baby in the U.S. in 1981.

1968 — Microbiologist Hamilton O. Smith discovers restriction enzymes, the proteins that can cut DNA at precise points in its genetic sequence, and microbiologist Daniel Nathans uses Smith's discovery to analyze the DNA of a virus that causes cancer in animals, achieving the first practical application of restriction enzymes. These accomplishments, along with Swiss microbiologist Werner Arber's initial theorizing on the existence of restriction enzymes, would earn Nathans, Smith and Arber the 1978 Nobel Prize in Medicine.[248]

1969 — Cardiac surgeons Vincent L. Gott and Harvey Bender Jr. perform the first heart transplant at Hopkins Hospital. Gott, a primary researcher in the development of the pacemaker, co-inventor of the Gott-Dagget Valve, an artificial heart valve, and the Gott Shunt, used to bypass sections of the aorta on which surgery is being performed, heads the Division of Cardiac Surgery for 17 of his 34 years on the Hopkins faculty, during which time he trains more than 60 cardiac surgeons. Bender leaves Hopkins in 1971 to become head of cardiothoracic surgery at Vanderbilt.[249]

1970s — The husband-and-wife allergist team of Kimishige and Teruko Ishizaka experiment on themselves, incurring inflamed, itching, burning arms and backs, to identify, isolate and purify IgE, the immune molecule that forms the basic ingredient of allergic reactions. They then isolate substances that turn gE antibody levels up and down, influencing the course and outcome of allergic reactions. Their research has made it possible for allergists to measure with unerring accuracy a patient's response to particular allergy-triggering proteins and the success of therapy.[250]

1971 — Orthopedic surgeon Lee Riley Jr. performs a landmark, full total knee replacement using a complete artificial knee Riley designed with four other surgeons, supplanting the previously used hinge-like replacements that had a very high failure rate. Later in the 1970s and early 1980s, orthopedic surgeons David Hungerford and Kenneth Krackow make major contributions to the field of knee surgery, including designing new instruments for carrying out the surgery, developing a new type of artificial knee, and implementing new training techniques. Both the instruments and the training techniques are adopted universally by orthopedic surgeons elsewhere. The innovations earn Hungerford and Krackow the first Annual Knee Society Lifetime Achievement Award in 2010.[251]

1972 — Neuroscientist Solomon Snyder, working with his student, Candace Pert, devises a unique radioactive tagging process to discover the long-suspected but elusive opiate receptors in the brain, the chemical messengers by which brain cells "talk" to one another and determine every brain-controlled process, from thinking and breathing to addiction, pain control, anxiety, mental illness, even love. The discovery earns him the 1978 Albert Lasker Basic Medical Research Award.[252]

1973 — Pediatric surgeon J. Alex Haller Jr., children's surgeon in charge since 1964, establishes the nation's first regional trauma center for children at Hopkins and creates a training program for pediatric surgeons.[253]

1979 — The nation's first Swallowing Center is opened. Its founder, Martin Donner, is head of the Department of Radiology and brings together radiologists, physiatrists, speech/language pathologists and other professionals to diagnose and treat swallowing disorders (dysphagia) of all kinds that afflict people with neurologic disabilities, head and neck cancers, and other medical conditions.[254]

1980 — Cardiac surgeon Levi Watkins Jr. installs the first implantable automatic heart defibrillator in a patient with repeated life-threatening episodes of ventricular fibrillation (irregular heartbeats). The implantable defibrillator had been developed by Mieczyslaw Mirowksi of Hopkins' Department of Medicine, working at Sinai Hospital.[255]

1981 — Neurosurgeon Dolin Long develops the Human Tissue Stimulator (HTS), an implantable electrical device similar in design to certain spacecraft batteries, that stimulates nerves to relieve patients of severe pain and uncontrollable tremors. Worn internally, it can be maintained indefinitely, unlike past such stimulators, being recharged by matching a special magnetic field generator next to the area where the device is implanted.[256]

1983 – Cardiologist Kenneth Baughman and cardiac surgeon William Baumgartner re-initiate Hopkins' heart transplant program, developing it into the one of the nation's leading centers for the surgical treatment of heart disease.[257]

1983 — Epidemiologist William Greenough III, Hopkins professor of medicine and international health, and fellow researchers Michael Field of the University of Chicago and John Fordtran of Baylor University, receive the King Faisal International Prize in Medicine for their two decades of research on the mechanism of action of the cholera toxin, which led to an effective oral vaccine and a unique probe into cell functions. [258]

1984 — Urologist Patrick Walsh performs the first of his nerve-sparing prostatectomy operations, using procedures he devised to remove a cancerous prostate gland while protecting the patient against post-operative incontinence and impotence. In the thousands of such operations he since has performed, nearly three-quarters of the patients have retained potency, more than 95 percent are continent, and only about 1 percent have had a local recurrence of their cancer. Adapting his nerve-sparing procedures to bladder cancer cases has produced similarly dramatic results in survival and function.[259]

1987 — Heart surgeons Bruce Reitz and Bill Baumgartner lead a Hopkins medical team to successful completion of the nation's first "domino donor," three-way transplant operation. The marathon 17-hour surgery involved Reitz's removal of a cystic fribrosis (CF) patient's healthy heart and diseased lungs; the harvesting of an accident victim's heart and lungs to give to the CF patient; and Baumgartner's "piggyback" placement of the CF patient's healthy heart on top of the diseased heart of a second patient. Fortified with immunosuppressive drugs that Reitz also helped pioneer, the patients went home to new lives a few weeks later.[260]

1987 — In the first successful operation of its kind, a 70-member Hopkins team of surgeons, physicians, nurses and technicians led by neurosurgeon Benjamin Carson undertake a 22-hour operation to separate a set of conjoined (Siamese) twins, Patrick and Benjamin Binder of Ulm, Germany, who were joined at the head. *Newsweek* magazine calls the operation "quite possibly… the most complex surgical procedure performed in this century."[261]

Hospital Directors/Presidents, 1940–1988[264]

Edwin L. Crosby (April 1946 – July 1952) succeeded his long-time boss, Winford H. Smith, following Smith's retirement after serving since 1911 as the Hospital's superintendant (a title changed to "director" during Smith's 35-year term). Crosby left Hopkins to direct the Joint Commission on Accreditation of Hospitals, the national, independent hospital-monitoring body based in Chicago. He later became head of the American Hospital Association.

Russell A. Nelson (July 1952 – October 1972), a 1937 graduate of the School of Medicine, spent his entire career at Hopkins. He saw the Hospital through a period that included an explosion in medical research, much of it funded with federal money, expansion of hospitals and medical schools, and government participation in medical care. Nelson shepherded the Hospital through a major financial crunch in the 1950s, and despite discussions about moving the medical institutions to the suburbs to follow the population shift, he held firm to Hopkins' original commitment to remain in East Baltimore as other local hospitals moved out of the inner city. Among the top issues during Nelson's time as Hospital director (the title was changed to "president" in 1963)[265] was the complex new partnership with local, state and federal governments in the financing and regulation of medical care. The Medicare and Medicaid programs came into being, vastly increasing the number of patients and full-time faculty. In turn, this solidified the Hospital's financial base, enabling it to do the kind of rebuilding that otherwise would not have taken place. During Nelson's tenure, Robert Heyssel, then-director of outpatient services and later Nelson's successor, made Hopkins the first academic medical center on the East Coast to launch community health maintenance organizations, the Columbia Health Plan in Columbia, Md., and a Medicaid HMO in Baltimore, which ultimately became part of the Hopkins Health Plan. Nelson, like Crosby, also served a one-year term as president of the American Hospital Association. He maintained his interest in clinical research, publishing papers on penicillin therapy for syphilis, as well as numerous articles on hospital management and the delivery of care.

Steven Muller (February 1972 – October 1982) became the first person since Daniel Coit Gilman in 1889 to serve as head of both the University and the Hospital. Having become president of the University in 1972, Muller recognized the glaring contrast between the high caliber of the School of Medicine faculty and the Hospital's patient care and the increasingly outmoded, cramped facilities in which they treated patients and conducted research. With a mammoth fundraising effort about to be launched to rectify this situation, as well a separate capital campaign to improve the financial health of the entire University, Muller thought it best not to duplicate such an undertaking and assumed the presidency of both the Hospital and the University.[264] During his tenure as Hospital president, a massive building program led to construction of a 60-bed Cancer Center, teaching and patient-care towers, the Meyer Center for Psychiatry and the Neurosciences, and expansion of the Wilmer Eye Institute. New emergency room facilities were among the most sophisticated in the nation. As chairman of an Association of American Medical Colleges' panel on medical education, Muller also oversaw production of a report that called for a new approach to the nation's way of training physicians, with a renewed emphasis on the personal physician-patient relationship. The enormous growth of both the Hospital and University during Muller's first decade as president of both, plus Hopkins Medicine's impending expansion as an increasingly diverse health system, led him to relinquish the Hospital presidency and appoint Robert Heyssel, the executive vice president and director of the Hospital, as its new president.

Robert M. Heyssel (February 1972 – January 1987; July 1988 – July 1992), had been responsible for the day-to-day operation of the Hospital and its substantial rebuilding as executive vice president and director of it during Muller's presidency. He encouraged a construction program that emphasized interrelations between specialties rather than separate, free-standing structures for each. Hailed nationally for his successful efforts to contain hospital costs at a time when they were skyrocketing elsewhere, Heyssel accomplished this in large part by decentralizing much of the Hopkins Hospital budget, making each clinical department director responsible for allocating and controlling the departmental budget. Known nationwide as a strategic business planner, Heyssel became convinced that the future of Hopkins Hospital lay in a vertically integrated system that would ensure the flow of patients at a time of increasing regulation and competition. In 1986, he oversaw incorporation of the Johns Hopkins Health System (JHHS), of which he also became president and soon included four hospitals, a health maintenance organization (HMO), a preferred provider organization (PPO), and a full-service insurance package offered through employers. He briefly relinquished the presidency of the Hospital but reassumed it in 1988, also maintaining the presidency of JHHS. Like so many of his predecessors, Heyssel was a national figure in academic medicine. He served as chairman of the Association of American Medical Colleges and its Council of Teaching Hospitals, as well as of the Commonwealth Fund Task Force on the Future of Academic Health Centers. Trained in hematology and nuclear medicine, he also served on many national commissions concerned with health care delivery. His pioneering role as the founder of JHHS made him an architect of the link between academic medical centers and HMOs.

Albert H. Owens Jr. (January 1987 – June 1988), who was considered instrumental in establishing medical oncology as a certified subspecialty, began his Hopkins career in 1954 as a fellow in the Department of Pharmacology. Subsequently chosen to become head of a new oncology division in the Department of Medicine, but lacking space for it at Hopkins Hospital, he established a model oncology unit in 1961 at what now is the Johns Hopkins Bayview Medical Center. In 1973, in response to Owens' successful proposal, Hopkins was awarded a grant from the National Institutes of Health to establish one of the newly authorized National Cancer Centers. He moved this center back to Hopkins Hospital in 1977.[265] Plans for expansion of the oncology department, of which he had remained head, and a construction of a new $40 million oncology building prompted Owens to relinquish the Hopkins Hospital presidency in 1988 in order to turn his full energies to the design of the new center and raising the funds for it. The auditorium in what became the Sidney Kimmel Comprehensive Cancer Center is named in his honor..

School of Medicine Deans, 1940–1988

Alan M. Chesney (1929 – 1953) As noted in the previous chapter, Chesney's 24-year term as dean of the School of Medicine was the longest in its history, encompassing the challenges of guiding Hopkins through the Great Depression and World War II.[266]

Philip Bard (1953 – 1957) was a brilliant physiologist whom Hopkins had recruited from Harvard in 1933, when he was only 34, to become director of the School of Medicine's physiology department. He headed it until 1964, even while simultaneously serving as the medical school dean for four years. Bard was a natural leader but hated administrative work and reluctantly accepted the deanship. His colleagues would recall how paperwork piled up around him until he couldn't stand it anymore. He then would stash it in a glass-doored bookcase where he still could see it without feeling buried underneath it. During his four years as dean, those who had business with him in that capacity entered his office through one door; those with physiology to discuss entered through another door. Bard's primary research interest was in the central nervous system's involvement in emotional expression. Along with colleagues Clinton Woolsey and Wade Marshall, Bard was one of the first to use an oscillograph to map the functions of the primate cortex. He was president of the American Physiological Society from 1941 to 1946, and became president of the Association for Research in Nervous and Mental Disease in 1950.[267]

Thomas B. Turner (1957 – 1968) was an authority on venereal diseases, polio, tetanus and other infections diseases, as well as on alcoholism. A graduate of St. John's College in Annapolis in 1921 and the University of Maryland Medical School in 1925, he came to the Hopkins School of Medicine on a fellowship in 1927 and then joined the faculty. In 1932, Turner left Hopkins to become a member of the Rockefeller Foundation's international health division. He returned to in 1936 and in 1939 was named professor and director of the Department of Bacteriology in what now is the Bloomberg School of Public Health. Renamed the Department of Microbiology in 1952, it was made a joint department with the School of Medicine in 1957, the same year Turner became the dean. During his decade as dean, the School of Medicine doubled its physical plant, added new departments, increased enrollment—including its first African American medical students—and received a record amount of federal government grants and contracts. After leaving the dean's office, Turner was named the first archivist of the Johns Hopkins Medical Institutions. While serving as archivist, he wrote a history of the medical institutions, *A Heritage of Excellence*. In 1982, he stepped down as archivist to head the newly-formed Alcoholic Beverage Medical Research Foundation based at Johns Hopkins. He was one of the last on the faculty to have known all of the School of Medicine's previous deans except Osler. He died in 2002 at the age of 100.[268]

David E. Rogers (1968 – 1971) was the 41-year-old chairman of Vanderbilt University's Department of Medicine when he was tapped to become the Hopkins School of Medicine dean. Initially, he held a dual appointment, also serving as medical director of the Hospital, a move spurred by the unprecedented growth in the number of full-time physicians and surgeons. This experiment in administrative consolidation was ended because the Hospital and School of Medicine believed at that time that they should maintain their independence of each other. Rogers encouraged Robert M. Heyssel, who had been at Vanderbilt with him, to come to Hopkins, where they both became involved vigorously in the issue of providing health care for the poor, particularly in the poverty-stricken East Baltimore neighborhood surrounding Hopkins Hospital. Rogers left Hopkins to become the first president of the Robert Wood Johnson Foundation, a major philanthropic medical organization.

Russell H. Morgan (1971 – 1975) was a pioneer in radiation safety and became renowned for perfecting ways not only to diminish radiation hazards but improve X-ray images in the 1960s, including wedding television technology with fluoroscopy. This enhanced the image without increasing the dose and led to angiography. Morgan also pioneered the evaluation of X-ray exposure to patients from diagnostic tests. Although planning to retire as director of the Department of Radiology, Morgan instead accepted the request that he become dean during a difficult period in the School of Medicine's history. He gracefully helped it weather financial troubles during the recession of the early 1970s. The radiology department now is named in his honor.

Richard S. Ross (1975 – 1990) was Hopkins' chief of cardiology when he was chosen to become the School of Medicine's 11th dean. A graduate of the Harvard medical school, he came to Hopkins in 1947, intending to spend a year on the Osler residential service—and never left. Beginning as a protégé of pediatric cardiologist Helen Taussig, treating young patients with congenital heart disease, he subsequently focused on the heart problems of adults. In the 1960s and early 1970s, he directed the Wellcome Research Laboratory and collaborated with his predecessor as dean, radiologist Russell Morgan, to introduce cineangiography to Hopkins and take the first motion pictures of a beating heart ever filmed here. He and cardiology colleagues Gottlieb Friesinger, J. Michael Criley and O'Neal Humphries were the first at Hopkins to produce coronary arteriograms, revealing how heart valves function, how heart sounds and murmurs are generated, and measuring coronary blood flow in patients by using radioactive gases. Much of Ross' tenure and accomplishments as dean will be covered in the next chapter. It is noteworthy, however, that several months before he was offered the job of dean, while he was serving as president of the American Heart Association, Ross turned down the directorship of the National Heart and Lung Institute. He did so because it would have taken him away from patients and research. During his entire 15-year tenure as dean, he continued seeing patients one morning a week, in large part because of the opportunity this gave him to interact with students and house staff.[269]

1989–1993:
Celebrations and Changing of the Guard

Even as elaborate preparations were being made to celebrate the centennial of The Johns Hopkins Hospital in 1989 and of the School of Medicine in 1993, long-simmering tensions between the two threatened to undermine their future.

Addressing what he feared would become an increasingly fractious relationship between the next generation of Hospital and School of Medicine leaders, Dean Richard Ross began drafting a proposal that he hoped the University and Hospital trustees would adopt to end the seemingly constant struggle over the two entities' shared direction.

In a nine-page draft memorandum to the joint trustee policy committee of the University and Hospital dated March 2, 1989, Ross wrote that they soon would have "a once in a 'trustee lifetime' opportunity to significantly strengthen the Medical Institutions," given the impending retirement of himself, the Hospital and Health System president, Robert Heyssel, and the University president, Steven Muller.[1]

The trustees could now "examine the governance relationships of … [the] health care entities and put in place a structure which is best for Johns Hopkins without considering the individuals currently occupying the several offices," Ross wrote.

He suggested creating a management organization for the medical institutions with a single chief executive officer. "I see the potential to become number one in the United States by a wide margin if we can pull the Hospital and the School of Medicine together under a single system of leadership," Ross predicted. He noted that Hopkins' local competition "in the health care delivery arena is also a significant threat. Our diverse and competing organization provides an opening for our local competitors."[2]

Ross would revise this memorandum over the next two years—even after his retirement from the School of Medicine deanship in 1990—and circulate it among numerous trustees and other leaders. Heyssel, with whom Ross often had clashed, voiced objections to Ross's overall proposal, but in a September 1989 memorandum to the executive committee of the Hospital and Johns Hopkins Health System, he acknowledged that many of Ross's points about the cumbersome management structure were "absolutely valid."[3]

Heyssel maintained his opposition until his own retirement in 1992, but when a deluge of internal conflict four years later led to the most dramatic restructuring in a century—unifying leadership under a single CEO as Ross had proposed—Heyssel

One unique aspect of The Johns Hopkins Hospital's 100th anniversary celebration in 1989 had a long-lasting impact on its own special, non-medical field. As part of the Hospital's centennial, then-U.S. Postmaster General Anthony Frank unveiled a $1 postage stamp bearing Mr. Johns Hopkins' portrait. The Hopkins stamp proved worthwhile for the post office, given its popularity among those in the special field of philately—stamp collecting. On the first day of issue, nearly 18,000 stamps were sold to collectors who would never put them on an envelope, making the stamps' sale pure profit for the Postal Service. Four years later, in January 1993, 125,000 of them were used on the oversized envelopes containing invitations to Bill Clinton's first inauguration as president. Above, Frank, left, shakes hands with Hopkins President Steven Muller after unveiling the Hopkins stamp. U.S. Secretary of Health and Human Services Louis Sullivan is on the right.

■ Adolf Meyer (1866–1950), director of the Phipps Psychiatric Clinic from 1912 to 1941, enjoyed practical jokes. At dinners he and his wife hosted at his home for first-year students, a large, dome-covered silver platter would be brought to the table. Meyer would ceremoniously lift the lid—revealing a flustered, live turkey.

was the consultant to the special 10-member group of University and Hospital trustees who drew up the new organizational plan.[4]

Notwithstanding the future turmoil bubbling beneath the surface during those five years, several impressive celebrations of all that was special about the Hospital and School of Medicine took place, significant new buildings were completed, a new medical school curriculum was introduced, and landmark research centers in genetics and cardiology were founded. In addition, advances were made in fostering diversity, as well as bettering relations with the community surrounding the medical campus. The difficulties did not prevent Hopkins from repeatedly being named the top hospital in the United States by *U.S. News & World Report*. And, in one small Hopkins research laboratory, a remarkable molecular discovery was achieved that would win a Nobel Prize a scant 11 years after it was made.

The Hospital's One Hundredth

Preparations for celebrating the 100th birthday of The Johns Hopkins Hospital began more than five years before the spring 1989 anniversary.

Under the direction of Elaine Freeman, the Office of Public Affairs (now known as Marketing and Communications) established an office to coordinate events. A flood of press releases and special supplemental materials was issued, all emblazoned with a dramatic "Centennial of Johns Hopkins Medicine" logo.[5]

Public Affairs persuaded the *Baltimore Sun* to devote an entire 39-page special issue of the *Sunday Sun Magazine* to Hopkins. (The willingness of businesses to purchase congratulatory ads in the supplement helped with the persuasion.)[6] Joann Rodgers, deputy director of public affairs and media relations director, also arranged with the editor of the *Journal of the American Medical Association* to have the publication devote virtually an entire issue to Hopkins—printing significant research papers written only by Hopkins faculty members. It was the first and remains the only time *JAMA* ever focused entirely on one institution.[7]

The New England Journal of Medicine also took notice of the centennial with a congratulatory editorial. "Although the *Journal* rarely acknowledges occasions of this kind," wrote Editor in Chief Arnold Relman, "Hopkins has played such a unique part in the history of American medicine that its hundredth birthday should not pass unnoticed." After summarizing key aspects of Hopkins' influence, Relman concluded, "Today, Hopkins remains one of our most distinguished centers of medical care and scholarship, still vital and innovative after an illustrious century of service. The *Journal* is happy to join what must be a great host of admirers in wishing Hopkins a happy birthday and another century of success."

In conjunction with the centennial festivities, the Johns Hopkins University Press published a two-volume history of the Hopkins medical institutions (upon which much of the earlier chapters in this book were based). Titled *A Model of Its Kind*, the book was a collaborative effort by A. McGehee Harvey; Gert H. Brieger, head of Hopkins' Institute of the History of Medicine; Victor McKusick; and Susan L. Abrams, a writer and editor at Hopkins. In *Uniquely Johns Hopkins*, another book illustrated with fine photographs of renowned Hopkins physicians and researchers of recent years, Joann Rodgers and Elaine Freeman updated Richard H. Shryock's 1953 monograph, *The Unique Influence of The Johns Hopkins University on American Medicine*, providing concise descriptions of the special achievements of these contemporary medical luminaries.

On May 7, 1989, one hundred years to the day after the Hospital's dedication, a procession of limousines took some 125 collateral descendants of Johns Hopkins and a large group of invited dignitaries from the original Hospital building to Baltimore's Green Mount Cemetery for a ceremony at the founder's modest resting place.[8]

The limousines then transported those at the graveside gathering back to the Billings Building—site of the 1889 dedication—for a reception under the dome.

A far more extensive celebration occurred a month later, between June 7 and 11. In a letter read at the opening event in the Baltimore Convention Center, President George H.W. Bush wrote that it was an anniversary "for all of us to celebrate" and called Hopkins' century of medical accomplishments "an inspiring chapter in human history."[9] His letter was read after a mammoth academic procession had filled the hall. It was followed by speeches by U.S. Secretary of Health and Human Services Louis Sullivan, Maryland Governor William Donald Schaefer, Baltimore Mayor Kurt Schmoke and others, as well as the unveiling by U.S. Postmaster General Anthony Frank of a $1 postage stamp bearing Johns Hopkins' portrait.

In an era when three television networks still dominated the media, the morning shows on NBC, CBS and ABC all broadcast segments on the centennial.

The celebration also featured a fireworks extravaganza in the city's Inner Harbor, a Baltimore Symphony Orchestra performance with famed flutist Jean-Pierre Rampal, two free-to-all-employees concerts by singer Gladys Knight, and a lavish ball and banquet with Pulitzer Prize-winning humor columnist Art Buchwald as the main speaker. Multiple symposia on critical issues in medicine, nursing and public health were held, and science fiction guru Isaac Asimov delivered a forecast on "Health in the Year 2000."[10]

1989

May 7: The 100th anniversary of opening of Hopkins Hospital is celebrated.

June 7: The U.S. Postal Service issues a $1 stamp bearing a portrait of Johns Hopkins. Nearly 18,000 are sold on first day of issue; 125,000 later are used on oversized envelopes with invitations to Bill Clinton's inauguration in January 1993.

The Center for Medical Genetics opens on the 10th floor of Blalock Building, with 25 full-time faculty, 100 staffers and research grants of more than $2.5 million. Its opening is the culmination of work begun 30 years earlier by Victor McKusick. Plans begin for creation of a "Genetic Data Base," scheduled to be unveiled in 1990.

November 18: The Johns Hopkins Asthma and Allergy Center is dedicated. It is the first building designed for the Hopkins Bayview Research Campus and represents a new initiative in patient care, physician training and research.

Jean-Pierre Rampal (1922–2000), considered by many to be the premiere flute soloist of the 20th century, was one of the acclaimed performers who participated in five days of festivities in June 1989 marking the centenary of The Johns Hopkins Hospital.

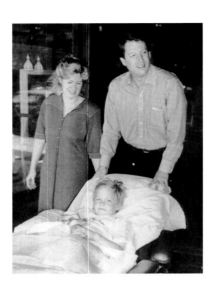

Invited to the School of Medicine's Centennial dinner but unable to attend was then-Vice President Al Gore, whose family had an intensely personal connection to Hopkins that was forged when his son, Albert III, was successfully treated in the Children's Center for serious injuries he sustained in 1989 after being run over by a car while leaving an Orioles game. In a videotaped message, Gore said: "Thinking back on those days when we were in Johns Hopkins, I remember how sincere and reassuring and how professional you all were. Your efforts and determination, your care, your love—they made the difference in Albert's full recovery and in our family's healing afterward. We will never forget your kindness or your prayers, and you will always have a special place in our hearts."[11]

1990
The Ciccarone Center for the Prevention of Heart Disease is founded. It is a "unique in the nation" research arm of the Division of Cardiology and named for Henry Ciccarone, a renowned former Hopkins lacrosse coach who died of heart disease at 50.

The Task Force on Women's Academic Careers in Medicine is founded. It becomes a model for medical schools across the country and in 1993 it issues a report on improvements for women in medicine.

The School of Medicine's Milestone

While the Hospital centennial brought alumni of all the health division schools together for the first time—and was supposed to eliminate the need to have separate celebrations for each one—four years later, with a new dean in place, the School of Medicine celebrated its own centennial with much panache, coordinating it with the traditional June biennial meeting of the Johns Hopkins Medical and Surgical Association, which always brings hundreds of former house staff, faculty and alumni back to East Baltimore. The school scored a news coup by having then-First Lady Hillary Rodham Clinton as the keynote speaker at its Centennial Symposium on June 10. Clinton was the main advocate of her husband's efforts to reform the nation's health care system, an issue then embroiling Congress and the public in heated debates, and her 25-minute speech to an enthusiastic, overflow crowd in Hopkins' 700-seat Turner Auditorium drew extensive press coverage.[12]

"You are a beacon in this region, and in the nation when it comes to clinical and tertiary care," Clinton said, "and, most importantly, you are an anchor in East Baltimore, providing quality treatment for poor and needy citizens who might otherwise go without."[13]

Health care reform was one of the main topics covered at the daylong symposium in June, as were future prospects for biomedical research and the relationship between academic medicine and business. Later, some 900 partygoers attended a lavish banquet.[14]

Research Makes Strides and Gets New Homes

The centenary events begun in 1989 not only celebrated past achievements but heralded the promise of great accomplishments to come. Participants at the centennial symposia had voiced the view that medicine's future would lie in genetics—and no medical center was better prepared to lead in that field than Hopkins. It was home not only to Victor McKusick, the "father of medical genetics," but to microbiologists Daniel Nathans and Hamilton Smith, whose Nobel Prize-winning discovery of restriction enzymes led the way to gene-splicing. Also still active on the faculty was pediatrician Barton Childs, who received the American Pediatric Society's 1989 Howland Award, the top honor in the field, for his establishment of pediatric genetics as a specialty.[15]

Ensuring that it remained in the forefront of genetic research and application, Hopkins opened its Center for Medical Genetics on the 10th floor of the Blalock Building in 1989. With 25 full-time faculty, 100 staff members and research grants of more than $2.5 million, the new center was considered the natural culmination of the work McKusick had begun 30 years earlier.[16] Results from the center's research would be stored and shared via the new Genomic Data Base, a high-speed, high-performance communications network project of the Howard Hughes Medical Institute in cooperation with the School of Medicine and the William H. Welch Medical Library. Decades of genetic data generation and collection at

Hopkins—beginning with McKusick's *Mendelian Inheritance in Man*, which relatively quickly became the *Online Mendelian Inheritance in Man*, better known to scientists worldwide as OMIM—made Hopkins the logical home for the database.[17] (A decade later, it would be consolidated with nine other Hopkins genetic research centers to form the McKusick-Nathans Institute of Genetic Medicine.)

During this period, genetic research at Hopkins received another important boost with the creation of the William S. Smilow Center for Marfan Syndrome Research, which opened in 1993 through the generosity of Joel and William Smilow and their families. The gift came two years after Hopkins scientists, led by Harry "Hal" Dietz III and working with colleagues at the Shriner's Hospital in Portland, Oregon, discovered the gene that causes Marfan. The discovery, reported in the July 25, 1991, issue of *Nature* magazine, had immediate impact on the early diagnosis of Marfan patients.[18] The multidisciplinary biomedical researchers and clinicians in the Smilow Center have done much to improve the medical care and lives of those with Marfan syndrome and related disorders.[19]

In another genetic breakthrough, a team of Hopkins researchers led by molecular oncologists Bert Vogelstein and Kenneth Kinzler, collaborating with scientists at the National Institutes of Health and the University of Helsinki, succeeded in 1993 in isolating a gene whose mutation causes colon cancer. Initially, Vogelstein had called the search for the cancer-causing gene "mission impossible," but working in an old East Baltimore supermarket converted into a laboratory, then in a sparkling new research lab that opened in 2000, Vogelstein and his team ultimately would succeed in isolating five of the seven genes linked to colorectal cancer and sequence the genes associated with breast, pancreas and brain cancers.[20]

"Now one can look at all of the genes at once in cancers and look in a much more comprehensive and unbiased way at what's wrong with the cancer cells," Vogelstein said in 2009. "That would have seemed like science fiction 20 years ago. And the first time that was done, it was done here at Hopkins. Hopkins is the only place so far that has the capability of doing this, although that obviously will change in the future and more and more institutions will be doing this, but so far, Hopkins is unique in having done this with any cancer. That has provided a completely new understanding of what is wrong with the cancer cell … And that has really altered our view and the view of cancer researchers in general about the nature of what cancer is and how it's formed."[21]

In the course of accomplishing such achievements, the affable, soft-spoken Vogelstein became the most cited medical researcher in the world, according to the Philadelphia-based Institute for Scientific Information. A 2008 article in the

Internationally acclaimed molecular oncologist Bert Vogelstein, a 1974 graduate of the School of Medicine, has spent his entire career at Hopkins conducting research that is unraveling the mysteries of cancer—why people get it and how obtaining that knowledge can help prevent its occurrence or cure it. His devotion to the field goes back to his days as a Hopkins intern in 1974.

"I have a very vivid memory of my first patient, when I was an intern here. She was a little girl with leukemia … and she was a beautiful little girl. She was four years old and right out of the blue, one day she was fine and the next day she got leukemia. I obviously identified with her father, and he asked, "Why did my little girl get this? What happened?" And all you could do was throw up your hands. I had no idea. No one had an idea. And it's very difficult to be hopeful when you have no idea of what's going on. It's like a plague from outer space. What's the next step you take? There's no pathway to follow. So that was certainly one of the things that stimulated me to go into cancer research, to try to find out some of the answers. And, you know, I think to a certain extent, based on efforts here and many other places, you can now answer those questions pretty precisely." Today, thanks to such advances, the kind of tumor that little girl had generally can be cured, he says.

1989
At the American Heart Association annual convention in November, three Hopkins cardiologists who headed the AHA attend: Myron "Mike" Weisfeldt, M.D., head of cardiology, the current president; Bernadine Healy, M.D., former assistant dean of the School of Medicine and immediate past president; and Richard S. Ross, M.D., dean of the SOM and AHA president in 1974.

December 18: Groundbreaking is held for the $140 million, 440,000-square-foot Outpatient Center across Broadway from Hopkins Hospital. It will be named for former Hospital President Robert Heyssel—but later becomes better known simply as JHOC—Jay-Hoc. It opens in 1992.

Cartoonist Doug Marlette, Pulitzer Prize-winning editorial cartoonist and creator of the comic strip *Kudzu*, enjoys inserting Hopkins' name in the comic, attributing wacky scientific "discoveries" to fictitious Hopkins researchers.

"Anybody who's been in the economy class in an airliner with a full bladder and the seat belt light goes on is keenly aware of aquaporin 2. Small differences in volume can make big differences in comfort."

—Peter Agre, 1974 graduate of the School of Medicine, co-recipient of the 2003 Nobel Prize for Chemistry in recognition of his discovery of aquaporins, the proteins that form the "water pore" channels through which water enters and exits cells

institute's newsletter, *Science Watch*, noted that between 1983 and 2002, Vogelstein's research findings were cited 106,401 times by other scientists in their papers, easily placing him at the top of the list of the 50 most influential scientists. (Number three on the institute's list of most-cited scientists was Hopkins neuroscientist Solomon Snyder, with 63,106 citations; Kinzler's 48,277 citations placed him at No. 19. Other Hopkins faculty listed in the top 250 were immunologist Lawrence Lichtenstein, neurologist John Griffin, urologist Patrick Walsh, neurologist Daniel Drachman, infectious disease expert Thomas Quinn, neurologist Ted Dawson, molecular biologist Andrew Feinberg, neuropathologist Donald Price and oncologist David Sidranksy.)[22]

Not mentioned in *Science Watch* because his research was not often cited between 1983 and 2002, but nonetheless pursuing extraordinarily important inquiries at Hopkins during this period, was Peter Agre. In 1992, working in a modest lab on the fourth floor of Hopkins' Preclinical Teaching Building, Agre and his research assistants discovered aquaporins, the previously elusive "water channel" proteins that enable water to move in and out of cells. The finding was so important that Nobel Prize officials, who often take decades to honor a scientific achievement, bestowed the Nobel for chemistry on Agre in 2003, a mere 11 years after his initial discovery.[23] More on Agre's discovery and its impact appears in Chapter 5.

Working in a modest laboratory on the fourth floor of Hopkins' Preclinical Teaching Building, Peter Agre (above) and his colleagues achieved a stunning scientific breakthrough in 1992, discovering aquaporins, the elusive "water channel" proteins that enable water to move in and out of cells. The finding would earn Agre the Nobel Prize for Chemistry in 2003.

The critical shortage of on-campus research laboratories that compelled Vogelstein and his colleagues to work in a converted supermarket and forced researchers in other fields to set up laboratory benches in hallways had already prompted a concerted drive to raise funds for a major research building.[24]

By late 1988, groundbreaking took place for an $85 million, 10-story building on Rutland Avenue, across Monument Street from the Hospital. With 22 state-of-the-art laboratory suites planned for each floor, it would increase Hopkins' research space by more than a third.[25]

Citing the institution's "preeminent position" as a recipient of federal research funding, Richard Ross said that to maintain its standards, Hopkins "must be able to offer the best scientists research facilities which match them in terms of quality. The expansion offered by the new research building is critical to our ability to compete successfully for these scientists and to allow all of our scientists to compete for research grants."[26]

Fittingly, the building that Ross had championed was named for him shortly before he retired in 1990. The occasion was Hopkins' annual Commemoration Day celebration, held in February to mark the University's official opening on February 22, 1876. Also present for the festivities that year was President George H.W. Bush, who received an honorary degree.

Ross also fostered a $5 million endowment drive for what became the Richard Starr Ross Fund for the Physician Scientist, a grant-bestowing program for young Hopkins researchers. When the Ross Building was dedicated in 1991, the results of the fundraising drive were announced.[27] The money collected included a substantial sum raised by those who worked on Ross's staff in the dean's office; and in 1992, the first Ross Scholar was Harry "Hal" Dietz III, the clinical fellow in pediatric cardiology who had led the Hopkins research team that found the Marfan syndrome gene. By 2009, 100 researchers had received more than $6.4 million in Ross Scholarships—and Hal Dietz had become the Victor A. McKusick Professor of Genetics in the departments of Medicine, Pediatrics, and Molecular Biology and Genetics; director of the Smilow Center for Marfan Syndrome Research; and discoverer of the ability of an existing drug, losartan, to dramatically improve the lives of Marfan patients.[28]

The researchers who received Ross grants subsequently returned that investment by obtaining more than $164 million in external funding to continue their research.[29]

At Hopkins' 1990 Commemoration Day celebration, held annually in February to mark the anniversary of the University's official opening on Feb. 22, 1876, President George H.W. Bush received an honorary degree from University President Steven Muller (right) and School of Medicine Dean Richard S. Ross (left). The naming of the School of Medicine's new research building in Ross' honor also was announced at that event.

Harry "Hal" Dietz (far left) is joined by Victor McKusick (center) and Dietz research colleagues (left to right) Clair Ann Francomano, Reed E. Pyeritz and Garry Cutting for the 1991 announcement that Dietz and his team had discovered the gene that causes Marfan syndrome, an affliction that McKusick had studied for decades.

Additional research space was found not on the East Baltimore campus but the grassy knolls of Johns Hopkins Bayview Medical Center on Eastern Avenue. In November 1989, the Johns Hopkins Asthma and Allergy Center, the first building designed for what was called the Research Campus at Hopkins Bayview, was dedicated.[30] The home for a new initiative in patient care and teacher training, as well as research, the $45 million, five-story facility housed 275 faculty, fellows, lab technicians, nurses and computer specialists. In addition, Hopkins Bayview—long the focal point of Hopkins' geriatric research and patient care—opened a new $17 million Geriatrics Center in 1991.[31] Dramatically adorning the five-story atrium of the Asthma and Allergy Center is an eclectic assemblage of painted maple, steel and aluminum by sculptor Larry Kirkland, who entitled it *Aspirato*, the Latin word for "I breathe." Hopkins employees gave it the unofficial but more commonly used name, "The Big Sneeze."

New Buildings Also Enhance Care for Inpatients and Outpatients

Expansion of facilities for patient care also was a priority in East Baltimore. In December 1989, ground was broken for a 440,000-square-foot, $140 million outpatient center across North Broadway from the Hospital, to which it would be linked by an underground concourse.[32] The center would have eight operating rooms for ambulatory surgery, nearly 200 examination rooms, a diabetes center, and suites for patients needing services in urology, neurosciences, orthopedics, otolaryngology, dermatology, pediatrics, plastic surgery, and gynecology and obstetrics. Four nuclear medicine suites and suites for general imaging were included with X-ray and EKG equipment, as well as two MRI machines and two CT units.[33]

The 400,000-square-foot, $140 million outpatient center, built right across North Broadway from the original Hopkins Hospital, was dedicated in 1992 and named for Robert Heyssel, soon-to-retire president of the Hospital. The building later became better known by the acronym "JHOC," pronounced "Jay-Hoc," for the Johns Hopkins Outpatient Center.

When the outpatient center was dedicated in 1992, it officially was named for Robert Heyssel, who would soon retire as president of the Hospital after a 24-year career at Hopkins.

Expanding Hopkins' outpatient activities far beyond the Hospital's immediate community was the Johns Hopkins Home Care Group, begun in 1983 as a nonprofit corporation owned jointly by the Health System and the University to provide ongoing, follow-up care for former Hospital patients and others. In 1992–1993, it underwent a major reorganization to improve its operations substantially. By the end of the first decade of the 2000s, the Home Care Group's nurses, therapists, home health aides and social workers were furnishing medical and respiratory supplies, home infusion therapy, pharmaceuticals and home hospice care during more than 35,000 annual patient visits throughout Baltimore City and its surrounding counties.[34]

Even with the dedication of the Ross Research Building on June 7, 1991, and the 1992 opening of the Heyssel Building, which became better known as JHOC, pronounced "Jay-Hoc" (the Johns Hopkins Outpatient Center), Hopkins' need to upgrade or expand its facilities remained great. An estimated $335 million in other construction projects were on drawing boards or already had shovels in the ground.[35] In ophthalmology and oncology, two areas in which Hopkins long had excelled, new structures were planned.

Morton Goldberg, a former house staff and faculty member of the Wilmer Eye Institute who had gone on to head ophthalmology at the University of Illinois in Chicago for 19 years, certainly had building on his mind when he returned to Hopkins early in 1989 to take over the helm of Wilmer from his mentor, Arnall Patz. By the following year, construction was under way on a two-floor expansion of the institute's Maumenee Building, providing additional space for patient care and research. Goldberg later would expand Wilmer's reach by establishing satellite patient centers at Hopkins Bayview and at surrounding suburban locations.[36]

■ Hopkins' impact on cardiology was apparent at the 1989 convention of the American Heart Association, when three Hopkins cardiologists who had headed the AHA stood side by side for a photo: then-president Myron "Mike" Weisfeldt, a 1965 graduate of the School of Medicine who would later become head of the Department of Medicine and physician in chief for Hopkins Hospital; the late Bernadine Healey, the immediate past president, who had been an assistant dean of the School of Medicine and later would become the first woman to head the National Institutes of Health; and Richard Ross, who had headed the AHA in 1974.[40]

Oncologist Martin Abeloff (left), a 1966 graduate of the School of Medicine and 20-year veteran of the faculty, succeeded Albert Owens (right) as head of Hopkins' 15-year-old oncology center in 1992. Here they review preliminary plans for what became Hopkins' new $184 million oncology clinical and research center, which opened in January 2000.

1990

January: William Richardson, provost of the Pennsylvania State University, is selected to assume The Johns Hopkins University presidency on July 1, succeeding Steve Muller.

February 22: President George H.W. Bush receives an honorary degree at the University's Commemoration Day ceremony at Shriver Hall on the Homewood Campus.

At the Commemoration Day ceremony, the new basic research building is named for Richard Ross, retiring dean, and will be dedicated in June 1991. The Richard Starr Ross Fund for the Physician Scientists endowment drive is launched, seeking $5 million to provide funds for young researchers.

July 1: Michael Johns, head of the Department of Otolaryngology—Head and Neck Surgery, becomes Dean of the School of Medicine, succeeding Richard S. Ross.

In first annual *U.S. News & World Report* nationwide survey of physicians, Hopkins is named the second best U.S. hospital after the Mayo Clinic. The following year, Hopkins is named the best hospital. Hospital President Robert Heyssel jokes that he'd always figured that first survey result "was a mistake." Hopkins will retain the top ranking for decades.

1991

June 6: A new $17 million, 190-bed Geriatrics Center opened at Johns Hopkins Bayview.

July 25: A Hopkins research team led by Harry (Hal) Dietz report their discovery of the gene that causes Marfan syndrome. The finding improves prenatal and early diagnosis of patients before symptoms of the condition appear.

Oncologist Martin Abeloff, a 1966 graduate of the School of Medicine and a 20-year veteran of the faculty, had even bigger building aspirations. Named director of the Hospital's 15-year-old oncology center in 1992, Abeloff succeeded its first director, Albert Owens. Shortly after Abeloff's appointment, Hopkins announced plans to replace its increasingly outdated cancer center in the Jefferson Street Building by constructing a $120 million care facility that would become one of the top oncology institutions in the country. The state of Maryland, recognizing that it had one of the highest cancer rates in the nation, initially pledged $30.5 million to the project, while Hopkins aimed to raise the rest of the money, primarily through donations.[37]

"This is more than another oncology center that we are building," Abeloff said. "It will be a comprehensive program that will put under one roof every Hopkins department involved in treating patients with cancer. The word 'interdisciplinary,' I know, sometimes can be overworked, but when we think of the strength of the departments we have here and imagine bringing them together, the potential for this center is incredibly exciting."[38]

Exciting though the vision was, some eight years elapsed before it became reality. The location at Broadway and Orleans Street turned out to have been the site of two long-lost, 18th and 19th century cemeteries. Some 400 unmarked graves had to be excavated carefully and removed—an archaeological process that took months. Rapid-fire changes in computer technology, a nursing shortage and rising construction costs that necessitated additional fundraising also delayed what became a $184 million project, which ultimately consisted of the Harry and Jeanette Weinberg clinical cancer center and the Bunting ◆ Blaustein Cancer Research Building.[39]

Advances in Education, Research, Outreach

The physical growth of this period was mirrored by Hopkins' continued influence on research, with the School of Medicine beginning its long reign in 1991 as the recipient of the largest amount of National Institutes of Health research funding in the country.[42] Significant initiatives were launched in such fields as cardiology research and in the battle against sickle cell anemia, in medical education with the adoption of a new curriculum in the School of Medicine, in patient care with Hopkins Hospital's first of many designations by *U.S. News & World Report* as the nation's best hospital, and with ongoing outreach to its community.

For example, providing immense technological assistance to the effort to diagnose and treat heart disease was Hopkins' development in 1990 of a new, noninvasive method to examine malfunctioning hearts. Led by Elias Zerhouni, then-director of Hopkins' MRI facilities, researchers devised a technique for improving the diagnosis of heart wall dysfunction by using a tag made with a radio-frequency pulse. The procedure enabled one team to locate a tumor in the heart wall of a newborn, and other clinical uses for it soon were found.[43]

Treating—and one day, perhaps, curing—sickle cell disease (SCD) had long been a key interest of pediatric hematologist George Dover, who joined the faculty in 1977 and became head of the Division of Pediatric Hematology in 1990 (and head of the Department of Pediatrics and the Hopkins Children's Center in 1997). In 1992, Dover and fellow Hopkins pediatrician Saul Brusilow discovered that 4-phenylbutyrate, a drug previously used to treat metabolic disorders, also would likely reduce the number of excruciatingly painful episodes—appropriately called "crises"—endured by the one in 500 African Americans who suffer from SCD. Three years later, Dover and another Hopkins colleague, Samuel Charache, discovered that a cancer drug, hydroxyurea, diminished the pain and lung disease caused by SCD. They and other Hopkins physicians then developed the hydroxyurea regimen that's still in use.[44]

Today, a team of Hopkins hematologists and oncologists headed by Robert Brodsky and Richard Jones is pursuing treatments with bone marrow transplants and cyclophosphamide that actually cure SCD.

Major advances in treating sickle cell disease (SCD) have been achieved by George Dover (above left), who became head of the Division of Pediatric Hematology in 1990 and director of the Department of Pediatrics in 1997, and (left to right) hematologists and oncologists Richard Jones and Robert Brodsky.

In 1990, a new field in cardiology research was opened with the founding of the Ciccarone Center for the Prevention of Heart Disease. Bringing together researchers and clinicians for comprehensive collaboration on preventing and curing heart disease, the center was named for Henry A. "Chic" Ciccarone, a Hopkins lacrosse coach who had died in 1988 at the age of 50 after his third heart attack. The center was the brainchild of Roger Blumenthal, then a fellow (now a professor) in the Division of Cardiology, who had been a statistician for Ciccarone's lacrosse teams. In the ensuing two decades, funds for it would be raised by annual "Heartfest" gatherings, featuring heart-healthy cuisine, celebrity honorees, and rock music provided by "Stevie V and the Heart Attackers," a group of Hopkins cardiologists led by Steve Valenti of Howard County General Hospital. To the left, Hopkins Hospital and Health System President Ron Peterson is shown making a guest appearance on the drums. Since its founding, Ciccarone Center fellows have published dozens of research papers in top journals, expanded its patient outreach to Hopkins' suburban locations, and helped educate practitioners on how to better identify and treat patients at risk of developing heart disease.[41]

Hopkins' first physician in chief, Willliam Osler, is shown in this *circa* 1891 photo in a small study on the third floor of Hopkins Hospital, working intently on his enormously influential textbook, *The Principals and Practice of Medicine*, initially published in 1892 and still in print a century later.

"There is no more difficult art to acquire than the art of observation, and for some men it is quite as difficult to record an observation in brief and plain language."

—William Osler

For more than two decades, The Johns Hopkins Hospital has been named the country's best in annual, nationwide surveys of physicians conducted by *U.S. News & World Report* magazine. In the first such ranking in 1990, Hopkins came in second to the Mayo Clinic. In 1991, when it began its two-decade record as No. 1, Hospital President Robert Heyssel observed wryly, "We were No. 2 last year, and I always figured that was a mistake."

A New Curriculum for the New Century, a New Hospital Accolade

Emblematic of Hopkins' profound and enduring impact on medical education was the 100th anniversary in 1992 of Sir William Osler's landmark textbook, *The Principles and Practice of Medicine*, which marked the centenary with its 22nd edition.[45] Appropriately, in the spirit of Osler's once-revolutionary yet still-relevant guide to the treatment of patients, the School of Medicine sought to reshape its curriculum to train students for medical practice in the 21st century.[46]

Under the leadership of Catherine DeAngelis, associate dean for faculty and academic affairs (and later the first woman editor of the *Journal of the American Medical Association*), and funded by a $2.5 million grant from the Robert Wood Johnson Foundation, faculty and student groups independently devised similar suggestions for the first major revision of the medical school curriculum in 75 years.[47]

The most significant aspect of the curriculum change implemented in 1992 was in the way students would be taught. They would spend less time attending large, lengthy lectures and more time reading and learning on their own, observing Baltimore-area private practitioners, solving problems independently, and discussing their findings with faculty in 10- to 15-member groups.[48]

Instruction in the use of sophisticated, three-dimensional computers and databases for patient care and research was added, as was a new, four-year "Physicians in Society" course. Led by epidemiologist Leon Gordis, associate dean of admissions, it focused on ethics, legal and political issues, finances, regulations, the history of medicine and even the fine arts as they relate to medicine.[49] For the first time, in response to the growing need for more primary-care physicians, a

generalist rotation later was added to the students' clinical training, requiring them to spend four weeks working with a family practitioner, general internist or pediatrician. The rotation also paired students with primary care physicians as mentors at Hopkins or other local hospitals or in private practice.[50]

Even as it strove to update its curriculum, the School of Medicine continued to be a magnet for an ever-increasing number of applicants. In 1993, a near-record 3,500 would-be physicians applied for the school's 120 places. The incoming Class of 1997 contained the most women in Hopkins' history—44 percent—and its portion of underrepresented minority students ranked well above the national average.[51] Both of those benchmarks would improve in the years ahead.

The spirit of innovation and excellence that led to the 1992 curriculum change also earned Hopkins Hospital recognition in 1991 as the best in the nation, according to a *U.S. News & World Report* survey of nearly 1,000 physicians across the country. The doctors were asked to rate hospitals in 15 specialties—and Hopkins excelled in 13 of them: AIDS, cancer, cardiology, endocrinology, eye surgery, gastroenterology, gynecology, neurology, otolaryngology, pediatrics, psychiatry, rheumatology, and urology. The first *U.S. News* hospital ranking had been made in 1990, with Hopkins coming in second to the Mayo Clinic. When Hopkins attained the top spot in 1991, Hospital President Robert Heyssel observed with a grin, "We were No. 2 last year, and I always figured that was a mistake." (Perhaps so, since Hopkins would retain the top ranking well into the 21st century.)[52]

Pediatrician Catharine DeAngelis (above), then-associate dean for faculty and academic affairs, led the 1992 effort to begin the first major revision of the medical school curriculum in 75 years. In October 1999, DeAngelis was named the first woman editor of the *Journal of the American Medical Association*, a post she held until 2011. She returned to Hopkins to develop a multi-disciplinary Center for Professionalism in Medicine and the Related Professions.

1991
August 9: Hopkins cancer researchers Bert Vogelstein and Kenneth Kinzler lead a team of scientists that isolates the gene APC, which, when defective, causes cells in the large intestine to form polyps that may become cancerous. The discovery improves screening of patients for potential colon cancer.

The Johns Hopkins Hospital is named No. 1 in the nation in second annual *U.S. News & World Report* survey of 1,000 physicians nationwide. Hopkins receives a $2.5 million grant from the Robert Wood Johnson Foundation to reform medical education.

Hopkins Hospital begins the creation of electronic patient records for all Hospital services to streamline and improve patient care.

Expanding Community Outreach

Although nationally acclaimed for the specialty treatment within its walls, Hopkins Hospital also undertook substantial initiatives during this period to address the basic health care needs—as well as economic aspirations—of the surrounding community, along with outpatients throughout the region.

Hopkins' Dwight Lassiter (left) chats with a few of the hundreds of Paul Laurence Dunbar High School students who participate in the Dunbar/Hopkins Health Partnership.

Community outreach efforts included such Hopkins-backed programs as the Center for Addiction and Pregnancy, launched in 1991 on the Hopkins Bayview campus. It remains one of the few programs in the country combining drug treatment with comprehensive prenatal and child care services, in addition to providing instruction in addiction relapse prevention, parenting, family planning and domestic violence issues.[53]

Also opened in 1991 was the Office of Community Health (OCH), with the help of a $150,000 commitment from the Hospital and the cooperation of East Baltimore community groups, the Mayor's Office, the city Health Department, and the schools of Medicine and Public Health. The OCH initially was designed

to be a clearinghouse of information about all the health care programs in the area, help groups obtain financial support for health-related programs, and refer groups and individuals to health care providers. In time, representatives from Hopkins' OCH would also become major participants in millions of dollars' worth of community-initiated job-training and other economic development and crime-prevention programs.[54]

A nearly decade-long career-development partnership between Hopkins and nearby Paul Laurence Dunbar High School, a citywide, magnet school offering programs for promising students interested in health-related careers, hit a major milestone in 1990 with the graduation of the first group of top Dunbar

A Hopkins Center for Nursing Research hypertension study was part of a broad initiative to tackle the persistent health problems of East Baltimore's residents. Here, project director Mary Roary (right) greets one of the 300 inner-city men being helped to control his high blood pressure.

students in an advanced college preparatory program developed by the Hospital, the University and the high school. All were headed for college at such schools as Hopkins, the Massachusetts Institute of Technology, Stanford,

Notre Dame, Loyola, Coppin State and Morgan State—while that spring another Dunbar alumnus, Thomas McFarlane, became the first Dunbar graduate to receive his M.D. from Hopkins.[55]

Icons Depart

As the 1990s got under way, those who had led the School of Medicine, some of its academic departments and the Hospital for some two decades began handing the reins over to their successors. The medical leadership changes that had begun in 1989 with Morton Goldberg's assumption of the directorship of the Wilmer Eye Institute continued with the 1991 retirement of Richard Johns, the founding director of the Department of Biomedical Engineering.

Johns, a 1948 graduate of the School of Medicine, is one of those rare individuals for whom excellence in whatever he does seems to come naturally and whose collegiality is contagious. "He is probably the smartest and most innovative person I've ever met," former Dean Richard Ross, a close friend since their days as fellow residents, told *Hopkins Medicine* magazine in 2009. "He can move into any situation and say this is the way we're going to go. He should have been dean of this medical school instead of me."[59]

Trained in physics at the University of Oregon, Johns became an exceptional internist at Hopkins before his passion for mechanical tinkering led him to become a laboratory technician for Samuel Talbot, then head of a Department of Medicine biomedical subdivision where tools for research and clinical practice were designed. There Johns swiftly became proficient in developing devices for making physiological measurements; and beginning in the 1960s, he also helped develop a revolutionary, three-dimensional radiography system. In 1966, he was appointed the first professor of biomedical engineering when Talbot left Hopkins for the University of Alabama; and in 1969, Johns was named director of what a year later would become the new Department of Biomedical Engineering.[60]

Sharing a serene moment aboard their sailboat, Richard Johns (left), founding director of the Department of Biomedical Engineering, and his wife, internationally renowned lung expert Carol Johnson Johns, met when they were Hopkins residents. He developed what is acknowledged to be the country's finest biomedical engineering program; she was a national advocate for women's careers in medicine.

Over the next 21 years, Johns expanded the department to foster research in hearing, speech, cardiovascular control and myocardial mechanics. When Arnall Patz at Wilmer told Johns of his dissatisfaction with the ruby laser that ophthalmologists used for eye surgery, Johns called on scientists at Hopkins' Applied Physics Laboratory to tackle the problem. Ultimately, they developed the helium-argon laser, which soon became the standard tool for performing retinal surgery on diabetic patients.[61]

Because Hopkins' Department of Biomedical Engineering was "born" in the medical school, not in the engineering school as has been the case at most universities; and because Johns himself was a Hopkins-trained physician, he strove to ensure a close, multidisciplinary collaboration with other departments in the School of Medicine. He also played a significant role in expanding research and training programs for Hopkins' undergraduate biomedical engineering majors, medical students and postdoctoral scientists in other fields while establishing what one colleague called "a deep sense of collegiality" in his own department, where decisions were reached through consensus. By the time *U.S. News & World Report* began its ranking of graduate programs in the mid-1990s, biomedical engineering at Hopkins had developed its reputation as the No. 1 program in the country—a distinction it still has.[62]

A deep commitment to Hopkins medicine was central to Johns' private as well as professional life. His wife, Carol Johnson Johns (whom he met when they both were Hopkins residents) became a renowned expert in lung disease, served as an assistant dean in the School of Medicine, and was a founder of the school's Task Force on Women's Academic Careers in Medicine. (From 1979 to 1980, she also was interim president of her undergraduate alma mater, Wellesley College.) "We really understood each other and the obligations of being a faculty member. We were best friends," Johns told *Hopkins Medicine* magazine in 2009, nine years after his wife's death.[63]

1991
September: The Office of Community Health is established to address the specific health problems of residents in the neighborhoods surrounding Hopkins Hospital.

The Center for Addiction and Pregnancy opens at Johns Hopkins Bayview. In future years, it has an indelible impact on the lives of women and children, earning statewide recognition.

1992
Peter Agre discovers aquaporins, for which he will win the Nobel Prize in Chemistry in 2003. In due course, he will discover that every question posed to a Nobel laureate is not seriously scientific: *Hopkins Medical News* reports in 1992 that *Spy* magazine asked Nobel laureates how they eat Oreo cookies. Previous Hopkins Nobel laureates Daniel Nathans and Hamilton Smith respond in typically unique ways.

July 1: James Block, M.D., becomes president of The Johns Hopkins Hospital, succeeding Robert Heyssel.

Neuroscientist and otolaryngologist Murray Sachs became the second director of the Department of Biomedical Engineering in 1991 and continued fostering a close relationship between it and a broad range of clinical departments. His efforts maintained and enhanced Hopkins' exceptional biomedical engineering program.

Murray Sachs, a 1966 Ph.D. graduate in electrical engineering from the Massachusetts Institute of Technology, took over from Johns and maintained and enhanced the distinction of Hopkins biomedical engineering created by his predecessor. Sachs came to Hopkins as an assistant professor of biomedical engineering in 1970—the year the department was created.

At Hopkins, Sachs' groundbreaking analysis, or modeling, of the workings of the cochlear nucleus—the inner ear's nerve system—combined with the work of his former Ph.D. student, Eric Young (now a Hopkins professor), to have a significant impact on the development of cochlear implants. These devices—which act as a kind of bionic ear, providing electronic stimulation to the auditory nerves—now enable many otherwise profoundly deaf individuals to hear.[64]

Becoming head of the department in 1991, Sachs—a professor of neuroscience and otolaryngology–head and neck surgery, as well as biomedical engineering—not only continued his work on how the brain processes sound, but expanded the department's studies into how other complex stimuli, such as visual images, are received, transmitted and comprehended. Believing, as did Johns, that much of the department's strength was due to its origins in the medical school, he continued fostering close interaction between Biomedical Engineering and colleagues in a broad range of clinical departments. Under Sachs, the department would become a leader in the computational modeling of physiological systems; researching mechanisms for delivering drugs directly into the cells of patients; and exploring tissue engineering, or the replacement of damaged human tissues and organs with engineered substitutes, such as artificial skin.[65]

Drive for Diversity

In 1990, the Department of Medicine formed an eight-woman Task Force on Women's Academic Careers in Medicine in response to University findings that female faculty were not being promoted in a timely fashion—and national surveys showing that women were less likely to become department heads.[56]

After questioning faculty and analyzing salary and promotional data, the task reported "major problems" in the department's treatment of its female members. It was a stunning wake-up call that prompted immediate efforts to change the department's practices and profile. Among the initiatives were a women's mentoring network, annual evaluations of the faculty, and a review by the promotions committee of the women's latest achievements to make sure they were on track for appropriate recognition.[57]

Within a few years, the number of women associate professors in Medicine soared from just four to 26 and salary equity was achieved. By 1993, some 44 medical schools in the United States and several in Europe sought the task force's advice on how to improve the status of women on their faculties, and the group's methods became a model for medical schools around the country. Much remained to be done, but a significant breakthrough had been achieved.[58]

In the forefront of efforts to advance the role of women in Hopkins medicine have been, left to right, Janice Clements, vice dean for faculty; Susan MacDonald, associate director of the Department of Medicine; Cynthia Wolberger, a Howard Hughes Medical Institute investigator; and Joan Bathon, director of the Johns Hopkins Arthritis Center.

A Trio of Key Administrative Changes

The first in the trio of non-clinical administrative guard changes that Richard Ross had suggested would facilitate a major alteration in the organization of the medical institutions occurred early in January 1990 when William Richardson, executive vice president and provost of Pennsylvania State University—and a recognized expert on the economics of health care—was chosen to succeed Steven Muller as president of The Johns Hopkins University, effective that July 1.[66]

July 1 also was the date for Ross' retirement to take effect. His successor was Michael Johns, a 1969 graduate of the University of Michigan medical school who had been at the University of Virginia before becoming head of Hopkins' Department of Otolaryngology–Head and Neck Surgery in 1984. Beginning in 1986, Johns simultaneously served as president of the faculty's Clinical Practice Association.[67]

Ross's Legacy

As *Hopkins Medical News*, the School of Medicine alumni magazine that Richard Ross had founded in 1976, observed in a headline, he would be "A Tough Act to Follow." During his 15 years as dean—a tenure second only in length to that of Alan Mason Chesney (1929–1953)—Ross was credited with an impressive list of accomplishments.

Looking back two decades after his retirement, Ross says the "great developments of my deanship were the growth of research and the growth of the clinical practice." When he became dean in 1975, he formed a committee of himself, microbiologist Daniel Nathans and surgeon George Zuidema that created the Clinical Practice Association (CPA) to increase patient volume as a source of funds. "Ten percent of the earnings came off the top to the dean. That's called the dean's tax, and it was a tremendous—and probably the most important—advantage I had over my predecessors, because I had that stream of unrestricted funds," Ross says.[68]

With that money—in Ross's day amounting to about $500,000 —he "did everything," from creating the Dean's Fund to underwrite research by young faculty members to recruiting top physicians to join the Hopkins faculty. "It made it possible to recruit people, because you need a dowry," Ross says.

Some faculty members objected that the CPA's efforts to increase patient volume "distracted them from their academic duties" and required them to spend more time making money, Ross recalls. "But I used to say, look, I don't make the rules. That's the way it is out in the marketplace, and if we want the best here at Hopkins, we have to match the marketplace."[69]

He encouraged more broadly educated young people to enter medicine, in part by dropping the Medical College Admissions Test (MCAT) requirement for admission to Hopkins, expanding the enrollment of minority students, and

■ Actor, author, screenwriter and journalist Evan Handler was treated successfully at Hopkins for acute myeloid leukemia, often considered incurable. He began his four-year battle against the illness at a well-known New York medical center, but switched to Hopkins when the disease recurred. In a 1993 one-man show, "Time on Fire," he praised Hopkins' physicians, support personnel and the fact that "the wheels on the IV carts all rolled really well there."

Richard S. Ross' 15-year tenure as dean of the School of Medicine (1975–1990) was the second longest in its history, and his accomplishments made him, as one publication put it, "A Tough Act to Follow."

Keeping a sharp eye on the progress of the research building that will bear his name is Medical School Dean Richard Ross (second from the left, pointing); along with Fred Reckenwald (far left), project manager, Whiting-Turner construction company, David Blake, associate dean for administration and planning, School of Medicine, Richard Grossi, assistant dean for administration, and Victor Litucsy, principal, Hansen Lind Meyer Inc., architects.

adopting a flex-med program with greater study options, such as a year of foreign study before entering medical school. He also strongly defended the right of private medical schools to select their own students, vigorously opposing a federal government effort to compel U.S. schools to admit underqualified third-year American medical students who had received their initial training overseas.[70]

In addition, under Ross, significant rebuilding of the medical school improved conditions for patients, students and researchers. Major construction projects were undertaken, including (in just a partial list) the Preclinical Teaching Building, a new Hunterian Building, the Denton A. Cooley Recreation Center for students and faculty, the Ross Research Building, and the A. McGehee Harvey Building, an administrative and lecture facility.[71]

Ross also oversaw establishment of 29 new endowed professorships for senior faculty and an exponential increase in federal biomedical research funding.[72]

Focusing intently on improving the School of Medicine's reputation, Ross displayed a penchant for what later would be called marketing—initially over the objections of more tradition-minded trustees and faculty.

For example, Ross was instrumental in the 1982 launch of *Hopkins Medical News*, a publication for the alumni, faculty, trustees and benefactors of the School of Medicine.[73] Renamed simply *Hopkins Medicine* in 2004, it became a glossy-covered publication that went on to win numerous awards under successive editors Terry Fortunato, Lisa Schroepfer, Janet Farrar Worthington, Edith Nichols and Sue DePasquale.[74]

"The Dean's number one job is to recruit people and to make sure that the people heading the departments are of the proper mold," Ross says. "I used to say that Halsted, Osler, Welch and Kelly recruited a second layer of faculty, and the second layer recruited the third layer, and we're now about the fourth layer…. When I came here [in 1947], there were a lot of older men who had known Osler, so I had contact with people who had contact with Osler. And then people who had contact with me had contact with people who had contact with Osler. So the lineage is clear. I say that my job was to try to do as good a job as they did to perpetuate the spirit of intellectual activity and academic medicine. So I felt that was part of marketing, to increase the image. If you burnish the image, it may be easier to recruit people."[75]

Michael Johns Becomes Dean

In selecting Michael Johns to succeed Ross and become the 12th dean of the School of Medicine, the trustees chose a popular faculty member who had done much to make his Department of Otolaryngology–Head and Neck Surgery one of the nation's finest. He also was considered an effective advocate for the faculty as head of the clinical practice group. An accomplished surgeon, researcher and teacher, he was seen as someone who understood "the careful balance a modern medical school must strike between research, teaching and patient care," said incoming university president, William Richardson.[76]

The Detroit-born son of Lebanese immigrants (whose family name of Marieb had been changed to Johns by an Ellis Island clerk), Johns initially planned to become a priest, but, he wryly explained decades later, "my interest in biology exceeded my interest in celibacy."[77]

Ross had appointed Johns to head Otolaryngology, as well as to be an associate dean in 1986, putting him in charge of the Clinical Practice Association.[78]

During his six years as dean, Johns increased the number of women and minorities who worked in his office, sought to improve Hopkins' contacts with the East Baltimore community, and oversaw implementation of the new curriculum. He created both a technology transfer program to facilitate the successful commercialization of Hopkins' research discoveries and an office of corporate liaison to encourage contacts between Hopkins researchers and industry. He oversaw the raising of more than $235 million in contributions and helped create 26 endowed faculty chairs. The Clinton administration even appointed him to head a committee of medical school chiefs to devise recommendations for the federal government's policies concerning medical schools.[79]

James Block Heads the Hospital

Michael Johns also had a hand in the hiring of James Block, who would succeed Robert Heyssel as Hopkins Hospital's president in 1992. Ultimately, Block and Johns would come to clash so forcefully over the direction that the medical institutions should take that their entire governing structure would be changed in the wake of their conflict.[80]

As noted by cardiologist John A. Kastor, a professor of medicine at the University of Maryland, in his 2004 book about academic medical centers, *Governance of Teaching Hospitals: Turmoil at Penn and Hopkins*, Johns met Block at a 1991 retreat. Block was then CEO of University Hospitals of Cleveland, the main teaching hospital for Case Western Reserve University School of Medicine. Born in Dayton, Ohio, and a 1966 graduate of the New York University School of Medicine, he was a pediatrician with a reputation as a forward-looking hospital and medical services administrator. Block had been president of the nine-member Rochester Area Hospitals Corporation from 1979 to 1985, when he moved to Cleveland. At the retreat, Block gave a speech that impressed Johns, who subsequently recommended that he be interviewed for Heyssel's position. Block similarly impressed the Hospital trustee search committee and was chosen to succeed Heyssel, beginning July 1, 1992.[81]

Left to right: James Block, appointed Hopkins Hospital and Health System President in 1992; John Stobo, director of the Department of Medicine since 1985; and Michael Johns, dean of the School of Medicine since 1990, were said to be forming a "seamless partnership" by a writer for the in-house publication, *Dome*. The relationship between Block and Johns eventually would come apart at the seams, leading to a complete restructuring of the governance of the Hospital and School of Medicine.

1992
A new School of Medicine curriculum is adopted to equip students for the practice of 21st century medicine. It is described as "the first major revision" in the curriculum "in 75 years." Members of the Class of 1996 will be the first to study under it. It is characterized as "generalist friendly," since for the first time in its history, the School of Medicine adds a required generalist rotation to medical students' clinical training.

The School of Medicine becomes the No. 1 recipient of federal research grants from the National Institutes of Health.

November: Plans for a new $120 million cancer center are announced.

The Smilow Center for Marfan Research is founded.

Geriatrician Linda Fried launches a study to find out why aging people go downhill physically. For the next decade, she will send researchers into the homes of 7,000 older men and women to study the causes of frailty.

William Osler's *The Principles and Practice of Medicine* marks its centennial with its 22nd edition.

1993

May: Bert Vogelstein and Kenneth Kinzler head a team of Hopkins researchers who report isolating the gene that causes colon cancer, in collaboration with researchers at the NIH and the University of Helsinki. Vogelstein initially called their effort "Mission Impossible."

June: The centennial of the School of Medicine is celebrated. First Lady Hillary Clinton visits on June 10 to help mark the milestone and speaks to an overflow crowd in the school's Turner Auditorium, calling Hopkins "a beacon in this region and in the nation." On October 28, President Clinton joins her on a subsequent visit, with Clinton giving a speech on the Homewood campus about his health care reform efforts. He also praises Hopkins, saying: "I want Americans all over this country who look to the Johns Hopkins Medical School ... [to] know that this medical center is a shining beacon of everything that is best about our health care."

The School of Medicine launches "The Physician and Society" course, part of the revised first-year curriculum, to "give tomorrow's physicians a new perspective on their roles and responsibilities in society." The course's co-directors are Henry Seidel, professor of pediatrics, and Leon Gordis, professor and head of epidemiology in the School of Public Health and also a professor of pediatrics in SOM.

Applications for the School of Medicine are booming, with a near-record 3,500 applications for schools' 120 places. The incoming Class of 1997 contains the most women in Hopkins' history (44%), and its portion of underrepresented minority students ranks well above the national average.

Hopkins' Nobel Prize-winning microbiologist Daniel Nathans receives the National Medal of Science, the country's highest scientific honor.

Heyssel's Legacy

At the time of his retirement, Robert Heyssel was hailed as a "visionary" hospital leader—an accolade he scorned. "It really didn't take a visionary person to see that managed care and HMOs were coming; and it didn't take anyone who's visionary to see that ambulatory care was going to be a bigger part of the scene, that cost pressures were going to accelerate," he told *Hopkins Medical News*. "What somebody in a place like this has to do is begin adjusting the institution to what you know is coming."[82]

Nevertheless, *BusinessWeek* magazine had named Heyssel one of the nation's top five hospital executives in 1990. His 24 years at Hopkins—with all but two years of the preceding two decades as head of the Hospital and, since 1986, also CEO of the Johns Hopkins Health System—were replete with achievements.[83]

When he became head of the Hospital in 1972, it was running a $1.2 million deficit and had many badly deteriorating buildings. Hospital trustees estimated that renovations and new buildings would cost $100 million, so Heyssel had to ensure that Hopkins did something it had never done before: make money. For eight decades, due to its commitment to founder Johns Hopkins' mandate to care for those who were unable to pay, the Hospital always had operated in the red. Its annual deficits had been covered by financial gifts from trustees and occasional dipping into the endowment.[84]

In part by citing the substantial number of impoverished patients Hopkins treated without compensation, Heyssel managed to persuade Maryland's Health Services Cost Review Commission (HSCRC), the agency that sets hospital charges in the state, to increase the fees Hopkins could charge those able to pay. He reduced the Hospital's staff and improved efficiency by devising an unprecedented system of decentralized management under which the heads of the clinical departments took control of their own budgets—an innovation that subsequently became a case study in management at the Harvard Business School. He searched for waste and cut costs wherever he could—saving nearly $25,000 alone by adopting the use of disposable eating utensils. By 1977, the Hospital was making money. He also launched the careers of a number of young hospital administrators, including Martin Diamond, who went on to become CEO of Mt. Zion Hospital in San Francisco; Ronald R. Peterson, his eventual successor at Hopkins; and Steven Lipstein, currently president and CEO of BJC HealthCare in St. Louis. Emblematic of his interest in mentoring, Heyssel created a formal, two-year administrative fellowship program in the late 1980s that continues to this day.[85]

Heyssel was the guiding force in Hopkins' acquisition of the old Baltimore City Hospitals in 1984, as well as the former U.S. Public Health Service hospital across from the Hopkins Homewood campus, and the nearby North Charles Hospital, a community facility. (Although the latter two acquisitions eventually led to hospital closures, they were harbingers of successful hospital mergers in the 1990s and early 2000s.) In addition, Heyssel became the first chief of The Johns Hopkins Health System (JHHS) created by the Hospital trustees at his instigation in 1986 to oversee Hopkins Hospital and Francis Scott Key Medical Center (now Johns Hopkins Bayview Medical Center). Over the next two decades, JHHS would come to control more than 20 health care corporations and facilities.[86]

Heyssel also was willing to borrow money—an action previous administrators had eschewed—to launch the building projects needed to replace some of the "deplorable" clinical facilities he inherited at Hopkins Hospital. Because the Hospital made a substantial financial commitment to the ensuing construction boom, Heyssel could share the credit with Richard

Ross for the new buildings that went up, as well as the updating of existing structures.[87]

Although Ross and Heyssel often battled over issues affecting the Hospital and the School of Medicine, they had enormous respect for each other's institution and a fervent commitment to the Hopkins medical institutions. Edward Halle, senior vice president of the Hospital under Heyssel, "used to say that Dick Ross and Bob Heyssel didn't see eye to eye, but they got the job done," Ross recalls. [88]

"Dick was Harvard-trained, rather distinguished, a prim-and-proper kind of guy, buttoned-up," says Ron Peterson. "Bob Heyssel was more gruff, barrel-chested, a big, tough guy on the exterior [who] had a heart of gold if you got to know him, but he had this exterior persona that was much tougher and he was just a lot different in his approach to dealing with people. Dick was a gentler kind of a guy in terms of how he conducted himself as the dean.[89]

"Although they sometimes knocked heads, they always figured out a way to resolve their differences and ultimately make decisions that were in the best interest of these institutions," Peterson says.[90]

The Challenges Ahead

Despite their respective accomplishments, Robert Heyssel and Richard Ross left challenges for their successors to meet and problems for them to resolve.

The Hospital complex in East Baltimore still needed major renovations. It also faced increasing competition from expanding health systems, particularly in the suburbs. Managed care programs and insurance companies kept a tight rein on fees and imposed restrictions on the services for which they would pay, and finances were strained by an ever-growing population of poor patients on Medicaid, which paid physicians even less than Medicare or other insurers. The Health System as a whole lost $4.1 million in the fiscal year before Block became its head.[91]

The School of Medicine also had a substantial deficit when Johns became dean—in part because of the way the University structured construction costs of the new Ross Research Building and even more because of a $20 million, five-year "tax" imposed on it by University President Steven Muller.[92] The University's finances had been severely undermined by a shaky stock market in 1987–1988, prompting the trustees to reject tapping into the school's endowment to pay for Muller's numerous non-medical initiatives. Muller's solution was to impose an overall, "one-time tax" on the revenues that the University's schools and centers had generated for themselves. Because the School of Medicine was the largest generator of revenue, it was assessed the biggest tax. The school was in the midst of paying that $20 million assessment—which amounted to 80 percent of the sum required of all the University's schools and centers—when Johns became dean in 1990.[93]

Although Michael Johns and James Block began their association as dean and hospital president on friendly terms, within a few years their dramatically different personalities, management styles and long-term goals led to an ongoing conflict that ultimately made its way into the press.[94] When that happened, Ross' idea of creating a single head for the medical institutions began to appeal to the University and Hospital trustees—but, as Ross notes, they didn't make the governance change "until the battle became public."[95] Once it did, the trustees undertook the most radical reinvention of Hopkins medicine since its founding.

Robert Heyssel (1928–2001), head of the Hopkins Hospital from 1972 to 1992, scoffed at being called a "visionary" hospital leader, but during his 20-year tenure he guided the Hospital and Health System through turbulent decades of change, substantially rebuilt and expanded the Hospital's East Baltimore campus, forged partnerships with corporations and the East Baltimore community, and earned national acclaim as one of the country's top health-service executives. Upon his retirement, the self-described "country boy" from Missouri was lauded by Hopkins officials and trustees as the "chief architect of a sophisticated . . . health care organization that is a national model," and as a man whose "booming laugh fills the duck-hunting lodge and whose Missouri accent and aphorisms—and love of heated discussion—keep the party going far into the night."

1994–1998:
Overcoming Turmoil and Addressing Challenges

Between January 6 and January 8, 1996, a mammoth blizzard buried Baltimore under nearly two feet of snow. At the height of the storm, 40-mile-an-hour winds reduced visibility to near zero. Occasional thunder and lightning flashes heightened the drama of the nor'easter, which blanketed most of the East Coast.[1]

No matter how badly the blizzard walloped Baltimore, it didn't faze The Johns Hopkins Hospital, which stayed open throughout the storm and its immediate aftermath. Even as most of Baltimore was snowed in, close to 400 Hopkins Hospital employees stayed on the job. Working 12- to 15-hour shifts, they grabbed what sleep they could in conference rooms and offices, and on portable cots, foam mattresses and examination room recliners—all to ensure that patients were cared for without missing a beat.[2]

The Hopkins community's reaction to the Blizzard of '96 was emblematic of its response to the extraordinarily stormy, five-year period between 1994 and 1998. Its entire governing structure first was buffeted by unprecedented interpersonal turmoil, then completely reorganized. Its finances were severely strained, requiring an equally unprecedented $55 million in budget cutbacks to address unsustainable growth in costs, all while competition from suburban hospitals and managed care groups intensified.[3]

Yet despite the upheaval and challenges, the Hopkins medical institutions continued making important clinical and research advances and refurbished and revitalized some aging facilities while opening new ones. Hopkins attracted ever-higher numbers of top applicants to its medical school, improved its finances, expanded its reach into the counties surrounding Baltimore—and the Hospital continued to be acclaimed as the best in the nation in *U.S. News & World Report*'s annual ranking.[4]

The Hopkins community's reaction to the Blizzard of '96 was emblematic of its response to the extraordinarily stormy, five-year period between 1994 and 1998.

1994

January: The Department of Emergency Medicine is created with Gabor Kelen, M.D. named as its head and the first professor of emergency medicine in Hopkins' history.

Hopkins Hospital works to improve its surrounding neighborhoods by collaborating with the newly incorporated Historic East Baltimore Community Action Coalition and nearly 100 community programs, as well as contributes $2 million to initiate new community outreach efforts to revitalize the area around Hopkins Hospital.

April 16: The Johns Hopkins Bayview Medical Center becomes the new name for the Eastern Avenue campus once known as Baltimore City Hospitals and then as the Francis Scott Key Medical Center; the new six-story Francis Scott Key Pavilion opens for inpatient care.

Retiring School of Medicine Dean Richard Ross (left) was pleased by the appointment of Michael Johns (right) as his successor in 1991, praising him as "capable of dealing well with people in all parts of this organization, because he has intelligence coupled with a genuine understanding of people and a good sense of humor." Ross had recruited Johns from the University of Virginia in 1984 to become director of the Department of Otolaryngology–Head and Neck Surgery, which under his leadership was perhaps the most successful department of its kind in the country. Ross also had appointed Johns head of the Clinical Practice Association in 1986. His leadership of it earned admiration from other clinical directors.

"I've never been a believer in unorthodox treatments; my idea of alternative medicine is a doctor who didn't go to Johns Hopkins."

—Calvin Trillin, author and essayist for *The New Yorker*, in his 1998 book, *Family Man*

The Seeds of Discord

As Hopkins maintained and extended its record of achievements between 1994 and 1998, those involved closely in running Hopkins Hospital and School of Medicine during that period have their memories of it forever colored by the epic conflict that arose because of the differing personalities and management styles of Michael Johns, the medical school's dean from 1991 to 1996, and James Block, the Hospital and Health System president from 1992 to 1996.

Both immensely capable, accomplished men, Block and Johns ended up being unable to work together under the organizational structure that long had governed their two institutions.

"To the credit of the trustees, when they discovered that there was both a structural and a [personal] chemistry problem between the dean and the Hospital president, they stepped in and did in retrospect what I think boards should do and managed to develop a structural realignment that has served us extremely well," says Ronald R. Peterson, who succeeded Block as head of the Hospital and Health System.[5]

For three years or so, however, the situation had been grim.

Although the Hospital and the medical school remained separate corporate entities, conflicts between the two always seemed near to the surface because their activities were so intertwined that they required close collaboration. Under Johns Hopkins' will, the Hospital was established as a stand-alone corporation with a board of trustees, while the University, of which the School of Medicine was a part, had its own corporate charter and trustee board.[6]

When Mr. Hopkins donated the 13-acre site for the Hospital to its board of trustees less than a year before his death in 1873, he wrote that it was his "wish and purpose" that the Hospital "should ultimately form a part of the medical school of that university for which I have made ample provision by my will." That was a passage Richard Ross

(and some of his predecessors as School of Medicine dean) liked to cite when the issue arose over whether the School of Medicine or Hospital was top dog, as Robert Heyssel would recall with amusement.[7]

The issue had prompted disagreements of varying intensity from the beginning of Hopkins' history. In 1889—right after the Hospital had opened, but four years before the School of Medicine admitted its first class in 1893—the University's first president, Daniel Coit Gilman, opposed a Hospital plan to offer postgraduate courses to physicians. Gilman wrote the Hospital trustees to remind them that "all that belongs to medical instruction should be under the control of the University; all that belongs to the care of the sick and suffering … belongs to the Hospital." The Hospital ignored Gilman and offered the courses until the arrival of medical students made teaching them more important.[8]

Every few decades, beginning in the 1920s, various presidents of the University and some trustees proposed revising the relationship that the Hopkins medical institutions had with each other and the University. Few proposals were adopted, and none had an impact on the inherent tension between the Hospital and medical school.[9]

James Block, president of the Johns Hopkins Hospital and Health System from 1992 to 1996, was in many ways a brilliant and visionary administrator, but he couldn't adjust to the collegial, teamwork style of Hopkins' culture and was "like a fish out of water," clashing repeatedly with Dean Michael Johns.

Clash at the Top

Although Heyssel was a far-sighted, tough, opinionated and controlling (some thought bullying) Hospital leader, he adapted ideas from a 1970s board-of-trustees proposal and a separate, faculty-prepared study that gave considerable management responsibility to the Hospital's clinical functional units' directors, who as faculty also head the equivalent academic department in the medical school. (Hopkins' department heads officially are called "directors," not "chairs," as in many other medical schools; the heads of departmental divisions are called "chiefs.")[10]

Internal rift is testing Hopkins' will

Medical school, hospital chiefs' clash runs deep

By John Fairhall and David Folkenflik
Sun Staff Writers

Dr. Michael E. Johns is dean of the Hopkins School of Medicine.

Dr. James A. Block is CEO of Hopkins Hospital and Health System.

While non-physician managers oversee the departmental budgets, the directors have immense control over their departments' clinical and research affairs. They consider the School of Medicine dean to be their main leader.[11]

Heyssel's "department-centric" management system worked well in the Hopkins Hospital/School of Medicine culture that—while immensely collegial—also prizes decentralization. It is a culture in which decisions tend to be made slowly, deliberately, following multiple meetings, discussions and melding of ideas. It was not the way James Block liked to decide things—and that would be his undoing.[12]

"The problem, in retrospect, was that Dr. Block, although a perfectly brilliant guy and in some respects a visionary, found this culture to be a bit foreign and did not choose to figure out how to embrace the culture at Hopkins, so he was like a fish out

1994
The School of Medicine's Women's Leadership Council is created.

Johns Hopkins at Green Spring Station opens. Its $12 million, four-story pavilion, establishes a "suburban outpost" for Hopkins Medicine. In its first year of operation, it serves 100,000 patients, 30 percent of whom are new to Hopkins.

For the first time in the School of Medicine's 101-year history, more women than men report for the freshman class, which began studies on Sept. 6. In 1993, 43% of freshman class were women; in 1994, 53% are women.

1994
Hopkins Hospital's 200th heart transplant is performed in December.

December 28: William Richardson announces he is leaving the Hopkins presidency to become head of the Kellogg Foundation. Daniel Nathans, Nobel laureate in the School of Medicine, is named interim president.

1995
February: A group of 10 trustees, with representatives from the University, the Hospital and the School of Medicine, begin an intense, four-month study of the institutions' governing structure and on June 19 announce creation of a revised "Office of Johns Hopkins Medicine," designed to resolve conflict between the School of Medicine and the Health System.

The Johns Hopkins University launches a $900 million fundraising campaign.

The Weinberg Foundation gives the Hopkins Hospital a $20 million gift for its new, comprehensive cancer center.

William Richardson, president of The Johns Hopkins University from 1990 to 1995, tried unsuccessfully to resolve the ongoing conflict between School of Medicine Dean Michael Johns and Hospital President James Block. He left Hopkins to become head of the W.K. Kellogg Foundation, where he once had been a fellow. He felt it was a job he couldn't decline; and although he found the Block-Johns conflict "an irritant," he insisted it "didn't drive me away" from Hopkins.

of water," reflected Ron Peterson in a 2009 interview. Peterson was president of Johns Hopkins Bayview Medical Center when Block asked him to assume a second job as chief operating officer of the Hopkins Health System in 1995.[13]

The culture "is not about top-down management," Peterson says. In the Hopkins Hospital president's office as he runs it, "it's about facilitating for those who are involved directly in the provision of patient care.

"What we do is help to create the environment that enables the great, bright clinicians and scientists and staff to carry on their important work each and every day," he explained. "And I would not underestimate the term *teamwork*. We take a great deal of pride in assembling people who are willing to work as teams of people, and I think long gone are the days when it's about any one or two people."[14]

That was not really James Block's style, according to many observers. He might engage in consultations, but preferred to act swiftly on his plans. Even if a group of University, Hospital and School of Medicine leaders postponed action on a proposal, "Jim simply would proceed as he wished," recalled former University president William Richardson.[15]

A number of leading clinical faculty members admired Block's dynamism. Others were unhappy with him, voicing concern that he gave too little attention to Hopkins' commitment to research and development of new medical procedures.[16] They worried that his entrepreneurial focus diluted the faculty's standing and would harm the medical school. It was said by some that he "understood the business of medicine. What he didn't understand was the culture of a research-intensive place like Hopkins."[17]

Michael Johns came to distrust him. As the animosity between the Hospital president and School of Medicine dean intensified, they often were not on speaking terms. The Hospital's influential board grew concerned, and some members of the board and of the faculty formed factions: "The Block People" versus "The Johns People."[18] Others on the faculty simply concentrated intently on doing what Hopkins does best—patient care, teaching and research.

"I was a junior faculty member then, and when you first start out, you kind of stay away from all the political stuff that goes on, although it does affect us," recalls oncologist, cell biologist and pathologist Chi Van Dang, who later served as vice dean for research.[19]

Even a senior faculty member well aware of the turmoil, then-director of surgery John Cameron, believes the Block-Johns battles ultimately did not harm either the Hospital or the School of Medicine. As Cameron told author John Kastor, "it was business as usual here. The place went on automatic pilot. The governance battles did not upset the daily routine."[20] Although Hopkins did maintain its high standards—and ranking—throughout this period, other senior faculty members were less sanguine about the toll the conflict was causing, later referring to a "siege mentality" that developed between the leadership of the institutions.[21] Yet one former top faculty member whose career before Hopkins had been at several other major institutions observed,

"the Block versus Johns days was the one aberration in [Hopkins'] collegiality that I can remember, but that's the way it is at Harvard all the time!"[22]

Resolving the Block and Johns Imbroglio

As president of the University, William Richardson was acutely aware of the ongoing battles in East Baltimore and wanted to end them. In June 1994, he began holding weekly meetings with Block and Johns, along with a few other senior University officials, to go over issues.[23]

In late December 1994, however, Richardson resigned to become head of the W. K. Kellogg Foundation, the Michigan-based philanthropy for education and health care where he once had been a fellow. It was, he later said, an offer he "couldn't turn down," adding that although he had found the medical institutions' difficulties "an irritant," the fights there "didn't drive me away."[24]

Until a permanent replacement for Richardson could be found by the University's board of trustees, they asked Daniel Nathans, the Nobel Prize-winning microbiologist, to serve as interim president. Reluctant to leave his laboratory but devoted to Hopkins, Nathans agreed. A soft-spoken, thoughtful man of keen judgment and immense integrity, he had the respect of everyone in Hopkins Medicine and was determined to resolve its governance difficulties.[25]

"Dan was this incredibly self-effacing, modest person who I always thought of as 'Mr. Hopkins,' as a lot of people would say," recalls Edward Miller, who then had only recently joined the School of Medicine faculty as head of the Department of Anesthesiology and Critical Care Medicine.[26]

"His moral compass was as straight as anybody's moral compass could be. He would listen and then he'd give you a piece of advice, and you'd listen very carefully when he spoke, because he just didn't ramble. Whatever he said was pithy and to the point and, I think, in many ways, cut through all of the bullshit that surrounds a lot of stuff and got to the essence of the issue."[27]

With Block and Johns still at loggerheads, news of the difficulties became public when a detailed story about the "fierce conflict" between them appeared in the *Baltimore Sun* on June 15, 1995. Other unflattering front-page stories followed.[28] The heads of both the University and Hospital felt they needed to act.

In 1994, the two boards had created an informal collaboration by forming "Johns Hopkins Medicine," but it did not quell the increasingly fractious situation. In June 1995, the board members decided to make Johns Hopkins Medicine a more structured entity to speak as one voice in responding to the marketplace. They established a new "Office of Johns Hopkins Medicine" with Nathans as its head, as well as an 11-member board made up of University and Hospital trustees, which Nathans would chair. The trustees hoped this would lessen the conflict between the Hospital and School of Medicine leaders—who actually were told to share an office and, when possible, support staff. Block and Johns both attended the subsequent weekly meetings of the Johns Hopkins Medicine board—but never did share an office.[29]

Daniel Nathans (1928–1999), Hopkins' Nobel Prize-winning microbiologist, served as interim president of the University from 1995 to 1996. A man of immense integrity, wisdom and a self-effacing, quiet manner, he was determined to resolve the governance difficulties of the School of Medicine and The Johns Hopkins Hospital and Health System. His leadership was key to the creation of Johns Hopkins Medicine.

■ As part of the 1995 refurbishment of the Hopkins Hospital's deluxe Marburg pavilion, some antique furniture that once belonged to Hopkins' pioneering first surgeon in chief, William Halsted (1852–1922), was used to give the entrance to the patient accommodations "the look and feel of a four-star hotel." Halsted, who had no children, left Hopkins much of the antique furniture that once graced his mammoth, three-story Bolton Hill mansion.

Edward D. Miller in 1994, shortly after he arrived from Columbia University—where he was unhappy with the constant bickering between the hospital and school of medicine dean—to become director of Hopkins' Department of Anesthesiology and Critical Care Medicine. Recruited by Hopkins School of Medicine Dean Michael Johns, with whom he had been a young faculty member at the University of Virginia medical school, Miller soon discovered that discord also reigned between Johns and Hospital President James Block. In 1996, he unexpectedly found himself called upon to begin righting the Hopkins Medicine ship.

"Virtually everyone who spends time here develops an uncommon affection for the place. My hypothesis is that it's due to a peculiar virus, call it JHV, that gets into the chromosomes and there expresses its genes for the rest of one's life."

—Daniel Nathans, 1978 Nobel Prize-winner, 1993 recipient of the National Medal of Science, interim president of The Johns Hopkins University (1995–1996), director of the Department of Molecular Biology and Genetics, senior investigator of the Howard Hughes Medical Institute at Hopkins

Johns earlier had urged that the University appoint a chancellor to oversee the medical, public health and nursing schools in East Baltimore, but the trustees had opted for the Office of Johns Hopkins Medicine instead. Six months later, in December 1995, Johns stunned the School of Medicine by announcing that he was leaving to become executive vice president of Emory University's medical center in Atlanta—essentially accepting a chancellor's position there like the one he had recommended be created at Hopkins.[30]

Emory first had approached Johns the previous August, just two months after the Office of Johns Hopkins Medicine was created. He felt the organizational structure in Atlanta was more suitable to his professional goals—and personal satisfaction. He also knew that as one of the two people most involved in what had become very public acrimony, he never would receive the top position at Hopkins. When he dropped what Ed Miller later called his "bombshell" about leaving, Johns told the medical school department directors that he would remain at Hopkins through June 1996 because his son was going to graduate from the School of Medicine and he wanted to be the dean who placed the graduation hood over his son's commencement gown.[31]

Many in the faculty were deeply saddened by Johns' decision to leave. They also feared that his departure would enable Block to become the dominant power on the East Baltimore campus, to the detriment of basic scientific research and teaching. Early in January 1996, a group of basic science and clinical department directors met and wrote a letter to the University trustees, requesting a meeting. On January 29, 1996, the University board chair, New York City-based banker Morris Offit, a 1957 graduate of Hopkins' Krieger School of Arts and Sciences, and the then-new Hospital board chair, George Bunting, former chief of the Noxell Corporation, joined Nathans in meeting with the top School of Medicine faculty members. Most of the faculty in attendance complained about Block, a few supported him, and the trustee chiefs concluded that a major change in Hopkins governance was essential.[32]

Nathans Taps Miller

Dan Nathans shared the concerns of the School of Medicine faculty. Even though Michael Johns would remain at Hopkins for another six months, Nathans believed he would no longer be engaged fully in protecting the School of Medicine's interests. Nathans thought the School needed a new dean on an interim basis and had his eye on what urologist Patrick Walsh later would call "an outside, inside" man—anesthesiology chief Ed Miller.[33] Having come to Hopkins from Columbia University only two years earlier, Miller was an "outside" man without a long history (or possible opponents) at Hopkins. As an exceptional leader of his department, he was an "inside" man, an administrative chief clearly at ease with the Hopkins culture.

"I got a phone call on a Friday afternoon from Dr. Nathans and he said to me, 'Are you going to be home Sunday morning?'" Miller recalls.[34]

"So, sure enough, Sunday morning comes and it's just terrible. It's a slushy, snowy, February 11th, and Dr. Nathans comes in, and we sit and talk about some of the issues that were involved. And he said, 'You know, Mike is going to stay but his heart is not here any longer, it's at Emory. And I'm very concerned that the system not get tilted away from the School of Medicine and all be in the Hospital, and I'm concerned about Jim Block as a leader. So I think you should be the interim dean. Send me your CV and give me your answer on Tuesday.' That's how it happened."[35]

Miller had not expected this turn of affairs.

"I was dumbfounded. I'd been here a very short period of time. I guess one way to look at it is that I hadn't time to make enemies but had some administrative experience. And so I called my friends around the country and said, should I do this or not? Because, you know, it's a pretty big departure from being a chief of anesthesia—which I really had mastered, to be quite truthful. So all my buddies said you need to do it; it's an opportunity you can't turn down."

He didn't wait until the following Tuesday. On Monday, February 12, 1996, Dan Nathans announced the appointment of Miller as interim dean of the School of Medicine and the University's vice president for medicine, effective March 1. Mike Johns would retain the title of dean but transfer decision-making responsibility to Miller "during an orderly transition of authority," Nathans said.[36]

Nathans observed that at Hopkins, the "ideal" leader was described "as having triple-threat strengths in research, teaching and patient care.

"Ed Miller adds a fourth strength, administration, to which he brings wisdom, decisiveness and generosity of spirit. I am grateful to Ed for accepting this responsibility," he said.[37]

As someone for whom becoming dean of the Johns Hopkins School of Medicine had never been an ambition, Miller appeared to take to the job with surprising smoothness and ease. An impressive figure at 6 feet, 5½ inches, with a thick mane of snow-white hair and a deep voice, Miller has variously been described as personable, friendly, open-minded, unassuming, fair and easy-going—but also as a blunt and no-nonsense leader.[38]

A native of Rochester, New York, where his father was an official of the Eastman Kodak company,[39] Miller is a graduate of Ohio Wesleyan and the University of Rochester medical school. He trained in surgery at University Hospital in Boston, then in anesthesiology at Peter Bent Brigham Hospital and in physiology at Harvard. He was on the faculty at the University of Virginia for 11 years before becoming head of anesthesiology at Columbia in 1986. Nationally known as a leading anesthesiologist, particularly in the fields of vascular smooth muscle relaxation and

1995
May 31: William Richardson leaves the Hopkins presidency to become president and CEO of the W.K. Kellogg Foundation. He is succeeded by Daniel Nathans, Nobel Prize-winning molecular biologist and geneticist. Nathans, a faculty member for 33 years, considered it his obligation to accept the post of interim president (the first such designation in university history) while the trustees searched for Richardson's replacement. "I think it's important to have a smooth transition to keep the momentum going," he said. He did so with exceptional skill. During his brief tenure, Nathans accepted for the university a substantial gift from trustee and alumnus Michael Bloomberg, which kept the then-ongoing Hopkins Initiative on track toward its goal of $900 million by the year 2000.

The Department of Pediatrics launches a unique, annual day-long training program, funded by the Cameron Kravitt Foundation, to train second-year pediatric residents in how to deliver bad news to the parents of critically ill youngsters who have died. The program trains more than 300 residents over the next 15 years and is expanded to other hospitals.

The Berman Bioethics Institute is founded.

■ Pulitzer Prize-winning editorial cartoonist and comic strip artist Doug Marlette (1949–2007) enjoyed inserting Hopkins' name in his comic *Kudzu*. "Whenever I'm doing outrageous [health] claims, I like to attribute them to Hopkins," he said. "I'm just glad Hopkins isn't suing me . . . . Hopkins is in the news; the name rings with authority." Periodically, the comic strip showed Marlette's chocolate-loving parakeet, Doris, listening to the radio or watching TV and hearing such "news" items as: "Research at Johns Hopkins has shown conclusively that everything you enjoy eating is bad for y o u … but everything you can't stand is good for you."

William R. Brody, a physician, electrical engineer, inventor, concert-quality pianist, and director of Hopkins' Department of Radiology from 1987 to 1994, was appointed the University's 13th president in April 1996. He would remove the "interim" from Edward Miller's title as dean of the School of Medicine and appoint him as the first dean/CEO of Johns Hopkins Medicine in January 1997.

1995
May: Ronald R. Peterson, president of the Johns Hopkins Bayview Medical Center, assumes additional responsibilities as executive vice president and chief operating officer of the Johns Hopkins Health System, managing its day-to-day operations. Peterson remains president of Johns Hopkins Bayview. Judy Reitz is appointed vice president for operations integration of the Health System, while continuing as executive vice president and chief operating officer of Hopkins Bayview, which she joined in 1984.

anesthetics and hypertension, he had written more than 150 scholarly research papers, abstracts and textbook chapters before coming to Hopkins. He notes with appropriate irony that when his erstwhile colleague from the University of Virginia, Michael Johns, began recruiting him for Hopkins in 1994, he was eager to leave Columbia because the school of medicine dean there and the president of the Columbia-Presbyterian Medical Center were always fighting.[40]

He did not take the term *interim dean* to mean *figurehead*. He promptly undertook significant initiatives—working with Block in negotiating a partnership between Hopkins and Suburban Hospital, a major community hospital in Bethesda, Maryland, near Washington, D.C., fostering the merger of Johns Hopkins Imaging with three community-based radiology networks, appointing some department directors, completing a Michael Johns initiative to revise the faculty's compensation plan, and forging ahead with a major reorganization of the clinical faculty's practice and billing organization, the Clinical Practice Association. Under Kenneth Wilczek, hired in 1996 as its executive director, the 1,500-member CPA would become one of the finest clinical practice groups in the country.[41]

Johns Hopkins Medicine Solidified

As Miller assumed the interim deanship with enthusiasm, a group of University trustees—responding to the letter and the January meeting they had with the School of Medicine faculty leaders—acted with equal decisiveness in making "Johns Hopkins Medicine" a more substantial, unified entity.[42] A 10-member group, with equal representation from the University, Hospital and Health System boards, worked with what one observer considered "breathtaking" speed to completely overhaul the century-old system by which the School of Medicine and The Johns Hopkins Hospital interacted and operated. Not too long after Miller became interim dean, the committee of board members announced a total revamping of governance. They created a "virtual corporation," not one legally formalized on paper, which was to be overseen by a 40- to 45-member board made up of University and Health System trustees and a 16-member executive committee that would be chaired by the University's president. Johns Hopkins Medicine, as it would officially be called, would be headed by a "dean/CEO"—a medical "czar" whose title emphasized the dual academic and business responsibilities of its holder and the increasingly complex needs of the institution.[43]

On April 8, the University's board ended its 16-month search for a successor to William Richardson and elected William R. Brody to become the University's 13th president. Brody, a physician and electrical engineer with a Ph.D. in engineering, had been director of the Hopkins Department of Radiology from 1987 to 1994 before leaving to become provost of the University of Minnesota's Academic Health Center. As a former School of Medicine faculty member and something of a renaissance man (he not only is a physician but also an inventor with several patents, an excellent writer, a concert-quality pianist, even a pilot), he understood the Hopkins culture well. He wanted to have whoever became the first permanent dean/CEO of Hopkins Medicine to be someone with whom he could work easily—not a person prone to conflict.[44]

Jim Block realized he would not have much chance to become the new dean/CEO, in part because he had spent most of his medical career as an administrator and did not have the academic credentials required of a medical school dean. Early in August 1996, he announced that he was resigning as president of the Hopkins Hospital and Health System.[45] (He subsequently became a private consultant on medical care, particularly well known for his work with the United Nations to eradicate polio and as an authority on palliative care.)[46]

Miller and Peterson: Two at the Helm

A month after Block's resignation, Ronald R. Peterson was named acting president of The Johns Hopkins Hospital and Health System. Already highly regarded as president of Johns Hopkins Bayview Medical Center and as chief operating officer of the Johns Hopkins Health System under Block, Peterson was a popular choice to head the Hospital, where he'd begun his Hopkins career as an administrative resident in 1973. The Hospital trustees removed the "acting" in front of his Hospital presidency title that December, making him the 10th president of Hopkins Hospital—the first non-physician to oversee the Hospital since Muller.[47] In February 1997, the "acting" also was removed from his title as president of the Health System. Jeremy Berg, former head of the Department of Biophysics and Biophysical Chemistry, calls Peterson "a hero," adding, "As much as Block didn't fit into the culture, Ron does. With Ron, you can count on an agreement with a handshake."[48] Peterson continued serving as president of Johns Hopkins Bayview until 1999.[49]

Ed Miller also was eager to have the "acting" removed from his title as medical school dean. He expressed interest in the new dean/CEO position. With the support of a number of department directors and Ron Peterson—as well as an impressive appearance before the trustee search committee—Miller got the job. His appointment as the first dean/CEO of Johns Hopkins Medicine and vice president for medicine of the University was announced on January 14, 1997—and greeted with cheers from faculty members, who gave him standing ovations wherever he went that day.[50]

The era of administrative angst had ended. With Miller and Peterson at the helm of a newly constituted Hopkins Medicine organization, the School of Medicine, the Hospital and the Health System would experience extraordinary expansion, advances and achievements—some of the groundwork for which had been established during the period of upheaval just concluded.

A native of New Brunswick, N.J., Ronald R. Peterson began his association with Johns Hopkins in 1966, when he entered the University as a freshman, hoping to pursue a career as a physician. Family circumstances prevented that, however, so upon graduation in 1970, he became a science and math teacher in the Baltimore City public schools and taught adult education at night. He even supplemented his income by working part-time as a catering firm manager. Yet he never lost his interested in medicine and by 1973 had obtained a master's degree in hospital administration from George Washington University. He leapt at an opportunity to do his administrative residency at Hopkins—and has never left.

He excelled as an administrator for the Phipps Psychiatric Clinic, the Hopkins Cost Improvement Program, and the Children's Medical and Surgical Center, winning national plaudits for his cost-saving skills and one newspaper's praise as "a fiscal surgeon." His masterful leadership of what now is the Johns Hopkins Bayview Medical Center, beginning in 1982, combined with adept service as executive vice president of the Johns Hopkins Health System, beginning in 1995, proved that he was supremely qualified to become president of The Johns Hopkins Hospital and Health System in 1997.

Johns Hopkins at Green Spring Station, as seen today, was Hopkins' first suburban outpatient facility, opened at a northern suburban crossroads in 1994. Beginning with a $12 million, four-story, 75,000-square-foot structure, it was immediately successful, serving 100,000 patients its first year. Its popularity led to expansion, first to an existing, adjacent four-story office complex, then to an entirely new, second pavilion—and thereafter to the creation of similar Johns Hopkins Health Care and Surgery Centers in White Marsh, northeast of the city; Odenton, south of the city along the Baltimore-Washington corridor; and Cedar Lane, to the southwest.

1995
June: A new Johns Hopkins radio news program, Health NewsFeed, is introduced, offering daily 60-second health and medicine updates; it goes worldwide in 1996, as it is distributed to the USA Radio Network's 1,300 stations and by the Voice of America, reaching more than 100 cities in Europe, Africa and Asia.

September 6: At Cal Ripken's 2,131st consecutive game, surpassing the 56-year-old record of Lou Gehrig, a $2 million pledge for the new Cal Ripken/Lou Gehrig Fund to study ALS is announced.

A Gallup poll commissioned by the Health System finds that two-thirds of Americans correctly associate the name "Johns Hopkins" with a medical center or university.

Expanding Hopkins' Reach: Envisioning the Future

After the *Baltimore Sun* broke the story about the conflict between James Block and Michael Johns in June 1995, a subsequent article succinctly described Block's overall vision:

"Dr. Block believes Hopkins' long-term interests lie in building a vast health system encompassing suburban facilities, alliances with community hospitals throughout the region and a network of community doctors who would generate revenue for the system by sending patients to Hopkins specialists and hospitals."[51]

The irony is that Block's goals—as outlined in the newspaper a little more than a year before he resigned—largely were adopted and achieved by Miller and Peterson, but within Hopkins' collegial "teamwork" culture and without diminishing its research or teaching missions. Miller thought Block had many good ideas but did "not always include others" in developing or executing them.[52]

That was an error neither he nor Ron Peterson would make. They also would give equal attention to what the *Sun* described as Michael Johns' and the faculty's goal of hiring "top scientists and equipping their laboratories to make breakthrough discoveries."[53]

Suburban Outposts: Green Spring Station and its Siblings

The general consensus is that Block's signal achievement was the creation in 1994 of Johns Hopkins at Green Spring Station, a $12 million, four-story, 75,000-square-foot outpatient facility built at a northern suburban crossroads.[54]

Johns Hopkins at White Marsh.

Over the strong objections of some senior faculty members who doubted that Hopkins physicians would want to traipse out to the suburbs to see patients, Block borrowed against the Health System's endowment to erect the Green Spring complex, initially keeping vague the scope of his plans for it. He believed establishing a presence in an area where well-insured patients lived was a key to ensuring the survival of the inner-city Hopkins Hospital and that the Baltimore County location would be more convenient for suburban patients who would be willing to go there for follow-up consultations and outpatient or inpatient procedures. He aimed to create a strong link between full-time faculty at the Hospital and referring, community physicians outside the city who possibly would qualify for status as part-time Hopkins faculty members.[55]

Green Spring Station proved an unequivocal success. In its first year, it served 100,000 patients—30 percent of whom were new to Hopkins.[56] By 1997, it was expanding into an adjacent, four-story office building to provide more space for its orthopedics and gynecology and obstetrics patients, as well as to add a comprehensive outpatient rehabilitation facility, a radiation oncology suite and a dialysis center.[57] Eventually, expansion at Green Spring included space in five different buildings—one of which had an ambulatory surgery center. By 2001, it

Johns Hopkins at Odenton.

had become the largest single source of new patients to Hopkins Hospital and School of Medicine, providing nearly $100 million in revenue, according to Gill Wylie, president of Johns Hopkins Medical Management Corp. (formerly Broadway Medical Management Corp.), the for-profit subsidiary of Hopkins Hospital that developed Green Spring and its subsequent siblings.[58]

Green Spring's success prompted plans, announced in 1998, to build a second, northeast suburban outpost in the White Marsh business center. The Hopkins Bayview Physicians faculty practice group had rented 5,000 square feet of space there in 1996 for primary and specialty care physicians. Two years later, more specialty services were added. In addition, other nearby Baltimore County sites in Essex and Perry Hall had small treatment facilities.[59]

The new outpatient center in White Marsh consolidated all of these operations at one site, which opened in 2000. By 2004, a second White Marsh building was opened and a third, comprehensive suburban outpatient center was planned for a location south of Hopkins Hospital in Odenton, Maryland, along the Baltimore-Washington corridor. It opened in 2005. Like Green Spring, White Marsh and Odenton eventually would have ambulatory surgery facilities and become known as "Johns Hopkins Health Care and Surgery Centers."[60]

■ In 1992, *Spy* magazine posed a critical question to a group of Nobel laureates: How do you eat Oreo cookies? Johns Hopkins' Daniel Nathans and Hamilton Smith, co-winners of the 1978 Nobel Prize for Medicine for their discovery of restriction enzymes, responded. Smith: "I love to take them apart. I don't know why.... I like to take one piece off and then the other, with the white filling still on it, [and] I eat that ... then I eat the single side separately...." Nathans said he hadn't eaten Oreos "for a while." Instead, "I eat animal cookies with my granddaughter. I generally bite the head off first, then proceed in a bit more random way to eat the rest."

Howard County General Hospital was born in 1973 as Columbia Hospital, part of a subsequently dissolved joint venture between Hopkins Hospital and University and the Columbia Medical Plan, a managed care group practice for Columbia, Md., a planned community located half-way between Baltimore and Washington, D.C. In March 1998, Hopkins renewed its ties to Howard County General, by then a 233-bed facility, welcoming it into the Hopkins Medicine family of health care centers. Adapting to a growing—and aging—suburban population, HCGH opened a new, four-story Patient Pavilion in July 2009. Housing a large outpatient facility and three floors of all-private inpatient rooms, the $73-million, 100,000-square-foot building almost doubled the hospital's outpatient space and is linked to the original HCGH by a distinctive, curved glass "knuckle" of corridors with floor-to-ceiling windows, seen in the center of this photo.

Adding to the Family: Suburban Hospital Affiliation, Howard County General Hospital Acquisition

Block, believing as did many contemporaries who then ran large hospitals that an academic medical center such as Hopkins needed to own "feeder hospitals" in the community, sought unsuccessfully to acquire Sinai Hospital in northwest Baltimore and to purchase Good Samaritan, another Baltimore community hospital where Hopkins long had based its inpatient rehabilitation program, rheumatology and certain orthopedic services because of a lack of space on the East Baltimore campus. Neither effort panned out.[61]

Block was more successful initiating a "foray into the suburban Washington market," which was how Baltimore's legal and business newspaper, the *Daily Record*, described a $2 million joint business agreement signed in July 1996—a month before Block resigned—between Hopkins and Suburban Hospital in Bethesda, Maryland. Located right outside the nation's capital and directly across the street from the National Institutes of Health's campus, the 238-bed Suburban and its NIH neighbor gave Hopkins a unique "bench to bedside" opportunity to translate findings from the laboratory back to the patient, particularly in cardiovascular disease, while also establishing an important foothold for Hopkins in one of the most prosperous communities in the country. In 2009, Suburban officially became part of the Hopkins Health System.[62]

Once leadership of Hopkins Hospital and School of Medicine had been unified under Johns Hopkins Medicine, efforts to expand Hopkins' reach beyond Baltimore had a smoother path. In March 1998, Howard County General Hospital, a 233-bed facility in Columbia, Maryland, became part of the Health System.[63]

Hopkins actually had a hand in creating Howard County General nearly 30 years earlier. In 1969, spurred by Heyssel and Hopkins pediatrician Henry Seidel, Hopkins Hospital and University joined with the Connecticut General Life Insurance Company to create the Columbia Medical Plan. It was a managed care-style group practice for Columbia, Maryland, a community being developed by James Rouse, a real estate magnate who had become a visionary of urban renewal and community planning. The Columbia Medical Plan built the Columbia

Hospital, the facility that eventually became Howard County General.[64]

The Columbia Health Plan initially proved unsuccessful, and by the mid-1970s, Hopkins had divested itself of its stake in what became the Howard Health System and its hospital. Two decades later, with health maintenance organizations and managed care groups largely determining how health care was paid for, and Howard County General Hospital wanting to grow to meet the needs of its now-burgeoning patient base, the time was right for Hopkins to acquire the hospital.[65]

Realizing that Howard County General couldn't achieve its potential on its own, and determined to gain more for its community than just physical expansion, the hospital's longtime president, Victor Broccolino, requested proposals from area health care systems that might be interested in its acquisition.

"We received 16 proposals," Broccolino recalled nearly a decade later. After several rounds of discussion, the hospital's board of trustees narrowed the field to three prospects—a religious hospital, a for-profit chain and Hopkins. "Then we asked our professional staff and our community for input," Broccolino remembers.[66]

"The two groups made their feelings crystal clear: The vote for Hopkins was unanimous."[67]

The overall deal cost about $140 million, although no cash was exchanged. Hopkins agreed to assume Howard County General's debt and fund its five-year strategic growth plan, as well as its plan to replace aging facilities. Hopkins also provided the funds to establish a community-based foundation to enhance the health of the county's population.[68]

1996
January: School of Medicine Dean Michael Johns announces that he will leave Hopkins to become executive vice president for health affairs and director of the Robert W. Woodruff Health Sciences Center at Emory University in Atlanta.

February 12: Edward Miller is chosen by interim President Daniel Nathans to serve as Acting Dean of the School of Medicine.

February: A 10-member group of Hopkins leaders, with equal representation from the University, Hospital and Health System boards, announces a completely new, unified governance structure for Johns Hopkins Medicine, with one senior executive, a "dean/CEO" at its head.

In July 1996, Suburban Hospital, a 238-bed health center located in Bethesda, Md., just outside of Washington, D.C. and directly across the street from the National Institutes of Health's campus, signed a $2 million joint business agreement with Hopkins Medicine. This arrangement gave Hopkins a unique "bench to bedside" opportunity to translate findings of research scientists to the care of patients. In 2009, Suburban officially became part of the Hopkins Health System.

1996
The Hopkins Hospital is named one of "America's 10 most computer-advanced healthcare facilities" by Healthcare Informatics. The Johns Hopkins Center for Information Services (JMCIS) invests heavily in new computer technology, adopting an $86 million, five-year plan to reorganize and put Hopkins in the forefront of electronic patient record technology. Electronic Patient Records, pioneered in pediatrics in July 1995, now are being deployed in other departments. Stephanie Reel, information services vice president, says the system will have a new "patient-centric approach."

June 24: Martha N. Hill, Ph.D., R.N., a research director and associate professor at The Johns Hopkins University School of Nursing, is named president-elect of the American Heart Association (AHA). With a background in disease prevention and behavioral science, she is the first nurse and the first non-physician ever to hold the position of president. Hill is the fifth Johns Hopkins leader to head the four-million-volunteer association. Previous Hopkins physicians who have been AHA presidents are E. Cowles Andrus, Helen Taussig, Richard Ross and Myron Weisfeldt.

July: Hopkins signs a $2 million joint business agreement with 392-bed Suburban Hospital, making its first foray into the suburban Washington market.

August 23: James Block, M.D., announces that he will resign as president of The Johns Hopkins Hospital, effective September 15.

■ Johns Hopkins' School of Medicine has been mentioned in many motion pictures and television programs—but in the 1994 comedy/murder mystery *Getting In*, attending Johns Hopkins became, as the movie's promotion said, "a matter of life or death." The film's plot involves a young man who is placed sixth on Hopkins' waiting list and tries to bribe those ahead of him to withdraw so he can get in. When they begin to die off mysteriously, he becomes the main suspect. One student who is bumped off was played by Calista Flockhart—the future "Ally McBeal" and wife of Harrison Ford; another of the waiting-list murder victims was played by Matthew Perry, who became a star of "Friends."

Bettering Bedside Manners

Hopkins leaders long had realized that the interaction between physician and patient—once central to medical care—was in danger of being supplanted by the technological advances that seemed to separate doctors from those whom they were striving to cure, as well as the economic pressures that sometimes curtailed the time spent with each ill person. It was a trend Hopkins was determined to buck.[69]

Schooling medical students in the fine art of bedside manners was part of the "Physicians in Society" course added to their first-year curriculum in 1992. Within a few years, new interns were advised at their orientation meeting that their effectiveness as healers would be key to their assessment as young physicians. In 1995, a program was launched to train second-year pediatric residents in how to deliver bad news to the parents of critically ill youngsters who had died. Supported by the Cameron Kravitt Foundation, created by Beverly Kravitt and her husband, Jason H.P. Kravitt, a Hopkins undergraduate alumnus, in honor of their stillborn son, the program would go on to train hundreds of young Hopkins pediatricians and be expanded to Cornell's school of medicine. In addition, formal group training in communications skills was instituted for Hopkins Children's Center interns in 1998.[70]

Bioethics: Facing Medical Care Dilemmas
Providing the finest, most sensitive patient care in complex cases can also require physicians to confront significant ethical conundrums. It is not uncommon for physicians to face these issues regularly, such as when to continue aggressive medical treatment versus the right of a critically or even terminally ill patient—or the patient's family—to either reject or demand it.[71]

As early as 1977, Hopkins internist Philip Wagley, a 1943 graduate of the School of Medicine, began lecturing first-year medical students on death-and-dying issues. In 1995, devoted former patients of Wagley's provided the money to endow a bioethics professorship in his name, and Hopkins appointed Ruth Faden, a Ph.D. in health psychology on the public health faculty, to the chair and named her director of the new Berman Bioethics Institute.[72] Created shortly after Faden's initial appointment when philanthropist Phoebe Berman died and left her estate to the School of Public Health, the Berman Institute is a freestanding entity that reports to the University provost and the deans of the Schools of Medicine, Nursing and Public Health.

Unlike any program of its kind in the country when it was founded, the institute now has a faculty numbering 30 scholars drawn from the School of Medicine, School of Nursing, the Bloomberg School of Public Health, and the Krieger School of Arts and Sciences. It has become one of the largest centers of its kind in the world, with its university-wide program focusing on five key areas: the ethics of clinical practice, biomedical research and discovery, public health ethics and health policy, research ethics, and global health ethics and research.[73]

Pediatrician Jeanne Santoli (center), a 1991 graduate of the School of Medicine, with Audrey Howard and her sons Sean, 8 (left) and Ricky, 7, all Medicaid recipients served by Hopkins' Medicaid managed care organization, Priority Partners, created in 1997, the year this photo was taken.

Expanding Care Locally, Nationally, Worldwide

When Maryland's legislature decreed in 1996 that the majority of the state's Medicaid patients had to be enrolled with managed care organizations to lower the costs of their medical care, Hopkins faced the prospect of losing this traditional patient population. In 1996 alone, The Johns Hopkins Hospital and Johns Hopkins Bayview had treated 12,000 Medicaid inpatients, more than a quarter of the agency's recipients in Baltimore. To protect these patients and remain true to Mr. Hopkins' mandate that his hospital always care for the poor, Hopkins joined in 1997 with the Maryland Community Health System to form its own Medicaid managed care organization, Priority Partners. It was an unusual collaboration between an academic medical center and a community health network.[74]

Priority Partners began operation under a state Medicaid rate structure unfavorable to teaching hospitals. Initially unable to cover the costs of its services and periodically challenged to stay out of the red, Priority Partners nevertheless expanded its care extensively over the next two decades. By the time of its 10th anniversary in 2007, it was providing services to more than 113,000 medical assistance beneficiaries at community health centers throughout Maryland. By 2010, that number had risen to 170,000.[75]

Another Hopkins entity, the Medical Services Corporation, became the largest provider of primary care services in the state by 1996. It was providing primary care to some 88,000 members, including the military services' personnel and their families who belonged to the Department of Defense's Uniformed Services Family Health Plan (USFHP).[76] Subsequently renamed Johns Hopkins Community Physicians (JHCP), the primary care network now has 31 sites across Maryland and Washington, D.C. Headed by physician/pharmacist Steve Kravet, it was providing care for about 250,000 patients throughout the state by 2010, including military families participating in USFHP, as well as Medicaid and fee-for-service patients.[77]

As Medicare, the other arm of the federal government's health insurance operation, became the fastest-growing segment of the health care industry and the largest purchaser of home care—generating an estimated $52 billion in services in 1996, two-thirds of it for non-hospitalized patients—the Johns Hopkins Home Care Group (JHHCG) expanded its own already-full spectrum of services and products.[78]

1996

September 1: William Brody, M.D., Ph.D., provost of the Academic Medical Health Center at the University of Minnesota, a former director of the Department of Radiology at Hopkins and former radiologist in chief at Hopkins Hospital, becomes president of The Johns Hopkins University.

November 4: The new Wilmer Eye Care Pavilion is opened. Outpatient visits have increased to 80,000 annually and surgeries broke the 5,000 annual case mark in 1995.

November: The Johns Hopkins Comprehensive Transplant Center (JHCTC) is created to consolidate and streamline Hopkins' extensive organ transplant services for all solid organ and bone marrow transplants for adults and children. In addition, the kidney transplant program originated at Johns Hopkins Bayview Medical Center is relocated to the East Baltimore campus. Pre- and post-operative care of kidney transplant recipients, including a transplant clinic, will remain at Bayview. The JHCTC establishes Hopkins as one of the largest providers of transplant services in the Mid-Atlantic region. With the largest and most experienced liver, heart and bone marrow transplant programs in Maryland, Hopkins physicians performed more than 300 transplant procedures including 176 bone marrow, 51 liver and 20 heart transplants in 1995.

November: Priority Partners, a Medicaid Managed Care Organization, is established by Johns Hopkins Health Care (JHHC) and the Maryland Community Health System (MCHS) to enroll beneficiaries throughout Maryland. It obtains state approval in May 1997 and begins accepting patients in June 1997.

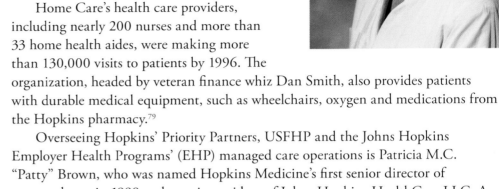

Patricia M.C. "Patty" Brown (right), was named in 1998 as Hopkins Medicine's first senior director of managed care. Now as President of Johns Hopkins HealthCare LLC, the managed care organization owned by the Johns Hopkins Health System and the School of Medicine, she is responsible for the administration of all Johns Hopkins managed care products for more than 300,000 individuals enrolled in self-funded employer, Medicaid, and Department of Defense health benefit plans. In addition, Brown coordinates managed care contracting, payer, and market strategy for all Johns Hopkins Medicine entities. An attorney, she also is senior counsel to the Johns Hopkins Health System, providing legal advice on managed care and regulatory compliance.

Home Care's health care providers, including nearly 200 nurses and more than 33 home health aides, were making more than 130,000 visits to patients by 1996. The organization, headed by veteran finance whiz Dan Smith, also provides patients with durable medical equipment, such as wheelchairs, oxygen and medications from the Hopkins pharmacy.[79]

Overseeing Hopkins' Priority Partners, USFHP and the Johns Hopkins Employer Health Programs' (EHP) managed care operations is Patricia M.C. "Patty" Brown, who was named Hopkins Medicine's first senior director of managed care in 1998 and now is president of Johns Hopkins HealthCare LLC. A former assistant attorney general for Maryland and a past president of the state bar association's health care law section, she joined the Johns Hopkins Health System in 1994 as senior counsel for patient care advice and regulatory issues. She now leads an organization that provides managed-care products for more than 300,000 patients served either by Priority Partners under Medicaid, the Department of Defense's USFHP, or the self-funded employer health benefit plans handled by EHP, which was launched in 1996 and now is headed by Keith Vander Kolk.

Attracting National and International Patients

Hopkins' outreach to patients went beyond those in its immediate community to include those throughout the state, across the country and overseas. Beginning in late 1994, Hopkins started ratcheting up its efforts to attract more foreign patients. Although international patients long had been coming to Hopkins (the King of Siam was one of ophthalmologist William Wilmer's patients), attracting their business had never been a goal.[81]

In the 1995 fiscal year, more than 1,500 patients from the Middle East, South America, the Far East and Europe had come to Hopkins for care—doubling to $12.5 million the previous fiscal year's billings for overseas visitors. The majority of the foreign patients sought treatment from Hopkins specialists in oncology, ophthalmology and orthopedics.

With Hopkins ratcheting up its efforts to attract foreign patients in the mid-1990s, its bustling international staff, assigned to provide four-star service to every patient from overseas, included program managers and translators from all over the globe—many of whom donned their native attire for this 1997 photo.

In 1998, Hopkins' overseas outreach entered a new phase with the signing of an agreement with the government of Singapore to provide care, lead collaborative research and offer medical education in that Southeast Asia city-state. The creation of the Johns Hopkins Singapore International Medical Centre led to the development of clinical facilities there for cancer and other seriously ill patients who came not only from Singapore but throughout the region. Unlike most arrangements between U.S.-based medical centers and foreign facilities, the uniquely comprehensive Hopkins-Singapore alliance followed Hopkins' "center of excellence" philosophy, placing emphasis on research and medical education as well as patient care.[82]

With patients from elsewhere in the U.S. constituting some 15 percent of Hopkins Medicine's business by 1997, a new office—Johns Hopkins USA—was created to help smooth their way through an often-labyrinthine path to appointments and treatment. The new service was also designed to help referring physicians arrange appointments for their patients.[83]

Research Advances

Genetic Breakthroughs Hold Promise

The mid-1990s witnessed advances in uncovering the genetic underpinnings of and potential cures for many illnesses. Significant breakthroughs in genetic research at Hopkins—and elsewhere by Hopkins-trained scientists—promised immense benefits in the future.

Early in the decade, neurologist and neuroscientist Ted Dawson, future scientific director of the Institute for Cell Engineering (ICE) and his wife, fellow neuroscientist Valina Dawson, future founding director of the Neuroregeneration Program in ICE, began work that would lead to discoveries revealing how the molecules nitric oxide (NO) and poly (ADP-ribose) within cells become "messengers of death." The Dawsons' research revealed the key role these molecules play in killing cells and nerves in the brain, heart and other organs during strokes, heart attacks and the progress of such neurogenerative diseases as Parkinson's and Alzheimer's. These seminal findings into the molecular mechanisms of neurodegeneration, enhanced through subsequent research by the Dawsons, led years later to drugs—now being tested in clinical trials—that would target these molecules, block their operation, and prevent the progressive destruction of cells and nerves by neurogenerative illnesses.[87]

In 1995 alone, three other major discoveries were made by Hopkins scientists:

- Molecular geneticist Hamilton Smith, co-winner with Daniel Nathans of the 1978 Nobel Prize for seminal work that led to gene splicing, once again made headlines as part of a team that completed the first sequencing of the entire genome of a free-living organism, a bacterium known as *H. influenzae*. Led by Smith and J. Craig Venter of the Rockville, Md.-based Institute for Genomic Research, the team, which included Hopkins research associate Jean-Francois Tomb and postdoctoral fellow Brian Dougherty, among others, accomplished its breakthrough in an astonishing 13 months. Deemed a "great moment in science" by Nobel laureate James Watson, co-discoverer of the structure of DNA, the achievement was attained by employing technical methods that ultimately hastened the sequencing of the entire human genome.[88]

- Hopkins nephrologist Gregory Germino sequenced the gene responsible for most cases of polycystic kidney disease, or PKD, an inherited disorder that afflicts more than a half-million Americans.[89]

- Pediatric oncologist and geneticist Gregg Semenza, future founding director of the Vascular Program in ICE, reported the purification and isolation of the coding sequences of HIF-1 (hypoxia-inducible factor-1), the protein that switches genes on and off in cells in response to low oxygen levels. Such low oxygen levels can be the result of conditions such as coronary artery disease, a stroke or cancer. HIF plays a prominent role in allowing cancer cells to adapt to low oxygen levels, while also affecting how the body responds to low oxygen levels related to heart attacks, angina and other cardiovascular

Hopkins' reputation for groundbreaking research and its important role in the Baltimore community were memorably combined by the crack of a bat and the cheers of a nationwide audience. On September 6, 1995, Baltimore Orioles shortstop Cal Ripken Jr. surpassed Yankee legend Lou Gehrig's 2,130 consecutive games streak and then presented Hopkins' researchers with a $2 million check to continue their quest to conquer amyotrophic lateral sclerosis (ALS), the neurodegenerative disease that killed Gehrig and has ever since been linked to his name.[84] Receiving the outsized check are neuromuscular nurse Lorn Clawson (left) and neuroresearcher Ralph Kuncl (center).

Creation of the Johns Hopkins ALS Cal Ripken Jr./Lou Gehrig Fund was spearheaded by Orioles owner Peter Angelos. The initial $2 million gift was raised by selling special, on-the-field seats to Ripken's record-breaking game at the Orioles' Camden Yards stadium. Retired from baseball since 2001—when special sky box seats for his final game also were sold to help support Hopkins' ALS research—Ripken continues contributing to the fight against that invariably fatal disease. The research is conducted in what now is known as the Robert Packard Center for ALS Research at Johns Hopkins, headed by neuroscientist Jeffrey Rothstein. Named in 2002 for a San Francisco investment banker who established a $4 million fund to combat ALS as Rothstein was treating him for it, the Packard Center is the only institution in the world dedicated solely to curing ALS.[85]

Cal Ripken and his wife, Kelly, also have been generous supporters of the Johns Hopkins Children's Center; and in 1997, Kelly herself founded the Kelly G. Ripken Program in Hopkins division of endocrinology to provide background information and education for patients with Graves' disease, a thyroid condition for which she has been treated successfully at Hopkins since 1993.[86]

conditions. Semenza's research is paving the way to treatments that could yield drugs that kill cancer cells by cutting off the supply of oxygen a tumor needs to grow, or increase the ability of HIF to ensure that tissues affected by such conditions as diabetes or arterial disease can survive on low oxygen levels. In 2010, Semenza's landmark achievement in opening the field of oxygen biology to molecular analysis earned him Canada's international prize for medical research, the Gairdner Award.[90]

Hopkins and the National Institutes of Health joined forces in 1996 to create a research headquarters for large-scale gene analysis, the Center for Inherited Disease Research (CIDR, pronounced "cider"), housing it on the Johns Hopkins Bayview campus. Beginning with a $21.8 million, five-year contract, CIDR since has become a key $115 million research center now under the direction of David Valle, head of Hopkins' Institute of Genetic Medicine. Supported by 14 NIH institutes, CIDR provides genotyping

In 1995, pediatric oncologist and geneticist **Gregg Semenza** (above) reported the purification and isolation of the protein that switches genes on and off in cells in response to low oxygen levels. That landmark discovery, and his continuing research based on it, is paving the way to development of treatments that could kill cancer cells by cutting off the supply of oxygen they need to grow, or increase the ability of tissues affected by such conditions as diabetes or arterial disease to survive on low oxygen levels. Semenza's achievement in opening the field of oxygen biology to molecular analysis earned him Canada's international prize for medical research, the Gairdner Award, in 2010.

and statistical genetics services for researchers seeking to identify the genes that contribute to human disease.[91]

In 1998, gynecologist, obstetrician and physiologist John Gearhart led the Hopkins research team that was among the first to identify, isolate and cultivate human embryonic germ cells—the primordial cells that develop into tissue for any portion of the human body. It was a pioneering discovery with immense potential for developing medications to treat such afflictions as ALS, diabetes, Parkinson's disease, strokes and spinal cord injuries, as well as the possible laboratory growth of tissues to replenish or replace diseased organs. Over the next decade, Gearhart became one of the world's most eloquent advocates for continuing stem cell research. In 2008, he was named director of the University of Pennsylvania's Institute for Regenerative Medicine.[92]

In addition, during the mid-1990s, physiologist William Guggino began overseeing new clinical trials on gene therapy for cystic fibrosis following the NIH's designation of Hopkins as one of just four institutions nationwide to be a Cystic Fibrosis Gene Therapy Center.[93]

Clinical Innovations

Just as research initiatives during this period led to long-standing achievements, so did the launching of new patient treatment centers—and creation of a brand-new department.

Emergence of Emergency Medicine

Recognizing the growing importance of emergency medicine as a major clinical specialty, Hopkins elevated the Department of Surgery's emergency division to departmental status in January 1994. Gabor Kelen, a widely published researcher best known for his studies of the HIV virus in inner cities, was named the School of Medicine's first professor of emergency medicine and the new department's first—and, to date, only—director. A member of the emergency medical faculty since 1984, the Canadian-born Kelen was known not only for establishing what became the nationally accepted methodology and precautions that health care providers employ in treating HIV patients, but also for innovative additions to the Hopkins medical school curriculum that led to emergency medicine research.[94] In 1998, Kelen collaborated with faculty from Hopkins' Bloomberg School of Public Health to obtain a grant to open a first-of-its-kind national emergency medicine research center that would produce substantive academic studies on the outcomes of emergency care and cost-benefit approaches to it.[95]

With the creation of the Department of Emergency Medicine in 1994, Gabor Kelen (above)—a member of the emergency medicine faculty since 1984—became its first (and, to date, only) director and the School of Medicine's first professor of emergency medicine. Kelen previously had developed novel methods to study HIV and other blood-borne pathogens, conducting research in this area that helped define the extent of the HIV epidemic and led to the adoption of universal precautions. He has been in the forefront of efforts to foster research into emergency medicine. "Controlling chaos," he says, is the essence of emergency medicine.

Transplant Central

The period also saw Hopkins surgeons perform their 200th heart transplant in 1994 and found the Johns Hopkins Comprehensive Transplant Center in 1996.[96]

In 1995, Hopkins physicians performed 20 heart transplants, 176 bone marrow transplants and 51 liver transplants, solidifying its reputation as one of the Mid-Atlantic region's most extensive organ and bone marrow transplant services for adults and children.[97]

To consolidate and streamline its transplant program, the Hospital created the Johns Hopkins Comprehensive Transplant Center. By centralizing the administration for all kidney, pancreas, liver, heart, lung, and bone marrow transplants, Hopkins became one of the few medical centers in the country to offer such a wide array of services in one location.[98]

From its inception, the center continued the innovative research and clinical procedures that had led to development of new organ transplant techniques. Hopkins transplant surgeons already had pioneered such procedures as minimally invasive laparoscopic donor kidney removal and a split liver procedure that increased the availability of organs for donation.[99]

1996

December 17: Ronald R. Peterson is named president of The Johns Hopkins Hospital. He was named acting president of the Johns Hopkins Hospital and Health System in September.

December 23: Judy A. Reitz, Sc.D., who has held numerous clinical and administrative leadership positions at The Johns Hopkins Hospital and Health System and Johns Hopkins Bayview Medical Center over the past 15 years, is named Senior Vice President-Operations for The Johns Hopkins Hospital. She had been Vice President of Operations Integration for the Health System since October 1995.

Henry Brem (above), now director of the Department of Neurosurgery, holds up the dime-size, chemotherapy-saturated porous wafer, known as Gliadel, that he and others developed over a 10-year period that concluded with its 1996 designation by the Food and Drug Administration as the first new major brain cancer therapy in 23 years. Originally limited to use in patients with advanced brain cancer, Gliadel received FDA approval in 2003 to be used on all patients with primary malignant brain cancer. Brem—also a professor of oncology and ophthalmology—has been instrumental in building a research and clinical center widely credited with changing the outlook for brain tumor patients. Raised in New York, the son of Polish immigrants who had survived internment in Nazi concentration camps, Brem was brought up in an atmosphere that nurtured hope. It is a philosophy that guides all of Brem's research and clinical endeavors.

John Cameron (far right), director of the Department of Surgery from 1984 to 2003, is the world's unequaled master of the Whipple operation, the complicated surgical procedure for treating pancreatic cancer. He has performed more of them than any surgeon in the world. This 1997 photo shows him in the midst of a pancreatic cancer surgery being assisted by his son, Andrew, a 1998 graduate of the School of Medicine. Born in the Hopkins Hospital, Andrew Cameron now is an assistant professor of surgery and surgical director of liver transplantation.

Caring for the Body and Mind in Both Clinics and Labs

Although Hopkins had been a primary center for the treatment of breast cancer since William Halsted performed the first radical mastectomies, creation of the Johns Hopkins Breast Center in 1994 brought together all of the Hospital's breast cancer services under a single roof. The center facilitated not only the scheduling of appointments with various specialists and central maintenance of patient records, but patient education, referrals, follow-up care and support.[100]

Providing unusually effective emotional grounding for patients of what now is known as the Johns Hopkins Avon Foundation Breast Center is its Breast Cancer Survivor Volunteers group, founded in 1997 by Lillie Shockney, a nurse, administrative director of the Center, and herself a breast cancer survivor who now is a University Distinguished Service Assistant Professor of Breast Cancer.

As devastating as breast cancer can be, it no longer is as potentially deadly as it was in Halsted's day. The same cannot be said for brain, pancreatic or colon cancer, all of which had high mortality rates that Hopkins specialists did much during this period to reduce.

In 1996, following more than a decade of effort by Hopkins neurosurgical oncologist Henry Brem and colleagues from Hopkins' biomedical engineering department and the Massachusetts Institute of Technology, the Food and Drug Administration gave final approval to the first new major brain cancer therapy in 23 years—a dime-size, porous polymer wafer that could be implanted in the brain to deliver chemotherapy drugs directly to a tumor.[101]

Known as Gliadel, the chemotherapy-saturated wafer now is used routinely in cancer centers worldwide. Although by no means a cure, it has extended the lives of brain cancer patients and, more important, made the once-scorned direct delivery of chemotherapy to the brain a rapidly growing area of medical research.[102]

By 1996, John Cameron, director of the surgery department, also had become world-famous as the unequaled master of the Whipple operation for pancreatic cancer, technically known as a pancreaticoduodenectomy but named for Allen Whipple (1881–1963), the Columbia University surgeon who devised it in the mid-1930s. Cameron performed half of the 150 Whipples at Hopkins in 1996—often operating up to six hours at a time, five times a week, to treat tumors in the pancreas, a large, vital digestive gland; the duodenum (small intestine), and adjacent digestive structures. As late as the 1970s, one in every four pancreatic cancer

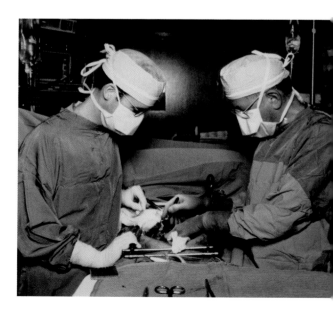

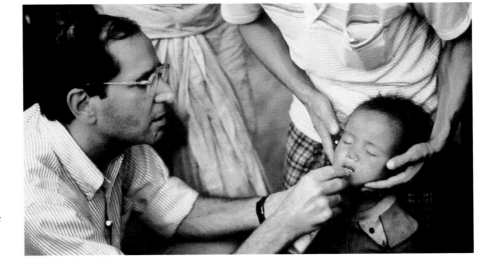

patients died during or after undergoing a Whipple at any hospital; by the mid-1990s, Cameron had so perfected the operation that the mortality rate for it at Hopkins had plummeted to less than 2 percent.[103]

Cameron's reputation as the world's top Whipple surgeon attracted patients with pancreatic cancer from around the globe and inspired such other Hopkins specialists as pathologist Ralph Hruban, geneticist Scott Kern, oncology surgeon Charles Yeo, and oncologist Connie Griffin to concentrate on studying pancreatic cancer and achieve significant scientific breakthroughs in the field.[104]

In 1997, molecular biologist Bert Vogelstein, along with his laboratory's co-director, Kenneth Kinzler, and other colleagues, announced the discovery of a subtle genetic weakness, known as I1307-k, in an otherwise intact DNA sequence that predisposed its bearer to develop colon cancer-causing mutations. The discovery meant that simple genetic tests would be the best way to determine early on if patients were prone to getting colon cancer.[105] Within weeks of the announcement, more than 1,200 people from throughout the country called Hopkins to find out more about the test.

By continually unveiling discoveries that revolutionized the view of cancer by defining it as a genetic disease, Vogelstein, Kinzler and others were achieving advances in cancer studies that some scientists likened to the discovery of polio's cause.[106]

In 1997, working in an old neighborhood supermarket near the Hopkins Hospital that had been converted into a research center, laboratory co-directors Bert Vogelstein (right), Ken Kinzler (center) and Ph.D. candidate Steve Laken (left) discovered the inherited, mutant gene that causes colon cancer and developed a test to detect it. Vogelstein and Kinzler remain among the most productive probers of cancer's mysteries, with their studies cited by other scientists more than any other researchers in the world.

Offices and Honors

As Hopkins researchers and clinicians were making advances, others were being recognized for exceptional past achievements that likely would herald more in the future.

Seminal achievements in genetics and public health earned two Hopkins medical school alumni and faculty members 1997 Albert Lasker Awards, typically called the "American Nobel." Geneticist Victor McKusick received the Lasker Award for Special Achievement in Medical Science for his lifetime of work developing genetics as a medical discipline, while ophthalmologist Alfred Sommer, then-dean of the School of Hygiene and Public Health (now the Bloomberg School of Public Health) and a member of the School of Medicine faculty, received the Lasker Clinical Research Award for pioneering studies which showed that simple doses of vitamin A could prevent blindness and various life-threatening infections in children.[107]

The award citation for McKusick read: "It is rare in the complex world of modern medicine for one man to have essentially founded an entire branch of medicine. It is rarer still when that field comes to occupy such a central place in the mainstream

Ophthalmologist Alfred Sommer, a one-time resident in the Wilmer Eye Institute, is shown here administering an oral high-dose of Vitamin A to an Third World toddler. Sommer spearheaded research which showed that even a mild Vitamin A deficiency could dramatically increase childhood blindness and death from such infectious diseases as measles and diarrhea.

In 1996, Martha Hill (right), a Hopkins-trained registered nurse with a Ph.D. in behavioral science from the Bloomberg School of Public Health, became the first nurse and non-physician elected president of the 4 million-member American Heart Association. Internationally known for developing and testing strategies to improve hypertension care and control among young, urban African American men, Hill became the dean of the School of Nursing in 2002.

Pamela Paulk, vice president for human resources for The Johns Hopkins Hospital and Health System, is among those at Hopkins who volunteer their time to help improve the community around the Hospital. Here she is working on a Habitat for Humanity house restoration project.

of clinical medicine." In honoring Sommer, the Lasker Foundation noted: "Throughout the developing world, in Indonesia and Tanzania, in South Africa and Nepal, in virtually all countries where vitamin A deficiency was once common, millions of children owe their eyesight and their very lives to a visionary, persistent doctor from Baltimore."[108]

Lasker officials were alluding to the fact that Sommer persisted even after his initial findings on the crucial importance of vitamin A had been ignored by scientists who refused to believe that such a far-reaching impact on public health could be achieved simply by a single vitamin. Only later did the medical establishment realize how correct—and life-changing—his discovery had been.[109]

In 1996, Martha Hill, a Hopkins-trained registered nurse with a Ph.D. in behavioral science from the Bloomberg School of Public Health, was elected the first nurse and nonphysician to be president of the 4 million-member American Heart Association. On the Hopkins faculty since 1980 and director of the Center for Nursing Research since 1994, Hill was the fifth Hopkins faculty member named to head the AHA. (E. Cowels Andres, Helen Taussig, Richard Ross and Myron "Mike" Weisfeldt preceded her; Bernadine Healy and Gordon Tomaselli would be later AHA presidents.) Internationally known for developing and testing strategies to improve hypertension care and control among young, urban African American men, she served as the AHA's president from 1997 to 1998 while continuing her research and academic work, which included directing of the postdoctoral program in Hopkins' School of Nursing. She would become dean of the nursing school in 2002.[110]

Investing More in the Community

Ironically, these extraordinary research and clinical achievements were taking place in an institution increasingly surrounded by urban blight.

Hopkins took on that challenge in several ways by providing increased support to community development initiatives, while effectively combating crime in its own neighborhood.

The Hospital and School of Medicine contributed $2 million to the Historic East Baltimore Community Action Coalition, which helped the organization provide nearly 100 community programs to foster revitalization projects in the area around the medical campus. By the end of 1998, Hopkins was directing more than $55 million in grants and in-kind services to dozens of community-based programs related to health, housing and crime prevention.[111]

In February 1994, as crime rose throughout the city and on the campus, Hopkins hired Joseph Coppola, a veteran of more than 24 years with the U.S. Secret Service, to fill the newly created position of chief of corporate security. Within a year of coming on board, Coppola had built an impressive security force from scratch. His determination and the efforts of his force dramatically lowered the crime statistics on campus, where a 38 percent reduction in on-campus thefts was recorded.[112]

Joseph Coppola (center), a 24-year veteran of the U.S. Secret Service, became the first chief of corporate security for Johns Hopkins Medicine in 1994. He quickly built a disciplined, effective security force that reduced on-campus thefts significantly and otherwise enhanced safety around the Hospital, in part by collaborating with community groups, city officials and the Baltimore Police.

Coppola realized, however, that simply reducing crime on the campus or its fringes was not enough. In 1995, he joined with leaders of HEBCAC, city officials and the police department to create the Eastern District Street Crime Unit and its Crime Reduction Task Force. Establishing a five-square-block target area surrounding the medical campus, the unit and task force—manned by both on-duty and volunteer police—succeeded in cutting area crime substantially, despite periodic setbacks. By 2000, the head of the national Association of Campus Law Enforcement Administrators praised as "unique and apparently very successful" what "Hopkins and the local police have put together to deal with criminal activity outside the medical campus."[113]

New Faculty Leadership

Even as the period saw major changes in the upper strata of Hopkins Medicine leadership, new department directors were being named.

John Stobo, known to friends as Jack, head of the Department of Medicine since 1985, relinquished the medicine directorship in 1994 to become head of the newly created Johns Hopkins HealthCare LLC. It had been founded to provide a single entity through which health care insurers could negotiate and carry out contracts with the hospital and providers, and unify Hopkins' clinical centers and programs throughout the state.[114] Within three years, however, Stobo accepted another challenge and left Hopkins in 1997 to become president of the University of Texas Medical Branch in Galveston. After a decade there, he became senior vice president for health sciences and services for the University of California.[115]

Stobo's successor as William Osler Professor of Medicine and director of the Department of Medicine—generally considered one of the premier posts of its kind in the country—was Edward Benz Jr., a renowned hematologist who had been chairman of the Department of Medicine at the University of Pittsburgh. Appointed in 1995, Benz had received international acclaim for research

John (Jack) Stobo, director of the Department of Medicine and physician in chief of Hopkins Hospital since 1985, left that position in 1994 to become the first head of Johns Hopkins HealthCare LLC, created to establish a unified approach to managed care contracting and to provide oversight to the fledgling Hopkins-sponsored health plan that would become the Employer Health Program. As head of medicine, Stobo had spearheaded what many considered a cultural revolution in the medical school, creating the Task Force for Women in Academic Medicine, appointing women faculty to previously all-male academic committees, and pushing for salary equity between men and women faculty.

To advance Hopkins' pediatric research, patient care and medical student education, George Dover became an inspired fundraiser, participating in annual telethons and radiothons in which he has gamely eaten bugs (above), been hit by cream pies and even let his trademark beard be shaved off to encourage donors to contribute generously.[119] Since 1995, the Center also has collaborated with Baltimore-based clothier Jos. A. Bank to create a unique line of men's neckwear—the Miracle Tie Collection—featuring colorful, abstract-like designs based on the molecular structure of children's medications, as well as drawings by Children's Center patients. Celebrities such as actors Will Smith, Matthew Perry and David Schwimmer, supermodel Cindy Crawford, country singers Tim McGraw and Shania Twain, and the ever-faithful Orioles legend Cal Ripken Jr. have been recruited to autograph some of the ties for special auctions[120]

1997
January 15: The "interim" is removed from Dean CEO Ed Miller's title and he is named Hopkins Medicine's first Dean/CEO.

February 15: The "acting" is removed from Ronald Peterson's title as president of The Johns Hopkins Health System, having served as its acting president since September 1996; president of the Hopkins Hospital since December 1996; and as president of The Johns Hopkins Bayview Medical Center since 1984.

An unprecedented $55 million in budget cuts are planned "to keep the institution competitive." Hopkins, on average, costs 30 percent more than suburban hospitals.

* Ultimately, Miller would appoint 31 of the 33 department directors. For details on those appointments, see Appendix A.

accomplishments that included the application of molecular genetics to the study of human diseases, providing essential information on the cloning of genes and determining the mutations in disease.[116]

Although Ed Miller had only been named interim dean of the School of Medicine in February 1996, within four months he made the first of what would be dozens of appointments of department leaders over the next 15 years by naming George Dover, one of the nation's top pediatric researchers, as head of the Department of Pediatrics and pediatrician in chief of the Hopkins Children's Center. Dover, a crinkly-eyed, full-bearded and energetic man with a voice that still bears a trace of his native Louisiana twang, succeeded the much-beloved Frank Oski, who had headed pediatrics for a decade. Oski, who remained on the faculty, would die of prostate cancer only seven months later at the age of 64.[117]

From the outset of his tenure as director of the Children's Center, Dover—best known for his work on sickle cell disease—met head-on the challenge of maintaining the Center's balance between top-flight research, superb teaching and excellent clinical care. He told reporters from the *Baltimore Sun* as early as two months after his appointment that he looked forward to having the opportunity to design what the children's hospital of the future would look like—a goal that more than 15 years later became a reality.[118]

In 2003, Dover announced receiving an anonymous $20 million gift, the largest individual donation in the history of the Children's Center, to support construction of the neonatal intensive care unit (NICU) and a nursery, and endow a fund for the faculty in the new Children's and Maternal Hospital at Johns Hopkins.[121] Five years later, the state-of-the-art, 12-story tower then under construction was named the Charlotte R. Bloomberg Children's Center, in honor of the mother of New York Mayor Michael Bloomberg, a 1964 engineering graduate of Hopkins, a former head of the University's trustees and the University's most generous benefactor ever.[122]

Other department heads among Miller's first administrative appointments were obstetrician Harold Fox, a former colleague from Columbia, named in 1996 as director of gynecology and obstetrics; pioneering glycobiologist Gerald Hart, chosen in 1997 to be director of the Department of Biological Chemistry; and neurologist John Griffin, named in 1998 as director of the Department of Neurology.[123] *

Fox, like Miller, an alumnus of the University of Rochester medical school, had been gynecologist in chief at George Washington University Medical Center prior to his time at Columbia, where he also headed obstetrics and gynecology, before being appointed the Dr. Dorothy Edwards Professor and head of the department at Hopkins.[124]

A well-known researcher who has focused on fetal health, particularly for mothers with HIV, Fox participates actively in the department's maternal-fetal

medicine division, among the nation's finest in caring for high-risk obstetrical patients, while overseeing the growth of other major departmental divisions. Among those he launched were the Division of Female Pelvic Medicine and Reconstructive Surgery, the Division of Gynecologic Specialties, and the Howard A. Kelly Gynecologic Oncology Service—named for the first gynecologist in chief at Hopkins, who was among the first to use radiation to treat cancer.[125]

Gerald Hart, appointed the DeLamar Professor and director of the Department of Biological Chemistry, is another international leader in research and a dedicated teacher known for his skill at the lecture podium. A Ph.D. in developmental biology from Kansas State University, Hart first had joined the Hopkins faculty in 1979. He left in 1993 to become head of biochemistry and molecular genetics at the University of Alabama-Birmingham. He returned to Hopkins when picked to head biological chemistry, one of the School of Medicine's oldest departments, founded in 1908 by John Jacob Abel.[126]

Harold Fox, an acclaimed expert on fetal health and high-risk obstetrical patients, was named director of the Department of Gynecology and Obstetrics in 1996.

Hart was a young faculty member at Hopkins when he discovered in 1983 that many proteins inside the nucleus of cells are often attached to a sugar—a type of carbohydrate called O-GlcNAc—that scientists previously thought was only found outside cells. This discovery and nearly two decades of subsequent studies overseen by Hart on the possible role this sugar plays in the activity of cells could have a profound effect on our understanding of and treatments for cancer, degenerative nerve diseases, diabetes and Alzheimer's.[127]

John Griffin was one of the world's leading experts in peripheral nerve disorders when he was appointed director of the Department of Neurology in 1998. Conducting research into how nerves respond to injury and how they regenerate and recover their function, he became a leading figure in studies of Guillian-Barré syndrome, a disease in which the immune system attacks nerves, leading to a rapidly evolving paralysis of the legs, arms, face and breathing.[128]

Winner of the School of Medicine's Professors Award for Excellence in Teaching, Griffin also served as president of the American Neurological Association from 2003 to 2005. In 2006, he gave up the directorship of neurology because of health problems, but became head of Hopkins' Brain Science Institute the following year. He continued his important research on peripheral nerves, including the roles of growth factors and molecules within the cells that protect the nerves' axons, or fibers, until his death in 2011.[129]

John (Jack) Griffin (1942–2011; left), a supernova in the field of neurology, was among the world's leading experts in peripheral nerve disorders when he was named head of the Department of Neurology in 1998. He later became head of Hopkins' Brain Science Institute, where he and neurologist/neuroscientist Amet Höke (right) used a confocal microscope to study nerve regeneration.

Elias Zerhouni (top right), named director of the Department of Radiology and Radiological Sciences in 1996, also served as executive vice dean, vice dean for clinical affairs, president of the Clinical Practice Association, and interim vice dean for research. "I rely on Elias to be my fireman, the person I could rush into a burning issue and get it resolved," dean/CEO Edward Miller said. Zerhouni left Hopkins in 2002 to become head of the National Institutes of Health.

Christine White (right) had been executive assistant to two previous Hopkins deans—D.A. Henderson in the School of Public Health and Michael Johns in the School of Medicine— when Edward Miller named her assistant dean for medicine in 1997. He added executive assistant to her title in 2007. "I love organizing, managing, figuring out a way to accomplish what people think can't happen," she says.

Karen Haller (above) was director of nursing for the Department of Medicine when Johns Hopkins Hospital President Ronald Peterson and Judy Reitz, senior vice president for operations, chose her to become vice president for nursing and patient care in 1998.

Beefing Up the Dean's Office

Along with the new department directors, Miller also began filling—and in some cases, creating—new administrative posts in the dean's office. Perhaps following the adage that the way to get something done is to give it to a busy person (long a Hopkins mantra), he persuaded several individuals who already had full plates to pile on some more responsibilities. It would not be the last time he tapped them.

In April 1997, Miller appointed Elias Zerhouni, a pioneer in advanced imaging techniques using computer tomography (CT) and magnetic resonance imaging (MRI), as executive vice dean. An Algerian-born physician who began his Hopkins career as a resident in 1975, Zerhouni was named director of the Department of Radiology and Radiological Sciences in February 1996. He had already been chosen by Miller to be vice dean for clinical affairs and president of the Clinical Practice Association, then desperately in need of reorganization. By 1999, Miller gave Zerhouni another job—interim vice dean for research (prompting him to relinquish the CPA presidency).[130]

Zerhouni performed immensely valuable services as an institutional leader until 2002, when he was recruited by President George W. Bush to become director of the National Institutes of Health, a post he held until 2008.[131]

Miller, who remained director of anesthesiology and continued his work in the operating room even after he became dean/CEO, also brought the anesthesiology department's administrator, Steven Thompson, into the dean's office as vice dean for administration in 1997. Like Zerhouni, Thompson proved a master of many trades. In 1998, he became CEO of Johns Hopkins Singapore Clinical Services; then head of Johns Hopkins International LLC, vice president of ambulatory services; then senior vice president of Johns Hopkins Medicine; and then head of Johns Hopkins International again.[132]

Miller also appointed Christine White as assistant dean. A seasoned veteran who had served as executive assistant to two deans—D.A. Henderson at the School of Public Health and Michael Johns in the medical school—White brought organizational, managerial and diplomatic skills to her new post.[133]

As president of Hopkins Hospital and Health System, Ron Peterson also had leadership posts to fill. In 1998, he and Judy Reitz, senior vice president for operations, appointed Karen Haller, a Ph.D. in nursing and director of nursing for the Department of Medicine for the past five years, as vice president for nursing and patient care services. Haller, who also had master's degrees in both nursing and public health, had arrived at Hopkins Hospital a decade earlier as director of nursing for research and education and held an associate professor's appointment in the School of Nursing. She also had won many grants and fellowships from the National Institute for Nursing Research—giving her the "triple-threat" credentials in research, teaching and patient care that Hopkins prizes.[134]

Diversity Strides—in Student Body and on Faculty

Enhancing the diversity of medical students and faculty at Hopkins, as well as improving the status of women faculty, were considered essential goals during this period, and the School of Medicine made advances on both fronts.

In 1994, for the first time in the School of Medicine's 101-year history, more women than men arrived on campus as first-year medical students. Without changing its admission policies or procedures, but clearly capturing the attention of an increasing number of well-qualified women candidates, Hopkins enrolled a freshman class that was 53 percent female, up from 43 percent in 1993.[135]

"Each year brings a greater number of exceptional women candidates," Catherine DeAngelis, then-vice dean for academic affairs and faculty, told *Dome*. "Our admissions committee simply chooses the best students. The process takes months, and no one stops for a gender tally."[136]

Efforts to increase minority students and faculty—African Americans, as well as Hispanics—had begun in earnest in the late 1960s and early 1970s and even served as a model for medical schools elsewhere. Overcoming a decades-long image of Hopkins as an institution uninterested in, even hostile to minority physicians and students proved challenging, but by the mid-1990s, there had been some progress, albeit modest—and more would be made in the future.[137]

By 1995, only 2.2 percent of the full-time faculty was African American—but that was better than the number of African American medical school faculty members at Harvard, Yale, the University of Pennsylvania and Stanford, among others, and only slightly lower than the national 2.4 percent average for all U.S. medical schools. Similarly, Hopkins had the highest percentage of black medical school graduates in the class of 1996—11.4 percent—compared, for example, to Harvard's 9.4 percent, Yale's 9.2 percent and the University of California at San Francisco's 6.5 percent.[138]

More impressive were the strides made in decreasing gender bias in promotion and improving salary equity for women faculty in the Department of Medicine. John Stobo, the department's head from 1985 to 1994, had spearheaded what many considered a cultural revolution in the medical school. Listening closely to geriatrician Linda Fried, who headed a women's faculty group, and Emma Stokes, a special assistant to the dean, Stobo insisted Hopkins had to become more female-friendly if it was to attract top women faculty and students. He not only had created the Task Force for Women in Academic Medicine but began appointing women faculty members to previously all-male academic committees and pushing for salary equity between men and women faculty.[139] So successful was Hopkins' Task Force on Women's Academic Careers in Medicine since its creation in 1990 that it also became a national model. By 1997, some 44 medical schools in the United States and several in Europe had asked for help in implementing a similar initiative on their own campuses. Within the Task Force's first five years, the number of women associate professors in the Department of Medicine had increased from four to 26, 86 percent of the women and 83 percent of the men questioned reported that gender bias had decreased, and from one-half to two-thirds of women faculty reported improvements in the timeliness of promotions. Mentoring of women faculty members also had improved.[140]

Third-year medical student Stephen Nurse-Findlay (above) prepares meticulously for his clinical rotations in 1998. A native of Jamaica, Nurse-Findlay remembered the precise moment he knew he wanted to go to Hopkins. In 1987, he was 16 years old and living in Trinidad when he saw a news report from the U.S. about an unprecedented surgery separating conjoined twins connected at the head. He saw an African American standing at a podium, describing the daring 22-hour operation. It was pediatric neurosurgeon Ben Carson of Hopkins. "I thought, wow, here's a place that lets someone who is brilliant be brilliant. And I thought, I've got to go there. I've got to get with that program." Now a pediatrician, as of this writing he is also a healthcare consultant working to improve the access to health care for women and children worldwide.

1997
A new $2.3 million, 13,000-square-foot Cardiology Care Unit is opened on the fifth floor of the Nelson Building. It is more than four times larger than the old quarters behind the Medical Intensive Care Unit on Osler 7 and is filled within two weeks.

Johns Hopkins USA is created. It is a new service designed to work behind the scenes to smooth the way for out-of-town patients to Hopkins Hospital.

March: Johns Hopkins Medicine launches its Web site, at http://hopkins.med.jhu.edu, offering a comprehensive overview of services, facilities and other resources for patients, health care professionals, the public and the press, with additional information soon to come for prospective students, alumni, and potential business partners.

September 22: Victor McKusick and Alfred Sommer each receive Lasker Awards, the "American Nobels," for their landmark achievements in genetics and public health, respectively.

1997

September: Johns Hopkins Medical Grand Rounds, a venerable, immensely popular presentation of a wide array of clinical topics, goes online for the first time, enabling physicians from Venezuela, Brazil, India, German, Korea, Malaysia, Canada and other nations to log on and learn about the latest in medical challenges and advances.

December: The School of Nursing opens its own building. The $17.8 million, five-story, 92,800-square-foot building had been under construction for two years and is named for philanthropist and University trustee Anne M. Pinkard. Instead of being spread all over the East Baltimore campus in six different buildings, the nursing school finally has a home of its own.

December 31: Benjamin S. Carson, Sr., director of pediatric neurosurgery, completes what is believed to be the first successful separation, without resulting neurologic deficits, of conjoined twins, in a 28-hour operation in South Africa. The infants are breathing on their own, drinking, and eating solid food, and apparently suffering no brain damage. Carson leads the team that separated 11-month-old Zambian boys Luka and Joseph Banda, who were born as type 2 vertical craniopagus twins—joined at the head but facing opposite directions. Carson has done two previous conjoined twin separation surgeries, one in South Africa and one at Hopkins. He credits the latest success to "surgical rehearsals" with a computerized, three-dimensional virtual "workbench" that allowed him to "see" computerized reconstructions of the twins' brains in spectacular detail.

■ Construction of the Harry and Jeanette Weinberg Clinical Cancer Center was delayed for months by the discovery of two long-lost, century-old cemeteries on its 2.3-acre site at Broadway and Orleans Streets. Some 400 previously unknown, unmarked graves had to be excavated carefully and removed before building of the $125 million, seven-story Weinberg Center could proceed. Hopkins paid for the dignified reburial of the remains. Because of additional delays caused by matters other than graveyards, the building, originally scheduled to open in 1996, finally opened for patients in September 2000.

Belt-Tightening, Improving Efficiency, Opening New Buildings

As soon as Ed Miller and Ron Peterson had the "interim" preface removed from their respective titles as dean/CEO of Hopkins Medicine and president of Hopkins Hospital and Health System, they acted swiftly to address a potentially devastating development: the growing gap between the rising cost of care at Hopkins and the lower fees charged by its area competitors.

On average, Hopkins' costs were some 30 percent higher than the nonacademic medical centers in its region, a discrepancy that threatened to price it out of the market for anything other than complex cases. "Payers might be willing to put up with an additional 15 percent to get our quality, but not 30 percent," Peterson said.[141]

Collaborating with Richard Grossi, Hopkins Medicine's vice president and chief financial officer, and Ronald Werthman, vice president of finance and chief financial officer for The Johns Hopkins Hospital and Health System, Miller and Peterson drew up plans for an unprecedented $55 million budget cut for fiscal year 1998—with up to $45 million slated to come from drastic expense reductions and increased efficiency initiatives in the Hospital.[142]

Miller made himself an example of what needed to be done. He noted that under his direction, the Department of Anesthesiology and Critical Care Medicine already had trimmed $1.2 million from its budget by eliminating duplication of efforts, negotiating better patient care and maintenance agreements, and reducing support staff. He also saved the expense of hiring a new department head by deciding to retain that job even after becoming dean/CEO. "Everyone is going to have to stretch themselves thin and make do with less," he said.[143]

The united front that Miller and Peterson presented in announcing their plans also enabled them to prevent a department director or administrator from getting around it, as Miller recalled with a chuckle a dozen years later.

"We cover each other's backsides, there's no question," he said. Early in his tenure as dean/CEO, Miller recalled, the head of one department came to see him "and wanted some preposterous amount of money for something that didn't make any sense. I remember I kind of said, well, that's just not going to work." Not long thereafter, Peterson told Miller that the same department director "wanted to come to see me about something or other." Miller said, "I think I know what it's about. Do you mind if I happen to be in your office when he comes? And sure enough, I was sitting there and [the director] walks in and his mouth just dropped open. He didn't know what to say, because in the old days, they went to the one side, then to the other side." Under the unified leadership, playing one side against the other had ended. "Here, they know they can't get around us," Miller said. "We're going to share information all the time."[144]

The stringent budget-cutting and cost-trimming—including a freeze on hiring any personnel other than those essential to patient care or required by grants and contracts, standardizing supplies and renegotiating contracts with vendors—paid off. By mid-1998, Ron Peterson could report at a "town meeting" for staff that the difference between the price of care at Hopkins and at its suburban competitors had been reduced from 35 percent in 1996 to 23 percent. In addition, Hopkins' prices were only 1 percent higher than other, nonacademic city hospitals, a remarkable feat considering that Hopkins had the additional costs associated with teaching and research.[145]

Moreover, Peterson reported, patient volumes actually had bucked a statewide trend, going up 3 percent while hospitals elsewhere in Maryland notched a 2 percent drop.[146]

The reorganization of the Clinical Practice Association under the leadership of Elias Zerhouni from 1997 to 1999, when cardiac surgery chief William Baumgartner became the group's president, and executive director Ken Wilczek also had a significant, positive impact on Hopkins' bottom line.[147]

Hopkins hardly was out of the woods, however. In May 1998, Peterson announced a second round of budget cuts (albeit less severe) for fiscal year 1999, citing increases in labor costs, a dramatic jump in health care insurance claim denials, a projected drop in patient volumes, and limits imposed by the state Health Services Cost Review Commission on reimbursements for costs associated with treating patients.[148]

In addition, Miller lamented how much money was being spent to maintain Hopkins' aging buildings. Persuaded by Peterson that Hopkins had to modernize, Miller sounded the first note of what would be a constant theme over the next dozen years: the "real need" for Hopkins to build new facilities, such as a critical care patient tower, a children's hospital and a research tower that would prepare all of Hopkins Medicine to enter the 21st century.[149]

Among Edward Miller's first acts as dean/CEO of Hopkins Medicine was to foster greater communication within the institution by initiating regular, monthly "Town Hall" meetings in Hurd Hall for faculty and staff. At one such gathering in May 1998, Hospital and Health Services President Ron Peterson could report that stringent belt-tightening had helped significantly reduce the gap between patient care costs at Hopkins and at its suburban competitors, and that patient volumes had gone up. At the same meeting, Miller (above) sounded the first note of what would be a constant theme over the next dozen years: the "real need" for Hopkins to build new clinical facilities.

Photographed in January 1997 on the day he was appointed dean/CEO, Edward Miller initially disliked one of the main responsibilities of his new position: fundraising. In time, however, he became a master of that art and found it to be "fun." He said, "I've got a great product. If you can connect potential donors to the institution, the 'ask' is easy."

Fundraising and Horn-Tooting

To achieve the goal of literally rebuilding the East Baltimore campus, Miller knew he would have to become a master fundraiser—a role he initially disliked. In time, however, he actually found it to be "fun." Reflecting a decade after becoming dean/CEO, he said, "I know how much money is out there. I've got a great product." Peterson observed that Miller's success at fundraising "is largely a function of the fact that raising money is about cultivating people. And Ed's very good at that."[150]

The Johns Hopkins University as a whole proved very good at fundraising, launching a university-wide "Johns Hopkins Initiative" in September 1994 to raise $900 million by the year 2000, earmarking about $355 million of that goal for the School of Medicine and $100 million for the Hospital. The funds would underwrite the kind of building projects Miller and Peterson envisioned, as well as ongoing programs and beefing up the University's endowment—which always had been much lower than its peer institutions.[151]

Meeting its $900 million goal in 1998, two years ahead of schedule, the Hopkins fundraisers upped the ante to $1.2 billion.[152] By the end of the six-year fundraising effort, Hopkins had outdone itself. In late 2000, the University announced that it had become only the sixth institution ever to raise $1.5 billion in a single campaign, and $703 million of it was committed to Johns Hopkins Medicine through an initiative known as the Partners in Discovery Campaign. Some 64 professorships and chairs, 19 fellowships, 77 research endowments, 47 program endowments and 76 scholarships would be funded.[153]

Hitting the Airwaves—and Internet

During this period, the expertise of Hopkins physicians was being disseminated more dramatically via the airwaves and, for the first time, online—spreading the institution's name while increasing the public's understanding of science.

Johns Hopkins' Health Newsfeed, a daily, 60-second health and medicine radio update, was introduced in June 1995 and quickly became a regular feature on the Bloomberg Business Radio Network. By 1996, the USA Radio Network's 1,300 stations also were receiving the radio reports and the Voice of America was distributing it to more than 100 cities in Europe, Africa and Asia. In 1997, both the CBS and UPI radio networks picked it up as well—increasing its reach to nearly 2,000 stations in the United States and overseas.[154]

In March 1997, Hopkins Medicine made its formal debut on the Internet with the launching of its first comprehensive Web site, offering an overview of services, facilities and other resources for patients, health care professionals, the public and the press. In September 1997, Hopkins' venerable—and popular—Medical Grand Rounds made its Internet debut. A weekly Hospital-based conference for medical students and physicians at which detailed presentations are given on the medical problems of one or more patients and how they were diagnosed and treated, Grand Rounds immediately caught on with physicians from Venezuela, Brazil, India, Germany, Korea, Malaysia, Canada and other nations, who logged on to "attend" the sessions on their personal computers and learn about the latest in medical challenges and advances. They even could submit questions to the presenters by email.[155]

Rebuilding Begins

Despite the administrative turmoil and financial difficulties that colored this period, Hopkins Medicine forged ahead with plans for new buildings, completing some of them even in the midst of the organizational restructuring and belt-tightening.

On the Hopkins Hospital campus, additional new buildings and units opened almost on a yearly basis, while ground was broken or plans drawn up for others.

In November 1996, the Wilmer Eye Institute opened a completely renovated Eye Care Pavilion, designed to more efficiently handle the institute's burgeoning outpatient caseload, now topping more than 80,000 annually, as well as surgical volume, which had surpassed 5,000 cases in 1995.[158]

As soon as the new Wilmer Pavilion opened, construction began on another $2 million renovation project within the original 1929 Wilmer building, creating the 14,500-square-foot Zanvyl Krieger Children's Eye Center, named for the Baltimore philanthropist who spearheaded the private fundraising drive for it. It opened in February 1998 with a star-

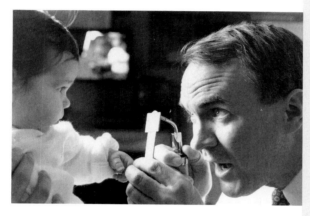

David Guyton (above), Director of the Krieger Children's Eye Center at the Wilmer Eye Institute, is internationally known for contributions to his specialty.

studded ribbon cutting that included remarks from U.S. Surgeon General C. Everett Koop and ABC News anchor Peter Jennings—both of whom had been patients of the center's director, David Guyton.[159]

In January 1997, even as Miller and Peterson were announcing the need to implement massive budget cuts, a new $2.3 million, 13,000-square-foot cardiology care unit, designed under the guidance of then-cardiology director Kenneth Baughman and his staff, opened on the fifth floor of the Nelson Building. With more intensive and intermediate care beds, and all the cardiology services grouped together on one floor, the new unit was more than four times larger than what it replaced—a claustrophobic facility that had been tucked behind the medical intensive care unit on the seventh floor of the Osler building for the preceding 27 years.[160]

Only a few months later, in February 1998, the School of Medicine's Department of Physical Medicine and Rehabilitation opened its first-ever inpatient rehabilitation unit in Hopkins Hospital—a 14-bed, $2.5 million

Eastern Avenue Rebirth

When Hopkins had acquired the old, debt-ridden Baltimore City Hospitals in 1984, it had renamed the facility the "Francis Scott Key Medical Center" in an early effort to alter its image—but not necessarily link it to Hopkins just yet. Philip Zieve, chair of the Department of Medicine and longtime head of the medical faculty there, remembered that the University president at that time, Steven Muller, "didn't want the Hopkins name on it until it had proven to be successful." As an alternative, University vice president Ross Jones had proposed the Francis Scott Key name to emphasize the hospital's lengthy history by associating it with the Baltimore attorney who had witnessed the British bombardment of Fort McHenry in 1814 and written "The Star-Spangled Banner" in honor of the fort's survival.[156]

By 1994, however, Ron Peterson, then Francis Scott Key's president, noted that the medical faculty and administration on the Eastern Avenue campus had "successfully completed … [a] major redevelopment and image transformation there." That April, they celebrated not only the opening of its new crown jewel, a $60 million, six-story, 221-bed critical care tower, but a new name: Johns Hopkins Bayview Medical Center. They christened the patient tower the Francis Scott Key Pavilion to perpetuate the legacy of the spectacular period of rejuvenation that the 221-year-old institution had experienced during the preceding decade. Combining the Hopkins name with "Bayview," which had been applied to the hilltop campus, on and off, since the 1860s, emphasized "its prominence in the Johns Hopkins family of medical institutions," Peterson wrote shortly thereafter. It conveyed "effectively that Hopkins medicine is practiced here by a medical staff that is largely full-time faculty of the Johns Hopkins University School of Medicine."[157]

A. Edward Maumenee Jr., a renowned corneal transplant and cataract surgeon who headed the Wilmer Eye Institute from 1955 to 1979, was known affectionately as "The Prof." He trained more heads of ophthalmology departments than anyone else in the world.

1998
February 12: The Zanvyl Krieger Children's Eye Center opens in the Wilmer Eye Institute with a ceremony featuring young patients, Baltimore Mayor Kurt Schmoke, former U.S. Surgeon General C. Everett Koop, M.D., and a taped greeting from ABC News anchor Peter Jennings. Krieger, a 1928 Hopkins graduate in political science, Baltimore businessman and philanthropist, had established a challenge grant that encouraged more than 100 individuals and foundations to contribute to the center. Krieger also gives generously to what becomes known as the Kennedy-Krieger Center for children with developmental problems.

February 13: the Department of Physical Medicine and Rehabilitation opens a new $2.5 million, 14-bed inpatient unit on Halsted 3. For the first time in Hopkins' history, patients recuperating from a serious illness can regain lost abilities without being transferred elsewhere.

March: Hopkins Medicine acquires Howard County General Hospital and the Howard County Health System. The merger is finalized on June 30.

November: Hopkins researchers, led by John Gearhart, are one of two teams to announce that they have isolated and identified human stem cells and proved them capable of forming the fundamental tissues that give rise to distinct human cells, such as muscle, bone and nerve.

facility on the third floor of the Halsted Building. (Because of previous space limitations, Hopkins' rehabilitation program had been located at Baltimore's Good Samaritan Hospital since its founding in 1970.)[161]

This flurry of openings was scheduled to be followed by more, with planning and construction under way for a new, $14 million ambulatory care center at Hopkins Bayview, $9 million in renovations for 17 basic science laboratories (some of which hadn't been updated since 1929), a new patient pavilion at the Green Spring Station suburban satellite center, and a new Comprehensive Cancer Center and companion cancer research building, at a combined, estimated cost of some $184 million.[162]

Exemplars

As the period drew to a close, two giants of the previous generation of Hopkins leaders died. Their extraordinary achievements—as superb physicians, inspiring teachers, accomplished administrators and international leaders in medicine—reminded the entire Hopkins community of the institution's immense impact and the need to nurture and advance it.

On January 18, 1998, A. Edward Maumenee Jr., the courtly, soft-spoken and world-renowned ophthalmologist who had headed the Wilmer Eye Institute from 1955 to 1979, died at the age of 84. During more than four decades of research, teaching and patient care, Maumenee had earned the enduring admiration of patients, colleagues and students who traveled to Baltimore from all over the globe to be treated by or work with him. One of the world's foremost corneal transplant and cataract surgeons, Maumenee also introduced the use of fluorescent dye injections to photograph the retina as blood flows through it—a technique that became standard practice. He studied congenital glaucoma, the blood vessel supply to the optic nerve, and uveitis, an inflammation of the eye's middle layer. He played a crucial role in the creation of eye banks nationwide and the founding of the National Eye Institute of the National Institutes of Health in 1968.[163]

To those who studied under Maumenee, he was known simply and affectionately as "the Prof"—and in the course of his career, he trained more heads of ophthalmology departments than anyone else in the world. Among his protégés was Morton Goldberg, one of his successors as head of Wilmer, who called him "one of the most famous names and towering figures in ophthalmology in the 20th century." Yet despite all the acclaim he received, Maumenee remained to his fellow physicians—and particularly his patients—an old-fashioned Southern gentleman of impeccable manners and quiet demeanor.[164]

Less than five months later, on May 7, 1998, A. McGehee Harvey, director of the Department of Medicine and physician in chief of The Johns Hopkins Hospital from 1946 to 1973—more than a quarter of the 20th century—died at the age of 86 in the Hospital he had done so much to transform.[165]

A 1934 graduate of the School of Medicine, Harvey became the mentor of such future Hopkins leaders as Richard Ross, Victor McKusick and Richard Johns. He had even taken a post-retirement interest in the then-budding career of a young administrator, Ron Peterson, who arrived at the Hospital in 1973, the year Harvey stepped down as physician in chief.[166]

In his 27 years as head of the School of Medicine's largest department, Harvey—nicknamed "Mac"—was credited with training more leaders in internal medicine than perhaps anyone else in the country: 2,151 students and 866 house officers, 16 of whom went on to become department chairs around the nation and eight of whom became medical school deans (among them, Richard Ross).[167]

A. McGehee (Mac) Harvey, who headed the Department of Medicine from 1946 to 1973—more than a quarter of the 20th century—did much to transform Hopkins Hospital and the School of Medicine. He achieved significant changes without fanfare, quietly initiating desegregation of the Hospital and working to eradicate racism and anti-Semitism in the School of Medicine. All the while, he was a key figure in the explosion of scientific medicine following World War II, a master diagnostician, prolific researcher, and a beloved teacher who trained more leaders in internal medicine than perhaps anyone else in the country, including 16 who became department chairs and eight who became medical school deans. At his death in 1998, he was praised by genetics pioneer Victor McKusick as "a sage of medicine, whose wisdom and insight were profound."

Praised by McKusick as "a sage of medicine, whose wisdom and insight were profound," Harvey not only was a master diagnostician, revered clinical teacher and prolific researcher but a key figure in the explosion of scientific medicine following World War II. He expanded the Department of Medicine from the three divisions it had maintained since the turn of the 20th century into an 18-division epicenter of advances in cardiology, medical genetics, gastroenterology, endocrinology, immunology and other disciplines.[168]

He also was the quiet but powerful force behind important cultural changes in both the Hospital, which was desegregated under his direction, and the School of Medicine, where he began the quiet process of weeding out racism and anti-Semitism.[169]

Emulating the professional and personal qualities exemplified by both Maumenee and Harvey would ensure the success of those who had just led Hopkins through an extraordinarily difficult period—and would face just as daunting challenges in the years ahead.

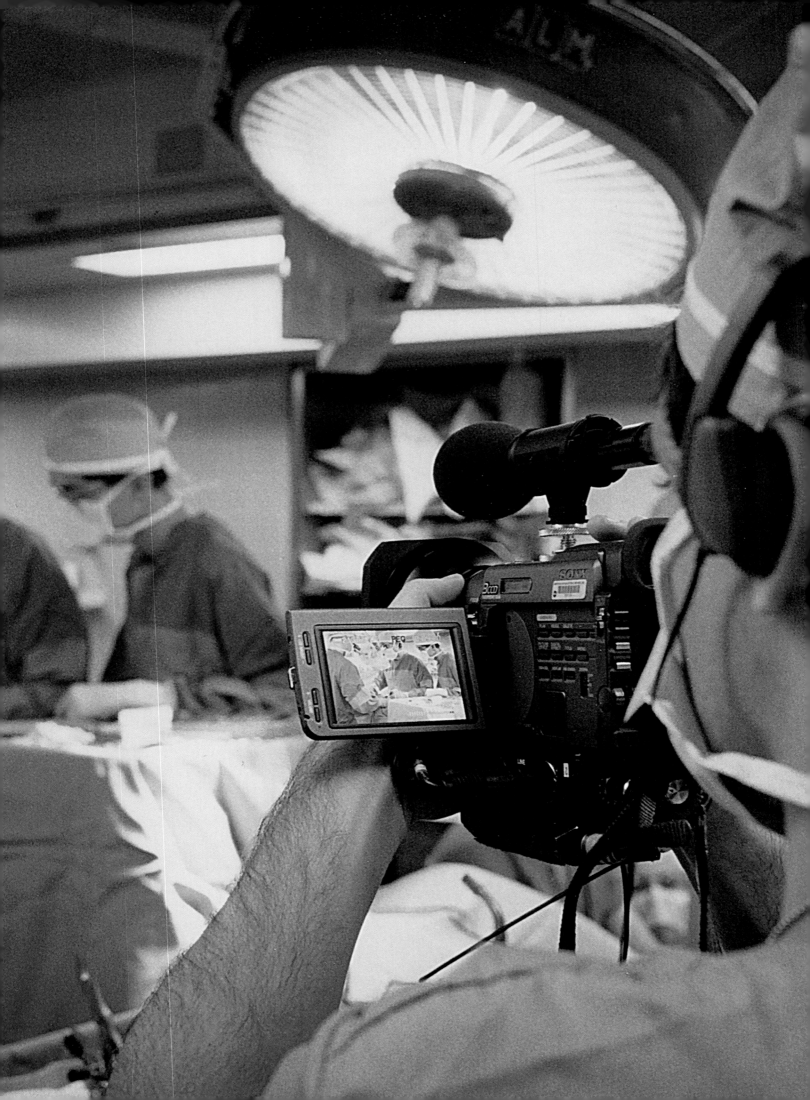

1999–2003:
Tested to the Limit: Surmounting Crises

Hopkins anesthesiologist Peter Pronovost (above, center), a 1991 graduate of the School of Medicine, is perhaps the world's most renowned expert on patient safety. He was involved intimately in addressing safety crises during this period. Facing Page: ABC-TV news crews were given unrestricted access to Hopkins Hospital for three months in 1999 to produce *Hopkins 24/7*, an acclaimed, six-part documentary that aired in 2000 and won Columbia University's prestigious duPont Award for Excellence in Broadcast Journalism in 2002.

On June 16, 1999, Joann Rodgers, Hopkins Medicine's director of media relations, received an unexpected telephone call from Severn Sandt, a producer for ABC News in New York.[1]

Sandt asked Rodgers if Hopkins would be interested in collaborating with ABC to create an unprecedented, six-part documentary-style series about the academic medical center. Taking three months to videotape, the series that ultimately was titled *Hopkins 24/7* would show how Hopkins tackles its multiple, sometimes intractable challenges: caring for the uninsured poor, addressing urban health issues, maintaining top research and education facilities, training future physicians and scientists to meet the demands of 21st century medicine, and confronting failures while also delivering medical miracles to patients in a financially constrained environment.[2]

The network's inquiry definitely interested Rodgers and her boss, Elaine Freeman, director of the Office of Communications and Public Affairs. When told of ABC's call, Hopkins Medicine's two top leaders, Dean/CEO Edward Miller and Ronald R. Peterson, president of the Johns Hopkins Hospital and Health System, also expressed interest. They both understood the perils in giving a network television news organization unfettered access to the entire medical complex over a three-month period. However, both felt confident that Hopkins not only could handle the close scrutiny but make the most of it. As Miller and Peterson later wrote in a "Dear Colleagues and Friends" letter, they believed "that participation carries the potential to give Hopkins, and academic medicine, a colossal platform for advocacy."[3]

"We have the qualities and the confidence to let the world see what we do, and to leverage what people see to explain better medicine's challenges and our solutions."

The faith that Miller and Peterson put in Hopkins' ability to handle the sometimes unforgiving, intense glare of penetrating coverage proved not only justified but prescient.

Over the next five years, Hopkins Medicine would be buffeted by unanticipated, soul-wrenching crises that critically affected all three key goals of its mission: patient care, research and education. The deaths of two young

"At Hopkins, we place a huge premium on discovering new ways to cure our patients. My goal is for our culture to place just as much emphasis on keeping patients safe."

—Peter Pronovost, 1991 graduate of the School of Medicine, Chief, Division of Adult Critical Care Medicine, Director of Quality and Safety Research Group, Medical Director, Center for Innovations in Quality Patient Care, Professor of Anesthesiology and Critical Care Medicine and Surgery, director of the Armstrong Institute for Patient Safety and Quality; recipient of a MacArthur Foundation Fellowship, known as a Genius Grant," and named one of *TIME* magazine's 100 most influential people in 2008

In this post-1999 aerial view of the growing East Baltimore campus, ground has been cleared to the south of the Weinberg Cancer Center (on the right) to make way for the new clinical towers, and the parking lot to the north (bottom, center) awaits removal for the Wilmer Eye Institute's new Smith building.

1999
January 6: Johns Hopkins International is founded to centralize and facilitate international initiatives and activities.

January 20: The McKusick-Nathans Institute of Genetic Medicine is founded, uniting nine centers, scores of physicians and scientists, and budgets worth tens of millions of dollars.

January: Hopkins' Office of Consumer Health Information publishes *The Johns Hopkins Family Health Book*, a nine-pound, 1,650-page text edited by Michael Klag, M.D., M.P.H., then director of the Division of Internal Medicine (and later dean of the Bloomberg School of Public Health). It is named a Book-of-the-Month Club main selection, with a first printing of 140,000 copies. Klag says: "It's meant to be like a call to your friend the doctor."

Johns Hopkins Bayview experiences a new building boom, with a new $13 million Bayview Medical Offices building providing a central location for ambulatory care.

Hopkins Hospital's Obstetrics and Gynecology Department opens a new Labor and Delivery unit, a $2 million project completed after 15 years of planning and postponements.

The Department of Emergency Medicine opens a Center for International Emergency Medicine Studies.

patients and a clinical research volunteer led to what Miller later would refer to as "among the darkest days many of us have ever known." A regulatory challenge to the accreditation of the Department of Medicine's fabled Osler residency program, world-renowned for physician training, threatened its existence.

Hopkins dealt with these crises swiftly, forcefully and forthrightly. The losses of life, avoidable and immensely tragic, led to dramatic initiatives in patient safety and research oversight that received national and international publicity, and since have served as models for improved guidelines that continue protecting people wherever implemented. Following intense discussions with the national body that accredits physician training programs, the Department of Medicine, in response to the citations against the Osler residency, preserved the program, a crown jewel of American medical education for more than a century.

Moreover, even during this period of profound internal difficulties and change, Hopkins Medicine maintained its forward momentum and extended its influence. Its leadership established the financial foundation for the construction of a completely new Hopkins Hospital. Its emergency medicine experts headed research and planning initiatives for national disaster preparedness following the September 11, 2001 terrorist attacks. New clinical and research buildings and units opened; two academic departments were created; three multidisciplinary research institutes were founded, heralding a significant trend in medical advances; and four interdisciplinary clinical centers were launched. More than 20 new department directors and division chiefs were named, and award-winning television programs, plus best-selling books, highlighted Hopkins' expertise and superior patient care.

Members of the faculty also continued garnering recognition for outstanding achievements, including two MacArthur Foundation "genius" awards, a National Medal of Science—and a Nobel Prize.

Y2K

As Hopkins got ready to give potentially troublesome, unprecedented access to ABC's cameras and crews, the Johns Hopkins Medicine Center for Information Services (JHMCIS) was working furiously to prepare for one crisis that never happened: the attack of the "Y2K bug."

As difficult as addressing this anticipated challenge was, it paled in comparison to that subsequent series of unexpected calamities which would have a far more profound impact on Hopkins Medicine.

Computer specialists worldwide had warned that once 2000 arrived, systems everywhere, long programmed to recognize a year only by its last two digits—for example, 1990 was denoted just as "90"—would be confused and unable to adapt.

The experts said computer systems had to be reprogrammed to expand year recognition from two digits to four, so that "00" would be recognized as 2000, not 1900—otherwise chaos might ensue. For example, what if a Hopkins computer was called upon to calculate a dosage of medicine for a patient born on 03/31/00? Was that patient a newborn—a millennium baby just arrived in 2000—or a centenarian, born in 1900 and blowing out 100 candles on a birthday cake?[4]

Under the direction of Stephanie Reel, the university's vice provost for information technology and chief information officer, as well as Hopkins Medicine's vice president for information services, dozens of workers spent hundreds of hours (at a cost of several million dollars) going over the operation of the health system's 65 central computer systems—which handled everything from patient registration and billing to pathology labs, pharmacies, medical records and employee payroll. In a process that Reel called "crucial," they identified, fixed and tested anything with a computer chip that stored years using just two digits.[5]

At the stroke of midnight that New Year's Eve, nothing broke down. All went quietly and well—not just at Hopkins, but around the world.

Although the Y2K bug proved to be a dud, Reel said that the anticipated crisis prompted Hopkins to purchase many new computer systems that it genuinely needed but might not have bought had concern over Y2K not forced its hand. "It gave us an opportunity to really modernize the environment. So I think it had lots of hidden benefits that we, at the time, may not have realized."[6]

Such an outcome would become a *leitmotif* for Hopkins as the 21st century got under way: the tougher the challenge, the stronger Hopkins would become after surmounting it.

To prepare for the potential Y2K crisis, Stephanie Reel, the University's vice provost for information technology and chief information officer, as well as Hopkins Medicine's vice president for information services, oversaw dozens of workers who spent hundreds of hours (at a cost of several million dollars) going over the operation of the health system's 65 central computer systems—which handled everything from patient registration and billing to pathology labs, pharmacies, medical records and employee payroll. They identified, fixed and tested anything with a computer chip that stored years using just two digits. Even the card swipers at Hopkins garages and parking lots were updated.

Although the feared computer breakdown never occurred, Reel said that making all the corrections to ensure that Hopkins' computers could read four-digit yearly numbers— something done by computer programmers all over the world— was "critical," and that Hopkins also purchased many new computer systems that it genuinely needed but might not have bought had concern over Y2K not forced its hand.

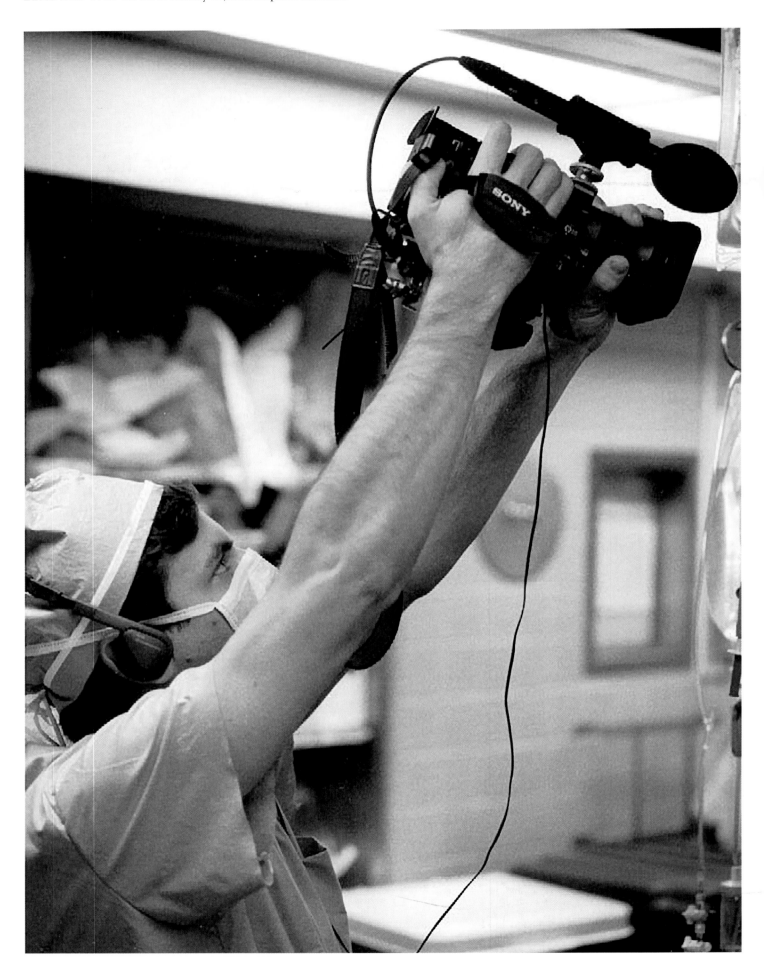

Hopkins 24/7: The Unblinking Eye

Arriving at the Hospital on September 29, 1999, the ABC News teams went everywhere with their unblinking TV eyes—from the patient wards to the operating rooms to the subterranean tunnels connecting various Hospital buildings. They even went home with some of their subjects—and grew close to them.

Months of preproduction negotiations and planning were required. "I believe we have the opportunity to create a truly unprecedented series," wrote Phyllis McGrady, vice president of ABC News.[7]

That is exactly what they did.

For most documentaries, producers used a single camera, perhaps two, noted ABC documentary producer Terry Wrong. For *Hopkins: 24/7*, ABC essentially set up a mini-bureau in the Hospital, deploying up to eight cameras simultaneously and treating the stories they found as breaking news.[8] By the time most of the crews left on December 19, they had shot nearly 900 hours of video on some 1,300 tapes.[9]

Once back in New York, the ABC producers, reporters and crew spent the next seven months culling their mammoth collection of material to form six one-hour, weekly episodes for the network to air, beginning on August 31, 2000. In addition, one of the videographers who worked on *Hopkins: 24/7*, Baltimore native Richard Chisholm, returned to Hopkins in May 2000 and spent eight additional weeks chronicling the extraordinary work of Hopkins' nurses for what became a subsequent, five-part series, "Nurses: Hearts of Mercy, Nerves of Steel," which aired on the Discovery Health Channel early in 2001.[10]

The risks that both ABC News and Hopkins took to create *Hopkins: 24/7* more than paid off. What the network teams produced was powerful, provocative television, capturing genuine issues, such as inner-city violence as seen nightly in the emergency room; the struggles with health management organizations and other insurers; and the constant training physicians undergo, exemplified by riveting footage of a Morbidity and Mortality conference, something no outsiders ever had seen, in which a physician's errors are reviewed—and assessed—with blunt frankness.[11]

Every episode featured an array of gripping stories, all united by the single fact that they each occurred at Hopkins. (ABC's brief episode summaries can be found in Appendix F.)

The series garnered acclaim from both critics and the public. More than 8.4 million viewers tuned in for the first episode. Subsequent episodes drew about 12 million viewers each. Over the series' six-week run, the new Hopkins Medicine website, www.hopkinsmedicine.org, launched just before the series began, had nearly 35,000 unique *Hopkins 24/7* visitors—which translated into more than 210 online requests for appointments with Hopkins physicians. The

One of the Hopkins physicians who featured prominently in **ABC-News'** *Hopkins 24/7* was Michael Ain, a 4-foot, 3-inch achondroplastic dwarf, an individual with an average-size trunk but shorter limbs, who has persevered to become a leading surgeon in spite of being a little person.

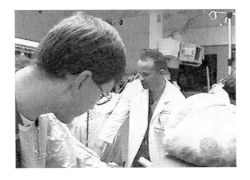

Another episode of *Hopkins 24/7* focused on Edward Cornwell, then Chief of Trauma Surgery, as he tried to save the wounded of East Baltimore's street wars, both in the operating room and through preventive outreach to local youngsters.

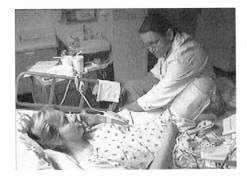

Yet another "star" of *Hopkins 24/7* was the late Rick Montz, whom some would call eccentric and irreverent, but who then was also considered one of the best surgical oncologists in the business. His vibrant, compassionate bedside manner captivated the TV audience.

1999
July: Victor McKusick's "Short Course in Medical and Experimental Mammalian Genetics" in Bar Harbor, Maine, marks its 40th anniversary.

September 29: An ABC-TV news crew begins a three-month assignment to film within the Hopkins Hospital for what will become the award-winning documentary series, *Hopkins 24/7*, aired in 2000. Nearly 900 hours of coverage are captured on some 1,300 videotapes. Says co-producer Severn Sandt: "No one ever has done six hours like this. Ever."

September: Hopkins transplant surgeons perform the first-ever kidney transplant using an organ provided by a volunteer donor who had never met the recipient.

October 12: Judy A. Reitz, Sc.D., chief architect of The Johns Hopkins Hospital's operations restructuring, is appointed executive vice president and chief operating officer of The Johns Hopkins Hospital—the sixth person and first woman to hold the post.

Josie King, whose February 2001 death at Hopkins was due to medical errors.

Hopkins telephone center was swamped with nearly 2,200 calls, leading to 700 more appointment requests. (ABC had some 250,000 visitors to its *Hopkins 24/7* website.)[12]

The arbiters of the Alfred I. duPont-Columbia University Awards for Excellence in Broadcast Journalism—the airwaves' equivalent of the Pulitzer Prize—were similarly impressed. In 2002, they awarded *Hopkins 24/7* a coveted Silver Baton, inscribed with the famous 1958 observation of legendary CBS newsman Edward R. Murrow about the potential power of television: "This instrument can teach, it can illuminate; yes, it can even inspire. But it can do so only to the extent that humans are determined to use it to those ends. Otherwise it is merely wires and lights in a box." [13]

Annus Horribilis

Barely six months after the glowing reviews of *Hopkins 24/7*, all of Hopkins Medicine was rocked by the first of those events that would make 2001 a truly *annus horribilis* (horrible year) for everyone in the institution. Only a few months later, it would be followed by yet another horrific tragedy.

Josie King

On February 22, 2001, Josie King, the engaging, blonde 18-month-old daughter of Sorrell and Tony King, a suburban Baltimore couple, died at Hopkins Hospital—not of some rare, perplexing disease, but of thirst and a common blood infection.[14]

A few weeks earlier, the blue-eyed child had received second-degree burns over 60 percent of her body in an accident at home. While playing in a bathtub, she turned on the hot water full-blast, and was scalded severely.[15]

Initially treated in the pediatric intensive care unit (PICU) at the Children's Center and responding well to care, which included some skin grafts, Josie was transferred to the intermediate care unit, with the prospect of being sent home within a few days.[16]

When she developed a blood infection from a catheter used to administer medications and fluids—then an unfortunately common occurrence not just at Hopkins but in hospitals nationwide—Josie was given antibiotics orally because it was difficult to insert an intravenous line in her damaged skin. She began vomiting and had diarrhea, both common reactions to antibiotics. Without a catheter or intravenous supply of liquids, her continuing diarrhea and vomiting dehydrated her. Although Sorrell King repeatedly told the nurses and physicians that her daughter was desperate for fluid and appeared to be weakening, she was assured that her child's vital signs were fine.[17]

Ongoing miscommunication between the teams of pain specialists, nurses and attending physicians handling Josie's case led to her continued deterioration, swift return to the PICU, the improper administration of a powerful pain killer, and what Hopkins patient safety expert Peter Pronovost later called her "senseless death."[18]

George Dover, head of the Hopkins Children's Center, immediately assumed full responsibility for the tragic outcome of the Josie King case, even though he had not been her physician. His conversations with those who had treated Josie convinced him that her death was due to medical errors. He went to see Sorrell and Tony King. He told them: "I am so sorry. This happened on my watch, at my hospital. I will help you get to the bottom of it." In pledging to help the Kings determine precisely what had happened to their daughter, and prevent such a tragedy from recurring, Dover had the full support of Dean/CEO Ed Miller, Hospital President Ron Peterson, pediatrician Beryl Rosenstein, who was then the Hospital's vice president for medical affairs, and the Hopkins Health System's then-managing attorney for claims and litigation, Rick Kidwell.[19]

Soon Peter Pronovost also began conferring with the Kings, explaining that he knew what they were experiencing—having lost his own father to a medical error at a New England hospital—and encouraging them to join with Hopkins in changing its culture, improving the safety of its patients, and fostering similar initiatives elsewhere. The Kings responded by donating a portion of their legal settlement with the institution back to Hopkins to create the Josie King Patient Safety Program and by establishing the Josie King Foundation to fund innovative safety programs that unite health care providers and consumers to foster a culture of patient safety.[20]

Sorrell King first spoke publicly about her daughter's death in September 2002. Her audience was not the press but a standing-room-only crowd of Hopkins Medicine staff and leaders in the Hospital's main auditorium, Hurd Hall. "Josie's death was the result of a combination of many errors, all of which were avoidable," she told the caregivers and executives. "You are the only ones who can solve this problem. The medical community must be open to the possibility that shortcomings do exist, and you must be prepared to make the necessary changes."[21]

Sorrel King (above, center), the mother of Josie King, has become a nationally known advocate of patient safety. She and her husband Tony used part of their settlement with Hopkins over the death of their 18-month-old daughter to create the Josie King Patient Safety Program at the Hospital. In this photo, Mrs. King is shown presenting the Josie King Nurses Award. Left to right: Karen Haller, vice president for nursing and patient care services, nurse Corrie Ann McKeen, Sorrel King, and nurses Joan Diamond and Susan Will.

Three months later, in December 2002, the Children's Center held its first patient safety summit, also in Hurd Hall. Discussions among staff included the medical situations they'd identified as the most error-prone and the solutions devised to address them. Hopkins since has made broad changes to improve communications among caregivers, to enhance resident training for calculating and documenting medication orders, and to replace potentially hazardous bottles of undiluted heparin, a blood-thinner, with pre-mixed bags of it. George Dover later observed:

"Sharing the teaching podium were a frontline nurse, a pharmacist, the chief of pediatric trauma surgery, a resident, a fellow, a neonatologist. And sitting behind me, in the audience, were Ed Miller, Ron Peterson, and most of the executive leadership of the Hospital and the School of Medicine.

"I sat in the front row and thought: Look who's teaching whom. *This hasn't happened before.*"[22]

1999
October 25; December 6: The Weinberg Oncology building opens on October 25; the Bunting ♦ Blaustein building opens on December 6. Bunting ♦ Blaustein is a "first in Hopkins' history, designed exclusively to bring oncology researchers together under one roof."

Hopkins Hospital's volumes are at an "all-time high" and its gain in market share is called "phenomenal" by Judith Reitz, newly named the Hospital's chief operating officer. The boom is the result of various moves to attract patients—including improving relations with referring physicians and opening suburban outposts such as Green Spring Station and aggressively pursuing international business. Also among the results is a bed shortage.

The magazine *Ophthalmology Times* names the 10 greatest living ophthalmologists – and five of them are from Hopkins' Wilmer Eye Institute.

November 16: Daniel Nathans, Nobel Prize-winning microbiologist and former interim Hopkins president, dies.

December 1: Nancy K. Roderer is chosen to become head of the William H. Welch Medical Library, one of the world's leaders in the management and delivery of health sciences information, with a particular focus on the development of innovative applications of new technologies.

"You have to have faith that asking fundamental questions about how things work ends up explaining disease processes in ways you'd never expect. Science is about discovering what you don't know, not confirming what you already think you know."

—Michael Caterina, 1995 graduate of the School of Medicine, Professor of Biological Chemistry, member of the Center for Sensory Biology, Johns Hopkins Institute of Basic Biomedical Science

What was happening at Hopkins—as often has been the case—also prompted changes elsewhere. Peter Pronovost's ideas for using checklists to improve not just the insertion of catheters (a key cause of hospital infections) but such other routine medical procedures as transporting gravely sick patients around the hospital, preparing medication instructions for patients being discharged, and improving communications between physicians and supporting caregivers, have been used widely in the U.S. and overseas.[23]

Ellen Roche

On June 2, 2001, not quite four months after Josie King's death, Ellen Roche, a perfectly healthy 24-year-old technician at the Asthma and Allergy Center on the Johns Hopkins Bayview campus, died after volunteering to participate in a study of how the lungs of normal individuals react to irritants.[24]

Alkis Togias, an associate professor of medicine and well-known asthma expert, was conducting the study, believing he could better understand how the lungs of asthmatics react during an attack if he tested what happened when irritants were inhaled by healthy people. He received funding from the National Institutes of Health for his research and asked three volunteers—Ellen Roche among them—to inhale hexamethonium, a drug that lowers blood pressure and long had been used to treat hypertension and decrease bleeding during surgery. The first volunteer developed a cough, but it cleared up. The second volunteer had no reaction. Roche, the third volunteer, responded differently. She developed a fever, a persistent cough, and progressive failure of her lungs and kidneys. Within five days, she was hospitalized in intensive care. After being placed on a ventilator, her condition continued to deteriorate drastically and she died less than a month after participating in the experiment.[25]

Roche's death, the first ever of a healthy research volunteer in nearly a century of studies at Hopkins, stunned the entire medical community. A little more than a month later, on July 19, the federal Office for Human Research Protections (OHRP) shocked Hopkins again by suspending all federally funded research involving human subjects at nearly every Hopkins division—halting some 2,400 investigations.[26]

Hopkins reacted angrily to what it considered the OHRP's "unwarranted, unnecessary, paralyzing and precipitous action." Critically ill patients undergoing experimental, potentially lifesaving treatments were being put in grave jeopardy, Hopkins officials argued. Moreover, the federal agency had never even replied to a December 28, 2000 letter responding to an earlier OHRP criticism of Hopkins' research procedures—a silence Hopkins officials interpreted as indicating that OHRP had been satisfied with the safeguards Hopkins had in place then. Within four days, the OHRP began lifting its restrictions on Hopkins' research after approving a corrective action plan Hopkins had devised to ensure the safety of future research efforts.[27]

"There must be a cultural change here," Dean/CEO Edward Miller told a Hopkins Medicine town hall audience only weeks after Roche's death. "We're going to have to raise the bar higher. There can't be any slippage. None."[28]

In the weeks immediately following Ellen Roche's death, the tragic incident raised questions about the thoroughness of pre-experiment research on the potential dangers of using hexamethonium in the study. The adequacy of the review of the original study proposal and the oversight of its procedures also came under scrutiny.[29] Among changes swiftly instituted by Miller was creation of additional Institutional Review Boards (IRBs) to handle authorization procedures for future research proposals and to do so more systematically. Hopkins also implemented policies to better define and communicate serious or unexpected adverse events that occurred during experiments. The institution additionally required far wider academic review of each research project involving humans and established stricter, more comprehensive standards for conducting pre-experiment research on drugs scheduled to be used on test subjects.[30]

To underscore Hopkins' enhanced commitment to the safest possible human subjects research, Miller also created the new post of vice dean for clinical investigation and appointed internist Michael Klag, an internationally known expert on epidemiology and prevention of cardiovascular and renal diseases, to the job. While overseeing all studies involving human subjects, Klag worked with Chi Van Dang, then-vice dean for research, to develop integrated programs for clinical investigation, ensuring that both biomedical basic science and patient-oriented research studies would be held to the highest standards. When Klag, who also was director of the Division of General Internal Medicine, became dean of Hopkins' Bloomberg School of Public Health in 2005, his successor as the School of Medicine's vice dean for clinical investigation was internal medicine specialist Daniel Ford, an experienced clinical investigator and a renowned expert in the field of depression and primary care.[31]

Left to right: Then-Vice Dean for Research Chi Van Dang, Dean/CEO Ed Miller, and cardiologist Lewis Becker, a specialist in stress testing, appear at a press conference after the death of volunteer participant Ellen Roche during a Hopkins research project. Calling the aftermath of that tragedy some of "the darkest days" anyone at Hopkins had known, Miller told reporters: "Our goal is to put in place a [research] review process that will serve as a national model. We are a mature enough institution to look at ourselves critically, and if we have faults, to fix them. And there can't be any slippage, none."

2000
The Harriet Lane Handbook marks its 50th anniversary. It is one of the most widely used publications ever to emerge from Hopkins.

February: Hopkins Hospital's medical board adds cardiac death to its organ donor protocol, supplementing its existing protocol that uses brain death as the criterion for recovering organs. Hopkins Hospital joins a handful of other medical centers that now use both criteria.

March 20: The Robert L. Packard Center for ALS Research is founded.

April 27: The Wilmer Eye Institute celebrates its 75th anniversary.

The Center for Complementary and Alternative Medicine (CAM) is established with a $7.8 million grant from the National Institutes of Health. Its non-traditional treatments boost patient care by working in tandem with conventional medicine. Acupuncture is among the treatments offered.

June 1: The Emergency Department has its 25th anniversary. Its founding-and-still Director Gabe Kelen is a man of many parts—including the builder of a harpsichord.

June 12: Groundbreaking is held for the 10-story, $14 million Broadway Research Building.

July 10: Johns Hopkins White Marsh opens and introduces one-stop ambulatory care to Baltimore's northeast corridor. On opening day, so many calls swamp the switchboard that administrators speedily have to add more phone lines and staff.

Johns Hopkins Bayview launches its Care-A-Van community outreach program.

The Tragedies' Impact

As was the case following the Josie King tragedy, Hopkins' prompt and powerful initiatives to right what had been wrong became national models for improving the safety of patients and research subjects.

"Those two deaths did galvanize the institution," Miller said in a 2009 interview. "First, in terms of clinical trials, I think we probably have the best oversight of clinical trials of any place in the country. We are a model. People have clearly now turned to us. The NIH has agreed that we have the best. The IRB panels do their job better than they ever have before."[32]

"Secondly, there's no question we have continued to focus ourselves on the patient safety measures. If you look at the number of bloodstream, surgical site infections, there's been a pretty dramatic decrease," Miller said. "Are we perfect? No way. We could still make that better. But I would say that without those two deaths, I don't think we'd be anywhere near where we are today."[33]

In a separate interview, Chi Dang voiced the same view that the deaths of Josie King and Ellen Roche had "transformed" Hopkins in fundamental ways—and credited the leadership of Hopkins, from University President William Brody to Miller, Peterson and other Hopkins Medicine officials, for fostering the change.

"Because of new leadership, we actually took the responsibility—that this was really our problem, that we're committed—to fix this place … It's a much safer hospital. We have one of the best safety programs, in terms of research [and patient care, with] Peter Pronovost being recognized with prizes and awards.

"Why is that? Because we invested in that. We think it's extremely important. The clinical directors, the clinical chiefs, their salaries are tied to safety. This place has taken it very seriously…. [O]ur IRB [process] is two-thousand-fold better; our faculty now can do clinical studies and we can track every little detail. We wouldn't have done that had it not been for these tragedies."[34]

The Center for Innovation in Quality Patient Care

The deaths of Josie King and Ellen Roche, plus a convergence of other factors—a November 1999 report by the Institute of Medicine (the National Academy of Science's think tank) on medical errors in hospitals, then a crackdown by hospital regulatory agencies—prompted University President Bill Brody, Dean/CEO Ed Miller, and Hospital President Ron Peterson to mandate in late 2001 that patient safety would be Hopkins Medicine's top priority.

In 2002, Hopkins created the Center for Innovation in Quality Patient Care, with funding provided by the University, its School of Medicine and the Hospital. The center serves as a learning laboratory that teaches frontline care providers how to troubleshoot potential problem areas and helps them obtain the resources to correct them.[35]

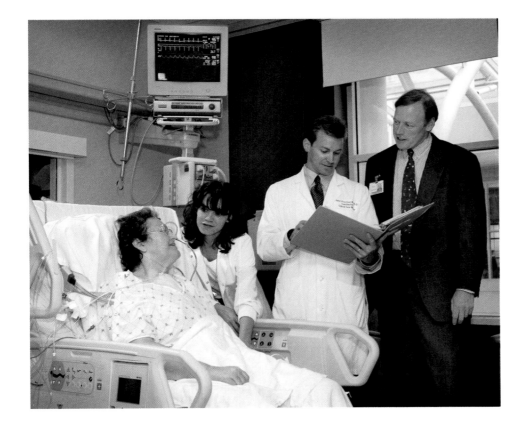

Following the deaths of patient Josie King and research volunteer Ellen Roche, Hopkins created the Center for Innovation in Quality Patient Care, funded jointly by the University, the School of Medicine and the Hospital, to spearhead safety initiatives throughout Johns Hopkins Medicine. It is led by executive director Richard "Chip" Davis (right), Hopkins Medicine's vice president for innovation and patient safety, and Peter Pronovost (with folder), the center's medical director.

Under the executive directorship of Richard "Chip" Davis, a Ph.D. in health policy and management who also is Hopkins Medicine's vice president for innovation and patient safety, and anesthesiologist and critical care medicine specialist Peter Pronovost, the center's medical director, the center launched initiatives that slashed the incidence of catheter-related bloodstream infections by more than 50 percent, fostered the adoption of a $20 million computerized system to reduce medication errors, and created the Executive Safety Rounds program, in which each of Hopkins Medicine's corporate officers adopts an intensive care unit, attends monthly staff meetings to encourage discussion of patient safety issues, and assists in helping address and correct them.[36]

Pronovost is perhaps the world's most visible patient safety advocate and analyst of where medical care communications break down. A boyish-looking, 50-year-old Hopkins Medical School graduate who has what a *New Yorker* writer described as a "fluttering, finchlike energy," the perpetually intense and engaging Pronovost also is director of the Quality and Safety Research Group, which he founded in 2003, and was named director of Hopkins' new Armstrong Institute for Patient Safety and Quality in 2011. In addition, he remains a practicing anesthesiologist, head of the Division of Adult Critical Care Medicine, and creator of now world-renowned checklists of patient safety procedures for intensive care units. He also has developed methods for measuring a health care facility's internal lines of communication and "culture of safety." [37]

"If a new drug were as effective at saving lives as Peter Pronovost's checklist, there would be a nationwide marketing campaign urging doctors to use it," wrote Atul Gawande, a physician and medical writer for the *New Yorker*, in 2007. Even without a nationwide marketing campaign, however, Pronovost's initiatives and insights have gained worldwide recognition— and adoption. In 2008, *Time* magazine named him one of the 100 most influential people of the year, and in 2009, he received a MacArthur Foundation "genius" award.[38]

2000

August 29: ABC-TV's *Hopkins 24/7* makes its debut and receives excellent reviews. "If 'ER' Were Real, It Would Be 'Hopkins 24/7'," reads the headline in *The New York Times*. Hopkins President William Brody later says: "I don't know that any other hospital could have survived the continuous on-camera surveillance that we did and come out so well under such close media scrutiny." The September 7 broadcast had audience of 12 million viewers, beating *ER* in ratings.

September 6: The seven-floor, 350,000-square-foot Weinberg Building for cancer patients opens.

September 13: Baltimore City approves a land swap, exchanging an eight-acre parcel of land at Orleans Street and Broadway, directly across from the Weinberg Building for former Church Home and Hospital property on Broadway. The deal gives Hopkins land it eventually will use for the Hackerman-Patz Patient and Family Pavilion for patients at the Sidney Kimmel Cancer Center, as well as new parking facilities for the Hopkins Hospital.

■ The 2003 death of two-year-old Briana Cohen, a cancer outpatient who received improper medication from Hopkins in an error that the institution quickly acknowledged, led to a major reorganization of pharmacy procedures at Johns Hopkins Home Care Group (JHHCG). The changes subsequently earned praise from a representative of the Joint Commission, the independent nationwide accrediting and certifying group for U.S. health care organizations. After a second, unannounced survey of JHHCG's operation, she wrote, "All I have been able to do is obtain ideas and best practices from JHHCG that I can share with others."

2000
September 29: Johns Hopkins Community Physicians is founded, combining the primary care practice of Johns Hopkins Bayview Physicians, which will be absorbed in 2002 by the Clinical Practice Association, and the Johns Hopkins Medical Services Corporation.

Howard County General Hospital begins construction on a major addition to expand emergency and obstetrical services.

October: The Website "Hopkins-Health" is launched in Oct. 2000 to distribute "the information on Hopkins' considerable consumer health database to the online community."

December 7: The Institute for Basic Biomedical Sciences (IBBS) is founded as the School of Medicine launches a $125 million fundraising campaign to support it, marking the largest single initiative for the basic sciences in the last 50 years at Hopkins. Thomas J. Kelly, M.D., Ph.D., is named its first director.

Brianna Cohen

Hopkins' ability to respond quickly and decisively to tragedies involving patients was tested again two years after the Josie King calamity with the December 2003 death of Brianna Cohen, a two-year-old cancer outpatient being treated at home with intravenously administered total parenteral nutrition (TPN). The TPN, which is liquefied nutrients for patients who cannot get their nutrition from eating, had been provided by Pediatrics at Home, a subsidiary of the Johns Hopkins Home Care Group (JHHCG).[39]

In a press release announcing "with deep regret, sadness and apology," Hopkins acknowledged that tests conducted immediately after Brianna's death showed that the TPN solution in the bag being used when she died "contained excessively high levels of potassium." The "most likely explanation for this tragic event," Hopkins said, was "human error in the manual formulation of the solutions at the Home Care pharmacy."[40]

Hopkins not only accepted complete responsibility for Brianna Cohen's death but fully cooperated with her family to find the exact reasons for the tragedy, notified and cooperated with the relevant regulatory agencies overseeing outpatient services, and immediately put into place substantive measures to ensure that TPN preparations from then on were safe. Among the measures, Hopkins Home Care implemented a fully automated system for preparing TPN solutions already in use for Hopkins inpatients.[41]

The intense efforts to improve not only the pharmacy procedures for Hopkins Home Care but every aspect of its operations became a national model.

By 2006, the Maryland Department of Health and Mental Hygiene (DHMH) found Home Care to be "deficiency free" following an unannounced, five-day survey of its patient care practices. Three years later, Home Care passed another unannounced, five-day DHMH survey with an identical "deficiency free" finding, a rarity for two consecutive surveys. The head surveyor for the state said most home care agencies tend to relax and loosen up their practices after one "deficiency free" finding, but Hopkins Home Care's director, Kim Carl, and vice president and chief operating officer, Mary Myers, instead had maintained an extraordinary commitment to patient safety and care.

That commitment was recognized again in 2009, when an unannounced survey by a team from the Joint Commission, the independent, nationwide accrediting and certifying group for U.S. health care organizations, resulted in an unusually strong commendation. Not only did the Joint Commission surveyors give Johns Hopkins Home Care the highest marks possible and offer no recommendations for improvement, but as one commissioner put it, "All I have been able to do is obtain ideas and best practices from JHHCG that I can share with others."[42]

A Storied Program in Peril

As shadows were being cast on Hopkins' patient care and research missions by the deaths of Josie King, Ellen Roche and Brianna Cohen, its education mission also found itself jeopardized by a July 1, 2003, mandate of the Accreditation Council for Graduate Medical Education (ACGME). A nonprofit group composed of physician, hospital and university representatives, the ACGME imposed new rules that strictly limited the number of hours that physician trainees in hospital residency programs could work.

Grueling 36-hour shifts and 104- to 120-hour work weeks for physicians-in-training had been common for decades—and long were considered a boot camp-like badge of honor among doctors. Yet studies on sleep deprivation among residents who worked longer than 24 hours at a stretch showed conclusively that physicians-in-training were more likely to exhibit lapses in decision-making—and more than twice as likely to fall asleep at the wheel and crash their cars on the way home.[43]

The ACGME's new rules prohibited residents from working more than 80 hours a week, averaged over four weeks, and no more than 30 hours continuously. Remaining on-call overnight in the hospital could occur no more than every third night, and residents must stop direct patient care after the first 24 hours, using the remaining six hours to study, do paperwork and prepare for the crucial "patient sign-off" procedure of turning over the patient's care to another physician. The residents were to get 10 hours off between work periods and a full day off once a week.[44]

For several years prior to implementation of the new residency work rules, the Department of Medicine had anticipated their adoption and made significant adjustments to prepare for them.[45] Within just a few weeks of the new rules' implementation, however, a resident on rotation in the general medicine service sent an email to the ACGME, charging that Hopkins was violating the just-imposed restrictions on residents' duty hours. The group's medicine residency review committee (RRC) looked over the resident's allegations, and soon thereafter, a three-member ACGME team arrived at Hopkins Hospital. They found that the frequency with which residents were on call in the medical intensive care unit exceeded the guidelines. They then slapped the heaviest penalty possible on Hopkins: Notice that the internal medicine residency program's accreditation would be withdrawn, effective in July 2004.[46]

Cardiologist Myron "Mike" Weisfeldt, a Hopkins alumnus who had returned to head the Department of Medicine in 2001, was stunned. "I felt like I'd been hit in the stomach," Weisfeldt later said. "It was rather like giving out a death sentence for possession of marijuana."[47]

The rules were the rules, however, and it was clear Hopkins had not complied fully with them—albeit briefly and with only a few residents. Weisfeldt met with the department's 110 worried internal medicine residents to assure them that everything would be done to regain the program's

Myron "Mike" Weisfeldt, director of the Department of Medicine and physician in chief of The Johns Hopkins Hospital, was vacationing at the beach in July 2003 when he learned that Hopkins' legendary Osler Medical Service residency program was threatened with losing its accreditation because of a few violations of new resident work rules imposed by the Accreditation Council for Graduate Medical Education (ACGME). "I felt like I'd been hit in the stomach," Weisfeldt later said. "It was rather like giving out a death sentence for possession of marijuana." After Hopkins instituted changes to ensure compliance with the new rules, the ACGME praised the procedures. Throughout the five-month period between the threatened accreditation loss and restoration of full accreditation, applications for the Osler residency never waned.

2001
Hopkins Medicine experiences two shattering events, the February 22 death of Josie King, an 18-month-old burn patient in the Children's Center who was treated improperly, and the June 2 death of Ellen Roche, a 24-year-old laboratory technician who was given a lethal dose of medication as a volunteer in an NIH-funded asthma research study. These tragedies lead to extraordinary—and successful—efforts to improve patient safety. On November 16, the new position of vice dean for clinical investigation is created to ensure the safest possible conduct of human subjects research; Michael Klag is appointed to the post.

January 28: The Discovery Health Channel airs *Nurses*, a five-part program about Johns Hopkins nurses, using footage from *Hopkins 24/7* that didn't make it into the documentary series.

January 30: The Institute for Cell Engineering (ICE) is created with a $58.8 million gift from an anonymous donor. The institute will be the first truly dedicated cell engineering research incubator of its kind in academic medicine.

2001

January: The iconic Dome on the Billings Building is lit with purple lights to support the Baltimore Ravens in Super Bowl XXXV, which they win on January 28, defeating the N.Y. Giants 34-7.

April 23: The School of Hygiene and Public Health is renamed the Bloomberg School of Public Health in honor of university alumnus, media entrepreneur, and head of the university's board of trustees, Michael Bloomberg. Bloomberg is elected Mayor of New York City the following November 6.

May 21: The Harriet Lane Primary Care Center for Children and Adolescents is created with the consolidation of the Children's Center's four primary care clinics—the Harriet Lane Primary Care Clinic, the Teen and Adolescent Clinic and the Intensive Primary Care Clinic—in one space on the ground floor of the now-vacated Jefferson Building, which once housed the oncology center.

Between May 21 and June 13, two former presidents of The Johns Hopkins Hospital, Russell Nelson and Robert Heyssel, die.

July 13: President George W. Bush visits Hopkins Hospital to make a pitch for reforming Medicare. While here, he sits briefly in the "President's Chair" at the Wilmer Eye Institute—the chair used by founder William Wilmer to examine every president from William McKinley to Franklin D. Roosevelt.

certification and remove any question about its reputation for excellence. Meanwhile, Ed Miller took the unusual step of immediately flying to Chicago to confer with leaders of the ACGME at their headquarters there.[48]

"I think in many ways the ACGME was looking for the alpha dog, to make an example of, to get everybody else in line," Miller said in a 2009 interview.[49]

"Now there were people who thought they were above regulations within the Department of Medicine. And I think I had to convince them that this was not something that we could say, 'We're Hopkins and we can do it differently.' I think, you know, there's a certain degree of arrogance about Hopkins, that we can do things because we're Hopkins. And I think we've kind of proved to ourselves that we're not always right. We've got to fix those things. And certainly the work hours was a big issue."[50]

Mike Weisfeldt needed no such convincing—and didn't consider Hopkins above the rules. Along with cardiac surgeon Levi Watkins, associate dean for postdoctoral studies; nephrologist Paul Scheel, the department's vice chair for clinical affairs; and anesthesiologist and pediatrician Dave Nichols, vice dean for education, he launched an exhaustive review of the residency program's policies, procedures and the initiatives that had been undertaken even before the new rules went into effect to ensure compliance with them. The goal was to ensure complete compliance with the new rules for all 75 of Hopkins' residency programs, not just the one in internal medicine. [51]

By mid-September, Hopkins was ready to send an official request to the ACGME, asking it to reconsider its withdrawal of accreditation from the internal medicine residency program.[52]

Within two weeks of receiving the document-packed reconsideration request, the ACGME replied that it would temper its action by granting the residency program probationary accreditation. That October 15, the ACGME's

For a decade, physiologist Charlie Wiener (right, next to the chalk board), professor of medicine, vice director of the Department of Medicine, and multi-tasker extraordinaire, headed the Osler Medical Service residency program while also leading the development of the School of Medicine's new curriculum and garnering numerous awards for teaching. After mentoring more than 1,000 students who rotated through the Osler Service, Wiener stepped down as its director in 2010 to become the interim dean and CEO of Perdana University Graduate School of Medicine in Kuala Lumpur. He remains on the Hopkins faculty as he oversees creation of this new medical school, established through a joint agreement between The Johns Hopkins University, the School of Medicine, Johns Hopkins Medicine International and the government of Malaysia.

residency review committee returned to Hopkins to reassess the residency program. Precisely two months later, the ACGME called Weisfeldt to say that it was fully restoring accreditation of the program without qualification—and even commended Hopkins for the significant changes it had made.[53]

Hopkins had handled another crisis with candor, commitment and dedication—and had emerged stronger. Pulmonologist Charles Wiener, then-director of the Osler Residency Training Program, later noted an ironic development during the anxious fall of 2003. As Hopkins awaited word on the program's fate, applications for the Osler residency never once let up—and the quality of those applying remained exactly as it always had been. What's more, as Weisfeldt was happy to report, the just-published *Osler Housestaff Manual*, entirely written, designed and produced by the Osler housestaff, had been praised in an outside review as "the best of its type…for education of medical students and residents."[54]

Instituting a Research Revolution

The *Hopkins 24/7* documentary series may have done everything Edward R. Murrow said television could accomplish—teach, illuminate and inspire—but it did so, understandably, by dealing largely with the highly dramatic aspects of health care and medical education, not the less visually stimulating but extremely important research work also under way at Hopkins 24 hours a day, seven days a week.

After completing *Hopkins 24/7*, producer Peter Bull said one of the things that most impressed him during his three months at the Hospital and School of Medicine were "the cross currents, the cross flow between disciplines and departments." The collegiality that he observed, which he said many people described as unique to Hopkins, always was present in the "remarkable ability for people to talk between departments. There isn't academic rivalry."[55]

Although interdisciplinary cooperation long had been a Hopkins hallmark, actively encouraging departments to join forces in the creation of unique research institutes became an exciting initiative—and source of vital research funding—on the cusp of the 21st century.

"What it has done for us is that all of these groups have been able to go out and compete for literally tens of millions of dollars of NIH funding that only comes to large programs," said Chi Van Dang, then-vice dean for research and a professor of medicine, oncology, pathology and cell biology, in a 2009 interview.[56]

"We've probably gotten close to $100 million in that type of money in the last five to seven years. And that's a new way of doing business here in terms of research. While we have maintained the strengths of departments, we now on purpose have stimulated interdisciplinary research through institutes," Dang said. "That's really a new wave."[57]

Christian A. Herter

■ The School of Medicine's Herter Lecture, the first and most prestigious of Hopkins' endowed lectureships (founded in 1903), and Johns Hopkins' School of Advanced International Studies have a significant connection. Christian A. Herter (1865–1910), a wealthy physician, biomedical researcher and protégé of pathologist William Henry Welch (1850–1934), the School of Medicine's first dean, donated the funds to create the Herter Lectureship. Its aim has been the dissemination of research by overseas scientists. The inaugural Herter Lecture was delivered by the renowned German bacteriologist Paul Ehrlich, a future Nobel Prize-winner already developing his "magic bullet" remedy for syphilis (an arsenic compound he called Salvarsan). He also is credited with coining the term "chemotherapy." The Herter Lectureship's connection to foreign research was prescient, since Herter's nephew and namesake, Christian A. Herter (1895–1966), a politician and diplomat, would become a founder of Johns Hopkins' School of Advanced International Studies (SAIS) and Secretary of State for President Dwight D. Eisenhower. (SAIS's Christian A. Herter Professorship in American Foreign Policy was established by his widow in 1967.)

Human geneticist Aravinda Chakravarti, famed for his studies of the genetic factors in diabetes, heart disease and mental illness, became the first director of the McKusick-Nathans Institute of Genetic Medicine in 2000. A native of Calcutta, India, he was trained at the University of Texas in Houston, where he'd been an integral member of the team that identified the gene responsible for cystic fibrosis. Highly regarded for his background in both experimental and computational genetics, he was recruited from Cleveland's Case Western Reserve University to head the McKusick-Nathans Institute. Perhaps not among the reasons for his selection is his skill in the kitchen. As a graduate student, he wrote a non-scientific book about Indian cuisine entitled *Not Everything We Eat is Curry.* In 2007, David Valle, a pediatrician, geneticist and molecular biologist, succeeded Chakravarti as head of McKusick-Nathans, where he also still directs the institute's Center for Complex Disease Genomics.

McKusick-Nathans Institute of Genetic Medicine

The wave began in January 1999 with the founding of the McKusick-Nathans Institute of Genetic Medicine. Uniting nine centers, scores of physicians and scientists, and combined budgets worth tens of millions of dollars, the new institute consolidated much of the genetic disease research, education and patient care activities that had been spread widely throughout Hopkins.[58]

Named for Victor McKusick and Daniel Nathans, the two faculty members whose pioneering work in the laboratory and at the bedside helped transform genetics into a driving force of medicine, the new institute was remarkable not only for the number of centers it housed within a single entity, but for the breadth of the clinical and research collaborations it supported. Its nine components included the Center for Inherited Disease Research (CIDR); the Online Mendelian Inheritance in Man (OMIM) Project, the internationally known Internet database used by geneticists worldwide to share findings; the Clinical Program in Genetic Medicine; and a Residency Program in Genetic Medicine.[59]

Other centers and programs within the Institute were the Greenberg Center for Skeletal Dysplasia, which unites research, diagnosis and care of patients with congenitally short stature; the Center for Craniofacial Developments and Disorders, whose physicians investigate normal skull and facial development and the genetic events leading to malformations; the DNA Diagnostic Lab, which specializes in testing for 14 genetic conditions and their carriers; the Predoctoral Training Program in Human Genetics; and the Genetics Resources Core Facility, a scientific "superstore" providing biochemical reagents and other products for researchers, as well as a cell culturing, DNA analysis and research planning service.[60]

Bart Chernow, then vice dean for research and technology, was named interim director of the genetic institute at its founding. In June 2000, Aravinda Chakravarti, a Ph.D. in human genetics renowned for his studies of predisposing genetic factors in such common and complex human illnesses as diabetes, heart disease and mental illness, was recruited from Case Western Reserve University to become the institute's first director. In 2007, pediatrician, geneticist and molecular biologist David Valle became director of the institute, succeeding Chakravarti, who remained head of the institute's new Center for Complex Disease Genomics.[61]

Valle, long a Howard Hughes Medical Institute investigator as well as Hopkins faculty member, had established himself as a *wunderkind* in genetics as far back as 1978, with work on an inherited form of blindness called gyrate atrophy. He oversaw the pediatric genetics clinic during his early years on the Hopkins faculty, focusing on understanding the basis for rare, inherited disorders. Citing a host of dramatic advances in genetic medicine, Valle says his goal is to ensure that every member of the Hopkins faculty feels welcome to become involved in and benefit from the institute's work.[62]

The Institute for Basic Biomedical Sciences

The move toward research unification continued with the founding of the Institute for Basic Biomedical Sciences (IBBS) in December 2000, consolidating under a single umbrella the efforts of several hundred scientists engaged in biomedical research and basic sciences teaching in eight departments. Thomas Kelly, director of the Department of Molecular Biology and Genetics, a physician and Ph.D. who had received both his undergraduate and medical degrees from Hopkins, was named the IBBS's first director.[63]

Begun with an anonymous donation of $30 million, the IBBS linked the research efforts of the existing basic science departments of Biological Chemistry, Biomedical Engineering, Biophysics and Biophysical Chemistry, Molecular Biology and Genetics, Molecular Cell Biology, Neuroscience, Pharmacology and Molecular Sciences, and Physiology. These departments already had garnered some $50 million of the $225 million in biomedical research grants received in 1999 alone from the National Institutes of Health, a sum that made Hopkins the nation's largest recipient of such research funding—a distinction it still has. Although the departments would remain independent, pooling their resources by sharing facilities and administrative costs would make research collaboration easier—and more efficient.[64]

Kelly observed that "nothing less than a revolution in molecular biology" had occurred in the preceding 20 years, "blurring the traditional boundaries of the basic sciences." To take advantage of that development required the "retooling" and "new scientific infrastructure" that the IBBS provided. "By enhancing resources for basic biomedical research, we'll keep gifted faculty at science's leading edge."[65]

Ironically, within two years Kelly himself left Hopkins. Keeping Hopkins' gifted faculty from being lured away by appealing opportunities elsewhere has long been a challenge—as well as a high compliment to the quality of the individuals and the work that they do at Hopkins. In 2002, Kelly ended his 30-year association with Hopkins to become director of the Memorial Sloan-Kettering Cancer Center's institute for basic research in New York.[66]

Kelly's successor as the IBBS's director was Jeremy Berg. A Ph.D. in chemistry, Berg had been on the faculty since 1986 and had become one of the nation's most distinguished basic scientists, renowned for his major contributions to our understanding of how zinc-containing proteins bind to the genetic material DNA and RNA and regulate genetic activity. Such findings may one day have medical applications in regulating genes involved in diseases. He led the institute from April 2002 until October 2003, when he, too, was snatched away to become director of the NIH's National Institute of General Medical Sciences.[67] The NIH director who appointed Berg to the NIGMS post knew him well. It was Elias Zerhouni, the School of Medicine's former executive vice dean and head of the Department of Radiology and Radiological Science

Biophysicist and biochemist Jeremy Berg (below, right), a researcher, teacher and administrator at Hopkins for 18 years, headed the Institute for Basic Biomedical Sciences from April 2002 to October 2003, when he was recruited away to become director of the National Institute for General Medical Sciences at the National Institutes of Health, overseeing a $1.8 billion budget for funding basic research. Berg's training as an inorganic chemist, combined with a profound interest in biology, aided his creation of a research program into zinc fingers—the small domains of proteins that bind zinc and interact with other biomolecules, such as DNA. These domains are active in many biological processes. In this 2000 photo, Berg and M.D./Ph.D. student Greg Gatto examine a model of a zinc-finger protein binding to DNA. Now at the University of Pittsburgh, Berg was elected president of the American Society for Biochemistry and Molecular Biology in 2011.

Molecular biologist and geneticist Stephen Desiderio (above), an M.D./Ph.D. graduate of the School of Medicine, became head of the Institute for Basic Biomedical Sciences (IBBS) in 2003. "Fundamental research touches upon everything from diagnosis to treatment and therapy for conditions ranging from cancer to autoimmune disease," says Desiderio, also a Howard Hughes Medical Institute investigator. The IBBS reinforces the unique and collaborative environment at Hopkins that bridges basic science and clinical research. With funding assistance from the IBBS, Hopkins' nine basic science departments "study all the fundamentals," Desiderio says, "from solving protein structures to dissecting cell movement, from analyzing chromosome structure to deconstructing biochemical pathways. We have so much exciting research going on here, it's really inspiring."

who had spearheaded the IBBS's creation. He had left Hopkins in 2002 when President George W. Bush appointed him the 15th director of the National Institutes of Health.[68]

Fortunately, Hopkins' talent bench is deep. Stephen Desiderio, a professor of molecular biology and genetics who received both his medical degree and Ph.D. from Hopkins, was named to succeed Berg within days of his departure. Desiderio's breadth of experience and ability is reflected by his appointment as a Howard Hughes Medical Institute investigator, as well as his membership in the American Society for Clinical Investigation, unusual for a member of a basic science department.

Under Desiderio's direction, the Hopkins research enterprise has fostered major research initiatives to use proteomics (the study of people's proteins) and genomics (the study of people's genomes) to make headway against sudden cardiac death, cancer, and psychiatric and neurological conditions. As he told the in-house Hopkins Medicine publication *Change* in 2001, "The hardest part of research is always maintaining faith that something important will turn up where we're looking."[69]

The Institute for Cell Engineering

Little more than a month after the founding of the IBBS, the Institute for Cell Engineering (ICE) was created in January 2001. It was launched with a $58.8 million gift from an anonymous donor who believed that Hopkins was the ideal place for advancing quests into a field of discovery that only a few years earlier would have been considered science fiction, but which now held profound potential.

The first dedicated cell engineering research incubator of its kind, ICE has researchers focusing on selecting, modifying and reprogramming human stem cells, aiming to mold them into therapeutic transplants for everything from Parkinson's, ALS and diabetes to heart failure, stroke and spinal cord injury.[70] Its first executive director was then-Vice Dean for Research Chi Dang, whose own research centers on the mechanisms underlying the abnormal growth of a gene that, when mutated, contributes to the formation of cancerous cells. In September 2011, after spending more than 30 years at Hopkins, Dang became director of the University of Pennsylvania's Abrahamson Cancer Center.[71] His successor as ICE's director is neurologist and neuroscientist Ted Dawson, renowned for his studies of the molecular mechanisms of neurodegeneration in such diseases as Parkinson's.

Hopkins President Bill Brody described ICE's creation as a herald of medicine's future, and Ed Miller noted that Hopkins' history in fostering stem cell research made it the ideal place for ICE. Miller observed that Hopkins was the home of the pioneering stem cell research of John Gearhart and Curt Civin, a hematologist, pediatrician and oncologist who had been on the faculty since 1979 and won international recognition for his 1984 discovery of a method for isolating stem cells from other blood cells. Such work, Miller said, formed the foundation on which ICE would build and "take the next great leap forward, deciphering some of the fundamental mysteries of how cells go awry in disease and behave in transplants."[72]

As the debate over stem cell research grew, however, Hopkins' role in the vanguard of this scientific revolution generated controversy, not just outside of Hopkins Medicine but within it. Gearhart, Civin, Miller, Brody and many others on the Hopkins medical faculty supported the effort to use stem cells to find cures for people suffering from debilitating diseases and injuries, but others on the faculty did not.[73] When an angry alumnus wrote a letter to the editor of *Hopkins Medicine* magazine in 2006, complaining about a Miller column in the previous issue that favored stem cell research and what the writer called its "misuse of human embryos," the dean/CEO replied: "Stem cell research will continue to be an issue that divides good people. Here at Hopkins, we remain firmly focused on the number of lives this amazing new area of medicine promises to save. Researchers also are keenly aware of the estimated 400,000 human embryos that are discarded each year" by laboratories having nothing to do with basic scientific research.[74]

Hematologist, oncologist and pediatrician Curt Civin (left), who joined the faculty in 1979, won international renown for his 1984 discovery of a method for isolating stem cells from other blood cells. Civin, whose scientific crusades include waging war on childhood leukemia, became one of the most vocal advocates of continuing federal and state funding for stem cell research. Civin left Hopkins in 2009 to become associate dean for research and director of the Center for Stem Cell Biology & Regenerative Medicine at the University of Maryland.

Pioneering stem cell researcher John Gearhart (below, center) with then-fellow Michael Shamblott, now an assistant professor and a research scientist in the International Center for Spinal Cord injury at the Kennedy-Krieger Institute. In 1998, Gearhart led the Hopkins research team that first reported identifying and isolating human embryonic stem cells, which can develop into any kind of cells in the adult human body or be expanded indefinitely for clinical use in treatments for a diseases and injuries. He became one of the most eloquent proponents of stem cell research and head of Hopkins' Division of Developmental Genetics. In 2008, he was appointed director of the Institute for Regenerative Medicine at the University of Pennsylvania.

Neuroscientist Jeffrey Rothstein's landmark studies of glutamate, a natural chemical released by nerve cells that transmits messages from one cell to another, led to development of riluzole, the first and still only drug on the market for slowing the progression of ALS, or Lou Gehrig's Disease. Rothstein heads the Robert Packard Center for ALS Research, founded in 2002. It has had a major impact on changing the course of ALS research around the world, distributing millions in grants to scientists in the United States and overseas.

2001

July 19: The federal Office of Human Research Protection notifies Hopkins that all federally supported medical projects involving human research are suspended, following a School of Medicine report three days earlier in which it took full responsibility for the death on June 2 of research volunteer Ellen Roche. Federal officials lift the suspension four days later, on July 23, after reviewing Hopkins' already-prepared plans for corrective action.

August 16: A paired kidney exchange program is launched by the Johns Hopkins Comprehensive Transplant Center. Its creation will lead to groundbreaking, multiple "domino" kidney swap operations in the years ahead.

August 20: Benjamin Carson, M.D., director of pediatric neurosurgery, and David Sidransky, M.D., professor of otolaryngology, oncology, pathology and urology, are cited by Time magazine as among 18 of "America's Best" in science and medicine.

In response to the Sept. 11, 2001 terrorist attacks, Hopkins Hospital mobilizes rapidly in case it needs to receive patients from Washington; in the aftermath, Hopkins focuses on preparing for bioterrorism, as well as disasters of every kind, by creating the Office of Critical Event Preparedness and Response (CEPAR) in September 2002. Hopkins prepares to be a major player in the event of a bioterrorist attack; D.A. Henderson, former dean of the School of Public Health, is tapped to head federal preparations for bioterrorism assaults; the Hopkins International Office is adversely affected by drop in patients.

The Hopkins Health Newsfeed—a one-minute medical news item heard on at least 500 radio stations in the U.S., as well as in Canada, Europe and Asia—marks its 1,000th show since its debut in 1985.

Research Centers

The consolidation of basic research also was accomplished during this period by the creation of multidisciplinary centers where studies could be conducted and new procedures and treatments tested on everything from ALS and cardiovascular disease to alternative medicine.

The Robert Packard Center for ALS Research at Johns Hopkins

In 2002, a $5 million donation from the Robert Packard Foundation, named for a San Francisco investment banker and ALS victim, led to the creation of an international consortium of scientists working in ALS research and based at Hopkins. The Packard Center has spearheaded an impressive expansion of ongoing efforts to defeat the invariably fatal, progressive neurodegenerative disease that causes loss of muscle control and death within two to five years of diagnosis.[77]

Heading the center since its inception has been neurologist and neuroscientist Jeffrey Rothstein. His pathbreaking studies of the neurotransmitter glutamate (a natural chemical, released by nerve cells, that transmits messages from one cell to another) led to the development of riluzole, the first—and still the only—drug on the market for slowing the progression of ALS.[78]

Rothstein and the Packard Center have had a major impact on changing the ALS research process around the globe. Within the center, Rothstein oversees the efforts of some two dozen postdoctoral fellows and neurology residents whose work has helped define ALS research, while also shedding light on the common paths of major diseases of the nervous system, from Alzheimer's to brain cancer.[79]

Frustrated by what he considered the slow, piecemeal pace of worldwide ALS research, Rothstein also guides the Packard Center's swift distribution of millions in grants to researchers both in this country and overseas, giving top scientists well-defined research goals and requiring them to share their findings. By the end of fiscal year 2011, the amount bestowed topped $21.5 million. Among the recipients of these funds are researchers who began their ALS work as fellows or

residents at the Packard Center and have—in classic Hopkins fashion—gone on to found their own research centers elsewhere. Because of Rothstein's leadership (and the center's largesse), they have remained affiliated with the Packard Center, giving it a globe-girding scope unlike any other research center in the field.[80]

The Donald W. Reynolds Cardiovascular Clinical Research Center

Within a seven-month period in 2002, the Hopkins Medicine family was stunned when three extraordinarily talented and admired young physicians—including one who had been featured prominently in *Hopkins 24/7*—all died of sudden cardiac arrest.

On April 21, David Nagey, 51, associate professor of gynecology and obstetrics, director of perinatal outreach, and a well-known authority on the management of high-risk pregnancies, collapsed and died while participating in a 5K fundraising race for his son's school.[81] On May 25, Jeffery Williams, 50, associate professor of neurosurgery and oncology, director of stereotactic radiosurgery and a 1977 graduate of the School of Medicine, collapsed and died while exercising at the school's fitness center.[82] Then, on November 21, Fredrick "Rick" Montz, known to millions of *Hopkins 24/7* viewers as the ponytailed, smiling and compassionate gynecologist, obstetrician, oncologist and surgeon who cared deeply for his patients (and they for him), collapsed and died while jogging. He was only 47.[83]

Having so recently been touched directly by sudden cardiac deaths, Hopkins was especially pleased in May 2003 to receive a four-year, $24 million gift from the Las Vegas-based Donald W. Reynolds Foundation to establish a multidisciplinary center that would focus exclusively on reducing the rate of these deadly heart attacks.[84]

To determine why specific patients have irregular heart beats, known as arrhythmias, or could be predisposed to developing potentially fatal, abnormal heartbeats, scientists at the Reynolds Cardiovascular Clinical Research Center aggressively pursue the creation of novel biological therapies, including the use of stem cells, to prevent abnormal heart rhythms and sudden death in patients recovering from heart attacks. They also use modern imaging techniques to better define the functional, structural and metabolic features of the heart that pose the greatest risk for life-threatening arrhythmias in post-heart attack patients. In addition, they seek to identify genetic and protein-related indicators of sudden cardiac death and develop new methods for studying the genetic factors in patients with varying levels of risk for the condition.[85]

Recognizing that by 2000, some 42 percent of Americans were using complementary and alternative medicines (CAM) for chronic conditions—especially in an effort to control pain when regular, prescribed medications prove ineffective—Hopkins researchers decided that it was time to conduct rigorous scientific studies to determine if these unconventional medical procedures and substances actually work, and if so, why.

With an initial five-year, $7.8 million grant from the National Institutes of Health, endocrinologist Adrian Dobs and biostatistician Steve Plantadosi launched the Johns Hopkins Center for Complementary and Alternative Medicine. It probes such subjects as the effects of alternative medicines and methods on prostate and breast cancer patients, trains new researchers in studying the impact of CAM, and educates the general Hopkins Medicine community on the use of such therapies.[75] It also has a care program, the Complementary and Integrative Medical Service, making Hopkins one of the few academic medical centers in the country to offer a comprehensive range of alternative treatments, including acupuncture, massage therapy and classes in meditation.[76]

In 2002, the Hopkins Medicine family was stunned by the sudden cardiac deaths of three remarkable talented, young Hopkins physicians—(from left to right) David Nagy, 51, director of perinatal outreach; Jeffery Williams, 50, director of stereotactic radiosurgery; and Fredrick "Rick" Montz, 47, a charismatic gynecologist, obstetrician, oncologist and surgeon featured prominently in the 2000 TV documentary *Hopkins 24/7*. Within months of their deaths, Hopkins received a $24 million gift to create the Donald W. Reynolds Cardiovascular Research Center, home to a multidisciplinary effort to focus on reducing the rate of these deadly heart attacks.

Steven Thompson (above), a one-time anesthesiology and critical care medicine administrator with an unflappable ability to manage multiple tasks with aplomb, was named the first director of what now is known as Johns Hopkins Medicine International in 1999. Characterized by Dean/CEO Ed Miller as his "strong right hand," Thompson and his team launched more than 50 innovative projects overseas between 1999 and 2003, generating more than $100 million in revenue for the clinical departments. In 2005, Thompson was promoted to senior vice president of Johns Hopkins Medicine, but in 2011 he returned to Johns Hopkins Medicine International as chief of its rapidly growing enterprises.

2001
September: The Johns Hopkins Early Learning and Child Care Center is opened in the former Church Home and Hospital Building, acquired in 2000 from the city several blocks south of the Hopkins Hospital on Broadway. With 13 classrooms, a crafts room, office space, a conference room and kitchen occupying the first and second floors of what now is the Church Home Professional Building, the center can accommodate 156 children ages 6 weeks to 5 years. Development of the center reflects Hopkins' commitment to becoming one of the nation's best places to work.

Exporting Hopkins

Early in 1999, Hopkins centralized all of its international initiatives and activities by creating Johns Hopkins International (later known as Johns Hopkins Medicine International). Noting that Hopkins already had shown that its "brand of quality can open doors to new opportunities and new activities around the world," Miller said that it only made "good business sense and good health care sense" to establish an entity to coordinate the Hopkins efforts in what was "a market place that is becoming truly global."[86]

At the time, more than 6,000 patients from overseas came to Baltimore each year for Hopkins medical services; the Hospital already had helped pioneer international telemedicine; and it was providing clinical care in Singapore through its outposts, which were then referred to as Johns Hopkins Singapore and Johns Hopkins Singapore Clinical Services. The potential for increasing all of these endeavors seemed substantial.[87]

To head this new enterprise, Miller chose Steve Thompson, a man whose apparently inexhaustible ability to deal calmly and efficiently with multiple responsibilities was quite familiar to him. Miller first hired Thompson to be his administrator of anesthesiology and critical care medicine in 1994.[88]

Soon after Miller had become dean/CEO of Hopkins Medicine in 1997, he tapped Thompson to become vice dean for administration; in 1998, Miller also chose Thompson to be CEO of Hopkins' clinical services operation in Singapore; in 2000, Miller added the vice presidency for ambulatory services to Thompson's portfolios. These multiple, institution-wide, globe-girdling assignments never seemed to ruffle the Minnesota-born Thompson's easygoing, Midwest demeanor or neatly combed and prematurely graying hair. By early 2005, he would be promoted to senior vice president of Johns Hopkins Medicine.[89]

For someone so skilled at handling a wide variety of complex jobs with a rare combination of intense concentration and low-key equanimity, Thompson actually had begun his health care career in a roundabout way. As an undergraduate at the University of Minnesota, he had majored in biology and considered becoming a physician. Summertime jobs in hospital operating rooms during his years in high school and college instead piqued his interest in the work of perfusionists, the technicians who operate the heart-lung machines during bypass and other cardiac operations.[90]

He undertook perfusion training at a time that just happened to coincide with a shortage of licensed practitioners in the field, and he quickly landed a job at Hopkins under cardiac surgeon in chief William Baumgartner, an equally calm, affable but intensely focused Kentucky native. Baumgartner later recalled, "During the long days we spent in the OR together, it was apparent that Steve had the innate ability to juggle a lot of responsibilities and work through a lot of issues."[91]

Thompson became the head of perfusion at Hopkins, then went to the University of Pennsylvania in 1993 at the invitation of a Hopkins cardiac surgeon who was going there to head its cardiothoracic and transplant program. Thompson never sold his Baltimore house, however—and within a year he came back, invited by Ed Miller, then the newly appointed head of the anesthesiology department, to become its administrator.[92]

Miller found Thompson to be his "strong right hand,"[93] and not surprisingly, when Miller became the first dean/CEO in Hopkins Medicine's history, he wanted Thompson to become his administrative vice dean. Thereafter, he continued to increase the number of jobs he asked Thompson to tackle. No matter how Thompson's responsibilities have grown, however, his institutional profile remains surprisingly muted. That's the way he prefers it. "He doesn't want credit," Vice Dean for Faculty Janice Clements once observed. "He just wants to get things done."[94]

Over the next decade, Thompson would get an immense amount done—much of which will be dealt with in the next chapter. During this period, however, he oversaw Johns Hopkins International's significant recovery from a sharp decline in revenues due to the precipitous drop in the number of overseas patients (many from the Middle East) following 9/11. Within a few years of the downturn, the influx of JHI's overseas clientele had rebounded, exceeding its budgeted revenues substantially. Between 1999 and 2003, Thompson and his band of managing directors also launched more than 50 innovative projects overseas, including a multi-departmental collaboration with the Sheikh Zayed Military Medical Hospital in Abu Dhabi, United Arab Emirates; a Department of Pathology diagnostic laboratory in Kuala Lumpur, Malaysia; and a high-amenity obstetric unit at the International Peace Maternity Child Health Hospital in Shanghai. More than $100 million in revenue, most of it going to the clinical departments, was generated by JHI during this period.[95]

Sharaf Saleh (above, left), Middle East patient services manager for Johns Hopkins Medicine International, said that patients from that region "were very afraid of blind reaction, of blind anger," in the wake of the Sept. 11, 2001 terrorist attacks, but swift action by University and Hospital leaders— who called for compassion toward Muslim patients and tolerance for all ethnic groups—helped minimize the temporary drop in the number of international patients.

Blossoming of New Hopkins Buildings

Even as new plans for rebuilding Hopkins were being formulated, enough construction projects were being completed or launched to give rise to the quip that the avian mascot of Hopkins' athletic teams should be changed from the blue jay to the crane.

Hopkins Hospital's gynecology/obstetrics department opened a new $2 million labor and delivery unit in June 1999, following 15 years of planning and postponements. It represented the conclusion of the first phase of plans to overhaul obstetrics completely. That same summer, the $13 million, 98,000-square-foot Hopkins Bayview Medical Center's outpatient center, called the Bayview Medical Offices building, opened on the Eastern Avenue campus. The new building was home to the primary care outpatient services of the medical center's departments of general internal medicine, gynecology, obstetrics and pediatrics. Among the specialty outpatient services in it were burn care, cardiology, dermatology,

2001
October 24: School of Medicine faculty members Kay Redfield Jamison, Ph.D., professor of psychiatry, and Geraldine Seydoux, Ph.D., associate professor of molecular biology, each win a MacArthur Foundation "genius grant" in October. Jamison, a psychologist and best-selling author of *An Unquiet Mind*, was recognized for her work to enhance mental health treatment, improve patient support and advocacy and increase public awareness of psychiatric disorders. Seydoux was recognized for her work to help illuminate some of the most complex processes in biology through her study of the molecular genetics of tiny roundworms called *C. elegans* (for *Caenorhabditis elegans*).

endocrinology, gastroenterology and imaging, including X-ray.[96] Out at Howard County General Hospital, construction began in October 2000 on the largest building project in the medical center's history, a $31.5 million wing that would make room for expanded emergency, pediatric, radiologic, neonatal and obstetrics services.[97]

In addition, in September 2000, Hopkins successfully negotiated an unusual land swap with Baltimore City, exchanging land the Johns Hopkins Health System had purchased from MedStar Health on the old Church Home and Hospital site several blocks to the south at Broadway and East Fayette street in return for an eight-acre parcel across Orleans Street from the just-completed Weinberg Building. The deal gave the city land on which it later built townhouses and apartments as part of a multi-million-dollar neighborhood redevelopment initiative, while simultaneously providing Hopkins Medicine with real estate conveniently situated in what would become a central part of its new master plan.

This site eventually became the location for a new parking garage; the Hackerman-Patz Patient and Family Pavilion, a residence for long-stay cancer patients and their families; the pediatric ambulatory building named for donor David Rubenstein; a new power plant; and a massive materials handling center, the largest in the city and "one of the largest receiving facilities in health care," says Kenneth Grant, vice president of general services. Its huge 22-bay loading dock replaced the often-overloaded, three-bay dock that was used to receive bulk orders of medical supplies, chemicals, lines and equipment. The creation of new parking and related projects south of Orleans street became the "enabler" for freeing up the five acres of land north of Orleans that were needed to proceed with construction of the two mammoth clinical towers that would become the new Johns Hopkins Hospital in 2012. [98]

The Bunting Family ◆ The Family of Jacob and Hilda Blaustein Building

In late 1999, several significant projects were completed. December saw the opening of the first building in Hopkins' history ever dedicated solely to cancer research, the $59 million, 122,000-square-foot structure known as The Bunting Family ◆ The Family of Jacob and Hilda Blaustein Building, named for two groups of Baltimore philanthropists. Ten stories tall, it contained working space for more than 400 researchers and staff who previously had been scattered from one end of the East Baltimore campus to another.[99]

Initially worried that the mammoth new building might disrupt the close-knit culture of his research cadre, molecular biologist Bert Vogelstein later said that once he had toured the nearly completed structure just prior to its opening, his "dread changed to anticipation." Impressed by the extremely well-designed labs and generous space for meeting rooms and offices,

The 10-story, 122,000-square-foot, $59 million new research center named The Bunting Family ◆ The Jacob and Hilda Blaustein Building when it opened in 1999, contains workspace for more than 400 researchers and staff previously scattered all over Hopkins' East Baltimore campus. Despite its size, it not only failed to disrupt the close, collegial interrelationships between Hopkins researchers but enhanced it.

"I realized the fantastic opportunities the new building will bring," he said.[100]

Vogelstein's sentiments were echoed by Drew Pardoll, the pioneering tumor immunologist whose laboratory had produced genetically engineered vaccines that activate the immune system against such malignancies as cervical cancer. Calling his research group "highly interactive," he noted that they'd "been hampered by our geographical dispersion."[101]

"It'll be a heck of a lot nicer to go up one floor in the new building than two-thirds of the way across campus."

The Harry and Jeanette Weinberg Building

A little more than a month later, in January 2000, Hopkins opened the $125 million Harry and Jeanette Weinberg Building, housing a new comprehensive cancer treatment center, plus state-of-the-art surgical unit and operating rooms. The opening followed nearly five years of frustrating construction complications and delays. Rapid developments in cancer-fighting technology had required redesigning portions of the structure; rising construction costs had necessitated additional fundraising; and Baltimore's distant past compelled a lengthy suspension of the project as two long-lost cemeteries were found on the site.[102]

The Weinberg Building, named for a philanthropic Baltimore-born real estate magnate and his wife, whose foundation covered much of its cost, is a seven-floor, 350,000-square-foot cancer treatment center consolidating cancer services previously dispersed throughout the Hopkins Hospital complex. In the lengthy course of its construction, the building's design was revised as more and more departments were added to its list of occupants. Cancer Center Director Martin Abeloff said that bringing all the specialties together meant "a lot less running around for patients." The inclusion of surgery provided the biggest convenience, since patients once needed to travel the length of three buildings to get to an operating room.[103] Sixteen operating suites and an expansive, 20-bed surgical intensive care unit occupied one floor, with pathology located one floor below.

Although hundreds of people had a say in the building's design, nurses had the biggest impact on the layout. The nurses also contributed significantly to the design of the building's top two floors, which had 134 inpatient beds in rooms large enough to allow family visitors to be seated comfortably, plus an area of lockers designed for long-term visitors to store belongings.[104]

Planners also paid considerable attention to infusing the building with natural light, giving it an uplifting atmosphere, as well as to painting its walls with bright colors and adorning them with more than 120 pieces of lively, original artwork. Funds for the art were the gift of a former patient, and many of the pieces were created by Maryland artists. "All patients deserve nice space," Abeloff said when the building finally opened, "but cancer patients spend huge amounts of time in the hospital, so it's especially important for them. I do believe in the power of good design."[105]

Following five years of frustrating construction complications and delays, the $125 million Harry and Jeanette Weinberg Building, housing Hopkins new comprehensive cancer treatment center, opened in January 2000. Designed to be infused with natural light and adorned with more than 120 piece of lively, original art, it provides an uplifting atmosphere for cancer patients who must spend extensive amounts of time there.

2001
November 14: Businessman Sidney Kimmel gives Hopkins $150 million, the largest single gift in Hopkins medicine history, for cancer research and patient care. The Kimmel Comprehensive Cancer Center is dedicated on May 4, 2002.

New York clothing industry billionaire Sidney Kimmel (above, left) never had lived in Baltimore, never had been a patient at Hopkins, and wasn't a University alumnus, but as one of the nation's leading individual donors to cancer research, he gave $150 million—the largest single philanthropic contribution in Hopkins' history—to create the Sidney Kimmel Comprehensive Cancer Center at Johns Hopkins in 2001. He said he considered himself "blessed" to be able provide a gift that would "support one of the leading institutions in the world and build on its momentum" in cancer research and treatment. The structures within the Kimmel Center—the Weinberg Building for patient care and the Bunting ♦ Blaustein Building for cancer research—retained their own names.

The Sidney Kimmel Comprehensive Cancer Center at Johns Hopkins

By November 2001, Abeloff's new building became home to the Sidney Kimmel Comprehensive Cancer Center at Johns Hopkins, created with a $150 million gift—the largest single philanthropic contribution in Hopkins' history—from New York clothing industry billionaire Sidney Kimmel. [106]

Never a patient at Hopkins and not an alumnus, Kimmel had become dedicated to cancer research when his best friend's daughter died of the disease. A layman without a scientific or medical background, Kimmel immersed himself in cancer studies. He quickly learned about the impressive breakthroughs of Hopkins' oncologists—such as their internationally recognized programs in the molecular genetics of cancer, bone marrow transplantation, radiation oncology, brain tumor treatment and research. He also was impressed by how the School of Medicine and Hospital had coordinated the use of federal, state and private funds to support ongoing research. He said he considered himself "blessed" to be able "to support one of the leading institutions in the world and build on its momentum" with his gift.[107]

The Weinberg and Bunting ♦ Blaustein buildings, which are the clinical and research arms, respectively, of the Kimmel Center and retained their own names, formed the cornerstones for Hopkins Medicine's ambitious new building plan. It included not only additional research buildings but literally a new East Baltimore hospital composed of two clinical facilities—one a children's hospital tower, the other an adult critical care tower.[108]

The Koch Cancer Research Building and Broadway Research Building

On the heels of the openings for the Weinberg and Bunting ♦ Blaustein buildings, Hopkins broke ground early in June 2000 for what would become the $140 million, eight-story, 372,000-square-foot Broadway Research Building, which opened in September 2003. Earlier in 2003, on a bright, bone-chilling afternoon in March, ground also was broken for a second 10-story, $88 million twin to Bunting ♦ Blaustein, which initially was known only as the CRBII—short for Cancer Research Building Two—and later was named for industrialist and philanthropist David H. Koch.[109]

The Broadway Research Building—with its six floors of biomedical research laboratories, each configured with equipment and utility requirements unique to individual research occupants—became home to the Institute for Basic Biomedical Sciences, the McKusick-Nathans Institute of Genetic Medicine, the

Institute of Cell Engineering, the laboratory programs in comparative medicine, a microarray facility featuring the latest technology for genetic analysis, and the Proteomics Center, where researchers conduct studies of how the body's proteins function.

The building also housed a 50,250-square-foot "vivarium," the new home for the thousands of rodents and other animals that play a vital role in medical research. Covering more than an acre of space, it featured computer-monitored heat, ventilation, lighting and purified water, as well as every measure possible to ensure that the animals it houses remain disease-free. The noxious job of routinely cleaning 1,000 pounds of soiled bedding from the animals' cages was automated, with two huge, Swedish-made industrial robots—appropriately dubbed "Lars" and "Inga"—saving humans the task of dumping out the cages, putting them on a conveyor belt to a gigantic washing machine, refilling them with new bedding and putting them on clean pallets.[110]

Upon its completion in 2006, the 267,000-square-foot, 10-story Koch Cancer Research Building devoted half of those floors to laboratories where research is being conducted on cancers of the prostate, brain, pancreas, skin, lung, and head and neck.[111]

Providing Superior Care as Patient Populations Surge

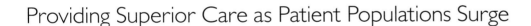

Naturally, the basic aims of all the building projects were to improve patient care, as well as to provide researchers with the facilities required to conduct studies, resulting in better treatments and possibly even the elimination of some illnesses.

The need for new clinical buildings was demonstrated dramatically in 1999 by an unprecedented surge in patient volumes at Hopkins Hospital. Chief Operating Officer Judy Reitz described the Hospital's gain in market share as "phenomenal" and its patient population as at an "all-time high."[114]

Reitz attributed the boom to various efforts undertaken to attract patients—clearly with considerable success. Among these was an initiative to improve relations with referring physicians; another was the opening of the suburban outpost at Green Spring Station (which had led to additional referrals to the Hospital and would be followed in July 2000 by the opening of Johns Hopkins White Marsh); a third was an aggressive pursuit of international business.[115]

Even with the demand for patient services increasing, not just at Hopkins Hospital but at Hopkins Bayview and Howard County General Hospital, and efforts being made to reduce hospitalization time, the provision of superior patient care remained a key Hopkins Medicine priority. Clinical departments and their faculty attained impressive achievements across Hopkins Medicine during this period—from cardiology and infectious diseases to neuroscience, oncology, ophthalmology, otolaryngology, pediatrics, physiology, and surgery.

Erecting new buildings was not the only way Hopkins Medicine responded to its need for space. In April 2003, the university bought five buildings on a 68-acre campus in the Mount Washington area, straddling the Baltimore City-Baltimore County line. For more than a century it was home to the Mount St. Agnes College for Women, then later a site for the United States Fidelity & Guarantee Company. It became home to a number of major administrative offices, as well as faculty and staff from the Department of Emergency Medicine's research, special operations, and CEPAR and PACER offices; Information Technology @ Johns Hopkins; the Alan Mason Chesney Medical Archives; and Johns Hopkins Medicine International.[112]

The signature building on the rolling, tree-covered hills of the campus was the 1855 Octagon Building, originally built to house the Mount Washington Female College. It was the residence of that college's first class, which included Maria Isabella "Belle" Boyd (1843–1900). She would be the most famous alumna of the short-lived school, which closed in 1861, soon after the outbreak of the Civil War in which Boyd would win renown as a Confederate spy—"La Belle Rebelle," in the words of a French journalist.[113]

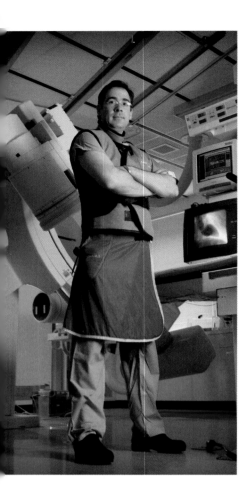

Interventional cardiologist Alan Heldman (above) joined with researchers at the Gerontology Research Center of the National Institute of Aging, located on the Johns Hopkins Bayview campus, to create a new type of medicine-laced coronary-artery stent that delivers a drug to prevent restenosis—the scarring caused by earlier, non-medicated stents that prompted arteries to narrow and become blocked again. Such "drug-eluting" stents "make it possible for us to offer patients treated with angioplasty much better long-term results," Heldman said in 2002. Arriving in 1989 as a resident in internal medicine, Heldman remained at Hopkins until 2007, when he was named clinical chief of cardiology at the University of Miami.

Cardiology

Groundbreaking research, culminating in 2001, was conducted by Hopkins' pioneering interventional cardiologist Alan Heldman on so-called drug-eluting, or drug-coated, coronary artery stents. Stents scar the inside of the vessel in about 20 percent of patients, causing a narrowing and reblocking known as restenosis. Working with scientists at the Gerontology Research Center of the National Institute on Aging, located on the Johns Hopkins Bayview campus, Heldman devised a process for coating the stents with paclitaxel, the active ingredient in the drug Taxol®, to prevent restenosis. A study he published in 2001 is one of the first to show that drug-coated stents inhibit restenosis in pig arteries, a finding that was crucial in the eventual approval of these devices for use in untold heart patients.[116] (Since 2007, Heldman has been the clinical chief of cardiology at the University of Miami.)

Infectious diseases

As this five-year period came to a close, the Hospital's Moore Outpatient Clinic marked its 20th anniversary as the centerpiece of the Johns Hopkins AIDS Service, the largest HIV care program in Maryland. The service provides nearly all the primary and specialty care required by some 3,300 patients annually and is internationally influential because of its exceptional research and incomparable patient database.[117]

Eighteen primary care physicians and eight nurse practitioners and physician assistants in the Moore Outpatient Clinic itself handle some 20,000 patient visits a year. The patients, in various stages of HIV infection, receive service tailored to their particular needs.[118]

The clinic originally was founded in 1915 to do much as it does today, study and treat what was one of that era's incurable, most feared diseases, syphilis. It evolved over time into a multifaceted chronic disease center and was named the Moore Clinic in 1957, following the death of its longtime director, venereal disease pioneer Joseph Earle Moore. Under Moore's successor, Victor McKusick, it became a center for such genetic conditions as Marfan's and dwarfism. In January 1984, however, indefatigable infectious disease expert John Bartlett and epidemiologist Frank Polk changed the clinic's focus to the newest sexually transmitted pandemic, HIV/AIDS. Polk—who had designed the Multicenter AIDS Protocol Study, or MACS, the first large epidemiological study of AIDS—died of a brain tumor in 1988, but the then-small Moore HIV/AIDS clinic grew rapidly along with the spread of the disease.[119]

Prior to concentrating on HIV/AIDS, the rumpled, bald, soft-spoken but intensely focused Bartlett was best known for discovering what he calls "the bug," *clostridium difficile* or *C. difficile*, which causes antibiotic-associated colitis, a chronic, debilitating diarrhea that sometimes afflicts patients on antibiotics.

As director of the Division of Infectious Diseases for 26 years and head of the Moore Clinic, Bartlett oversaw Hopkins' crucial participation in the decade-long, international research and testing effort that became what he later called "the best success story in medicine" since creation of the flu vaccine and the eradication of smallpox: the development of HAART (highly active antiretroviral therapy), the multidrug combination for combating HIV.[120]

Since the introduction of the complex but effective HAART "drug cocktail" in 1997, the HIV/AIDS death rate has plunged by 50 to 80 percent. By carefully calibrating each patient's HAART regimen, caring for HIV/AIDS patients, which Bartlett once had called a "temporizing" effort to forestall an inevitably "morbid death," has become a way to treat a manageable, chronic disease.

Bartlett stepped down as head of the division and the clinic in 2006 and was succeeded by David Thomas, a 13-year faculty veteran and specialist in chronic viral hepatitis, hepatitis C and AIDS.[121]

Oncology

Advances in cancer treatments also marked this period in Hopkins Medicine history. These achievements include:

Landmark work by interventional radiologist J. F. "Jeff" Geschwind, director of cardiovascular and interventional radiology, was central to developing and refining a new procedure called chemoembolization, which successfully kills advanced, inoperable liver tumors by using a catheter to cut off the tumor's blood supply while socking it directly with chemotherapy. By 2001, 80 to 88 percent of patients who had come to Geschwind with diagnoses that gave them just three to six months to live had survived for a year after he treated them with chemoembolization—and 60 to 75 percent had survived for three years.[122]

Saraswati Sukumar, professor of oncology and pathology and co-director of the Oncology Department's Breast Cancer Program, developed a new, molecular-based test that reveals cancers in patients who were shown to be tumor-free both by mammography and clinical examination—a vitally important breakthrough in early detection methods.[123] Sukumar's procedure was able to do what mammography often can't—pick up very early forms of the disease by analyzing the molecules in the cells of milk ducts.[124]

World-renowned pathologist Jonathan Epstein, acclaimed for his expertise in deciphering more potentially cancerous prostate tissue than anyone else, oversaw studies that changed the way cancer patients are treated nationwide.[125] Epstein's examination of a staggering 5,000 specimens of cancerous prostate tissue surgically removed by surgeons Patrick Walsh and Alan Partin formed the foundation for what became known as the Partin tables, a significant diagnostic tool developed by Partin and Walsh in 1997 and now used worldwide.[126] A 1999 report based on two Epstein-led reviews of the tissue slides of 6,171 patients who had been sent by outside physicians to Hopkins for cancer treatment showed that one or two out of every 100 people had received a "totally wrong" diagnosis from the original pathologist's biopsy, signaling the prospect of 30,000 medical mistakes in the United States every year. In the decade since the Epstein group's report about biopsy errors, the move toward obtaining second opinions on pathology specimens has become almost universal in American medicine.

John Bartlett (above, right), head of the Division of Infectious Diseases for 26 years, is seen discussing AIDS cases with his house staff. Regularly working 100-hour weeks, Bartlett conducted research, taught students, treated about 100 AIDS patients a week, wrote scientific papers, and oversaw Hopkins' crucial participation in the decade-long international research and testing effort that produced HAART (high active antiretroviral therapy), which has transformed HIV/AIDS from an invariably fatal affliction to a manageable, chronic disease. He calls the achievement "the best success story in medicine" since the development of flu vaccine and the eradication of smallpox.

■ A few years after the 2000 launch of "DermAtlas," an online Hopkins database of images illustrating case studies of skin conditions, its creators got an education in the law of unintended consequences. Christoph Lehmann, assistant professor of pediatrics and health information sciences, and Bernard "Buddy" Cohen, professor of dermatology and pediatrics, were stunned to discover that Web prowlers for pornography had begun visiting DermAtlas in droves, thinking it contained dirty pictures. It seemed an unauthorized link through a porn site called "gorgasm.com" provided access to it. Lehmann and Cohen wrote a scholarly article about the problem for the Proceedings of the Annual Symposium of the American Medical Infomatics Association— and devised an effective filter for DermAtlas to screen out those looking for what isn't there.

Pediatrics

Hopkins' impact on pediatric care worldwide was highlighted in 2000 with publication of the 50th anniversary, 15th edition of one of the most influential medical texts ever produced: the *Harriet Lane Handbook*, a guide to pediatric care used by virtually every pediatrician in the country for fast, accurate bedside consultation. As in the past, it was compiled by senior pediatric residents at Hopkins and jam-packed with the latest diagnostic guidelines, recommended tests and therapeutic information. Although now fatter than ever at more than 1,000 pages, it still was designed to fit into a lab coat pocket (albeit, barely). It was also made available online. Moving with the times, the *Lane* (as it's known at Hopkins) had new chapters on surgery, psychiatry, oncology—and biostatistics. Maintaining its traditional, three-year schedule, the *Lane*'s 16th edition came out in 2003. Each edition sells more than 100,000 copies in the U.S. and overseas.

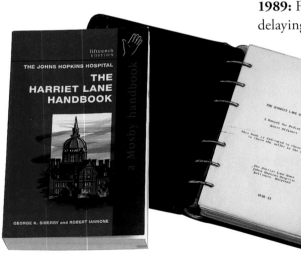

The 50th anniversary edition of *The Harriet Lane Handbook*, above left, and the first edition—both compiled by Hopkins' pediatric residents. The original was a stack of mimeographed pages containing basic clinical information compiled for fellow house officers.

Ophthalmology

The Wilmer Eye Institute ended the 20th century on a high note. In 1999, the magazine *Ophthalmology Times* named the 10 "greatest living ophthalmologists"—and five of them were from Wilmer; and in 2000, the Institute marked its 75th anniversary. Ophthalmologists from as far away as Japan and Australia gathered at Hopkins for a two-day scientific conference and a gala black-tie celebration. They extolled not only the institute's incomparable impact on eye care over the preceding three-quarters of a century, but its current achievements and the certainty of its future accomplishments.[127]

The roster of *Ophthalmology Times'* 10 greatest living ophthalmologists included Wilmer's then-director, Morton Goldberg; his immediate predecessor, Arnall Patz; Alfred Sommer, a Wilmer grad who by then was the dean of the School of Public Health; W. Richard Green, on the Wilmer faculty since 1968; and J. Donald Gass, a one-time resident at Wilmer who went on to be among the founding faculty of the Bascom Palmer Eye Institute at the University of Miami and then, in "retirement," an ophthalmology professor at Vanderbilt.[128]

In the decade just prior to Wilmer's 75th anniversary, its scientists and surgeons had accomplished at least a half-dozen notable firsts in ophthalmological research and care, including:

1989: First use of excimer laser energy to erase scars on the cornea, thereby delaying and, in some cases, eliminating the need for a transplant.

1991: Discovery that a drug, foscarnet, not only treats an infection threatening the vision of AIDS patients but also prolongs their survival.

1994: Worked with the National Aeronautics and Space Administration (NASA) to introduce the Low Vision Enhancement System (LVES, nicknamed "Elvis"), battery-powered, high-tech goggles that enhance contrast and enlarge and brighten images for the severely vision-impaired.

1998: Developed macular translocation, a surgical technique to save sight in some patients with age-related macular degeneration. By 2000, the procedure had been performed on more than 500 patients and found to work well in about a third of them—all of whom would have faced a certain loss of their ability to drive, read and see faces without the treatment.[129]

1998: Discovered that exposure to sunlight increases the risks of getting cataracts.

2000: Proved that the standard medical tests routinely performed before cataract surgery do not measurably improve outcomes or reduce deaths or complications from the surgery, and estimated that stopping such testing could save $150 million annually in direct costs to Medicare.[130]

Otolaryngology

During this period, Hopkins became one of only a handful of medical centers in the nation to provide a dramatic technique for restoring the hearing of patients who have sustained damage that blocks the conduction of sound to the cochlea, or inner ear, which in turn conveys it to the brain. Otolaryngologist Lawrence Lustig and his colleague, John Niparko, head of Hopkins' Listening Center and director of the Division of Otology, Audiology, Neurotology, and Skull Base Surgery, perfected a 30-minute outpatient operation under local anesthesia to implant a tiny titanium post into the mastoid bone behind a patient's ear. That gives them a means to amplify sounds to their inner ear. A tiny mounting bracket is placed on the titanium post, which rises above the surface of the skin, so that a hearing aid can be attached to it after three months of healing. Once the hearing aid is snapped into place and turned on, most patients experience a dramatic improvement in their hearing. Hopkins now has the largest cochlear implant program in the country.[133]

Physiology

In 1999, cellular biologist Douglas Murphy became director of the Optical and Electron Microscope Laboratory and used funding from the School of Medicine's basic science departments and the dean's office to build a center containing an armamentarium of enormously sophisticated—and expensive—microscopes that would have been far too costly for any single departmental laboratory to afford.[134] The simply named Microscope Facility, then located in the subterranean precinct of the Physiology Building's basement, employed amazing new technology during this period that enabled Hopkins researchers to make important discoveries by providing them with astounding images of what actually is taking place in cells, something they had never expected to see.[135] For example, in 2001, psychiatrist and neurologist Christopher Ross pinpointed in a cell the location of the critical proteins that cause Huntington's disease, a genetically programmed degeneration of certain brain cells leading to uncontrolled movements, intellectual impairment and emotional disturbance. Oncologist Stephen Baylin's lab also was able for the first time to understand how key proteins interact to silence genes abnormally in cancer cells.[136]

Transplants

1999: Hopkins transplant experts perform a historic operation that, for the first time, gave a kidney to a patient who had never met the organ's live donor, an unrelated individual who simply wished to offer the recipient a new chance at life.[137]

2001: Hopkins' Comprehensive Transplant Center and transplant surgeon Robert Montgomery launched a pioneering paired kidney exchange program, which helps

Hopkins has the largest cochlear implant program in the country. It counts among its former patients one-time Miss America Heather Whitestone McCallum (above, right), who in 1995 became the first hearing-impaired individual to win a national beauty crown.[131] McCallum had been deaf for nearly 30 years, after having lost her hearing at the age of 18 months during a bout of meningitis. She hadn't minded not being able to hear her name announced when she was proclaimed Miss America, but when one of her young sons fell in her backyard in 2001 and she couldn't hear his cries for help, she began exploring the possibility of getting a cochlear implant.

In August 2002, Hopkins hearing specialist John Niparko and his surgical team implanted the device for McCallum. Following six weeks of recovery from the surgery, she returned to Hopkins for a simple test. Audiologist Jennier Yeagle activated the computerized implant, sat in front of McCallum, who was staring at the floor, and clapped her hands once—hard and loudly. McCallum, startled, looked up, put her hand over her mouth and whispered, "I heard that," as she began to cry soundlessly. The implant had worked. A week later, McCallum told Yeagle and Niparko that she remained overwhelmed at finally being able to hear the conversation between her 3-year-old and 18-month-old sons at breakfast. She told reporters that the sounds she now heard were "a gift...every day."[132]

With his shoulder-length hair and long, flowing, Civil War-era mustache, master kidney transplant surgeon Robert Montgomery (above, left), is a quiet-spoken, unconventional and charismatic surgical pioneer. He has led the teams that performed the first successful "triple swap" kidney transplant operation, the first multi-hospital, three-way kidney transplant swap, and the first six-way, multi-hospital, domino kidney transplant, serving 12 recipients, among other breakthroughs. His work has helped make Hopkins "Kidney Swap Central." Here he chats with Kelly Finan (seated), transplant recipient, and her twin cousins, Andrew and Sean Hayden, who raised $4,105 for Hopkins' transplant center.

Urologist Patrick Walsh (below), director of the Brady Urological Institute from 1974 to 2004, revolutionized the treatment of prostate cancer by developing his landmark nerve-sparing method for removing a cancerous prostate, thereby helping patients avoid incontinence or impotence following surgery. His 45-minute instructional DVD on how he performs the operation—free to all urologic surgeons who request it—has been sent to more than 48,000 practitioners. By the time he set down his scalpel in June 2011, he had performed 4,569 radical prostatectomies. He continues to see patients, teach students and do research.

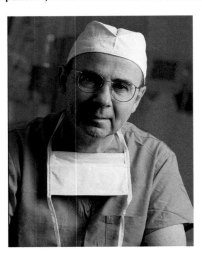

patients obtain a kidney transplant when they have a willing, designated donor whose blood type is compatible with theirs. In a paired kidney exchange, a kidney from such a designated donor is matched and transplanted into a compatible recipient in a second donor-patient duo that provides a kidney for the first patient—and vice versa. The transplant operations are performed simultaneously.[138]

2003: Montgomery oversaw what was believed to be the world's first successful "triple swap" kidney transplant operation, involving three pairs of would-be donors and recipients, six operating rooms, two surgeons, two anesthesiologists and two nurses. All the recipients had to be ready at the same time as the donors' kidneys were removed. After 11 hours of surgery, all the transplanted kidneys were in place, and a woman from Miami, a woman from Pittsburgh and a child from Washington, D.C., had new leases on life.[139]

Urology

In 1999, Patrick Walsh, who was then Hopkins' urologist in chief, noted that many accomplished athletes long had used film and later videotape to perfect their technique. Walsh, creator of the landmark, nerve-sparing method for removing a cancerous prostate, wondered why surgeons also didn't use video technology to improve what they did.[140] He decided to videotape dozens of his own operations and then undertake a meticulous, months-long review of the videos to see if anything he did affected his patients' recovery.[141]

With his patients' consent, Walsh videotaped the operations of 64 men undergoing radical prostatectomy at Hopkins.[142] He wanted to find out why some men retained control over their bladders and their sexual function immediately after surgery while others might not recover full sexual function or remained incontinent for a long time. Hours of intense scrutiny were needed to watch a single two-hour operation several times, often frame-by-frame.[143]

All of the hard work paid off. Although he couldn't uncover any difference in procedure that affected how quickly a patient regained bladder control, he discovered a significant, previously unknown difference in the anatomy of neurovascular nerve bundles that could affect postoperative recovery of sexual function, as well as identified four slight variations in his technique that appeared to make a difference in the men who recovered their sexual function the soonest.[144]

By 2004, he had developed a one-hour, 45-minute DVD providing a detailed description of the prostatectomy procedure he had perfected—and offered copies of it for free to "all urologists in the world who want it."[145] As soon as Walsh made the DVD available, requests for it poured in. So far, more than 48,000 free copies have been sent to urologists worldwide.[146]

In 2001, Walsh also published *Dr. Patrick Walsh's Guide to Surviving Prostate Cancer: Give Yourself a Second Opinion*, co-authored by Janet Farrar Worthington. It immediately became a best seller—and led to a second, updated and expanded edition just six years later. To date, it has sold more than 100,000 copies.[147]

On the Frontiers of Discovery

The construction of three new buildings devoted to research during this period reflected Hopkins' enduring commitment to turning discoveries on the laboratory bench into treatments for use at a patient's bedside. Even as the buildings were going up, Hopkins researchers were making significant advances in genetics, HIV/AIDS, neurology and pharmacology—reporting breakthroughs involving everything from stem cells to broccoli sprouts.

Genetics

2001: Pioneering tumor immunologist Drew Pardoll and his laboratory colleagues announced their development of genetically engineered vaccines that activate the immune system against such malignancies as cervical cancer. Clinical trials of these vaccines are continuing.

2003: Hopkins scientists led by oncologist and pathologist Scott Kern announced the discovery that three genes that were long associated with a rare, inherited form of anemia also appeared to play a role in many pancreatic cancer cases, offering researchers new targets for improving the treatment of patients suffering from this particularly deadly form of cancer. Their study found that the culprit genes could actually be the Achilles heel of pancreatic tumors and might make them more responsive to several existing cancer drugs for those patients who have the anemia-related, mutated genes.[148]

Neurology

2003: Hopkins neurologists Ola Selnes and Guy McKhann, along with Hopkins cardiology colleagues, announce research findings that show that post-operative cognitive impairment of coronary bypass patients, reported in earlier studies, generally was temporary and reversible, lasting for no more than three months. Selnes and McKhann had spent nearly a decade conducting studies to determine if coronary artery bypass grafting (CABG), a frequently used surgical procedure for unblocking clogged arteries, inadvertently leads to brain damage. Their earlier studies showed that nearly a third of CABG patients had experienced changes in their cognitive abilities following surgery, including hard-to-pinpoint alterations in their perception and memory.[153] In a subsequent study, conducted along with cardiac surgeon William Baumgartner and released in 2009, Selnes and McKhann report that evidence showed that any long-term memory losses and cognitive problems that heart bypass surgery patients experienced are due to their underlying coronary artery disease, not to after-effects from the use of a heart-lung machine during surgery.[154]

HIV/AIDS

Overcoming clinical and political obstacles, Hopkins pathologist Brooks Jackson concluded a two-year-long, Uganda-based clinical AIDS-prevention trial in 1999 with stunning results.

Jackson's research showed that the inexpensive antiretroviral drug nevirapine was nearly 50 percent more effective than the costlier drug AZT in cutting the spread of AIDS from an infected mother to her newborn child. The finding was projected to prevent HIV transmission in more than 300,000 newborns annually—for as little as $4 per mother-baby pair.[149] Providing HIV-infected mothers in the United States with AZT had dramatically reduced the instances of newborn babies being infected by the virus, which usually is passed along either during birth or through breast milk. The cost of such treatment was about $1,700 per case, however—far too expensive for the developing nations of Africa, where HIV was striking up to one in three people.[150]

Jackson was determined to find a less-costly way to protect newborns from HIV/AIDS and chose to conduct his clinical trial in Africa, where such a solution was most desperately needed. He picked Uganda because the government there, unlike in other African nations, was confronting the AIDS crisis and trying to educate its citizens on how to avoid the disease.[151]

Today, Jackson's HIV Specialty Lab at Hopkins provides a staggering number of scientific services every month, not just for Hopkins Hospital but also the National AIDS Clinical Trials Group (ACTG) and medical centers around the country. His colleague, Thomas Quinn, director of Hopkins' Center for Global Health, has shown that multidrug therapies can be used to treat HIV inexpensively in Third World countries.[152]

"My greatest fear is that on my tombstone, they'll say, 'He made broccoli famous.'"

— Paul Talalay

Pharmacology

A decade after Hopkins' Laboratory for Molecular Pharmacology, led by the feisty, energetic pharmacologist Paul Talalay, discovered the cancer-fighting properties of sulforaphane gluscosinate, a chemical compound abundant in broccoli sprouts, Talalay and his researchers issued a 2002 study showing that sulforaphane also may help prevent human retinal disease.[155]

Pharmacologist Paul Talalay has been investigating the links between diet and disease for more than 30 years. He coined the term "chemoprotection" to describe the work of his lab in finding substances, mostly from plants such as the broccoli sprouts he holds in this photo, that boost the body's own protective mechanisms. His study that revealed the cancer-fighting properties of sulforaphane gluscosinate, a chemical compound abundant in broccoli sprouts, garnered front-page coverage in the *New York Times*.

Talalay's research team tested three different cell types, including cancer cells and cells from the retina. When the cells were treated briefly with sulforaphane before exposure to an oxidant (potentially destructive molecules produced in cells or from outside sources, such as smoke, alcohol, strong sunlight or fatty foods), all cell types defended themselves against damage. What's more, the response triggered by sulforaphane protected the cells against oxidants for two or three days. The extent of the protection was tied to the amount of sulforaphane, as well as the type of oxidant and amount.[156]

These test results added to the already good evidence "that eating large quantities of vegetables—and cruciferous ones play a special role—is one thing that really works to fight disease," said Talalay, an exuberant octogenarian who had been investigating the links between diet and disease for more than 30 years. He coined the term "chemoprotection" to describe the work of his lab and others like it—finding substances, mostly from plants, that bolster the body's own protective mechanisms. His broccoli sprout breakthrough alone garnered front-page coverage in the *New York Times* and was proclaimed one of the top 100 scientific discoveries of the 20th century by *Popular Mechanics*.[157]

Internationally recognized as one of the world's premier biomedical researchers, Talalay came to Hopkins in 1963 to head the Department of Pharmacology and Experimental Therapeutics. He built the Ph.D. program in pharmacology, and brought to Hopkins and mentored some of the country's most promising young biomedical researchers, including oncologist Donald Coffey, who became director of research at the Brady Urological Institute.

Talalay's lab, now known as the Lewis B. and Dorothy Cullman Cancer Chemoprotection Center in the Institute for Basic Biomedical Sciences, continues to make important discoveries about the disease-combating properties of broccoli sprouts. With wry humor, Talalay told *Hopkins Magazine*, "My greatest fear is that on my tombstone, they'll say, 'He made broccoli famous.'"[158]

Addressing Community Needs

Hopkins' physicians, researchers and alumni may influence medical education, health policies and awareness nationwide, but the faculty, administrators and employees of Hopkins Medicine never have lost sight of the importance of addressing the needs of those who live in the troubled neighborhood surrounding the Hospital, as well as the greater Baltimore community as a whole. Initiatives during this period included programs to mentor students from city elementary and high schools; combat drug abuse and street crime; provide computer training for area residents; recruit employees from local neighborhoods; and supply new community patient services via a roving van and a nursing home collaboration.

Since 2000, Johns Hopkins Bayview's Care-A-Van has arrived four days a week at medically underserved Southeast Baltimore schools and community and family centers, offering well-child and acute care, as well as help with chronic problems such as asthma and obesity. It also provides 24-hour coverage for people with questions at night or on the weekends. Seen above on one of its early runs are (left to right) driver Charles Webster, certified medical assistant Angela Holmes, and physician assistant Patricia Letke-Alexander. Michael Crocetti, head of outpatient pediatrics, said the Care-A-Van's role is "to extend the hospital's doors," but not serve as "a medical home." It connects people with the services they need, such as nearby primary care providers—just as it did in an early encounter with a young Hispanic mother who couldn't speak English and had no idea how to find a physician. The Care-A-Van put her in touch with a Spanish-speaking primary care doctor.[162]

Hopkins Data Services, a unique employment-training and drug treatment program, was launched in 2000 by psychiatrist and behavioral scientist Kenneth Silverman (above, right) at Johns Hopkins Bayview. It is a nonprofit business that provides a therapeutic workplace where recovering addicts earn a living wage to learn and then perform data entry services for academic researchers.[160] It has proven remarkably successful at helping one-time addicts kick their drug dependence. Every Monday, Wednesday and Friday morning, employees of Hopkins Data Services take a urine test to determine if they have remained abstinent. If they pass, they go right to work; if they fail, they go home—until the next day, when they can return and try again. One longtime employee told the Hopkins publication *Dome*, "Now I can pay my rent, buy my kids clothes, buy myself clothes. This is the only income I have, and I'm doin' all right. It's still a struggle to stay clean because where I live, there are drugs all around—up the street, out the back, everywhere I walk is a drug strip. But I have it in my head that I just don't wanna use. So I don't pick up. I would have too much to lose."[161]

Inspiring youngsters to pursue academic achievement was the goal of several programs that by the early 2000s were linking volunteers from Hopkins Medicine to area students from the third grade through high school. At Hopkins Hospital, an annual "Science Day" program, begun in 2000, opens basic science and oncology laboratories to fourth and fifth grades from several East Baltimore elementary schools. About 100 students rotate through the labs and work on experiments. For example, here Tammy Morrish, a postdoctoral student in molecular biology and genetics, shows fifth graders that if they mush up bananas with salt mixed in dish soap, sprinkle on meat tenderizer, layer on rubbing alcohol and— presto!— a white, thread-like stuff appears. The threads are actual strands of DNA, visible to the naked eye.[159]

Hopkins Medicine's Office of Critical Event Preparedness and Response (CEPAR), created after 9/11, has been in the forefront of disaster planning throughout the institution and nationwide. In March 2003, it joined with the Baltimore City Health Department to co-sponsor a disaster preparedness exercise attended by then-Mayor (now Maryland Governor) Martin O'Malley and more than 100 key decision makers from the city's hospitals, public health organizations and fire and police response teams. The interagency exercise, shown here, was held at Hopkins' Applied Physics Laboratory's Warfare Analysis Laboratory (WAL) and dubbed the "WALEX." It was believed to be the first such citywide emergency response drill in the nation.

2002

January 1: Johns Hopkins Bayview Physicians (formerly Chesapeake Physicians), the pioneering faculty practice group, merges with the Johns Hopkins' Clinical Practice Association (CPA), uniting the faculty practices of both hospitals under the CPA. The merger combines 312 physicians from Bayview with nearly 1,200 on the East Baltimore campus, consolidates administrative and billing functions, and creates one of the largest academic group practices in the country.

Jan. 11: Coretta Scott King, widow of slain civil rights leader and Nobel laureate Martin Luther King, Jr., is the keynote speaker at Hopkins' 20th annual Martin Luther King, Jr. celebration, created by cardiac surgeon Levi Watkins, M.D., a friend of the King family.

January: A new master plan for the East Baltimore campus is unveiled, revising the plan created in 1992-93. In April, plans are revealed for a biotech park east of the Hopkins medical campus.

To soften its image as being "too competitive," the School of Medicine overhauls its grading system, eliminating alphabetical grades of A, B, or C.

Preparedness and Response

9/11

On the morning of 9/11, Hopkins physicians, administrators and employees were like Americans everywhere, going about their usual duties—admitting patients, performing surgeries, teaching students, doing research. Then word of the attacks in New York and Washington began filtering in.

Because then-federal Health and Human Services Secretary Tommy Thompson anticipated additional attacks in or near New York or Washington, all of Hopkins Medicine immediately was put on alert to prepare for a surge in patients as the wounded were referred to Hopkins, Bayview or Howard County General Hospital. Medical personnel, especially surgeons, nurses and burn specialists not already on duty, were recalled to their posts. The Baltimore Regional Burn Center at Bayview, the state's top facility in its field, was staffed to accommodate additional patients. An emergency blood drive was launched in anticipation of an increased demand for blood. All elective surgeries and elective admissions and transfers were postponed until further notice. Patients already cleared for discharge were sent home quickly to make available as many hospital beds as possible. Even patients admitted for surgery but not yet under anesthesia were discharged.[163]

Edward Bessman, director of Emergency Medicine at Hopkins Bayview was (and remains) an active member of one of the 28 Federal Emergency Management Agency (FEMA) Urban Search and Rescue teams. He was working in the Emergency Department, helping activate the hospital's disaster plan and preparing to receive burn patients, when he received word that his task force had been put on alert. Within an hour, he was dispatched to the Pentagon to join his 62-member team.[164]

At the Pentagon, he spent a lot of time caring for the pounding headaches suffered by the emergency responders who were continually exposed to jet fuel fumes—as well as the blisters that developed on feet encased in boots waterlogged by efforts to extinguish the flames.[165]

On September 12, the American Red Cross's then-president, the late Bernadine Healy, a former associate dean of the School of Medicine, personally called Ron Peterson to ask for backup help at the Red Cross's national headquarters in Washington. Within an hour, the first Hopkins volunteers were on the road to the capital 40 miles away, followed by three police-escorted van loads the next day.[166]

CEPAR

In the immediate aftermath of the 9/11 attacks and Hopkins' response to them, University President Bill Brody asked that a mass casualty task force be created to plan for any similar crises in the future. Brody looked to Gabor Kelen, veteran head of the Department of Emergency Medicine, as the logical choice to head it.

"It quickly became apparent this wasn't a task-force issue," Kelen recalled a year later. "Terrorism wasn't going to go away any time soon, perhaps for as long as a generation or more." He realized that the many components of Johns Hopkins Medicine—including the immensely diverse educational and research entities throughout the University—needed a unifying crisis structure and single voice.

In December 2001, he recommended that the task force become a permanent body that could tap every Hopkins asset and speak on behalf of all its entities in the event of a calamity. Bill Brody agreed, having already told the federal government that the services of every Hopkins institution would be available to assist in developing a national homeland security plan. In July 2002, the Office of Critical Event Preparedness and Response (CEPAR) was launched officially, with Kelen as its head, and James Scheulen, manager of the Emergency Department, as its executive director.[167]

Department of Emergency Medicine Director Gabor Kelen (above, right) was named head of the Office of Critical Event Preparedness and Response (CEPAR) in July 2002. CEPAR's executive director since its founding has been James Scheulen (left), manager of Hopkins' Emergency Department.

CEPAR combines the talents of Hopkins Medicine with the tactical and planning capabilities of Hopkins' Applied Physics Laboratory (APL), a major research resource for the federal government since World War II; the investigative and planning skills of the Bloomberg School of Public Health; and the faculty and facilities of Hopkins' Krieger School of Arts and Sciences on the Homewood campus.[168]

Since its creation, CEPAR has demonstrated its crisis preparation and response leadership both within Hopkins and nationwide. It spearheaded the 2005 creation of the national Center for the Study of Preparedness and Catastrophic Event Response (PACER), a Hopkins-led, 24-member consortium of academic, business and government entities. Funded with a $15 million grant from the federal Department of Homeland Security, PACER is developing projects to ensure that this nation's immense technological, medical and physical resources can be mobilized effectively in response to any disaster.

Along with the APL, CEPAR also created a special software program, Electronic Mass Casualty Assessment & Planning Scenarios (EMCAPS), which enables hospitals and public agencies to produce realistic projections of their needs should a disaster strike. Using EMCAPS, copies of which CEPAR distributes for free, disaster planners nationwide can estimate casualties arising from biological, chemical, radiological or explosive attacks, adjusting the figures to conform to their specific area's populations and conditions at the time of the disaster.[169]

2002
March: Elias Zerhouni, M.D., executive vice-dean of the School of Medicine, chair of the Russell H. Morgan department of radiology and radiological science and professor of radiology and biomedical engineering, is appointed the 15th director of the National Institutes of Health in May.

The Center for Innovation and Quality Patient Care is founded, ushering in a revolution in patient safety initiatives.

Judy A. Reitz

2002

May 6: The fundraising "Campaign for Johns Hopkins Medicine" is launched, sets a goal of $1 billion by 2007.

May 9: Victor McKusick, the "father of genetic medicine," receives the National Medal of Science, the country's highest scientific honor, at a White House ceremony.

The Neuroscience Critical Care Unit, already the largest in the country, is expanded, ushering in "the latest technology and a new spirit of collaboration."

A New Generation of Leaders

Administrative Changes

In 1999, Ronald R. Peterson, president of The Johns Hopkins Hospital and Health System, appointed Judy Reitz—already senior vice president of operations for Hopkins Hospital and chief architect of its ongoing operations restructuring—as executive vice president and COO of The Johns Hopkins Hospital. She became the first woman to hold that post.[170]

Reitz sought to bridge what had been a longstanding gap in understanding between the central hospital administration and the decentralized (and fiercely independent) functional units, which are made up of clinical faculty and headed by the medical school clinical department directors and their administrators.[171]

Reitz recognized that what often can be seen as Hopkins' maddening decentralization also is among its greatest strengths. She saw no reason, however, why Hopkins couldn't be run as an efficient organization while retaining its unique, invaluable attributes. She continues to strive to see that it does both.[172]

Also in 1999, Peterson relinquished the presidency of Johns Hopkins Bayview, having held it jointly with the presidency of Hopkins Hospital for two years. He was succeeded by Gregory F. Schaffer, the senior vice president of operations since 1995.[173]

New Hands on the Dean's Crew

Beginning late in 1999, Miller began reshuffling duties and job descriptions in the dean's office, making some new appointments that enhanced both the diversity and hands-on quality of the leadership there.

When one woman of dynamic impact—the charismatic and irrepressible pediatrician Catherine DeAngelis, vice dean for academic affairs and faculty since 1991—decided to leave Hopkins to become the first female editor of the *Journal of the American Medical Association* (*JAMA*), Miller chose another titan, Janice Clements, to succeed her in December 1999.[174]

Miller actually had decided it was necessary to take DeAngelis' old job and make it two: a vice deanship for the faculty and a vice deanship for education. He believed that the increasing complexity of issues to be addressed in both realms required the focus of a single individual for each one.[175]

In assuming the role as vice dean for faculty, Clements added multifaceted administrative duties to her academic activities. A Ph.D. in biochemistry and a professor of comparative medicine, neurology and pathology, she became the representative of the 3,000 physicians and basic scientists who make up the School of Medicine's full- and part-time faculty. She also succeeded DeAngelis as head of the School of Medicine's Women's Leadership Council, founded in 1993. In that role, she became a skillful advocate for the advancement of women in Hopkins Medicine. In addition, she successfully urged that her particular area of expertise, molecular and comparative pathobiology, merited elevation

Janice E. Clements

from a divisional to departmental status, a change that finally occurred in 2002.[176]

She graduated with a degree in chemistry from a small Catholic women's college, and after earning a Ph.D. at the University of Maryland, she arrived at Hopkins in 1974 as a postdoctoral fellow in the Department of Molecular Biology and Genetics. In 1976, she received a second fellowship in the Department of Neurology. Clements became a respected investigator of viruses, such as HIV, that use their RNA as a launching pad to incorporate themselves into genetic material. By 1999, she was interim chief of the Retrovirus Laboratory in the Department of Medicine's Division of Comparative Medicine, of which she would become the first director when it became a department.[177]

Having filled the vice deanship for faculty, Miller turned to the new post of vice dean for education and appointed anesthesiologist and critical care medicine specialist David Nichols to fill it in January 2000. To many, Nichols' appointment was another inspired choice.[178]

Nichols seems to have teaching and mentorship in his blood. The eldest of three sons, he is the offspring of two teachers and was born on the campus of the Hampton Institute, a historically black college in Virginia. His father, an English professor and Fulbright scholar, was invited to direct the Freie Universitat in Berlin and moved his family there when Nichols was 7. Growing up, Nichols had no contact with other American youngsters, much less African American children, and even briefly considered becoming a German citizen. He was intrigued by the civil rights and anti-war movements under way in America, however, and decided he wanted to return home to see what was occurring in his native land.[179]

He entered Yale in 1969, intent on becoming a physician, a goal he'd set when he was 6. Yet he also found that he loved working as a student counselor advising other undergraduates on how to pursue a medical career. He went to medical school at Mt. Sinai in New York and did his residency and a fellowship at Children's Hospital in Philadelphia. He never considered applying to Hopkins, he remembered, because he'd heard it wasn't friendly to African Americans.[180]

He found the reality at Hopkins to be the complete opposite of that when he applied for a faculty position in 1984, basically as a courtesy to a friend. Instead of being standoffish, the Hopkins that Nichols encountered was embracing.

"It was *the* most vibrant, *the* most exciting, *the* most stimulating, and, I have to say, *the* most welcoming place I'd visited. I was immediately smitten," he recalled.[181]

He joined the anesthesiology department in 1984. Nichols undertook significant clinical and research work in the bustling Pediatric Intensive Care Unit (PICU), of which he eventually was appointed chief. He became a full professor of anesthesiology and critical care medicine as well as pediatrics, and an extraordinary mentor to postdoctoral fellows who since have gone on to head their own pediatric critical care programs around the world, from Seattle, Washington to Great Britain to Australia.[182]

"I was attracted to Hopkins really for two related reasons. The first was the enormous excitement and energy that I felt when I came to visit [in 1984]. There was so much going on in pushing back frontiers and discovering new things and establishing new fields that you just wanted to be a part of this place.... Related to this was that everybody seemed open to collaborating and working with me.... And really that has been the way for the whole 25 years I've been here.... That quality is worth its weight in gold—to have more senior people who are willing to mentor along those lines."

—David G. Nichols, Vice Dean for Education, Professor of Education, Anesthesiology and Critical Care Medicine and Pediatrics

Switching his research attention in the 1950s from cardiology to genes—the biomedical basis for every living thing's uniqueness—made Victor McKusick the "Father of Genetic Medicine" and ultimately earned him the country's highest scientific honor, the National Medal of Science.[252] Established in 1959 and first awarded in 1962 in recognition of lifelong contributions in the life and physical sciences, social sciences and engineering, the medal is given to individuals nominated by their peers and selected by a 12-person committee, named by the president. McKusick said that receipt of the medal, presented to him by President George W. Bush at a June 12, 2002, ceremony in the White House, honored not just his contributions to the field of genetics, "but also those to colleagues and students over the years."[253] (That week in June proved to be a busy one for McKusick, then 80. The day after getting the medal from Bush, he went up to New York to get an honorary degree from the Rockefeller University—one of some 21 honorary doctorates he had received since 1974.)[254]

As vice dean for education, Nichols established the Office of Academic Computing to harness the rapid advances in educational technology to improve medical and graduate education; created an associate dean's position to ensure continual oversight of what medical students are being taught; and addressed the thorny issue of properly compensating faculty for teaching time. In 2001, he also launched what became an eight-year project to revise the medical school curriculum—a mammoth undertaking described in the next chapter.[183]

Adding another member to the dean's office lineup, Miller tapped Chi Van Dang, a 1982 graduate of the School of Medicine and chief of the Division of Hematology, to become vice dean for research.[184]

Born in Saigon (now Ho Chi Minh City), Dang was one of the 10 children of Viet Nam's first neurosurgeon and dean of the University of Saigon's School of Medicine. In 1967, at the height of the Viet Nam War, his parents obtained refuge for him and a brother by sending them to live with an American family in Flint, Michigan.[185]

After obtaining his undergraduate degree at the University of Michigan and a doctorate in chemistry from Georgetown, Dang arrived at Hopkins in 1978 as a medical student and remained for the next 32 years, becoming a professor of medicine, cell biology, oncology and pathology, as well as the highest ranking physician of Vietnamese descent in academic medicine—not just in the U.S., but in the world.[186] (He left Hopkins in late 2011 to become director of the University of Pennsylvania's cancer center.)[187]

Miller added one more new leadership position in the dean's office, born out of tragic necessity. After the death of human subject research volunteer Ellen Roche and the temporary shutdown of Hopkins' federally funded clinical trials, the institution made sweeping changes in its oversight of clinical trials to ensure the safest possible human subjects research efforts.[188] Among these initiatives was Miller's creation of the new post of vice dean for clinical investigation. He appointed internist Michael Klag, an internationally known

Oncologist, cell biologist and pathologist Chi Van Dang, a 1982 graduate of the School of Medicine, became the highest ranking physician of Vietnamese decent anywhere in the world when he was named vice dean for Hopkins' new Division of Research Affairs in 2000. Dang concluded a three-decade career at Hopkins in 2011 when he was named director of the University of Pennsylvania's cancer center.

expert on epidemiology and prevention of cardiovascular and renal diseases, to the job. While overseeing all studies involving human subjects, Klag worked with Chi Van Dang to develop integrated programs for clinical investigation, ensuring that both biomedical basic science and patient-oriented research studies would be held to the highest standards. When Klag, who also was director of the Division of General Internal Medicine, became dean of Hopkins' Bloomberg School of Public Health in 2005, his successor as the School of Medicine's vice dean for clinical investigation was psychiatrist Daniel Ford, a pioneer in researching the interrelations between mental disorders and such chronic medical conditions as heart disease.[189]

Student Admissions

Also in 1999, cardiologist James Weiss, a clinician, researcher and head of fellowship training in the Division of Cardiology, was appointed associate dean of admissions and academic affairs. He succeeded pediatrician and epidemiologist Leon Gordis, who had held the admissions post for six years while directing the School of Medicine's physicians and society course and its clinical scholars program.

Weiss, a 10-year veteran of the admissions committee, is a 1968 graduate of the Yale medical school and a fellow of the American College of Cardiology, and he also happens to be an accomplished oboist and art connoisseur. His exceptional scientific background, eclectic interests and calm demeanor were well-suited to the complex task of overseeing the selection of incoming Hopkins medical students.[190]

The year after Weiss' appointment, the admissions committee received 6,500 applications for 120 spots in the next class. Those selected represented 72 different colleges and universities, a wide variety of geographic regions and many different walks of life—everything from a professional harpist to an Alaskan fisherman. Admissions administrators continued striving to increase the appeal of the School of Medicine and the diversity of its students. To temper its image as "too competitive," the school overhauled its grading system in 2002, eliminating the alphabetical grades of A, B or C. In some ways, this was a return to a system that had been in place decades earlier, when students were not told what their grades were unless they were doing unsatisfactory work. The letter grading system had been adopted in the 1960s, at the behest of students who wanted more clarity on their standing.[191]

By 2003, the ongoing effort to recruit more women and underrepresented minority students appeared effective. The entering Class of 2007 was 53 percent women, continuing what would be a three-year trend in which women outnumbered men. (The trend temporarily stalled with the Class of '09, which had 52 women vs. 68 men.) The recruitment of underrepresented minority students continued to have an impact, however, with the Class of '09 containing 16 percent minority applicants, up from 11 percent in the Class of '08.[192]

Cardiologist James Weiss, head of the School of Medicine's admissions committee since 1999, oversees the exhaustive review of the thousands of applications for the 120 or so spots in each year's entering class. Each senior member on the committee spends about 200 hours of uncompensated time on the process, winnowing down the applications to the number of individuals deemed eligible for interviews.

Weiss has said he finds the job "a lot of fun, and the psychology and sociology of the process are fascinating." His greatest joy, he said, was the "gratification of playing a role in molding our future…. It's seeing the students when they first come and watching their progress over the years, their growth as they interact with faculty and each other. Some end up on the house staff or faculty, and it's very gratifying to see how critical it is to bring great students here."

2002
July 11: The Robert Packard Center for ALS Research is founded. It is the only one of its kind in the nation dedicated solely to discovering the causes and curing this fatal disease, and its subsequent achievements are impressive.

A former Miss America, Heather Whitestone McCallum, the first hearing-impaired individual to win the national beauty crown (in 1995), receives a cochlear implant at Hopkins. Hopkins has the largest cochlear implant program in the country.

For decades, all of Hopkins' medical students spent countless hours in the venerable William H. Welch Medical Library, among the nation's oldest and most comprehensive repositories of medical information. It obtained a new director in 1999 with the appointment of Nancy Roderer (right), former coordinator of the Associate Fellowship Program of the National Library of Medicine.

From the moment she arrived, Roderer knew that the days of rummaging through the library's stacks for books and journals was coming to an end.[193] Prior to her work at the National Library of Medicine, Roderer had distinguished herself as head of Columbia University's Integrated Advanced Information Management Systems (IAIMS), becoming a leading expert in its operation. She later served as director of the Cushing/Whitney Medical Library at Yale and managed its IAIMS operations.[194] Ever since its opening in 1929, the Welch Library and its directors had played a leading role in the management and delivery of medical information, focusing especially on the development of innovative applications of new technology. At the time of Roderer's appointment, the Library had 367,000 volumes and subscribed to an estimated 2,400 print journals and 1,000 electronic journals.[195] Beginning in 2001, the Welch Library worked with outside consultants and two in-house advisory committees to develop a long-range strategic plan, which ultimately called for a rapid and assertive move into all-digital collections.

As director of the Johns Hopkins Division of Health Sciences Informatics as well as head of the Welch, Roderer began in 2005 to dispatch some members of a new breed of librarians known as "informationists" to work within Hopkins Medicine's departments and divisions. They have become valued assets for research, teaching and clinical practice, and Roderer hopes to add more to the staff in the future. By the time the library hit its 81st birthday in 2010, the number of bound volumes had risen to some 400,000—but the online usage of books, journals and other materials was surging.[196] Roderer anticipates that eventually all of the library's materials will be available online and expects 98 percent of the library's budget by 2012 will go toward electronic materials.[197]

When Hopkins Medicine Speaks, Others Listen—and Learn

Hopkins Medicine's impact on physicians and educators—and the patients they both ultimately serve—remained as far-reaching as ever as the new century approached. Just as Sir William Osler's *The Principles and Practice of Medicine* and the Hopkins Children's Center's *Harriet Lane Handbook* had, over time, become among the most influential medical texts ever published, other Hopkins publications, continuing medical education programs, radio news broadcasts and websites during this period brought the latest in medical developments to physicians, researchers and patients alike.

Continuing Medical Education

By 2000, what had begun a quarter century earlier as a modest program through which a few Hopkins professors offered seminars on the latest developments in medicine to outside physicians and in-house colleagues had developed into the nation's largest, self-supporting continuing medical education enterprise.[198]

Literally every branch of medicine—and even the business of medicine—were covered in hundreds of Hopkins' continuing education activities. Every year, tens of thousands of physicians and other health care providers across the country and overseas were benefitting from the knowledge and insights of Hopkins experts, as well as occasional visiting instructors from other top medical schools.[199]

According to Hopkins' CME's current leader, Associate Dean Todd Dorman, in 2006 alone, upward of 75,000 health care professionals enrolled in the more than 380 educational activities provided by Hopkins' ever-growing CME program. It also earns Hopkins Medicine millions of dollars in gross revenues per year.[200]

In Print, On the Air, Over the Web

The success of Hopkins' continuing medical education program was just one example during this period of the power of the Hopkins Medicine name and the public's desire for the knowledge it disseminated.

In January 1999, Hopkins' then-Office of Consumer Health Information (since absorbed by Hopkins' Health Information Management Group) published the *Johns Hopkins Family Health Book*, a nine-pound, 1,650-page compendium of consumer health information, spreading Hopkins' expertise to a wide audience. More than 100 members of the faculty participated in its writing and editing—and three medical artists from the Department of Art as Applied to Medicine produced its illustrations.[201]

Such consumer health publications were not new to Hopkins. A decade earlier, its Office of Communications and Public Affairs (now the Office of Marketing and Communications) had partnered with an outside publisher, Rebus, to launch the newsletter *Health After 50*, which by 1999 had a half-million subscribers. A year before the *Family Health Book* was published, another Rebus product, the *Johns Hopkins Complete Home Encyclopedia of Drugs*—edited by biological chemist Simeon Margolis, also the *Health After 50* editor—quickly sold out its initial 25,000-book press run. Geared especially to people over 50, it described in detail more than 700 generic and 2,300 brand-name prescriptions and over-the-counter drugs.[202]

Hopkins Medicine also was a regular presence on the radio and Internet by late 2000, when it marked the 1,000th broadcast of the popular, 60-second Hopkins Health Newsfeed reports. During the same time period, a regular Hopkins-prepared package of faculty commentaries on medical news and helpful Ask-the-Doc items was reaching between four and five million viewers of the InteliHealth Web site.

Michael Klag (above), an internationally known expert on the epidemiology and prevention of heart and kidney disease, was editor in chief of *The Johns Hopkins Family Health Book*, 1,650-page volume of consumer health information published in 1999. Shown here holding the hefty tome, Klag would be named the School of Medicine's first vice dean for clinical investigation in 2001, a post in which he oversaw research involving human volunteers. In 2005, he became dean of the Bloomberg School of Public Health, from which he had earned a master of public health degree in 1987, the year he joined the School of Medicine's faculty.

Advocacy and Influence

No matter how widespread the reach of the airwaves and Internet, a single physician with a major message can have a powerful impact, as the late endocrinologist Christopher Saudek, founder and director of the Johns Hopkins Diabetes Center, demonstrated vividly during his year as president of the American Diabetes Association between 2001 and 2002.[203]

To fight the mounting epidemic of diabetes, which by 2001 was affecting approximately 17 million Americans, Saudek launched a personal crusade during his year as ADA president, initiating a campaign to raise public awareness about diabetes' link to heart disease and stroke. His effort led to newspaper headlines, cover stories in both *Time* and *Newsweek*, and the *Journal of the American Medical Association*'s first-ever issue devoted entirely to the condition. Tommy Thompson, then federal Secretary of Health and Human Services, made diabetes prevention and management a top priority.[204]

At left, endocrinologist Christopher D. Saudek (1942–2010), founder and director of the Johns Hopkins Comprehensive Diabetes Center and a pioneer in the development of implantable insulin pumps, served as president of the American Diabetes Association from 2001 to 2002. During that time, he launched an immensely effective, nationwide campaign to raise public awareness about diabetes' link to heart disease and stroke. "People with diabetes need to have the motivation and the knowledge to treat themselves hour by hour, day by day and year by year," Saudek said. "This is why diabetes education is so important." Saudek traveled widely to lecture about diabetes, its complications and how to prevent them. His audiences included students, medical residents and other health care professionals at all levels.

Passages

During this period, some of Hopkins Medicine's most influential past leaders died, reminding their successors of the breadth of the shoulders on which they stood.

Daniel Nathans, whose brilliance in the laboratory earned him the Nobel Prize and the U.S.'s highest scientific award, the National Medal of Science, and whose soft-spoken intensity, unquestioned integrity and devotion to Hopkins led to his appointment as interim president of the University during the most tumultuous time in the School of Medicine's and Hospital's history, died of leukemia on November 16, 1999, at his Baltimore home at the age of 71.[205]

Carol Johnson Johns (1923–2000), a 1950 graduate of the School of Medicine and internationally renowned expert on lung diseases, was the mentor to generations of female medical students and junior faculty.

Nathans once said he had never so much as "fantasized" about winning a Nobel Prize, and he certainly hadn't anticipated being named interim president of Hopkins. He took the job, he said, because he'd "always considered it a privilege to be here, to lead the kind of life Hopkins has given me the opportunity to lead."

"Virtually everyone who spends time here develops an uncommon affection for the place," Nathans observed. "My hypothesis is that it's due to a peculiar virus, call it JHV, that gets into the chromosomes and there expresses its genes for the rest of one's life."[206]

Only three months after Nathan's death, Hopkins lost lung disease expert Carol Johnson Johns, who died on February 24, 2000, of melanoma at the age of 76. For nearly 50 years, Johns had carried the torch for Hopkins women. A member of the School of Medicine's Class of 1950 and married to Richard Johns, Class of '48 and founder of the Department of Biomedical Engineering, she epitomized the satisfactions—and struggles—of being a woman in medicine.[207]

A longtime mentor to female medical students and junior faculty, she was a staunch champion of women's education. Actively involved in the Johns Hopkins Women's Medical Alumnae Association, serving as its director three times, she also was director of the Stetler Research Fund for Women Physicians, co-chaired a national conference on Women Physicians in Contemporary Society, and founded the Women's Task Force for Faculty Careers in Medicine at Hopkins.[208]

An international authority on tuberculosis and sarcoidosis, a disease of unknown cause in which inflammation occurs in the lymph nodes, lungs and other bodily tissues, Johns' reputation grew even as she raised three sons, worked part-time as an unpaid volunteer director of the Hopkins sarcoid clinic she had founded, and constantly struggled, as she once said, to keep a "toe in the academic door."

Her ability to juggle all these roles and excel in each one of them "made her a mythical figure among medical students and junior faculty," recalled oncologist Georgia Vogelsang, a 1980 School of Medicine graduate.[209]

In 1974, Johns was named "Medical Woman of the Year" by the Medical College of Pennsylvania, and in 1979, her undergraduate alma mater, Wellesley College, named her acting president, a position she held for 18 months. At Hopkins, she would serve as assistant dean and director of continuing medical education from 1981 to 1993. She became the first female member of the century-old American Clinical and Climatological Association and then its first woman president in 1994. A year before her death, she published a widely acclaimed article in the journal *Medicine* titled "The Clinical Management of Sarcoidosis: A 50-year Experience at the Johns Hopkins Hospital." She was appointed a full professor of medicine just four months before she died.[210]

"For more than 20 years, Carol Johns was always there for me," pathologist Karen King, a 1986 graduate of the School of Medicine, told *Hopkins Medical News* after Johns' death. "Now that she's gone, several of us women on the faculty have wondered: Who will look after us now?"[211]

During a three-week span in the spring of 2001, two former presidents of The Johns Hopkins Hospital died, both having left enduring legacies.

Russell Nelson, who headed the hospital during a two-decade period of substantial expansion between 1952 and 1972, died on May 19, 2001 at his retirement home in Naples, Florida. He was 88. Less than a month later, on June 13, Robert Heyssel, another 20-year head of the Hospital, as well as the founder and first CEO of the Johns Hopkins Health System, died of lung cancer in his retirement community of Seaford, Delaware. He was 72.[212]

The 20 years of Nelson's Hospital leadership demonstrated not only his talent as a hospital administrator but also as an innovator in the field of health care delivery.[213] His influence ranged well beyond Hopkins. He was a consultant on the design and operation of medical complexes in Germany and the Netherlands, as well as a member of a five-person U.S. delegation that was sent in 1965 to study hospital services in what then was the Soviet Union. He was president of the American Hospital Association from 1959 to 1960, chaired the board of commissioners of the Joint Commission on the Accreditation of Hospitals from 1961 to 1962, and also was chairman of the Executive Council of the Association of American Medical Colleges.

A 1937 graduate of the School of Medicine, Nelson retained an interest in clinical research even as he pursued an administrative career. He published papers on penicillin therapy for syphilis, in addition to numerous articles on hospital management and the delivery of health care. His leadership in hospital administration won him widespread recognition. He was the recipient of the Distinguished Service Award from the American Hospital Association, as well as honorary degrees and other awards from hospitals in the U.S. and overseas.[214]

Heyssel's 20-year tenure at the helm of Hopkins Hospital, as described earlier, was an even more turbulent two decades than those through which

As president of The Johns Hopkins Hospital from 1952 to 1972, Russell Nelson (1913–2001) demonstrated his talents not only as a hospital administrator but also as an innovator in the field of health care delivery. Under Nelson's stewardship, the Hospital greatly expanded its physical plant and made sweeping changes in its financial structure to accommodate pioneering prepaid comprehensive health care programs in East Baltimore and Columbia, Md.

2002
September: The Johns Hopkins Office of Critical Event Preparedness and Response (CEPAR) is created.

Executive Safety Rounds are begun, with leaders such as Hopkins President William Brody and Dean/CEO Ed Miller "adopting" intensive care units around the Hospital and visiting with them regularly to emphasize that patient safety is paramount.

Hopkins Medicine marks the 20th anniversary of its partnership with the U.S. Military Health System.

The Hopkins Medicine family is immensely saddened when three extraordinarily talented and admired young physicians—including one who was featured prominently in *Hopkins 24/7*—die of sudden cardiac arrest during a seven-month period: Frederick (Rick) Montz, M.D., a nationally recognized authority on gynecologic cancer who became known to millions via *Hopkins 24/7*; David Nagy, M.D., Ph.D., director of perinatal outreach; and Jeffrey Williams, M.D., one of the world's foremost radiosurgeons. Nagy dies on April 21; Williams, class of '77, dies on May 26; Montz dies on November 21.

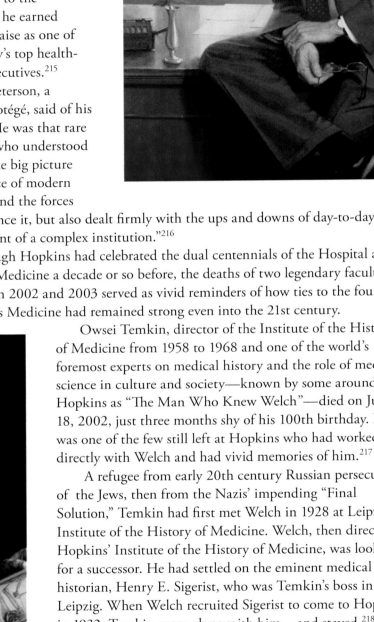

This oil portrait of Hopkins Hospital President Robert Heyssel (1928–2001) is on display in the 440,000-square-foot, $140 million outpatient center named for him and formally dedicated in 1992 across North Broadway from the Hospital. During his 20-year tenure as head of The Johns Hopkins Hospital, Heyssel earned a nationwide reputation as one of the country's savviest, most forward-looking health care executives.

Medical historian Owsei Temkin (1902–2002) was recruited by William Welch himself to come to Hopkins in 1932 and ultimately would hold the William H. Welch Professorship in the history of medicine. Known as a captivating lecturer, supportive teacher, and prodigious writer, publishing hundreds of articles on the history of medicine, Temkin served as president of the American Association for the History of Medicine from 1948 to 1968 and as editor of *The Bulletin of the History of Medicine.* He also wrote a dozen books. His last one, *On Second Thought and Other Essays in the History of Medicine and Science,* was published the year he died.

Nelson had steered the institution. Successfully responding to the challenges, he earned national praise as one of the country's top health-services executives.[215]

Ron Peterson, a Heyssel protégé, said of his mentor: "He was that rare physician who understood not only the big picture and practice of modern medicine and the forces that influence it, but also dealt firmly with the ups and downs of day-to-day management of a complex institution."[216]

Although Hopkins had celebrated the dual centennials of the Hospital and School of Medicine a decade or so before, the deaths of two legendary faculty members in 2002 and 2003 served as vivid reminders of how ties to the founders of Hopkins Medicine had remained strong even into the 21st century.

Owsei Temkin, director of the Institute of the History of Medicine from 1958 to 1968 and one of the world's foremost experts on medical history and the role of medical science in culture and society—known by some around Hopkins as "The Man Who Knew Welch"—died on July 18, 2002, just three months shy of his 100th birthday. He was one of the few still left at Hopkins who had worked directly with Welch and had vivid memories of him.[217]

A refugee from early 20th century Russian persecution of the Jews, then from the Nazis' impending "Final Solution," Temkin had first met Welch in 1928 at Leipzig's Institute of the History of Medicine. Welch, then director of Hopkins' Institute of the History of Medicine, was looking for a successor. He had settled on the eminent medical historian, Henry E. Sigerist, who was Temkin's boss in Leipzig. When Welch recruited Sigerist to come to Hopkins in 1932, Temkin came along with him—and stayed.[218]

Nearly 25 years after Welch's death in 1934, Temkin would be named to the William H. Welch professorship and become head of the History of Medicine Institute that Welch had founded. Known as a captivating lecturer, supportive teacher and prodigious writer, he served as president of the American Association for the History of Medicine from 1948 to 1968 and as editor of the *Bulletin of the History of Medicine.*[219]

Almost exactly a year after Owsei Temkin died, Hopkins also lost Benjamin M. Baker Jr., another physician who had "known Welch" and then become the physician to such celebrity patients as the Duke of Windsor (the former King Edward VIII of England), actor Clark Gable, writers F. Scott Fitzgerald and H. L. Mencken, as well as a medical consultant to General Douglas MacArthur during World War II. A one-time Rhodes Scholar and accomplished athlete, Baker died on July 14, 2003, at the age of 101.[220]

The son of a "horse and buggy" doctor in Virginia, Baker was a star athlete at the University of Virginia. He then received a Rhodes to attend Balliol College at Oxford, where he became acquainted with the widow of Sir William Osler. He also ran track against famed Olympians Harold Abrahams and Eric Liddell, both portrayed in the 1982 movie *Chariots of Fire*.[221]

A 1927 graduate of the School of Medicine, Baker completed his studies with honors in just two years. He then served his internship and residency at Hopkins before going into private practice—while also volunteering six mornings a week as a clinician in the Hospital and instructor in the School of Medicine.[222] He would become a full professor even though he was only a "part-time" member of the staff for more than 35 years.

During World War II, Baker was chief of medical services for Hopkins' 18th General Hospital in the Fiji Islands. He led a team that investigated the use of atabrine as a preventive measure against malaria and also served as a combat physician during the invasion of Okinawa. Following three years of overseas duty, he was headed home when General MacArthur intervened, recruiting him to become chief consultant in medicine as U.S. forces were preparing to invade Japan in 1945, prior to the atomic bombings of Hiroshima and Nagasaki that ended the war. Baker received the Legion of Merit and five combat stars for his service.[223]

Known as a master diagnostician, Baker was among the first to study the link between diet and coronary heart disease, and he pioneered the use of the balistocardiogram, a sensitive device that measures the force of a patient's heartbeat, in the diagnosis and evaluation of heart disease.[224]

In the early 1980s, Baker's interests turned to colon cancer. He first became convinced that the mysteries of the disease could best be solved by a team approach, then recruited top Hopkins oncologists, pathologists, geneticists, molecular biologists and gastroenterologists to join the Hopkins Bowel Tumor Working Group, which he established and funded through charitable contributions that he raised. More than 100 scientific articles resulted from the group's work, including landmark studies that explained the molecular genetics of colon cancer.[225]

Master diagnostician Benjamin M. Baker Jr. (1902–2003) was a Rhodes scholar, accomplished athlete, physician to such celebrities as actor Clark Gable and author F. Scott Fitzgerald, chief of medical services for Hopkins' 18th General Hospital unit overseas during World War II, and creator of Hopkins' Bowel Tumor Working Group for the study of colon cancer. Former School of Medicine dean and chief of cardiology Richard Ross said of him: "He was a magnificent man. He had it all. He was endowed with the charm of a Virginia gentleman, the easy grace of an athlete and the intellect of a medical scientist and physician. He was a role model for generations of Hopkins students and residents."

In "Partners of the Heart," the award-winning Public Broadcasting System's 2003 documentary about the collaboration between surgeon Alfred Blalock and African American surgical technician Vivien Thomas, actor Chris Haley (above, left) portrayed Thomas, shown here as other employees in the segregated world of the 1940s Johns Hopkins Hospital look askance at a black man in a white coat.

Accolades

The remarkable breadth of experience and achievement in the lives of many such Hopkins physicians could provide source material for television programs or films as gripping as *Hopkins 24/7* or *Nurses: Hearts of Mercy, Nerves of Steel* early in this period. Yet it was one particular story—about the unlikely collaboration in the segregated 1940s between a renowned Hopkins surgeon who was white and an African American laboratory technician—that inspired a 2003 TV documentary and a subsequent cable TV feature film. Both won national awards, just as a half-dozen currently active Hopkins physicians garnered impressive accolades between 1999 and 2003—including MacArthur Foundation "Genius" awards, the National Medal of Science, and the granddaddy of them all, the Nobel Prize.

Partners of the Heart and *Something the Lord Made*

In February 2003, the Public Broadcasting System's series, *The American Experience*, began airing "Partners of the Heart," a four-hour documentary about the remarkable, medical history-making collaboration between Alfred Blalock and Vivien Thomas, described in Chapter 2, that led to the 1944 "blue baby" operation and the birth of cardiac surgery.[226]

The documentary's title came from Thomas' autobiography, published only two days before his death in 1985. That book in turn led to a 1989 article in *Washingtonian* magazine. Written by Katie McCabe, it was titled "Like Something the Lord Made" (the expression Blalock had used to describe the results of a surgical procedure devised by Thomas) and earned her the 1990 National Magazine Feature Writing Award. Among the article's fascinated readers were documentary producer and director Andrea Kalin, who knew the Blalock-Thomas partnership would make a riveting TV program, and a Washington, D.C. dentist, Irving Sorkin, who thought it would make a terrific movie.[227]

The Organization of American Historians gave "Partners of the Heart" its award as the outstanding documentary of 2003, and later that year, GlaxoSmithKline, the multinational pharmaceutical company, launched the Vivien Thomas Scholarship for Medical Science and Research, administered by the Congressional Black Caucus, to encourage African American students to pursue degrees in medicine or science. Each year, the fund awards a total of $100,000 in Thomas scholarships. Kalin said she was most proud of the scholarship program's recognition of Thomas. "That's how he should be remembered," she said.[228]

Almost simultaneously, as Kalin was rounding up the funding to turn the Blalock-Thomas collaboration into a powerful documentary, Sorkin was trying to persuade movie studios to dramatize the tale. By the fall of 2003, he had succeeded, and movie crews and actors from Home Box Office were readying themselves to descend on Hopkins Hospital and other Baltimore locations to begin the January 2004 filming of an HBO movie that, harkening back to the original title of the *Washingtonian* article, was called *Something the Lord Made*. Its stars were almost as unlikely a set of partners as Blalock and Thomas: British actor Alan Rickman (best known as Professor Snape in the popular *Harry Potter* movies), who successfully transformed his Queen's English into Blalock's broad Georgia drawl, and rap star Mos Def, whose restrained, convincing and dignified portrayal of Thomas impressed surviving family members.[229]

British actor Alan Rickman (above, left) and rap star Mos Def earned universal acclaim for their portrayals of Alfred Blalock and Vivien Thomas, respectively, in Home Box Office's Emmy award-winning 2004 film, *Something the Lord Made.*

As they had with the ABC News documentarians, many at Hopkins helped the HBO company make its "docudrama" a film that, while juxtaposing fact with fiction, would prove to be as true to the men, the medicine and the times as possible. The effort did not always go smoothly. "It sometimes was a real clash of cultures between an institution with a heritage to protect and a network that wanted to make an entertaining film," observed Gary Stephenson, then-associate director of media relations for Hopkins Medicine, who served as Hopkins' coordinator for the project.[230]

A number of former colleagues of Blalock and Thomas, including retired chief cardiac surgeon Vincent Gott and retired chief pediatric surgeon Alex Haller, were among those who read the initial script for the film and grew uneasy over its overall tone and inaccuracies. They were ready to distance themselves and Hopkins from the production unless some of the most serious flaws were fixed. To their surprise, the HBO decision-makers not only listened to their criticisms but agreed to make numerous changes for the sake of accuracy.[231]

The HBO people learned from their encounters with Hopkins personnel. Producer Robert Cort later said, "In the course of making the film, I came to understand the importance of the heritage of excellence at Johns Hopkins."[232]

The film, which aired in May 2004, was critically acclaimed. It received three Emmy Awards, including Best Made for Television Movie; and in 2005, it received the University of Georgia's prestigious George Foster Peabody Award for broadcasting excellence in both news and entertainment. For his portrayal of Thomas, Mos Def won both an Image Award from the National Association for the Advancement of Colored People and a Black Reel Award.[233]

Less than a week before *Something the Lord Made* debuted on HBO, Hopkins established the Vivien Thomas Fund for Diversity to increase the number of minorities in the academic medicine talent pool.[234]

Veteran Hopkins surgeons Vincent Gott (above, left) and R. Robinson Baker (center), who knew and worked with both Alfred Blalock and Vivien Thomas, acted as consultants on *Something the Lord Made* and were impressed by the dedication and skill with which its stars such as Mos Def (right) captured the personalities of the men they portrayed.

David Sidransky, director of Hopkins' Head and Neck Cancer Research Laboratory and a professor of otolaryngology–head and neck surgery, oncology, pathology, urology, and cellular and molecular medicine, was dubbed the "Cancer Spotter" when named to *Time* magazine's list of 18 exemplars of "America's Best" in 2001. He was praised for developing new tests that can detect cancer more accurately and at an earlier stage than previously possible.

■ In 1988, a 2,500-year-old Egyptian mummy, nicknamed "Boris," was the oldest individual ever examined at Hopkins Hospital. He was brought to the Hospital for radiologic scans. CT scans had been used on mummies before, but Hopkins was the first to employ a sophisticated computer program to enhance the images into three dimensions.

"America's Best"

In the August 20, 2001 issue of *Time* magazine, a list of 18 of "America's Best" individuals in science and medicine included two Hopkins physicians, "Cancer Spotter" David Sidransky and "Super Surgeon" Benjamin Carson. Both, the magazine's Michael Lemonick wrote, were among the "brilliant individuals with a passion for understanding the world and the ability to concentrate obsessively on a problem until they have solved it." They were, Lemonick said, "at the forefront of asking the crucial questions and finding the breathtaking answers."[235]

A graduate of Brandeis University in New York and Baylor College of Medicine in Houston, Sidransky was by no means a stranger to those in the cancer research and clinical fields. Arriving at Hopkins as a clinical oncology fellow in 1988 and later an oncology research fellow under Bert Vogelstein, he had joined the Hopkins faculty in 1992 and was named director of the Hopkins Head and Neck Cancer Research Laboratory in 1998.[236]

Time magazine was especially impressed with Sidransky's pioneering work in developing what writer Alice Park called "a new generation of tests that can detect cancer more accurately and at a much earlier stage in the progress of the disease." Sidransky's research targets were blood, urine, saliva, stool and DNA, in which he and his 20-member laboratory seek to identify traces of the molecular changes that indicate a cell is becoming cancerous. In one study of bladder-cancer patients, a screening test that Sidransky developed picked up more than 90 percent of the tumors, "a hit rate that could revolutionize the early detection and treatment of bladder cancer," Park wrote.[237]

The DNA of other cancers can be harder to detect, however, and Sidransky continues searching for additional genetic clues that could improve the methods for catching cancer when it's just beginning and launching the fight against it then. "This approach may one day lead to a simple blood test to detect cancers at an early stage," Sidransky explains. "No matter what area we research, we always try to bridge basic research into the clinical setting."[238]

Ben Carson was no stranger to the limelight, either, when *Time* decided to single him out as one of "America's Best." As early as 1993, the *New York Times* had called Carson "arguably the most famous surgeon on the Hopkins staff" in a profile it headlined, "For Many, Pediatric Neurosurgeon is a Folk Hero." By 1999, he also was a professor of oncology, pediatrics and plastic surgery, as well as neurological surgery. As *Time*'s Christine Gorman observed, as dramatic and inspiring as Carson's surgical achievements were in successfully separating conjoined twins joined at the head; in developing important new surgical techniques for performing a hemispherectomy, the dramatic removal of half a child's brain to end crippling seizures; or in successfully completing the first intra-uterine procedure to relieve pressure on the brain of a hydrocephalic fetal twin—"they almost pale before the story of his life." Had he not gone to medical school, Gorman wrote, Carson "might just as easily have landed in jail."[239]

Born in inner-city Detroit and raised by a devoted single mother who often worked three jobs in order to support him, Ben Carson initially received poor grades in school and struggled with a violent, uncontrollable temper. His mother,

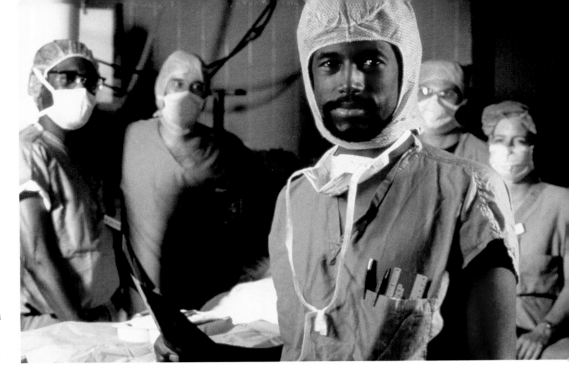

Sonya, had only a third-grade education, could barely read, married at 13, and had been left to bring up Carson and his older brother on her own after she and the boys' father divorced when Ben was eight.[240]

A devout believer in education because of her own lack of it, Sonya Carson succeeded in at first compelling, then inspiring her sons to read and excel in school. Yet Ben still was plagued by a fierce temper that led to furious, potentially dangerous outbursts. After nearly stabbing a friend to death following an argument over the choice of a radio station (only the friend's belt buckle prevented a possibly mortal wound), Ben was so shattered by the experience that he ran home, locked himself in the bathroom with a Bible, and emerged with a resolve to control his temper.[241]

Graduating with honors from high school, Carson worked summer jobs and won a scholarship that enabled him to attend Yale, where he earned a B.A. degree in psychology. At the University of Michigan's medical school, he realized that his skill at visualizing the brain in three dimensions and his excellent hand-eye coordination would serve him well if he pursued neurosurgery. Upon graduation in 1977, he came to Hopkins, where he did his internship, clinical and surgical residencies and fellowships, and joined the faculty. By 1984, at the age of just 33, he was named director of the Division of Pediatric Neurosurgery. In 1991, appointment as co-director of the Johns Hopkins Cleft and Craniofacial Center followed.[242]

Carson's impact went far beyond the operating room. His best-selling 1996 autobiography, *Gifted Hands*, and his 2001 philosophical and motivational book, *The Big Picture* (followed in 2006 by *Think Big*), made him a much-in-demand public speaker. He remains especially interested in addressing high school audiences, delivering up to 50 speeches a year. "I've been trying to get young people, especially from disadvantaged neighborhoods, to know that there are other things to aspire to be than being basketball players and rappers," he says.[243]

Carson and his wife, Lacrena (Candy), also founded the Carson Scholars Fund in 1994 and the Ben Carson Reading Project in 2000 to encourage students in grades four to 11 to maintain high academic achievement and demonstrate strong humanitarian qualities. In the succeeding years, Carson scholars have received more than $800,000 in scholarship funds and attended more than 200 colleges and universities, while more than 45 Carson reading rooms have been opened in schools nationwide.[244]

Time magazine wasn't the only one to recognize Carson in 2001. That year, CNN also named him as one of the nation's top doctors and scientists, and in celebrating its own 200th birthday, the Library of Congress chose him as one of 89 "Living Legends." Even more honors—and achievements—for Carson would follow in the years ahead.[245]

Pediatric neurosurgeon Ben Carson, professor of oncology, pediatrics, plastic surgery and neurological surgery, has become what the *New York Times* called a "folk hero," with an impact that goes far beyond the operating room. His best-selling books, including his autobiography, *Gifted Hands* and several motivational texts; his extensive public speaking; and the student scholarship programs he and his wife Lacrena (Candy) Carson founded, all have earned him numerous, well-deserved accolades.

"When I was a kid in Detroit and I would hear a story about medicine on the radio or the television, so often those stories had something to do with Johns Hopkins. So even as a little kid, I was focused on Johns Hopkins...."

—Benjamin S. Carson Sr., Director of Pediatric Neurosurgery, Co-director of the Craniofacial Center, Professor of Neurological Surgery, Oncology, Plastic Surgery and Pediatrics, Recipient of the Presidential Medal of Freedom

The best-selling 1995 memoir of Kay Redfield Jamison (above), *The Unquiet Mind*, recounting her own battles with manic depression, and her subsequent books, scientific papers and advocacy efforts, have played an important part in increasing public awareness of psychiatric disorders—while earning her a MacArthur Foundation "genius" award in 2001.

Insights into how life systems work, obtained by developmental biologist Geraldine Seydoux through her long, meticulous study of the tiny, transparent roundworm *C.elegans*, garnered a MacArthur Foundation "genius" award in 2001. Seydoux (center) was surrounded by her research team in this photo, taken shortly after she received the unrestricted MacArthur grant.

MacArthur "Geniuses"

Since 1981, the John D. and Catherine T. MacArthur Foundation has been presenting unrestricted, five-year MacArthur Fellowships—promptly dubbed the "genius" awards—in recognition of and support for the creative efforts of those committed to discovering and advancing knowledge and improving lives. In 2001, two Hopkins School of Medicine professors, psychologist and psychiatry professor Kay Redfield Jamison and developmental biologist Geraldine Seydoux, received the award.[246]

Jamison's best-selling 1995 memoir, *An Unquiet Mind*, recounting her own battles with manic depression, or bipolar disorder, was credited with increasing public awareness of psychiatric disorders, enhancing mental health treatment, and improving patient support and advocacy. Co-director of Hopkins' Mood Disorders Center, Jamison also is the author of numerous scientific papers, an influential medical text on manic-depressive illness, and of several other best-selling books for a general audience. Her work has been instrumental in helping to increase the understanding of suicide and serious mood disorders.[247]

Geraldine Seydoux attained fascinating insights into how life systems work not by studying as complex an organism as a human being but instead focusing her microscope on teensy roundworms called nematodes, which have a lengthy scientific name, *Caenorhabditis elegans* (*C. elegans* for short). Although a tiny, transparent creature no bigger than Lincoln's nose on a penny, the crystalline *C. elegans* has cells that are perfectly visible, even as they divide. By studying its life cycle, especially how its embryo develops, Seydoux helped illuminate some of the most complex processes in biology, the MacArthur Foundation said.[248]

Seydoux, a native of France who received her undergraduate degree in biochemistry at the University of Maine and her Ph.D. in molecular biology from Princeton, joined the Hopkins School of Medicine faculty in 1995 after completing postdoctoral work at the Carnegie Institution of Washington's branch on Hopkins' Homewood campus. She brought her *C. elegans* fascination, first fired when she saw it through a microscope at Princeton, with her.[249]

At Hopkins, her pioneering work with the worm has led to identification of the molecular mechanisms that generate germ cells, the precursors of adult reproductive organs. She also identified how the initial asymmetry of the embryo, necessary for development of specialized tissues, derives from the interaction of specific structures within the ovum and sperm and fertilization. Her discoveries have "had a major impact on understanding fundamental problems that have challenged developmental biologists for over a century," said Thomas Kelly, then-head of the Department of Molecular Biology and Genetics, when Seydoux received her "genius" award.[250]

To those who might scoff that studying a worm can't be of much help to curing human ills, Seydoux has the ready, incontestable fact that big problems can be solved by studying tiny creatures. Many of *C. elegans'* 19,000 genes are conserved across species, she notes, and humans share a full one-third of them. For example, some of the worm work could help illuminate how cancer develops. "I like logic," Seydoux said shortly after receiving her MacArthur fellowship. "When you study a simple organism, it's easier to make logical arguments because there are few variables."[251]

Agre's Aquaporins

The telephone call from the Royal Swedish Academy of Sciences, bestowers of the Nobel Prize, came at 5:30 a.m. on October 3, 2003. Pajama-clad Peter Agre thought it "didn't seem to be a joke," so he believed it when the man on the other end of the line told him that he'd won the Nobel Prize for chemistry for having solved the age-old mystery of how the body regulates water. Agre later enjoyed relating that when his wife, Mary, called his mother in Minnesota to tell her that he'd just received the world's most prestigious award in science, she replied, "Oh, that's very good, but don't let it go to his head."[255]

For easygoing, modest, Minnesota-born and Johns Hopkins-trained Peter Agre, a swelled head was unlikely. He was—and remains—resolutely down-to-earth in describing his achievement. In 1992, he and Hopkins physiologist Bill Guggino reported Agre's discovery of a protein, which he dubbed "aquaporin," for the "water pore" channels it forms in cells that enable water to flow in and out of them. The Nobel arbiters quickly recognized that his initial discovery and subsequent findings of many other aquaporins were "of great importance for our understanding of many diseases of the kidneys, heart, muscles and nervous system." Agre concedes the accuracy of that assessment but has a less grand way of explaining what he found—and characterizing himself.[256]

"Anybody who's been in the economy class in an airliner with a full bladder and the seat belt light goes on is keenly aware of aquaporin 2," he said in a 2009 interview. "Small differences in volume can make big differences in comfort." As for receiving the Nobel Prize, he says it "clobbered me over the head.[257]

"And life has never totally returned to normal. In part that's because there's an expectation of some grand individual like Alexander Fleming [the discoverer of penicillin]. Sorry, it's just me."[258]

That may be true—but the discovery by Agre (pronounced AHG-ray), followed by the finding of Rockefeller University biophysicist Roderick MacKinnon of the three-dimensional shape of the water channels, for which he shared the Nobel Prize with Agre—ushered in a golden age of biochemical, physiological and genetic studies of these proteins in bacteria, plants and mammals. In a mere dozen years, the inquiries provided a fundamental understanding, at the molecular level, of how the malfunctioning of these channels plays a role in many diseases. Working from this basic knowledge, scientists throughout the world now are searching for drugs that can specifically target water channel defects.[259]

Peter Agre (above, left) a 1974 graduate of the School of Medicine who has spent most of his career at Hopkins, where he discovered aquaporins, receives the 2003 Nobel Prize in Chemistry from King Carl XVI Gustaf of Sweden at the Nobel award ceremony in Stockholm.

2003

February 10: The Public Broadcasting Service's series "The American Experience" airs a documentary, *Partners of the Heart*, about the remarkable collaboration between African American laboratory technician Vivien Thomas and white surgeon Alfred Blalock to pioneer cardiac surgery at Hopkins. It is based on a *Washingtonian* article, "Like Something the Lord Made," which would inspire the title of an HBO TV movie that airs on May 30, 2004, starring Alan Rickman and Mos Def.

February: Julie Freischlag becomes the first woman to be named director of the Department of Surgery and surgeon in chief of Hopkins Hospital.

March 3: Groundbreaking is held for another 10-story, $80 million cancer research building (CRBII). It will be a twin to the existing Bunting/Blaustein Building and a companion to it, as well as to the Harry and Jeanette Weinberg Building, as part of the Sidney Kimmel Comprehensive Cancer Center.

April: Hopkins acquires its Mt. Washington campus, a historic, 68-acre, tree-covered property with its signature mid-19th century Octagon building (once a girls' school whose alumnae included a future female Confederate spy), from Mt. St. Agnes College for Women. Eventually, this new campus becomes the home of Johns Hopkins International, CEPAR, PACER, the Chesney Medical Archives, the information technology offices and other Hopkins Medicine entities.

April 14: The federal Health Insurance Portability and Accountability Act (HIPAA) takes effect, imposing strict new rules to protect the privacy of patient medical information.

May 9: The School of Medicine receives a four-year, $24 million grant from the Donald W. Reynolds Foundation to establish a multidisciplinary center focused exclusively on reducing the rate of sudden cardiac death.

July 28: What is believed to be the first three-way, paired kidney swap is performed by a Hopkins surgical team led by Robert A. Montgomery, director of the incompatible kidney transplant programs at Hopkins Hospital.

■ Actor Cary Grant (1904–1986) was a patient at Hopkins in the 1950s and stayed in isolation—but not for medical purposes. Back then, the Hospital would close an entire floor in the Marburg pavilion for a complete cleaning. Grant was admitted to one of those floors and had the whole place to himself.

When he received the Nobel, Agre said he was honored that it "not only recognizes the discoveries, but also their usefulness to the advancement of fundamental science."

"It is amazing and gratifying that the Nobel committee feels our work has accomplished that milestone in just 12 years. That's warp speed in molecular chemistry, and it could never have happened as fast as it did without the wonderful resources and collaborators available at Johns Hopkins. This is an honor for the entire Hopkins family. This could have happened to any of my colleagues, but I was up at bat when a fast ball was thrown right down the middle of the plate."[260]

The baseball metaphor was typical for the self-effacing Agre, born in Northfield, Minnesota, "an idyllic little Norwegian town," in 1949 and raised with five siblings amid the southern Minnesota farmland. Agre's father was a professor of organic chemistry at St. Olaf College and later at Augsburg College in Minneapolis and "would invite us kids to his lab and have us do little experiments." Although that certainly captured young Peter's interest, he's quick to point out that it did not make him a scientific genius overnight. He got a D in chemistry at Theodore Roosevelt High School in Minneapolis. By the time he went to Augsburg as an undergraduate, however, his chemistry skills had improved, and in 1970, he received his bachelor's degree in it—with honors. He had decided by then to go to medical school and had been accepted at Hopkins.[261]

Initially interested in clinical work, he wasn't certain he was cut out to be a basic scientist, but "as a medical student at Hopkins, the lights really went on," he says. "Here, I met people who were so interested in science that it made me think it would be fun."[262]

Receiving his medical degree in 1974, Agre spent another year at Hopkins as a fellow in pharmacology, then completed his residency in medicine at Case Western Reserve University in Cleveland and did another postdoctoral fellowship at the University of North Carolina in Chapel Hill. He became an assistant professor of medicine there but remained attracted by the lure of the laboratory. An offer from Hopkins to return with a dual appointment in the departments of medicine and cell biology and anatomy quickly brought him back in 1981.[263]

Specializing in hematology, Agre was working in a lab on the fourth floor of Hopkins' Preclinical Teaching Building in the 1980s, studying a sub-unit of red blood cells with technician Barbara Smith, when they found their samples repeatedly contaminated by a smaller molecule. At first he thought it might be a fragment of the substance he'd originally been studying, but soon he and his team discovered that it was a protein in its own right—and located not just in the membranes of red blood cells but in kidney cells and blood vessels, with related proteins also found in tear ducts, salivary glands, even plant cells. The sheer abundance of the protein fascinated him. "It's in the brain! It's in bacteria!" Agre recalls, describing the excitement of those early days of discovery. He especially was intrigued because it didn't resemble any other known biological molecules except a distant relative in the lens of the eye. What was it?[264]

For years, scientists had assumed that water simply slipped past the fats, proteins and cholesterols that make up cell membranes, gently diffusing to one side or the other through the principles of osmosis. Yet physiologists noted that certain cases defied that explanation. In kidneys, for example, water cruises through membranes far faster than it would by simple diffusion. The physiologists proposed the existence of a specialized structure that somehow facilitates the flow. The problem was, nobody could find such a thing in the membrane of cells, many of which themselves are 70 percent water. Scientists searched for proteins that might serve this purpose but repeatedly came up empty-handed. Agre, not looking for the protein—but having the insight to realize he may have found it—set out to determine if he had.[265]

Agre typically credits a friend and former colleague at the University North Carolina, "a superb basic physiologist," with suggesting that the tiny molecule gumming up his blood study might be the elusive water channel. After finding it and varieties of it in so many tissues of the body, Agre needed to devise an experiment that would prove that it was, indeed, the thing for which so many scientists had long been searching.

By 1991, Gregory Preston, then a postdoctoral fellow in Agre's lab, had succeeded in cloning the DNA of the mysterious molecule. To collaborate on an experiment, Agre enlisted the assistance of his colleague Bill Guggino, who since the 1970s had been exploring why fish eggs laid in salt water don't shrivel, and why frog eggs laid in a pond don't explode. Guggino decided that their test case would be a frog oocyte, an egg that is one big cell, visible to the human eye. Sure enough, each frog egg doctored with the gene that produced the mystery protein and then placed in fresh water would swell up and burst within three minutes, indicating plenty of water now was flowing into them. "The things explode like popcorn," Agre later told the *New York Times*.[266]

As soon as the paper by Agre and Guggino describing their experiment appeared in the journal *Science* in 1992, "we got calls from all over the world," Agre recalled. "We put in our thumb and pulled out a plum. It completely changed the focus of my lab … and suddenly, a lot of people wanted to work on this." Since then, hundreds of aquaporins have been found in plants, bacteria and other forms of life, as well as in mammals. Although much remains to be learned about them, Agre is certain that scientists will be able to use their growing understanding of aquaporins to benefit medicine, biotechnology and even agriculture.[267]

The easy-going, modest, Minnesota-born Nobel laureate Peter Agre speaks to his research team after learning that he'd received the 2003 prize for chemistry. Life after the award has "never totally returned to normal," he says, and affably asserts that those who now meet him and expect to find "some grand individual like Alexander Fleming," the discoverer of penicillin, will be disappointed. "Sorry, it's just me," he says.

2003
August: The Accreditation Council for Graduate Medical Education (ACGME), a non-profit group composed of physician, hospital and university representatives, threatens the famed Osler internal medicine residency program with loss of accreditation because of its alleged failure to comply with new 80-hour workweek rules for house staff members. Medicine Department Director Myron (Mike) Weisfeldt oversees a comprehensive review of the residency program's compliance with the new ACGME rules, and by December, the ACGME restores full accreditation of the program without qualification—and even commends Hopkins for the significant changes it has made.

Peter Agre celebrates his 2003 Nobel Prize for the discovery of aquaporins, the proteins that form the "water channels" through which moisture enters and leaves cells, with fellow aquaporin researcher Landon King (rear, left), now Vice Dean for Research, and post-doctoral members of his laboratory team.

"[Receiving the 2003 Nobel Prize for Chemistry] clobbered me over the head. And life has never totally returned to normal. In part that's because there's an expectation of some grand individual like Alexander Fleming [the discoverer of penicillin]. Sorry, it's just me."

—Peter Agre, 1974 graduate of the School of Medicine, co-recipient of the 2003 Nobel Prize for Chemistry for his discovery of aquaporins, the proteins that form the "water pore" channels through which water enters and exits cells

"I'd like to see more progress being made in terms of using these as diuretics," he said in 2009. "If we could lose more water and make more diluted urine, people with fluid overload would be protected. Or could we engineer, for example, plants that are relatively intolerant to drought because they have more efficient roots for water uptake? Some plants already do that. There are a lot of practical things that we hope will follow. There's not a guarantee."[268]

In the months following Agre's Nobel Prize, he was flooded not with water but speaking invitations. He was determined that one way he'd cash in some of his chips of newfound fame was to promote a concept dear to his heart: getting young people interested in science. He traveled to inner city Baltimore high schools—and back to his Minneapolis alma mater, Roosevelt High, where he flunked chemistry—toting his nearly half-pound, gold Nobel medallion in a crumpled plastic bag and allowing students to pass it around. "This is what all the commotion is about," he'd say. "I think it's more valuable when people touch it."[269] He'd also praise the schools' teachers. "I'm here to honor you," he told faculty at Baltimore's Dunbar High School. "The fact is, public education can be outstanding."[270]

Winning the Nobel also ensured him plenty of job offers. A particularly appealing one came from Duke University, where his lifelong friend and former Hopkins medical school roommate, George Vann Bennett, was on the faculty. Duke offered him the post of vice chancellor for science and technology, "with a bully pulpit for advancing science." Feeling he was "out of gas" after all the Nobel hoopla, he decided to accept it. After being at Hopkins for 24 years, in 2005 he returned to North Carolina.[271]

"I was kind of hoping I could spend a lot of time and get rejuvenated as a scientist—and that didn't happen," Agre says now. When Hopkins offered him an opportunity to return in 2008 and head the Johns Hopkins Malaria Research Institute, headquartered in the Bloomberg School of Public Health, he happily agreed.[272]

Becoming president of the National Association for the Advancement of Science the year after he returned to Hopkins, Agre today says, "I don't feel like I ever left."[273]

For such a large institution, Agre says, Hopkins still maintains a "very real family-type atmosphere." Although now based in the School of Public Health, not the School of Medicine, he retains a visiting professorship—and an office—in the School of Medicine's biological chemistry department. He enjoys recalling that one weekend shortly after he returned to Hopkins, he was entering the Hospital and one of the guards at the door gave him a warm greeting: "Dr. Agre, you're back! It's good to see you!" [274]

Building a New Hopkins

Remaking the best medical care and education programs in the wake of difficulties and challenges during this period was mirrored by the determination of Miller, Peterson and other institutional leaders to remake Hopkins Medicine physically as well.

Early in their relationship as colleagues, Peterson convinced Miller that Hopkins desperately needed to build a new medical campus. It became and remains a key focus for them both. "In some parts of the Hospital, we've been doing world-class medicine in third-world facilities," Peterson observed a decade into his collaboration with Miller.[275]

"To the skeptics who say we're giving too much attention to physical structures, I would only repeat that we've had to remain focused because we'd gotten so far behind," Peterson said in a 2007 interview. "Everything we do depends on our buildings. They provide the environment where we care for our patients; they're the workshops for our faculty, staff and nurses; and they determine which physicians and scientists Hopkins is able to attract and retain."[276]

By 1999, the average age per square foot for Hopkins Hospital was nearly 40 years, compared to the national average of 11 years. Some of its structures were "totally antiquated," Peterson recalled, referring to the Osler-Halsted building complex designed in the 1920s, built in the early 1930s, and no longer suitable for accommodating modern clinical care and research programs.[277]

■ Loudenslager's Hill is not a place name that most within Hopkins Medicine would recognize—but it is it important to the history of Johns Hopkins. Mr. Johns Hopkins originally wanted his hospital to be built on the grounds of his 330-acre estate, "Clifton," located on land that now is a golf course in northeast Baltimore City. He later was persuaded to purchase new land in the city for the hospital, a 13-acre site at the crown of what then was called Loudenslager's Hill (named for a prominent butcher and innkeeper, Jacob Loudenslager). Hopkins' handpicked hospital trustees convinced him that if he wanted to help the poor, it made better sense to build his hospital closer to where they lived. Perhaps he also decided that placing the hospital on high ground would provide fresher air and safety from floods. He paid $150,000 for the Loudenslager's Hill site, shown above in an 1869 lithograph. It is where original Johns Hopkins Hospital was built— adjacent to where the magnificent new Hopkins clinical towers now stand.

The girders forming the superstructure of what would become The Johns Hopkins Hospital's new Sheikh Zayed Cardiovascular and Critical Care Adult Tower symbolized the massive redevelopment project envisioned in Hopkins Medicine's 2002 master plan for the East Baltimore campus.

Guiding much of The Johns Hopkins Hospital's campus redevelopment was an in-house design team (left to right): architect Marge Siegmeister, mechanical systems expert Anatoly Gimburg, construction manager Howard Reel, facilities vice president Sally MacConnell, and architect Mike Iati, who with MacConnell orchestrated every aspect of the dual-tower megastructure that will transform the East Baltimore landscape. Each bolt, beam, window and pipe—among other things—on the five-acre plot at Orleans and Wolfe streets had to be accounted for. "It's like building a 60-ton watch," Iati said at the height of construction. "All these pieces have to fit together. Somebody is out there right now making sure that above the ceiling, ducts and pipes don't collide and that there's room for that electrical conduit to get between them. Between architects, engineers, contractors, the construction manager and the suppliers, all these things need to be coordinated."

■ By late 1983, the fifth floors of five adjacent buildings—Blalock, Halsted, Harvey, Nelson and the Children's Center—had been renovated to create what then was called the Clayton Heart Center. Now absorbed into the Johns Hopkins Heart and Vascular Institute, the center had been named for William L. Clayton (1880–1966), father-in-law of Hopkins' legendary professor of medicine, Benjamin M. Baker Jr. (1901–2003). Clayton had been the undersecretary of state who hand-wrote the first draft of what became known as the post-war Marshall Plan. By the time the Clayton Center was completed, 800 open heart surgeries were being performed annually at Hopkins and the new center's creation was designed to facilitate collaborations on patient care and research by adult and pediatric cardiologists, radiologists, cardiac surgeons and anesthesiologists, as well as make it easier to share the expensive medical equipment their specialties required.

Despite financial constraints that required painful budgetary cutbacks in the late 1990s, the outlines began taking shape for what would become probably the largest medical center building project in U.S. history—a decade-long, multibillion-dollar effort to make the buildings of Hopkins Hospital and its affiliates the finest in the country.

The key question, as always, was how to raise the money to do it. Told by health system budgeters that funds probably would be available for only one new clinical building—either a children's hospital or a cardiovascular/critical care center, not both—Miller balked.[278]

"I said this just doesn't make sense to me. They said, well, you just can't afford it. So I said, why don't you go away and stop telling me how I can't do it and come back and tell me how I can do it."

He credits master number-crunchers Richard "Rich" Grossi, now senior vice president and chief financial officer for Johns Hopkins Medicine, and Ronald Werthman, vice president of finance/treasurer and chief financial officer of the Hopkins Health System and Hospital, with being the genuine financial "geniuses" who figured out how to manage it. "They went away for six weeks and devised this 10-year plan of how much money we had to borrow, how much money we had to raise, how much money we needed from the state, how much money we had to get out of operations."[279]

With this master financial plan—the first one Hopkins Medicine ever had that established "very clear-cut objectives of what we needed to do to move to the next step each year"—Miller, Peterson and others "spent a lot of time getting everybody on board on why we needed to build the buildings. I spent a year going through this," Miller says.[280]

By early 2002, an entire new campus master plan had been unveiled, revising the master plan from 1992-93. It was the first long-range plan ever to deal comprehensively not just with additions to Hopkins Hospital and School of Medicine but with all of the East Baltimore area surrounding Hopkins, including the schools of Nursing and Public Health and the Kennedy-Krieger Institute.

Miller and Peterson knew that they had to make such an undertaking financially feasible—somehow.[281]

The budget for the 2004 fiscal year would be the linchpin for the 10-year, billion-dollar campus redevelopment plan.[282] How well Hopkins Hospital met its FY04 budgetary targets would determine the decisions that state regulators would make on its Certificate of Need application to erect the new buildings, influence the behavior of the bond-rating agencies when the Hospital tried to sell $400 million in revenue bonds, and ensure that Hopkins Medicine's trustees would continue to back Miller and Peterson's vision of the future.[283]

The message from Miller and Peterson to Hopkins' clinicians for FY04 was that they needed to increase patient volumes, conserve resources and not expect program funding increases unless their departments were self-sufficient. For the next seven fiscal years, the budget message would have that familiar ring. Miller warned: "We're running way behind in our patient volume projections, and we can't get a certificate of need and new rates without financial discipline. The trustees will pull the plug, and we'll have nothing left to show but a fancy parking garage."

The Campaign for Johns Hopkins Medicine

Another central element in the drive to build a new Hopkins Hospital was the launching in 2002 of a medicine-centered charitable fundraising campaign—Seeking Cures, Saving Lives— as part of the university-wide Knowledge for the World campaign begun in 2000.[284]

Both Knowledge for the World and the Campaign for Hopkins Medicine proved to be spectacularly successful. More than a quarter-million individuals and organizations committed to investing $3.74 billion in Johns Hopkins by the time the Knowledge for the World campaign finally concluded in 2008—and the Hopkins Medicine share of that amounted to $2.17 billion, more than double its original goal.[285] The fiscal foundation for Hopkins Medicine's new century of medicine was given an excellent start.

2003
September: Following five years of construction, the Broadway Research Building opens, providing six stories of research labs, two stories of administrative offices, meeting spaces and a large animal facility, or vivarium, for research animals. It has its formal "grand opening" on May 25, 2004, with NIH head Elias Zerhouni giving the keynote address.

October 8: Peter Agre, M.D., professor of chemistry, receives the 2003 Nobel Prize in Chemistry for his discovery of aquaporins, the previously elusive water channels through which cells discharge and receive water. His mother in Minnesota says," Oh, that's very good, but don't let it go to his head." To encourage schoolchildren to become interested in science, Agre later will carry his Nobel gold medal around in a plastic bag to school visits and let young students pass it among themselves.

Outreach to the East Baltimore community continues as employees throughout Johns Hopkins Medicine seek to mentor students at the neighborhood Tench Tilghman Elementary School. The Urban Health Institute receives praise (and some criticism). Computer training is offered to neighborhood residents.

The Hopkins Hospital reduces its length-of-stay (LOS) rate, an important sign of improved efficiency.

The Moore Clinic marks its 20th year as one of the nation's premier facilities for treating AIDS patients.

December 19: The Johns Hopkins Health System announces the death of Brianna Cohen, 2, a cancer patient who was being treated at home with Total Parenteral Nutrition (TPN) infusion therapy prepared by The Johns Hopkins Home Care Group. The TPN bag provided to the child contained excessively high levels of potassium. "Hopkins not only accepts full responsibility, but also continues to fully cooperate with the family in its quest" to determine the cause of the error. JHHCG undergoes significant improvement of its patient safety procedures and receives commendation from the state Department of Health and Mental Hygiene for being "deficiency free" in 2006 and again in 2009.

2004–2011:
Creating a New
Johns Hopkins Medicine

I n the fall of 2005, Dean/CEO Edward Miller waited for a prearranged, late-night telephone call from C. Michael Armstrong, the newly elected chair of the Johns Hopkins Medicine Board of Trustees. Since the trustees' adoption two years earlier of a comprehensive 10-year financial plan to fund an ambitious billion-dollar medical campus redevelopment, Miller's life had become a whirlwind on the fundraising circuit: "flyby" events around the country to shake hands and introduce himself to prospective donors, sit-down dinners with top "people of interest" in Washington, D.C., and train hops to New York for fêtes with potential philanthropists.[1]

Once set in motion, the redevelopment would become the largest hospital construction project in the country. It also would define the first decade of the 21st century for Hopkins Medicine, even with so many other extraordinary markers of the era: international accolades, National Medals of Science, Presidential Medals of Freedom, Lasker Awards, MacArthur "genius" grants, a revolutionary new medical curriculum, and recognition as a national leader in patient safety research and health care reform.

When he telephoned Miller, Armstrong knew well the enormous pressure bearing down on him and Hospital and Health System President Ron Peterson to raise money for building two massive clinical towers and adding research space to solidify the institution's competitive edge in scientific funding—even at the possibly unavoidable expense of Hopkins Medicine's teaching mission.

"I watched Ed wrestle with this dilemma for months," said Armstrong, a longtime, trusted advisor to the dean. Armstrong had for years held a keen interest in medical education. The two men had extensive talks about curriculum reform and the necessity for a state-of-the-art teaching facility. Miller frankly had told Armstrong and other trustees that, although he would like nothing better than to throw a new education building into the mix, he couldn't afford to lose his focus on raising money for the clinical and research priorities.

The call that the dean took from Armstrong that autumn evening didn't solve all of his problems, but it did lift one burden off his shoulders. "Mike started right out saying he knew how much education meant to me and that he understood that I couldn't afford to diffuse my fundraising." Then, the next thing Armstrong said nearly took the dean's breath away. "He asked me if $20 million would get us started on a new education building. I said, 'You bet.'"[2]

"I wish that other friends and supporters could experience Hopkins' collaborative excellence. Hopkins excels in congeniality in education, research and clinical practice. It's a very real and important attribute and what makes Hopkins unique as well as the best."

—C. Michael Armstrong, chairman of the board of Johns Hopkins Medicine (2005–2011), former chairman of IBM, Comcast, AT&T and Hughes Electronics

C. Michael Armstrong (right), the newly elected chair of the Johns Hopkins Medicine Board of Trustees, significantly eased the concerns of Dean/CEO Edward Miller in the fall of 2005 by offering a $20 million gift to ensure construction of what became the four-story Anne and Mike Armstrong Education Building, dedicated in October 2009 and specifically designed for teaching 21st century medicine.

The extraordinary gift from Armstrong and his wife, Anne, exemplified the fact that Johns Hopkins Medicine's unequalled record of excellence in medical education, as well as research and patient care, ensured that the always-challenging effort to find the money needed to advance its multiple missions would continue to succeed.

Even amid a period of severe, worldwide economic downturn, Hopkins was able not only to press ahead on its mammoth East Baltimore construction project but also to extend the imprimatur of Hopkins Medicine's quality locally, nationally and internationally. During the period, the institution would more fully integrate Johns Hopkins Bayview Medical Center and Howard County General Hospital into the academic enterprise, bring two premier Washington, D.C.,-area community hospitals, as well as a children's hospital in Florida, into the Hopkins Medicine family, and sign affiliation agreements with several suburban hospitals. Beyond the United States, Johns Hopkins Medicine International established management agreements and partnerships with leading hospitals in the Middle East, Asia and Latin America.

As Johns Hopkins Medicine grew in strength and scope, Hopkins Hospital continued its unprecedented run of being named by *U.S. News & World Report* as the best in the nation, attaining that distinction for the 21st year in a row, and the School of Medicine remained the top recipient of funding from the National Institutes of Health and one of the top three medical schools in the country.

Significant recognition for medical leadership also remained a hallmark of the Hopkins faculty. Between 2004 and 2011, the kudos included one National Medal of Science, one Lasker Award, two Presidential Medals of Freedom, two MacArthur Foundation "genius" awards—and another Nobel Prize.

Beyond the Buildings

Clinical Achievements Herald Medicine's Future

Hopkins physicians continued attaining clinical breakthroughs to improve the lives of patients, providing an advance look into medicine's future and Hopkins' exciting role as a leader in its vanguard.

Biomedical Engineering

In 2007, Murray Sachs, director of biomedical engineering, created the Center for Bioengineering Innovation and Design to enhance medical device creation and development, while also providing greater access to biomedical engineering for clinicians, other research scientists, and industry. At the center, students go into operating rooms and clinics, watch physicians practice, then "brainstorm for weeks about how those [physicians' clinical] goals could be met in a more efficient, better way through technology," explains Elliot McVeigh, successor to Sachs as head of biomedical engineering.[3]

■ In 2009, Carol Greider, the Daniel Nathans Professor and Director of Molecular Biology and Genetics in the Institute of Basic Biomedical Science, became the most recent Hopkins Medicine faculty member or alumnus to win the Nobel Prize in Physiology or Medicine. The first Hopkins-associated Nobel Prize winner was Thomas Hunt Morgan (1866–1945), who received his Hopkins Ph.D. in zoology in 1890. He won the 1933 Nobel Prize in Medicine for his discoveries concerning the role played by the chromosome in heredity. In addition to his genetics work, he made important findings in experimental embryology and regeneration.

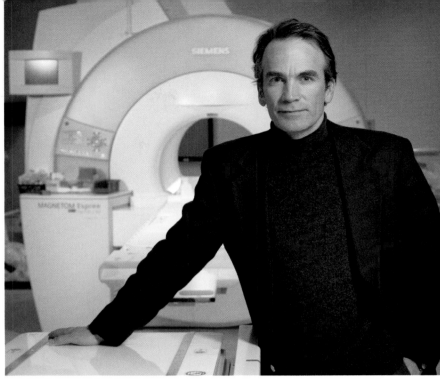

Since CBIC's founding, its student teams have completed more than 50 medical device projects. Their creations include:

- polymer cable ties that provide a less intrusive and safer way to close a patient's chest after heart surgery;
- a carbon fiber orthopedic ankle support;
- a pullout-resistant pedicle screw for the treatment of vertebral fractures in patients with osteoporosis;
- quick-dissolving, oral strips for delivering vaccines that prevent rotavirus infections, a leading cause of severe diarrhea in children; and
- a surgical thread embedded with a patient's own adult stem cells, designed to enhance the healing of serious orthopedic injuries, such as ruptured tendons.[4]

Elliot McVeigh, director of the Department of Biomedical Engineering since 2007, is maintaining the growth and excellence of the department, which spans both the School of Medicine and Homewood campuses. Repeatedly recognized as the nation's finest, BME now encompasses an undergraduate training program with more than 500 students, a graduate program with more than 200 master's degree and doctoral candidates, and more than 30 postdoctoral fellows.

Cardiology and Cardiac Surgery

The Johns Hopkins Heart and Vascular Institute

The September 2004 creation of the Johns Hopkins Heart and Vascular Institute (originally the Johns Hopkins Heart Institute) recognized that the future of cardiology and cardiac surgery increasingly will require the cooperation of physicians across many disciplines—not only from general surgery to cardiac surgery but from interventional radiology to pediatrics. The scope of the institute's reach was deemed substantial enough to require its own advisory board of governors, led by Art Modell, former majority owner of the Baltimore Ravens football team.[5]

From its inception, the Heart and Vascular Institute was slated to be based in the new cardiovascular and critical care tower then being planned as part of Hopkins Hospital's billion-dollar overhaul. Now, every aspect of advanced diagnostic and therapeutic services for the heart will be centralized on two floors of the institute's new home in the adult care tower.[6]

The institute is at the forefront of innovations in cardiac surgery, the treatment of arrhythmias (heartbeat irregularities), heart-related stem cell research, and the development of genetic biopacemakers. [7] Its pioneering efforts include using less-invasive methods of robotic, closed-chest open-heart surgery.

It also houses the United States' first training program for ventricular restoration, an operation that reshapes the heart and improves its ability to pump blood; and the institute responded to the growing problem of heart disease in women by founding a women's heart program in 2008.[8]

Hopkins is one of a handful of hospitals in the world authorized to implant the newest self-contained, total artificial heart (TAH), held in this photo by cardiac surgeon John Conte, director of the heart and lung transplantation program and the ventricular-assist device program. Although not suitable for everyone on a heart-transplant waiting list, the new TAH can prolong life—even double its expectancy—and improve its quality. The TAH operates on both internal and external batteries and is just one tool in an arsenal of ventricular assist devices that are marking their mark on the treatment of heart disease at Hopkins' Heart and Vascular Institute.[9]

Top right: Every year, the Johns Hopkins Emergency Department's Lifeline helicopter logs about 750 flights—averaging 70 a month—transporting critically ill patients brought to the hospital from within a 150-mile radius. Equipped as a sort of airborne intensive care unit, the helicopter has its missions limited only by intense fog or thunderstorms. "Otherwise, no one is too sick to transport," said flight nurse Lisa Denton. "We all thrive on helping the sickest patients."[10]

Under the direction of the Office of Critical Event Preparedness and Response, Hopkins' clinical reach was extended to victims of Hurricane Katrina in Louisiana and earthquake survivors in Haiti. Other teams of Hopkins doctors and nurses rushed to the aid of disaster survivors around the world. At right, "Team Echo" members await a flight back to Baltimore in a military hangar near New Orleans.

2004

A decade of success in reducing East Baltimore campus crime is marked by Joe Coppola, head of security. A 24-year veteran of the Secret Service, Coppola arrived in 1994 and has overseen a dramatic drop in on-campus crime. In 1993, 23 robberies were reported; in 2003, zero. In 1994, campus thefts totaled 700; by 2003, the number was down to just 200.

May 25: The "grand opening" of the Broadway Research Building features a speech by Elias Zerhouni, M.D., former SOM executive vice dean and now NIH head. The talk, entitled "Stimulating Science," is about "the making of the building."

May 30: The Home Box Office movie, *Something the Lord Made*, about the collaboration between Alfred Blalock and Vivien Thomas, airs. It would win the 2004 Emmy as best made-for-TV movie and a 2005 Peabody Award, as well as numerous other awards, including an NAACP Image Award and Black Reel Award. Four days before the movie's premiere, the School of Medicine announces establishment of the Vivien Thomas Fund for Diversity to increase the number of minorities in the academic medicine talent pool. The Fund honors the memory of the African American surgical technician whose pivotal contributions to the development of the "blue baby" operation at Hopkins 60 years ago ushered in the era of cardiac surgery.

Emergency Medicine

The threat of future disasters compelled the Department of Emergency Medicine and the Office of Critical Event Preparedness and Response (CEPAR) to undertake rapid responses and long-term planning initiatives during this period. Its efforts included overseeing a Hopkins enterprise-wide planning process to respond to a possible pandemic involving the H5N1 bird flu virus in 2005–2006, which also prepared the institution to respond swiftly to the 2009–2010 outbreak of the so-called swine flu, H1N1.

Geriatrics and Gerontology

With the Baby Boom generation of 1946-64 poised to become the Senior Tsunami of the 21st century, Hopkins' Division of Geriatric Medicine and Gerontology continues inaugurating clinical programs and research initiatives that promise to have a powerful impact on future health care of the elderly.

It is well-equipped to lead such efforts, having been a driving force in geriatric medicine for decades. Beginning in the 1980s, when gerontology was in its infancy and little medical literature was devoted to it, Hopkins was central to its development. (Indeed, as mentioned previously, the old Baltimore City Hospitals—now Johns Hopkins Bayview—became the birthplace in 1958 of the Baltimore Longitudinal Study of Aging, the unique, still-active research effort involving hundreds of city residents. In 1968 it also became home to the National Institute on Aging's Gerontology Research Center.)[11] Hopkins' most recent initiatives in the field include:

2008: Launching the nation's first-ever Biology of Frailty Program under co-directors Jeremy Walston and Neal Fedarko. Its aim is to understand the biological mechanisms that underlie frailty—a condition that afflicts some 7 percent to 10 percent of older people with such problems as weight loss, muscle

weakness, exhaustion and other symptoms that were not defined scientifically as the signs of "frailty" until Hopkins experts did so.[12]

2009: Completing the 20-year-long Cardiovascular Heart Study, begun by geriatrician and epidemiologist Linda Fried. One of the largest single NIH-funded research projects when it was launched in 1989, its findings on heart disease and stroke in more than 6,000 men and women over 65 have had a nationwide impact on the preventive approaches to those conditions.[13]

Continuing another unique Fried-initiated project, the Women's Health and Aging Study, begun in 1992 with a $20 million grant from the National Institute on Aging to unravel the myriad factors that lead to frailty, the physical decline that often results in disability.[14]

2009: Opening the Johns Hopkins Memory and Alzheimer's Treatment Center under neuropsychiatrist Constantine "Kostas" Lyketos, head of psychiatry at Johns Hopkins Bayview. It has developed a systematic approach to the disease that links Hopkins Hospital, Johns Hopkins Bayview, and the School of Medicine-affiliated Copper Ridge Institute in Sykesville, Maryland, a private research and education entity. They work together to develop the evaluation and treatment of Alzheimer's patients, providing a unique continuum of care from diagnosis to death, whether or not patients are participating in a research project.[15]

Genetics

Hopkins' long tradition as a center of human genetic medicine is being maintained and advanced. Its discoveries and initiatives during this period include:

2006: Pediatric cardiologist Harry "Hal" Dietz makes a "jaw-dropping" discovery by combining his previous identification of the gene that causes Marfan syndrome with his finding of the influence of the cell-growth factor TGF-beta (or TGF-ß) in enabling that defective gene to cause the disease. He and colleagues discover that a common, government-approved blood pressure

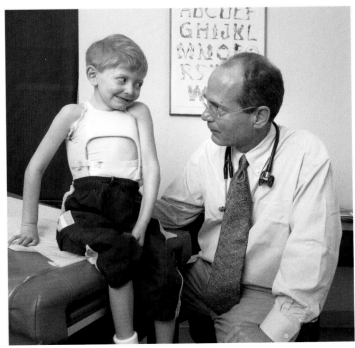

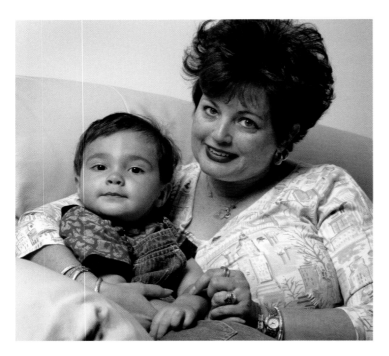

At Hopkins, the uniqueness of each patient with a genetic-based affliction—or the potential for one—is addressed daily by genetic counselors who work in a variety of disciplines, including obstetrics, pediatrics, oncology and cardiology.[18] From enabling a young mother to learn how to handle the multiple requirements needed to care for and raise an infant with a rare metabolic disorder, to offering background information to women found to have a mutation in the hereditary breast cancer gene, Hopkins' genetic counselors provide support and help patients and family members weigh the benefits, risks and limitations of learning their genetic predisposition for diseases. For example, genetic counselors are helping Debbie Ekonomides (above) and her son, Michael, learn how to cope with a rare metabolic disorder he has.[19]

2004

June: Arnall Patz, director emeritus of the Wilmer Eye Institute, receives the Presidential Medal of Freedom at a White House ceremony.

Cancer researcher Bert Vogelstein, M.D. tops the list of the world's 50 most influential scientists in the past 20 years, according to the Philadelphia-based Institute for Scientific Information. His research papers are the most cited by other researchers worldwide. Hopkins' neuroscientist Solomon Snyder is ranked No. 3 and oncology professor Kenneth Kinzler is No. 19.

medication—losartan—can block TGF-ß's actions. This provides a "silver bullet" capable of preventing many of Marfan's most devastating developments.[16]

2009: Pediatrician and molecular biologist David Valle, director of the McKusick-Nathans Institute of Genetic Medicine, launches the Johns Hopkins Individualized Medicine Program (JHIMP), which could lead to development of "individualized" medicine tailored to each patient's DNA. Harnessing the rapidly expanding knowledge being gained about the human genome, Valle and colleagues are collecting DNA samples from Hopkins patients who are donating them willingly. Researchers then analyze each individual's genome to identify those variants with special properties that ought to appear on a computerized medical chart of that person and could influence how that patient is treated. For example, even a tiny fraction of variation between the 3 billion DNA letters that make up one person's genome and that of another can account for the difference between loving peanuts and having a life-threatening allergic reaction to them.[17]

Hematology

Although Hopkins long has had a superb pediatric program for children suffering from sickle cell disease, adults who experience the excruciating "pain crises" that are the hallmark of this genetically inherited blood disorder often had to wait agonizing hours in emergency rooms for treatment. Recognizing the need to respond swiftly to such crises with the powerful pain-reducing drugs and fluids prescribed for adults:[20] Hopkins Hospital opens a Sickle Cell Infusion Center for Adults in 2008, providing services to some 300 patients regularly. Spearheaded by Sophie Lanzkron, assistant professor of medicine and oncology, and launched with $500,000 in funding from Hopkins Hospital, the center provides comprehensive care. This includes not only access to clinical trials of new treatments but counseling to address the depression and other psychological factors affecting the chronically ill sickle cell patients.[21]

Hematology Division head Robert Brodsky also develops a new procedure for the established process of swapping a sickle cell patient's bone marrow, the body's blood-making factory, with marrow from a healthy person, which produces only normal, healthy blood cells.[22] In the past, most sickle cell patients needed a suitable "matching" donor—usually a full sibling—whose immune cells can prevent rejection of the new marrow.[23] Brodsky and his colleagues devise a new, experimental procedure that gives patients "half-match" bone marrow transplants, allowing for a larger pool of donors that includes parents and even half-siblings. Patients additionally receive a high dose of a drug called cyclophosphamide (in a regimen called "Hi-Cy"). This helps reboot their immune systems.[24]

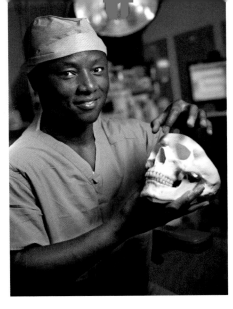

Neurosurgery

The drive to perform less invasive surgery whenever possible, thereby ensuring less costly procedures and more rapid recoveries for patients, is a goal of surgeons throughout Hopkins. It has avid advocates in neurosurgery, with an especially promising collaboration between brain surgeon Alfredo Quiñones-Hinojosa and facial plastic and reconstructive surgeon Kofi Boahene. Their innovation: Development in 2007 of an extraordinary new method for removing brain tumors through a patient's eyelid.[25] A small incision is made in the eyelid's crease and a coin-sized piece of skull is removed from above the eyebrow, using a computer-guided endoscope with a tiny camera to deploy the surgical instruments. When the surgery is completed, the piece of bone is replaced, secured by a small metal plate, and self-dissolving sutures on the eyelid leave no visible scar.[26] By 2010, the method, formally known as a transpalpebral orbitofrontal craniotomy, had been used successfully to remove brain tumors or repair brain fluid leaks in 18 patients. All recovered quickly, with less risk of infection than in such surgeries conducted through nasal passages and with none of the scarring left by traditional surgery.[27]

Ophthalmology

On the cusp of the Wilmer Eye Institute's 85th anniversary, and soon after the opening of the $100 million, ultramodern building that substantially expands its patient care and research facilities, a remarkable milestone was reached in the history of Hopkins ophthalmology. In 2009, Wilmer attained the unprecedented record of having 100 eye departments, institutes or foundations around the world headed by Hopkins-affiliated ophthalmologists—either former students, residents, fellows or faculty—since Wilmer was founded in 1925.[29] The achievement powerfully reflects not only the service Wilmer has provided to generations of patients who came to Baltimore, but the care given to untold millions who were and still are being treated by Wilmer-trained ophthalmologists worldwide.[30]

Facial plastic and reconstructive surgeon Kofi Boahene (above) collaborated with brain surgeon Alfredo Quiñones-Hinojosa to develop a remarkable new method for removing brain tumors through a patient's eyelid. Formally known as a transpalpebral orbitofrontal craniotomy, the procedure "is a very viable and practical option for thousands of surgeries done each year in the United States," Quiñones-Hinojosa told The Baltimore Sun in June 2010. It is effective for problems that are deeply seated behind the eyes or at the front of the brain, he said, and eventually could also be used on victims of automobile crashes, as well as patients who have undergone traditional brain surgery but need to have a follow-up operation.

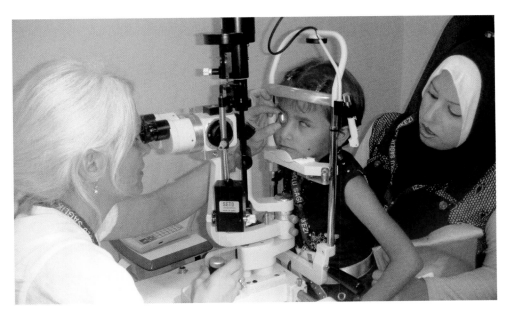

In August 2008, Hopkins eye surgeon Esen Akpek flew to Istanbul, Turkey, on a unique war relief mission: to restore the sight of eight Iraqi children who desperately needed corneal transplants. Working out the complex details for transporting the children from war-battered Baghdad to Turkey was accomplished by two of Akpek's Wilmer Eye Institute colleagues—technician Shanna Igrodi and coordinator Kim Pratzer—collaborating with Ilkay Baylam of the Johns Hopkins-affiliated Anadolu Medical Center in Turkey. In this photo, Akpek examines 7-year-old Noora, blind since birth but able to see following surgery. The Iraqi children and their adult companions initially were apprehensive, but once the Wilmer team began to work, the human chemistry changed dramatically. "They treated us like we were angels or something," Akpek recalls.[28]

From bubble-blowing to sleight-of-hand tricks, professional performers from the New York-based Big Apple Circus' Clown Care Unit regularly provide patients at the Hopkins Children's Center with laughter and relief. The clowns are impressed by the willingness of Hopkins faculty and staff to call on them for "large doses of clown," whether it's to encourage a determined 5-year-old to eat, prompt a recalcitrant teenager to accept an IV, or distract a burn victim while her bandages are changed.

Lower right: Elliot Fishman was named in 2007 as the nation's top radiologist by *Medical Imaging* magazine—which also cited the exceptional work of seven other Hopkins radiologists and proclaimed Hopkins' radiology department the best in the country. Fishman not only was among the pioneers in developing 3-D medical imaging—initially with Pixar, then a computer manufacturing firm that now is better known for the wizardry behind such computer-animated movies as *Toy Story*—but launched the first radiology-related podcasting and vodcasting on his website, www.ctisus.com (as in CT-Is-Us, a parody of the name of the popular toy store chain). The website includes his "Ask the Fish" column, which takes its title from his childhood nickname. He calls it "kind of a 'Dear Abby' for medical radiology stuff" and answers every radiology question personally or relays replies from colleagues, whom he credits.[34]

Pediatrics

From innovatively fixing children's damaged hearts to warming them with the antics of clowns, the Johns Hopkins Children's Center continues to earn recognition as one the top facilities of its kind in the country.[31] For example, in 2005, Hal Dietz, the pediatric geneticist and cardiologist whose breakthroughs in treating Marfan syndrome were described earlier, announced the discovery of a related but clinically and genetically different connective tissue affliction that he and fellow Hopkins research scientist Bart L. Loeys uncovered. Dubbed the Loeys-Dietz Syndrome, it is linked to two defective genes that can cause an even more rapidly developing weakness in the aortas of children than Marfan's, putting them at greater risk of suffering a sudden, fatal aortic rupture, known as an aneurysm.[32] Hopkins cardiac surgeon Duke Cameron—who leads the surgical team and has the world's longest record in repairing aortic aneurysms in Marfan patients—soon thereafter performed the first valve-sparing aortic root replacement surgery on a child diagnosed with Loeys-Dietz Syndrome.[33] Since then, many other patients have benefited from this operation.

Radiology

Throughout this period, the Russell H. Morgan Department of Radiology and Radiological Science maintained an extraordinary pace of technological advancement to improve its services to the patients of heart and brain specialists, oncologists and other clinicians—for whom it performs more than 360,000 radiology exams annually. Its research and teaching activities also benefitted immensely from the constant upgrading.[35] For example, in 2007, Johns Hopkins became the first hospital in North America to install and put into operation the most powerful X-ray imaging machine in its class—a 320-slice CT scanner. The 4,400-pound machine cost more than $1 million, which is shared by radiology

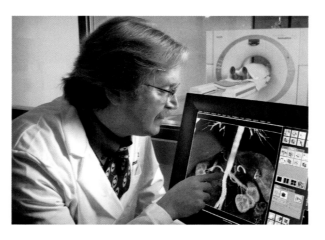

and cardiology. Located in the cardiology division, it is available immediately for general clinical use, having received approval from the U.S. Food and Drug Administration just a month earlier—based in large part on the initial testing at Hopkins of the similar, but slightly less powerful 256-CT scanner.[36]

Surgery

As they have since the beginning, Johns Hopkins' skilled surgery practitioners continue breaking new ground, devising innovative methods and operating on patients that surgeons elsewhere have declined to serve. That is especially true now as Hopkins pushes the surgical boundaries for seniors and redefines the surgery eligibility of others as well. Among Hopkins' surgeons and specialists, the word *can't* repeatedly is replaced by *let's see about that.*

Among Hopkins' current trailblazers is a team of gastrointestinal surgeons and biomedical engineers who performed the first successful experimental endoscopic suturing of the stomach lining, which had been perforated to gain access to such internal organs as the appendix, gallbladder, liver, uterus and ovaries. This 2007 breakthrough could usher in a more ideal kind of gastrointestinal surgery, enabling surgeons to enter the body through a natural opening such as the mouth without making a single cut through the skin or muscle. Tony Kalloo, head of gastroenterology and hepatology since 2005, had first proposed such a surgical procedure in 1997, shortly after coming to Hopkins. He and key associates published the first study advocating endoscopic transgastric surgery in a 2004 issue of the journal *Gastrointestinal Endoscopy*. Work on perfecting the technique continues.[38]

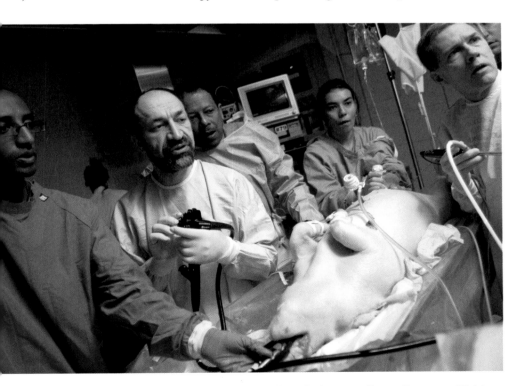

From left to right, endoscopist Samuel Giday, gastrointestinal surgeon Sergey Kantsevoy, Division of Gastroenterology chief Anthony Kalloo, computer integrated surgery expert Lia Assumpcao, and laparoscopic surgeon Michael Marohn. Their expressions indicate awe as a series of titanium sutures, deployed by the prototype of a tiny stapling gun attached to an endoscope, flash across a computer screen and instantly seal a 6-centimeter opening in a pig's stomach lining "like magic," says Kantsevoy, following this first successful experimental employment of the procedure.

Rheumatology

When Nancy Bechtle, above left, former head of the board of the San Francisco Symphony, contracted scleroderma, a mysterious, painful and incurable autoimmune disease of unknown origin, she went to the heads of three top medical schools—Harvard, Stanford, and the University of San Francisco—and asked for the name of the finest scleroderma doctor in the country. All three said Fredrick Wigley, above right, head of the Johns Hopkins Scleroderma Center. In three decades of treating more than 1,700 scleroderma patients, Wigley has collaborated in what he calls "the Hopkins way" with colleagues who are plastic surgeons, dermatologists, pulmonologists, immunologists, cardiologists, vascular biologists, pathologists and psychiatrists. Under his leadership, they have shown that while scleroderma may be incurable, it is not untreatable.

By 2004, after a year and a half of taking a medication prescribed by Wigley, the hot, stiffening skin on Nancy Bechtle's hands, feet and face, and her agonizing, disabling pain, had disappeared. She could return to skiing, scuba-diving and other favorite activities. She still has scleroderma, however, and when worrying once about the possibility of a relapse, she emailed Wigley, "What if I go over a cliff?" He replied, "I will have a net there for you." Wigley also is working with Rheumatology Director Antony Rosen, his wife, Livia Casciol-Rosen, and other Hopkins scientists to uncover the cell and molecular biology mechanisms underlying such autoimmune diseases as scleroderma, lupus and rheumatoid arthritis. These occur when the body's natural ability to attack invading viruses and bacteria goes haywire and mistakenly sees healthy body proteins as enemies. "Eventually, we'll be able to predict autoimmunity," Rosen says, "and find a way to turn it off." [37]

The husband-and-wife team of mathematician Sommer Gentry (left) and transplant surgeon Dorry Segev figured out the complicated organization required for matching multiple kidney removals with transplant recipients in "domino" surgical procedures.

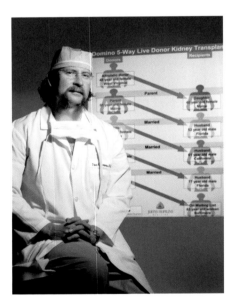

Transplant chief Robert Montgomery has pioneered "domino" transplant surgery swaps, overseeing breakthrough 10- , 12- and 16-patient surgeries often involving other hospitals and making Hopkins the United States' "Kidney Swap Central." It currently is estimated that of the 60,000 U.S. patients now on a waiting list for a matching kidney, some 3,000 could die before getting it. Montgomery has developed a "daisy chain" strategy for kidney swaps that, if fully implemented with a functioning national Kidney Paired Donation system, could increase the annual number of successful transplants by 2,000.

Transplants

Hopkins fast has become what one writer called "Kidney Swap Central"—with reason:

2004: Surgeon Dorry Segev and his mathematician wife, Sommer Gentry solve the convoluted organizational problems of how to match the multiple kidney removals and transplantations needed for what became known as "domino" surgeries.

2005: Hopkins performs the first double- and triple-domino transplants.

2006: Hopkins performs the first 10-patient domino transplants.

2007: Transplant chief Robert Montgomery oversees the first multihospital, transcontinental three-way swap transplant.

2008: Robert Montgomery leads the Hopkins teams that perform the first six-way, 12-patient domino transplant—a 10-hour procedure that simultaneously uses six Hopkins operating rooms and nine surgical teams.[40]

The example that Hopkins was setting soon led to the adoption of similar procedures around the world, extending the lives of more than 1,000 patients.[41] At Hopkins, the pace of advancement in kidney transplants only accelerated as the decade neared its end:

January 2009: Surgeon and biomedical engineer Mohamad Allaf uses a sophisticated, noninvasive method he has developed for removing a healthy

Leslie Persell (left) of California, a kidney donor, grasps the hand of Kristine Jantzi of Maine, a kidney recipient, following the world's first 10-patient domino transplant surgery, conducted at Hopkins in November 2006. Persell's kidney went to Gerald Loevner of Florida, while Jantzi received a kidney from Honey Rothstein of West Virginia, a "nondirected" or altruistic donor whose decision to donate one of her kidneys to anyone in need of it launched the landmark procedure.

female donor's kidney through the vaginal canal, a breakthrough likely to increase the number of prospective kidney donations. In what was believed to be the first-ever procedure of its kind, the need for a 5-to-6-inch abdominal incision was eliminated and the 48-year-old donor was left with only three pea-size scars on her abdomen, one of which was hidden in her navel. Her kidney was transplanted into her niece. [42]

February 2009: Robert Montgomery leads the first 12-patient, multihospital domino donor kidney transplant surgeries, in cooperation with surgical teams at Barnes-Jewish Hospital in St. Louis and INTEGRIS Baptist Medical Center in Oklahoma City.[43]

March 2009: Montgomery publishes an article in the *New England Journal of Medicine*, unveiling a strategy for a "daisy chain" of kidney swaps in which optimal pairings can be identified amid an even longer string of matches, with surgeries then executed one pair at a time. [44] An altruistic donor, unconnected to any patients on the American waiting list, can set in motion a chain of transplants that rolls across time zones and could, ostensibly, reach around the world, Montgomery writes.[45] The Hopkins-inspired, expanding kidney-swap concept clearly is poised to save more lives than ever, if implemented. By 2009, some 3,000 of the 60,000 U.S. patients on the waiting list for a matching kidney are likely to die before getting it. A fully functioning, national Kidney Paired Donation system could add up to 2,000 more successful transplants annually.[46]

April 2009: Hopkins' 1,000th kidney transplant from a deceased donor is performed since the 1996 merger of the Hopkins Bayview and Hopkins Hospital transplant programs.[47]

June 15 and July 9, 2009: Montgomery oversees the extraordinary effort of surgical teams at Hopkins, Barnes-Jewish Hospital in St. Louis, INTEGRIS Baptist Medical Center in Oklahoma City and Henry Ford Hospital in Detroit as they successfully complete the first eight-way, 16-patient, multihospital domino kidney transplant. [48] Nearly 100 medical professionals took part in these marathon transplants, including 10 surgeons-in-charge, anesthesiologists, operating room nurses, nephrologists, transfusion medicine physicians, critical care doctors, immunogeneticists, nurse coordinators, technicians, psychologists, social workers, pharmacists, financial coordinators and administrative support personnel.[49]

December 2009: Hopkins Hospital completes its 100th kidney paired donation swap.[50]

Pamela Paulk, right, vice president of human resources at The Johns Hopkins Hospital and Health System, was one of the donors in the historic July 2009 eight-way, 16-patient, multi-medical center domino kidney transplant procedure led by Hopkins. Paulk volunteered to donate a kidney to a needy recipient. The ultimate beneficiary of Paulk's generosity was Robert Imes, left, a painter and mechanic on Hopkins Hospital's facilities staff, who received someone else's kidney because of Paulk's donation. Surgical teams at Hopkins, Barnes-Jewish Hospital in St. Louis, INTEGRIS Baptist Medical Center in Oklahoma City and Henry Ford Hospital in Detroit successfully completed the multiple operations involving eight donors—3 men and 5 women—along with eight organ recipients, also 3 men and 5 women.[39]

Biological chemist and physician Landon King, at right, director of the Division of Pulmonary and Critical Care Medicine since 2005, became the new Vice Dean for Research in September 2011. He succeeded Chi Van Dang, who had been named director of the University of Pennsylvania's cancer center. Since 1994, the School of Medicine has received more biomedical funding from the National Institutes of Health than any other institution in the country. (In 2010, the amount was $438.8 million.) King is eager not just to maintain that record but improve it. A 1989 graduate of the Vanderbilt medical school, King came to Hopkins as an intern that year and later worked as a postdoctoral fellow studying aquaporins with Hopkins' future Nobel Prize-winner Peter Agre, the discoverer of these proteins that form the channels through which water goes in and out of cells. King's current research focuses on the role of these water channels in lung function, as well as immunologic mechanisms of recovery from lung injury.[52]

Research: The Perpetual Engine Powering Medical Miracles

Behind all these stories of clinical achievements and the medical miracles they entail is the research that brought such wonders to the patients' bedsides.

Throughout this period, Hopkins maintained its long record as the nation's top recipient of federal research grants. In fiscal year 2008—its 30th consecutive year as the leading U.S. academic institution in total research and development spending—Hopkins performed $1.68 billion in medical, science and engineering research.[51] From 2004 to the present, the growth of its multidepartment, interdisciplinary inquiries into medicine's newest frontiers continued unabated with the founding of four institutes and nine centers for basic research.

Biomedical engineer Raimond Winslow is the director of Hopkins' Institute for Computational Medicine (ICM), believed to be the first such research institute in the world. It uses information management and computing technologies to improve the understanding of how human illnesses originate and explore ways to enhance treatment of them. The ICM, founded by Winslow in 2005, has made impressive gains in unraveling the underpinnings of cancer, heart disease and brain afflictions.

The Institute for Computational Medicine (ICM)

Created jointly by the School of Medicine and Hopkins' Whiting School of Engineering in October 2005, ICM is believed to be the first research institute of its kind, designed to produce a better understanding of the origins of human illnesses by using information management and computing technologies to identify diseases in their earliest stages and to look for new ways to treat them.[53] Launched with more than $8 million in grants from the National Institutes of Health, the D.W. Reynolds Foundation and the Falk Medical Trust, ICM is focusing on three key research areas: the use of computers to shed light on the molecular basis of disease, computer analysis of the structure of healthy and diseased sections of the body, and development of new ways to mathematically represent and manage biomedical data, an area known as bioinformatics.[54]

In the five years since it was founded by Raimond Winslow, a Hopkins-trained biomedical engineer and the center's director, ICM researchers have achieved impressive gains in unraveling the underpinnings of cancer, heart disease and afflictions affecting the brain. For example, they have been:

- collecting and synthesizing vast amounts of data about healthy and unhealthy hearts, thereby making it easier for physicians to quickly and accurately identify heart abnormalities, assess risks, and determine treatment and intervention options for individual patients; and

- conducting computer analysis of how diseases initially appear in the brain and change its biological structure. This will help physicians detect autism, Alzheimer's, and schizophrenia earlier than ever before, so patients can benefit more quickly from the revolutionary therapies being devised to slow the progress of these devastating conditions.[55]

The Brain Science Institute

The continuing movement toward more interdisciplinary research at Hopkins received additional emphasis in February 2007 when faculty from the School of Medicine's basic and clinical neuroscience departments, other School of Medicine disciplines, the Whiting School of Engineering, the Krieger School of Arts and Sciences, and the Applied Physics Laboratory all jointly founded the Brain Science Institute (BSi).[56]

BSi is setting out to explore the brain in more depth than ever previously attempted. Under its founding director, the late John Griffin, former head of the Department of Neurology, his successor Richard Huganir, head of the Department of Neuroscience, and co-director Jeffrey Rothstein, head of the Packard Center for ALS Research, researchers are seeking to solve fundamental questions about how the brain develops and functions, then use those insights to understand the mechanisms of brain disease.[57]

Neurologists Jeffrey Rothstein (above), head of the Packard Center for ALS Research, and Richard Huganir (left), director of the Department of Neuroscience, jointly head Hopkins' interdisciplinary Brain Science Institute.

Funded by a family that wishes to remain anonymous, the BSi has used its resources to award more than $13 million in research grants since its founding to more than 20 departments across the School of Medicine and University.[58] The BSi also has underwritten the creation of a unique, academically based drug-discovery unit that's working on development of new medications for a wide variety of brain diseases, including cancers, ALS and schizophrenia.[59]

2004
The School of Medicine's reign as the nation's No. 2 medical school ends after 13 years as Washington University edges it out by one point (96 to 95) in the *U.S. News & World Report* survey of 125 accredited medical school. It regains the No. 2 spot (second only to Harvard) in the 2005 rankings, only to lose it again in 2009 to the University of Pennsylvania.

The Institute for Clinical and Translational Research

A five-year, $100 million grant from the National Institutes of Health's National Center for Research Resources launched the Johns Hopkins Institute for Clinical and Translational Research (ICTR) in September 2007 under the direction of Daniel Ford, vice dean for clinical investigation.[60]

The ICTR is designed to speed promising research findings from the laboratory to medical clinics and the community at large through the Johns Hopkins Clinical Research Network, which includes Hopkins Medicine's community hospital affiliates, the Anne Arundel Health System and the Greater Baltimore Medical Center.[61]

Daniel Ford (left), vice dean for clinical investigation, heads the Johns Hopkins Institute for Clinical and Translational Research, dedicated to accelerating the transfer of promising research findings to clinical treatments. Ford, a professor of medicine with joint appointments in the Department of Psychiatry and the Bloomberg School of Public Health's Departments of Epidemiology and Health Policy and Management, is considered a pioneer in what he calls "the interface of medical and mental health." He's won acclaim for documenting depression as an independent risk factor for heart disease and for describing the long-term risks of sleep disorders.

2004

September 21: As part of a major initiative to revolutionize care for cardiac patients in the 21st century, The Johns Hopkins Heart Institute is created. It will have its own advisory board of governors, headed by Arthur B. Modell, former majority owner of the Baltimore Ravens football team and former President of the Cleveland Clinic Foundation.

October 18: Hopkins' Institute of the History of Medicine, founded by William H. Welch, M.D. as the first department of its kind in the nation, celebrates its 75th anniversary.

Johns Hopkins at Green Spring Station marks its 10th anniversary with plans to add additional space and specialists. Hopkins Medicine's suburban centers grow, with expansion planned at White Marsh and Odenton labeled the "next big thing."

The Office of Critical Event Preparedness and Response (CEPAR), citing reports of the spread of so-called avian (bird) flu, convenes an institution-wide, 64-member Pandemic Influenza Steering Committee to plan for the possible onset of a pandemic and its impact on every segment of the Johns Hopkins enterprise. The comprehensive plan developed by this committee enabled Hopkins to respond effectively when the H1N1 (swine) flu virus pandemic began in 2009.

Molecular biologist Andrew Feinberg and biophysicist and biophysical chemist Cynthia Wolberger, co-directors of Hopkins' Center for Epigenetics in the Institute for Basic Biomedical Sciences, don hard hats in 2006 as they stand on the construction site of the $54 million John G. Rangos Sr. Building. Opened in 2008, it became the home of their new center, as well as the first structure built just north of Hopkins Hospital as part of an 80-acre urban redevelopment called the "Science + Technology Park at Johns Hopkins." Scientists in the Center for Epigenetics, founded in 2004 as the first university-based entity to concentrate on researching how cells establish and maintain control over genetic activity, study information unrelated to DNA that is inherited during cell division. Such epigenetic (meaning "above the gene") traits could make it possible to identify patients at risk for certain cancers and possibly explain the roots of illnesses as widespread as Alzheimer's disease and bipolar disorder.

The Institute for Basic Biomedical Sciences' Offspring

The Institute for Basic Biomedical Sciences helps coordinate fundraising for basic science and define the general way in which research efforts can be directed toward clinical work. It nevertheless recognizes that ensuring the departments' independence and scientific focus is essential. Such an arrangement gives Hopkins the maneuverability to adjust its financial and laboratory space commitments to certain kinds of research should a sudden change in knowledge or technology warrant enhanced concentration on a particular department's field.[62]

As part of its coordination and focusing efforts between 2004 and 2006, the IBBS became the birthplace of five interdisciplinary research centers.

The Center for Epigenetics

Founded in June 2004, this is the first university-based research entity to concentrate on the study of how cells establish and maintain control over genetic activity.[63] It currently accommodates 27 researchers associated with departments across Hopkins.[64] It is directed by pioneering epigeneticist Andy Feinberg, a 1976 graduate of the School of Medicine and since 2005 head of the Division of Molecular Medicine, and biophysical chemist Cynthia Wolberger. Its researchers are studying the impact of environmental factors on genetic activity. They also are concentrating on how the chemical changes in a process called "methylation," which occurs when a cell divides, can later affect the workings of the genes in a person's DNA. By allowing some genes to act as they should or causing others to go haywire, such "epigenetic" chemical changes—so called because they act "above the genome" to control the genetic activity within it—can impact both normal human development and the onset of such diseases as cancer, autism, schizophrenia, macular degeneration and glaucoma.[65] Epigenetics offers hope for devising new therapies that can reverse or prevent the genetic changes that underlie so many illnesses.

The Center for High Throughput Biology

Built around a technological question—how best to analyze hundreds, sometimes thousands of biological data samples all at once—this center (nicknamed "HiT") was founded in 2005 by molecular biologist and geneticist Jef Boeke. Along with neuroscientist Min Li, he oversees five independent laboratories that use a wide range of disciplines—including physics, chemistry, mathematics, engineering and computer science—to push the frontiers of research technologies by applying so-called high throughput techniques. These combine vast amounts of biological material, selected by both human researchers and robotic devices programmed to do so, with computers designed to analyze the material's contents at an accelerated pace.[66]

The Center for Sensory Biology

Under directors Paul Fuchs and Randall Reed, this is the first such entity in the world to conduct interdisciplinary research focused on understanding the fundamental processes underlying all of the primary senses—vision, touch (including pain), taste, smell and hearing.[67]

The Center for Cell Dynamics

Like the scientists working with its sibling, the Center for High Throughput Biology, the researchers in this group are technologically oriented, concentrating on developing the microscopic and chemical tools needed to advance inquiries into the functions within and between cells—as they are happening, in real time. Unraveling those mysteries can lead to new ways to diagnose and treat a wide range of diseases.[69] Headed by neuroscientist Denise Montell, the Center for Cell Dynamics aims to develop the high-powered microscopes and the chemical probes required to study such phenomena as cancer metastases and the body's immune response.[69]

The Center for Metabolism and Obesity Research

With an epidemic of obesity sweeping not just the United States but other western countries, scientists in this research center aim to identify the genes and the compounds, both chemical and those naturally produced in the body itself, that control human metabolism. These compounds conceivably could be used as treatments for metabolic disorders—and to head off development of such obesity-related diseases as diabetes.[70] Under co-directors Gabriele Ronnett, Timothy Moran and Fredric Wondisford, researchers focus on how metabolism works at the cellular level, looking at factors influencing cell survival, growth and aging. They also address how nutrients, hormone levels and energy usage affect feeding behaviors, exercise capacity, cognitive function, reproduction and longevity.[71]

Several more research groups are in the planning stage, including the Center on Drug Addiction, the Center for Chemoprotection, which will study the therapeutic benefits of edible plants, and the Center for Transport Biology, which will focus on discovering new drugs and gene therapies to treat conditions arising from faulty transport of water and salts within the body.[72]

Molecular biologist and geneticist Jef Boeke (above, left), founder of the Center for High Throughput Biology (HiT), discusses its work with a group of visitors to the center's office in the Broadway Research Building (BRB) shortly after the building's official opening in May 2004. Among those listening are then-Director of the National Institutes of Health, Elias Zerhouni (center), who had helped plan the BRB when he was the School of Medicine's executive vice dean and vice dean for research; neuroscientist Min Li (second from the right), who co-directs HiT with Boeke, and Dean/CEO Edward Miller, (right). In a speech prior to the BRB's official opening, Zerhouni said that as head of the NIH, he could describe dazzling research discoveries to the public all he wished, but what people invariably wanted to know was the answer to one question: "Where's the cure?" The technology used at HiT to unravel the mysteries of the molecular events that lead to disease should help the provide answer.

"We need to speed up the discovery and application of medical breakthroughs. We're at a critical juncture—both as an institution and as a society. We must extend our reach by investing everything we can in the art of what is possible."

—Ronald R. Peterson, President of The Johns Hopkins Hospital and Health System

Richard Jones (left), head of Hopkins' bone marrow transplant program, and Robert Brodsky, chief of hematology, have helped more than 100 aplastic anemia and multiple sclerosis patients by re-booting their autoimmune system with a treatment known as HiCy, which uses substantial doses of cyclophosphamide, a powerful synthetic drug used in low amounts as an anticancer chemotherapy.

Research by cardiologist Ilan Wittstein (above) received considerable publicity in 2005 when he and colleague Hunter Champion released a study revealing that "broken heart syndrome" (BHS), or stress cardiomyopathy, a severe but reversible weakness in the heart muscle prompted by emotional distress, is more common than previously believed. Their findings should help physicians distinguish between BHS and genuine heart attacks.

Promising Research Achievements

Throughout this period and into the present, Hopkins researchers have been conducting studies that could lead to new treatments for a host of illnesses. From finding a new way to use an old medicine to successfully reverse—perhaps even eliminate—such devastating autoimmune diseases as multiple sclerosis, lupus and aplastic anemia, to exploring techniques and devising experimental trials in cardiology, infectious diseases and oncology, the scientists in Hopkins' laboratories are achieving remarkably promising results.

Autoimmune Diseases

Hopkins physicians developed a way to reboot the immune systems of patients who are suffering from such autoimmune diseases as aplastic anemia and multiple sclerosis. Led by Hopkins oncologist Robert Brodsky, chief of hematology, and Richard Jones, head of the bone marrow transplant program, Hopkins physicians have bucked the medical establishment and significantly helped more than 100 patients by administering huge amounts of cyclophosphamide, a powerful synthetic drug devised decades ago to use in low doses as an anticancer chemotherapy.[73] Known as HiCy when given in high doses, cyclophosphamide can inflict serious side effects when it essentially shuts down a patient's immune system—which, in the case of autoimmune diseases, has mysteriously gone out of whack and begun attacking the body's own cells. Yet Hopkins researchers have found that as the immune system begins to recover slowly from the use of HiCy, it somehow reboots and cleanses itself of whatever caused it to attack the body originally.[74] HiCy essentially is a stem cell therapy, because what causes the dramatic improvements in autoimmune disease sufferers *are* stem cells—the patients' own stem cells. They just need to be protected from attack by a haywire immune system and given the chance to regenerate. That is exactly what HiCy accomplishes.[75]

Cardiology

Cardiologist Ilan Wittstein grabbed headlines in 2005 with research into cases of stress cardiomyopathy—colloquially known as "broken heart syndrome" (BHS)— in which patients react to emotional distress or shocking news with severe but reversible heart muscle weakness that mimics a heart attack but does not cause lasting damage.[76] Such cases often are misdiagnosed as a massive heart attack when what the patient actually has suffered is a surge in so-called catecholamines, such as adrenalin (epinephrine) and other stress hormones, which temporarily stun the heart.[77] Wittstein and his chief associate, Hunter Champion, subsequently conducted additional studies showing that BHS was more prevalent than first believed. To further help physicians distinguish between BHS and heart attacks, Wittstein and Champion also expanded the category of events that cause similarly transient chest pain in many more patients, including severe migraine headache, an asthma flare-up, major bone fractures, stroke, seizure, acute internal bleeding or sudden withdrawal from medicines.[78]

Cell Engineering

Between 2007 and 2011, Hongjun Song, director of the Stem Cell Program in the Institute for Cell Engineering (ICE) and his wife, fellow neuroscientist Guo-Li Ming, made seminal discoveries about the mechanism of demethylation, the process by which cells can be reprogrammed to develop differently by affecting how their genes interact. Methylation is the genetic process by which the development of stem cells is directed—determining if the cells become part of a particular organ or neurological activity. By removing the methylation signals by demethylation, changes can be effected in the development of cells. This could lead, for example, to treatments that would deactivate the development of brain cells that have become addicted to drugs, turning them into cells unaffected by opiates.[79]

In 2010, radiologist Jeff Bulte, director of cellular imaging for the Institute for Cell Engineering, made advances in stem cell therapies by developing non-invasive, MRI scanning methods to track and monitor the use of stem cells when they are being used to treat diseases. His laboratory has pioneered methods for labeling cells magnetically to make them visible by MRIs and already has used this technology in the clinic to determine the effectiveness of experimental cancer vaccines based on dendritic cells, which play a key part in the body's natural effort to combat disease.[80]

In 2010, Stephen Desiderio, director of the Institute of Basic Biomedical Sciences, also made important discoveries regarding the body's own system for developing antibodies in response to previously unknown organisms that invade it.[81]

Husband-and-wife neuroscientists Guo-Li Ming (center) and Hongjun Song, director of the Stem Cell Program in the Institute for Cell Engineering (ICE), were high school sweethearts in China. Today they have research laboratories side by side at ICE, where they work on finding ways to regenerate brain cells after they are damaged. It is not uncommon for them to have their children Maggie and Max in the labs with them. On the faculty since 2004, they say that one reason they chose to come to Hopkins was because of its family-friendly environment.

Radiologist Jeff Bulte (left), director of cellular imaging for the Institute for Cell Engineering, is pioneering methods for labeling cells magnetically so the use of stem cells in treatments such as experimental cancer vaccines can be tracked and monitored using non-invasive magnetic resonance imaging (MRI) scanning.

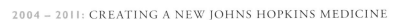

"The reason I came here is because this [tuberculosis] center is the most complete in the world. I came to Hopkins because they asked me to. I rushed to come."

—Jacques Grosset, French-born tuberculosis researcher, the world's foremost scholar in the field, now at the Johns Hopkins Center for Tuberculosis Research, founded in 1998

Infectious Diseases

By the early 1950s, physicians and scientists thought they finally had conquered tuberculosis with antibiotics such as streptomycin and other medications. They were wrong.[82]

By the mid-1980s, the pernicious, fiercely adaptable TB pathogen, *M. tuberculosis*, had resurfaced at an alarming rate. Fueled in part by the appearance of multidrug-resistant strains, in part due to an environment made hospitable by HIV infection, in part by a host of such negative social factors as homelessness and intravenous drug use, tuberculosis was once more the largest single cause of infectious disease deaths.[83]

Hopkins researchers are on the front lines in the battle against TB—just as they were in 1898, when William Osler obtained a grant of $750 for each of five years (a substantial sum in those days) to study TB containment.[84]

In 2003, under the direction of infectious diseases specialist Richard Chaisson, Hopkins' TB research efforts obtained more than $60 million in research grants by assembling an extraordinary group of exceptional scientists. Among them is Jacques Grosset, a septuagenarian French physician and researcher who once suffered from TB and now is perhaps the world's best-known tuberculosis researcher. Other principal researchers include center co-director, molecular biologist William Bishai, infectious disease

Left to right: Infectious disease expert Richard Chaisson, director of the Johns Hopkins Center for Tuberculosis Research (CTR); legendary TB researcher Jacques Grosset; CTR co-director, molecular biologist William Bishai, and infectious disease expert Eric Nuermberger are among the Hopkins scientists whose passion and vision for conquering TB has brought more than $60 million in grants to Hopkins' worldwide TB research efforts. Chaisson also is the founding director of the Consortium to Respond Effectively to the AIDS-TB Epidemic (CREATE), an international research collaboration based at CTR and funded with more than $76 million in grants from the Bill & Melinda Gates Foundation. It is undertaking campaigns to curb the spread of these two related scourges in Zambia, South Africa and Brazil. "It's hard to act on a grand level most of the time, but that kind of thinking is what inspires me," says Chaisson.

experts Paul Converse, Susan Dorman, Petros Karakousis, Yukari Manabe, Gueno Nedeltchev, and Eric Nuermberger, microbiologist Gyanu Lamichhane, epidemiologist Jonathan Golub, and pediatrician Sanjai Jain.[85]

In July 2004, the Bill & Melinda Gates Foundation recognized the depth and worldwide scope of the anti-TB battle of the Johns Hopkins Center for Tuberculosis Research by giving it $44.7 million to fund CREATE (Consortium to Respond Effectively to the AIDS-TB Epidemic), a Hopkins-led research group that is working on urgently needed strategies to control TB in communities with high HIV infection rates. Nelson Mandela, the former president of South Africa and himself once a TB patient, spoke on behalf of CREATE's mission at the international AIDS conference in Bangkok where the Gates Foundation's grant was announced.[86]

By October 2008, The Gates Foundation had decided to give CREATE another $32 million in funding. CREATE, considered the largest TB-related research effort in the world, now encompasses more than 250 scientists and dozens of health policy experts in Africa, South America, Europe and the United States.[87]

Neuroscience

With a daylong November 2005 symposium that featured lectures by two Nobel laureates and other top neuroscientists from around the country, the Department of Neuroscience marked its 25th anniversary by celebrating a century's-worth of neuroscience at Hopkins.[88]

That anomaly can be explained by nomenclature—essentially the difference between the extraordinary neuroscientific accomplishments of Hopkins researchers for 100 years and the founding of a formal neuroscience department in 1980.

Beginning in 1906, Hopkins leaders in three clinical areas—neurosurgery, psychiatry and physiology—had been conducting critically important studies of the functions and abilities of the brain. Their achievements, recounted earlier, spanned everything from discovering the role that the pituitary gland plays in growth to developing procedures for the first brain X-rays, and from describing the molecular and anatomical regulation of our "biological clock" to uncovering the precise, column-like arrangement of brain cells, now acknowledged as a universal organizing principle of the brain.[89]

Even apart from neurosurgery, psychiatry and physiology, there "were enclaves of scientists and physicians studying the brain in various departments at Hopkins well before the neuroscience department was formed," explains Solomon Snyder, who—as noted previously—made the pivotal discovery of the brain's receptors for opiates. He headed the department from its founding until 2006 and still remains active in its laboratories.[90]

By the time it reached its 25th anniversary, the Department of Neuroscience had become the largest basic science department in the School of Medicine. Four of the department's members rank among the world's most cited people in the field—more than any other institution anywhere.[91] For example:

Multi-departmental researcher, neuroscientist/neurologist/psychiatrist Christopher Ross, directs the Huntington's Disease Center—possibly the world's largest HD clinic. Housed in the Psychiatry Department, Ross and his team of researchers identified the protein, named huntingin, that causes the Huntington's Disease gene to mutate and run amok. Huntington's—so named because it first was described by physician George Huntington (1850–1916)—is a rare, still-incurable and invariably fatal disease that kills nerve cells, bringing on not only ceaseless, jerky movements but an asylum-ful of significant psychiatric disturbances: apathy, depression, obsessions, psychosis, substance abuse, paranoia.[92] Ross' team has demonstrated that the earlier the disease begins, the more severe its consequences. They have used this genetic knowledge to breed mice to have HD so that potential therapies can be tested. Doing this, they have found ways to slow the disease's progression in lab animals.[93] For HD sufferers today, Ross and his team seek to delay the disease's onset as long as possible by looking for the earliest indications that treatment should begin. "We use the most advanced molecular and cell biology techniques, along with mouse genetics, to tackle clinical problems," Ross says. "That's not so common in psychiatry departments."[94] He adds: "Because Huntington's resembles Parkinson's and Alzheimer's in major ways … if we find an HD treatment, something should be there for the other illnesses."[95]

Multi-departmental researcher Christopher Ross (above) directs the Huntington's Disease Center at Hopkins—possibly the world's largest Huntington's Disease (HD) clinic. HD is a rare, still-incurable and invariably fatal disease that kills nerve cells, leading to ceaseless jerky motions and numerous psychiatric problems. Ross and his research team identified the protein, huntingin, that causes the HD gene to mutate and run amok, a key finding that is fostering further studies aimed at developing potential therapies.

2005

Nobel laureate Peter Agre leaves Hopkins—as it turns out, only temporarily—to become vice chancellor of science and technology at Duke. He returns in 2008, saying, "I never really left Hopkins."

The master plan for the East Baltimore campus's multi-million dollar revitalization and expansion marks a milestone in the spring when all the "users" of two new clinical towers settle on what will go in the buildings and where.

Hopkins is the acknowledged national leader in continuing medical education under Todd Dorman, M.D., associate dean for CME.

March 14: Pioneering neuroscientist Solomon Snyder belatedly receives the 2003 National Medal of Science, the nation's highest honor for scientific achievement, at a White House ceremony. The Department of Neuroscience celebrates 25 years of groundbreaking research under Snyder, who over the past two decades has become the third most-cited scientist in the world. His work has been referred to by other researchers more than 63,000 times in journal articles and books. Snyder later announces his plans to retire as head of the Department of Neuroscience, which subsequently is named in his honor. Neuroscience itself at Hopkins can be traced back a century to the studies of Harvey Cushing, Walter Dandy and Adolf Meyer.

Neurologist Richard O'Brien, director of the Department of Neurology at Johns Hopkins Bayview, has led research uncovering important distinctions between the brains of elderly individuals who did not develop dementia despite significant Alzheimer's pathology discovered in their brains during autopsies, and those who did develop dementia. The findings indicate that cerebrovascular disease is the key culprit in determining whether a person develops dementia.

"Hopkins has established itself as a leader in what's called 'translational' research, which is just a fancy word that means applying the fruits of basic research to patients through either improved diagnosis or therapy. A lot of places give lip service to that ideal, but it's very difficult to realize. Hopkins has achieved singular success in it, I think, because of the environment we have.... People want to help—whereas I know at other institutions, a lot of the time, the first questions you hear are, 'Well, what's this going to do for me or my career or my department?' I've never heard that at Hopkins. It's always, 'It sounds exciting. What can we do to help?' That's an environment that Hopkins has fostered and that's really paying dividends."

—Bert Vogelstein, 1974 graduate of the School of Medicine, Professor of Oncology and Pathology, Director of the Ludwig Center for Cancer Genetics and Therapeutics

Neurology

Alzheimer's disease, the affliction poised to increase exponentially as the number of elderly Americans explodes, has long been a focus of Hopkins researchers. As director of the Alzheimer's Disease Research Center in the late 1980s and 1990s, neuropathologist Donald Price was instrumental in development of donepezil, the medication marketed as Aricept, one of the most widely used drugs for the treatment of Alzheimer's symptoms. Alzheimer's also is the main interest in the laboratory of Johns Hopkins Bayview-based neurologist, Richard O'Brien, the department chief there. His research has provided striking evidence that reducing the risk of stroke could be the key to turning back the anticipated wave of mind-destroying dementia cases.[96]

Using the autopsy reports of individuals who participated in the renowned, half-century-old Baltimore Longitudinal Study, launched in 1958 by geriatrics pioneer Nathan Shock at what now is Johns Hopkins Bayview, O'Brien and fellow researchers reported in 2008 that they had discovered two chief distinctions among the so-called "normals"—those whose brains showed significant Alzheimer's pathology without any dementia. First, Hopkins pathologist Juan Troncoso found that these individuals had many enlarged neurons, meaning they somehow retained "high plasticity" in their brains and managed to form new tissue to bypass the obstacles formed by Alzheimer's.[97] The second distinguishing trait for the normals is that they showed a very low incidence of cerebrovascular disease, chiefly demonstrated by a lack of strokes of any kind. "It's the combination of the two pathologies that matters most," O'Brien says. "Even a single stroke doubles your risk of dementia when combined with Alzheimer's pathology.... Eliminate the strokes and you eliminate a third of all dementias."[98] Researchers haven't found a way to stop Alzheimer's yet, but progress is being made in tackling its partner in crime—cerebrovascular disease. The chief culprit is unmanaged hypertension.

Oncology

Laboratory experiment after laboratory experiment, scientific journal paper after scientific journal paper, Hopkins' cancer researchers continue their record as among the world's most productive—and influential—in their field. They have been in the forefront of those scientists who have demonstrated the genetic basis for cancer, making "genetic testing and counseling part and parcel" of the care of families with a genetic predisposition to the disease, says Bert Vogelstein.[101]

The genetic testing "can accurately determine who is at risk for cancer and who is not, and for those who aren't, it removes a lot of the anxiety and the need for routine testing," Vogelstein observed in 2009. "Twenty years ago, none of those causes were known. None of the genes had been discovered. And now they all have been, without exception."[102]

Much of that discovery has been done at Hopkins.

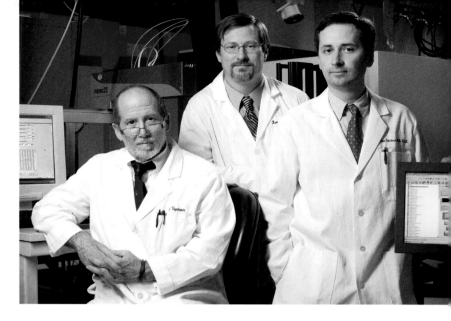

2005–2006: Hopkins researchers became the first to examine simultaneously all of the genes in a tumor to obtain a more comprehensive and unbiased determination of what specifically is wrong with the cancer cells. Vogelstein notes that "since that time, four major cancers—the first was colon and now breast, pancreas and brain tumors—have been comprehensively investigated in the same way. All at Hopkins."[103]

Vogelstein and his team found five of the seven genes linked to colon rectal cancer. Indeed, "Hopkins is the only place so far that has the capability of doing this," says Vogelstein, "although that will obviously change in the future and more and more institutions will be doing this, but so far, Hopkins is unique in having done this with any cancer, and that has provided a completely new understanding of what is wrong with the cancer cell" and its many different manifestations.[104]

2008: Cancer researchers Victor Velculescu, Ken Kinzler and Bert Vogelstein sequenced the DNA of numerous brain cancers and made the seminal discovery that in some of them, a specific mutation in an enzyme (a protein that accelerates chemical reactions in a cell) affects its metabolism, causing the genome of the cell to become highly flexible and thereby foster the development of various cancers. The discovery of this mutation in the metabolism of cancer cells led many pharmaceutical companies to begin developing medications that will combat that mutant enzyme.[105]

2010: Oncologist, pathologist and molecular biologist Drew Pardoll and oncologist Cindy Sears reported finding that humans have far more bacteria—clustered together in communal groups known as microbiomes—than the body has cells, a discovery that could have an important impact on developing methods to improve our immunity to cellular mutations that lead to cancer.[106]

2010: Molecular biologist Josh Mendell reported that research he conducted with Chi Van Dang showed that oncogenes—normal genes that, if they mutate, can lead to cancer—may be affected by altering the metabolism of cells through manipulation of the molecule RNA (ribonucleic acid) within them. Mendell and Dang's findings showed that manipulation of the RNA molecules can be used to treat malignancies such as liver cancer.[107]

Having made the genetic discoveries, Hopkins' cancer researchers and their clinical colleagues now are switching their focus to figuring out how to use these insights clinically. "Hopkins has established itself as a leader in what's called 'translational' research," Vogelstein notes, "which is just a fancy word that means applying the fruits of basic research to patients through either improved diagnosis or therapy."[108]

Vogelstein's own lab has turned its attention almost entirely from doing basic inquiries into the mechanisms that underlie cancer to research that can show how to apply the information they already have acquired. "I think we have something like a dozen clinical trials going on that are directed by members of our lab," Vogelstein says. "Ten years ago, we had none. It would have been unthinkable for a basic research lab

Hopkins cancer researchers continuously are praised by their peers as among the best— and most frequently cited—in the world. Left to right, Bert Vogelstein, Ken Kinzler and Victor Velculescu were responsible for deciphering the genetic codes of breast, colon, pancreatic and brain cancer, findings that could lead to earlier detection of these diseases and a better chance of treating them successfully.

■ Scientific journals are researchers' way of disseminating their key findings throughout the medical community. Researchers elsewhere, rather than starting from scratch, use the discoveries of others as building blocks for their own inquiries—and cite those earlier findings in their papers.[99] The January/February 2006 issue of *Science Watch*, a newsletter published by Thomson Scientific, then a Philadelphia-based publisher of college textbooks and information publications in science, technology, financial and other fields, proclaimed Hopkins' Kimmel Cancer Center "a research powerhouse in the field of oncology," noting that five of its faculty were the most frequently cited oncology researchers of the 1995 to 2005 period. Between them, Bert Vogelstein, Kenneth Kinzler, James Herman, Stephen Baylin and David Sidransky were cited more than 90,000 times, the newsletter said.[100]

2005

May 15: Surgeons at Hopkins' Comprehensive Transplant Center perform what is believed to be the world's first "domino" three-way kidney transplant, prompted by an altruistic donor's decision to give a kidney to a stranger.

October 12: The Institute for Computational Medicine is founded, with Aravinda Chakravarati, director of the Institute of Genetic Medicine, named co-director with Raimond Winslow, a professor of biomedical engineering in the Krieger School of Arts and Sciences. The ICM is believed to be the first research institute of its kind, aiming to use information management and computing technologies to form a better understanding of the origins of human illness.

to take that route, because there wasn't enough knowledge, really, to do it. But now we can and we are, and lots of other labs here are doing that as well."

"That's clearly the wave of the future, and I think something that Hopkins is particularly good at and well-suited for. When I ask people, within our department but not in our lab, and in the institution in general, for help or to join a project, there's usually, almost without exception, great enthusiasm to do so. People want to help—whereas I know at other institutions, a lot of the time, the first questions you hear are, 'Well, what's this going to do for me or my career or my department?' I've never heard that at Hopkins. It's always, 'It sounds exciting. What can we do to help?' That's an environment that Hopkins has fostered and that's really paying dividends."[109]

Among such dividends are not just promising improvements in diagnosis and treatment. The impact of this work has inspired several multimillion-dollar gifts in recent years for the establishment of important interdisciplinary research centers to combat pancreatic cancer, among the deadliest of malignancies. Only five percent of patients survive five years after diagnosis. Two Hopkins centers aim to change that percentage dramatically.

Pathologist Ralph Hruban (above, right), director of the Sol Goldman Pancreatic Cancer Research Center, joins with the center's benefactors (left to right) Amy Goldman and Jane Goldman, the daughters of Sol Goldman; Hopkins cancer researchers Ken Kinzler and Bert Vogelstein; and New York cardiologist Benjamin Lewis, the husband of Jane Goldman, following the $10 million gift from the Sol Goldman Charitable Trust and the Lillian Goldman Charitable Trust (named for the Goldman sisters' parents) to found the multi-disciplinary research center in 2005. It already has had a powerful impact, says Hruban. "Just five years after the human genome project was completed in 2003, the team at Hopkins sequenced 24 pancreatic cancers in this landmark study. All this happened within a year and a half of receiving the funding from the Goldman family. It is truly extraordinary."

The Sol Goldman Pancreatic Cancer Research Center

In 2005, a $10 million gift from the New York-based Sol Goldman Charitable Trust and the Lillian Goldman Charitable Trust created the Sol Goldman Pancreatic Cancer Research Center. The Goldman Trust had no previous association with Hopkins, but its leaders determined that Hopkins' scientists were the field's superstars.[110] Representatives from five departments—pathology, surgery, oncology, radiation oncology and radiology—have joined forces in the center under the direction of pathologist Ralph Hruban. They are using the Hopkins-based National Familial Pancreas Tumor Registry, which contains more than 1,400 family samples, as a major research source. Their research philosophy is to "go out on a limb," looking for unconventional ways to fight the disease, Hruban said at the time of the center's founding. "And we need to swing for a home run—that's what pancreatic cancer research needs."[111]

In 2008, Hopkins researchers announced the complete deciphering of pancreatic cancer's genetic blueprint, an accomplishment that has forever changed the study and treatment of the disease—the fourth leading cause of cancer death in the United States, killing nearly 31,000 Americans each year.[112] The breakthrough was made possible by another $4 million gift from the Goldman Trust. Hopkins researchers used the funds to launch the Goldman Pancreatic Cancer Genome Initiative, the world's first and most comprehensive pancreatic genome study. "Just five years after the human genome project was completed in 2003, the team at Hopkins sequenced 24 pancreatic cancers in this landmark study," said Hruban. "All this happened within a year and a half of receiving the funding from the Goldman family. It is truly extraordinary." The findings could pave the way to genetic blood tests for earlier tumor detection, more efficient medications, and innovative therapies that will stop cancer-forming genes in their tracks.[113]

The Skip Viragh Center for Pancreas Cancer Clinical Research and Patient Care

In 2010, The Kimmel Cancer Center received the largest gift for pancreas cancer research in its history. Albert P. "Skip" Viragh Jr. bequeathed $20 million for establishment of the Skip Viragh Center for Pancreas Cancer Clinical Research and Patient Care. Viragh, a mutual fund leader, had been a pancreas cancer patient at Hopkins who died in 2003 at the age of 62.[114] Co-directors of the Viragh Center are oncology researcher Elizabeth Jaffee and medical oncologist Daniel Laheru. Since 2004, they have made substantial strides in developing a vaccine to trigger the body's own immune system to fight pancreatic cancer and suppress it. "With the results we're getting, we'll be able to think about making pancreas cancer a chronic disease rather than a death sentence," Jaffee has said.[115]

In just one of Hopkins' key advances in combating cancer, Kimmel Cancer Center oncologists Elizabeth Jaffee and Daniel Laheru (above, holding a model of a pancreas), have been fine-tuning a treatment vaccine for pancreatic cancer since 2005 and reporting positive results from preliminary tests. Jaffee and Laheru believe it might well make possible a broader reversal of pancreatic cancer and improved long-term survival.

Opthalmology

In 2009, molecular biologist and geneticist Jeremy Nathans discovered that the color receptors in the eyes of mice and men are different, providing a richer understanding of the visual process. A year earlier, Nathans and his colleague, King-Wai Yau, shared the $1.4 million Champalimaud Vision Award, the world's richest accolade for eye research, for their groundbreaking laboratory discoveries that enhance our knowledge and comprehension of sight. Nathans was recognized for determining the genetic code of human visual pigment molecules within the eye that capture light and mediate our sense of color. He also discovered some of the ways in which genes control eye development and contribute to inherited eye dysfunctions, such as color-blindness, and diseases, such as macular degeneration. Yau was cited for discovering how the absorption of light by these visual pigments generates the electrical signals that initiate vision and also regulates the natural rhythms of our bodies. His work

has led to major advances in understanding many hereditary blinding diseases, such as retinitis pigmentosa. (Yau's research interests go beyond the eyes to the nose. Over the past decade, he also uncovered how the nerves in our noses respond to odor molecules, unraveling the mystery of how we smell. This could lead to treatments for people with disorders of smell and taste.)[116]

2005
October 22: Michael Armstrong, head of the Hopkins Medicine board of trustees, and his wife, Anne, donate $20 million for a new education building, the first such facility on the campus in nearly 25 years.

October 27: the School of Medicine celebrates attaining the milestone of having 100 women promoted to professor since the SOM was founded in 1893. As recently as 1979, only seven women had been promoted to full professor. By 2005, the number had reached 115—three of whom are department heads.

On the left, molecular biologist and geneticist Jeremy Nathans (left) and neuroscientist King-Wai Yau each have made landmark discoveries about the visual process, earning them the $1.4 million Champalimaud Vision Award, the world's richest accolade for eye research, in 2008.

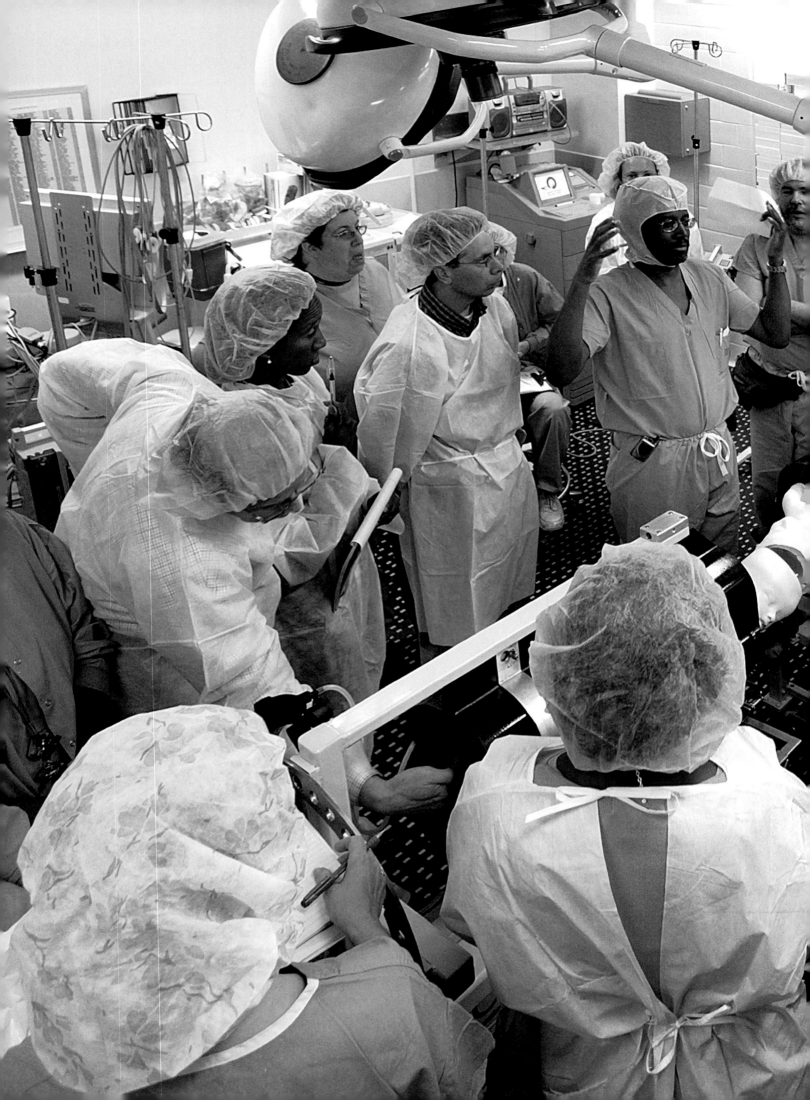

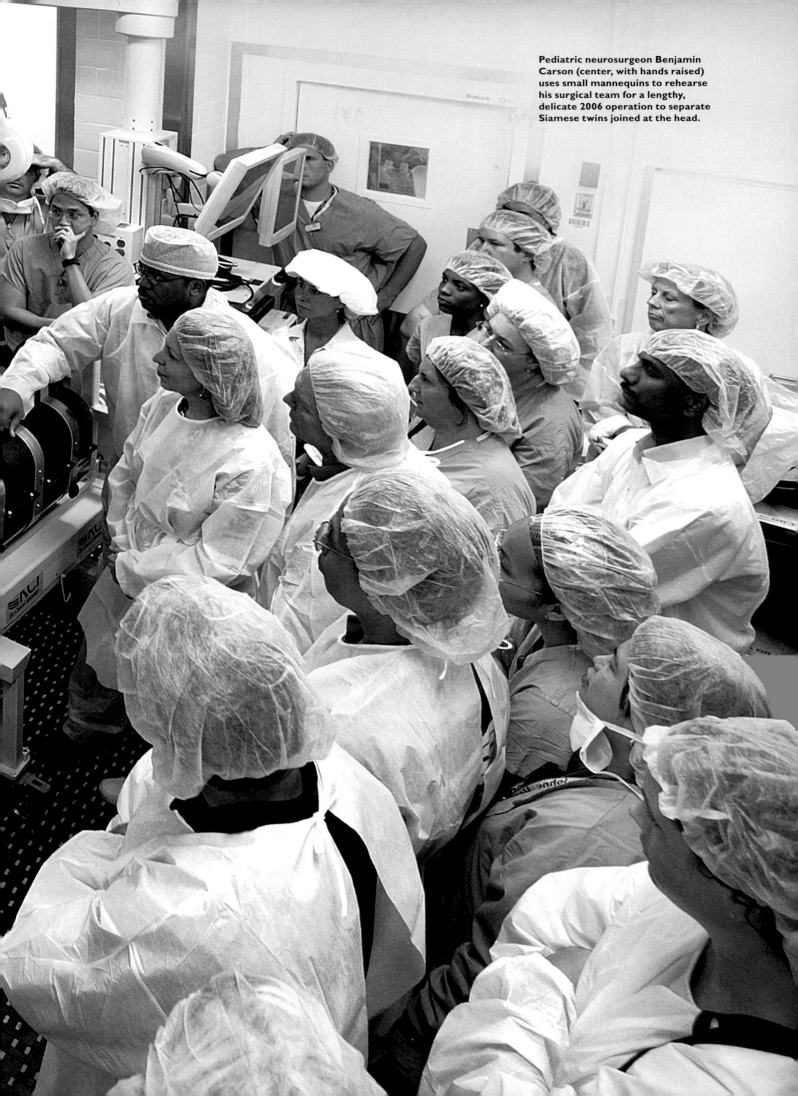

Pediatric neurosurgeon Benjamin Carson (center, with hands raised) uses small mannequins to rehearse his surgical team for a lengthy, delicate 2006 operation to separate Siamese twins joined at the head.

Otolarygology

**A self-described "music addict,"
otolaryngologist Charles Limb has
the rare distinction of holding faculty
appointments at both the School
of Medicine and Hopkins' Peabody
Conservatory of Music. In collaboration
with fellow researchers at the National
Institute on Deafness and Other
Communication Disorders, he has
conducted innovative studies that use
functional magnetic resonance imaging
(fMRI) to track how jazz performers'
brains react when they're improvising.
Music, Limb notes, represents the
pinnacle of hearing—an art form as
complex in structure as language. It
can provide important insights on how
the human brain processes auditory
information. As a cochlear implant
specialist, Limb foresees that his studies
of "the brain on jazz" could help him
improve how deaf individuals with
cochlear implants perceive music. He
also believes a musical ear exam—
rather than the current use of simple
sounds such as blips and pings—might
be sensitive enough to detect hearing
loss at its earliest stages, or detect
defects in pitch or tone that standard
tests may miss. Limb's expertise in
both hearing and music led to his
collaboration with Baltimore Symphony
Orchestra music director Marin Alsop
and others in a special February 2008
program, "Beethoven: CSI." The
orchestra performed excerpts from all
nine of Beethoven's symphonies, after
which Limb and the other participants
discussed what might have caused the
composer's hearing loss and how it may
have affected his music.[117]**

Milestones

During the period recounted in this chapter, Hopkins Medicine's
dedication to excellence, exemplified by such clinical and research
achievements as those just described, led to many of the highest accolades in
science and medicine being bestowed repeatedly on Hopkins physicians and
scientists.

White House Honors

During a span of four years, beginning in 2004, three Hopkins Medicine
luminaries—ophthalmologist Arnall Patz, neuroscientist Solomon Snyder and
pediatric neurosurgeon Benjamin Carson—traveled to the White House to
receive presidential medals for distinguished work in their fields.

Patz, director of the Wilmer Eye Institute from 1979 to 1989, went to
Washington on June 23, 2004 to receive the Presidential Medal of Freedom,
the nation's highest civilian honor, for his notable achievements in
ophthalmology. President George W. Bush praised Patz as "a great healer
[who] has been the gold standard in the field of researching the causes
and treatment of eye disease."[118]

Prior to Patz, who died in March 2010 at the age of 89, the only other
Hopkins recipients of the medal had been pediatric cardiologist Helen Taussig
and former University president Detlev W. Bronk in 1964 and former School
of Public Health dean D.A. Henderson in 2002.[119] (The ranks of Hopkins'
Presidential Medal of Freedom recipients soon would grow.)

The next Hopkins Medicine notable to head to the White House was
neuroscience legend Solomon Snyder, who received the National Medal of
Science, the nation's highest honor for scientific achievement, from President
Bush at a March 14, 2005, ceremony.[120]

Snyder insists that he is a "klutz" in the laboratory, inept at handling its
equipment, and credits his graduate and postgraduate students with figuring
out how to put his ideas for experiments into practice. When word arrived of
his National Medal of Science, his response was typical: "I am grateful that the
work of my students over the past 40 years has received recognition," he said.
Snyder also said he was honored to join the list of distinguished neuroscientists
who had received the award, "especially my friend and colleague Vernon
Mountcastle," a 1986 honoree.[121]

Pediatric neurosurgeon Benjamin Carson made not one but two trips to
the White House in 2008—which turned out to be quite a year for him. On
February 10, he joined retiring Supreme Court Justice Sandra Day O'Connor
at 1600 Pennsylvania Avenue to receive the 2008 Ford's Theatre Lincoln Medal
from President Bush. The award, bestowed by the Executive Committee of the
Ford's Theatre Board, honors individuals whose professional accomplishments or
personal attributes exemplify the qualities of Abraham Lincoln. Carson returned
to the White House on June 19 to receive the Presidential Medal of Freedom
from Bush, who praised his "groundbreaking contributions to medicine and his
inspiring efforts to help America's youth fulfill their potential."[122]

Hopkins pediatric neurosurgeon Benjamin Carson shakes hands with President George W. Bush after receiving the Presidential Medal of Freedom in a June 2008 White House ceremony. Bush praised Carson's "groundbreaking contributions to medicine and his inspiring efforts to help America's youth fulfill their potential," alluding to the Carson Scholars Fund, co-founded by Carson and his wife, Candy, which recognizes young people of all backgrounds for exceptional academic and humanitarian accomplishments. The $1,000 Carson Scholarships, awarded annually to hundreds of students in grades 4–11 nationwide, are invested toward their college education. The fund also supports the Ben Carson Reading Project, which creates school-based reading rooms in which youngsters can discover the joy of reading.

The 2008 accolade that meant the most to Carson, however, came from Hopkins, where an endowed professorship was named in his honor. It was made possible by the generosity and concerted fundraising efforts of medical equipment entrepreneur Ernie Bates, a 1958 graduate of Hopkins' Krieger School of Arts and Sciences, the second African American in the United States to become a neurosurgeon, and a long-time vice-chair of the university's board of trustees.[123]

Bates, a great admirer of Carson, eagerly had joined the campaign to raise the endowment money needed for the professorship. Recruited by neurosurgery director Henry Brem to participate in the drive, Bates capped the $2.5 million goal by persuading retired Oppenheimer Funds executive Don Spiro and his wife, Evelyn, to donate $1 million toward the endowed chair, which also bears Evelyn Spiro's name.[124]

Legendary ophthalmologist Arnall Patz (right) director of the Wilmer Eye Institute from 1979 to 1989, received the Presidential Medal of Freedom, the nation's highest civilian honor, from President George W. Bush in June 2004. Bush lauded Patz as the "gold standard" in eye disease research and treatment. Also receiving the medal at the same White House ceremony was golfing icon Arnold Palmer. For Patz, an avid duffer, having the opportunity to meet Palmer was almost

At a tribute dinner held to celebrate Carson's official receipt of the endowed chair, the soft-spoken honoree expressed his appreciation not just for the honor but for all that his Hopkins colleagues and support staff had done to help him.[125]

as impressive as receiving the presidential accolade. "To just be sitting next to him…," Patz later marveled.

"The reason I think it's so easy to work at Hopkins is because, you know, in other places, you get these really smart people and they're special—but here they're not special because everybody's smart," Carson said, eliciting a huge laugh from the audience. "It makes it a whole lot easier to work with people, let me tell you. And, yet, you know that if you get in trouble, you know there's going to be somebody who has an answer. I can't thank my colleagues enough."[126]

At left, neuroscientist Solomon Snyder receives the National Medal of Science, the nation's highest scientific honor, from President Bush in 2005. Crediting his graduate and postgraduate students with devising the laboratory methods for putting his experiment ideas into practice, Snyder said upon receiving the medal, "I am grateful that the work of my students over the past 40 years has received recognition."

Liberian-born internist and epidemiologist Lisa Cooper (above) received a $500,000 MacArthur Foundation fellowship—popularly known as a "genius" award— in 2007 for her landmark studies into the disparities between the health care provided to minority patients and what non-minorities receive. She continues developing programs aimed at overcoming racial and ethnic disparities in medical care and research.

2005
Howard County General Hospital plans a $73 million expansion that will include a five-story patient tower with 90 private rooms and two new operating rooms.

Over an intense Labor Day weekend, Hopkins disaster response experts worked almost 'round-the-clock to assemble and dispatch a team of physicians, nurses and others to help a Katrina-ravaged community in Louisiana, not far from a devastated New Orleans.

The Hopkins Hospital's signature edifice, the Billings Administration Building, undergoes an extensive, exterior renovation that includes a completely new tile roof on its iconic Dome. Months later, original, 116-year-old tiles from the Dome will be sold to Hopkins employees to help fund restoration of a home for a low-income family in East Baltimore.

Geniuses

Among the many "smart people" whose presence at Hopkins was praised by Ben Carson are two professors in entirely different fields—internist and epidemiologist Lisa Cooper and critical care specialist and patient safety advocate Peter Pronovost. They were recipients in 2007 and 2008, respectively, of John D. and Catherine T. MacArthur Foundation fellowships—popularly known as "genius grants"—for their exceptionally creative work and likely achievement of more.[127]

The Liberian-born daughter of a surgeon and a research librarian, Cooper lived in Africa until she was 14, went to high school in Switzerland and came to the United States for college and medical school. Having spent her formative years outside the U.S., she brought a unique perspective to American medical care, the MacArthur Foundation observed.[128]

After obtaining her medical degree at Emory University in Atlanta, Cooper did her residency at the University of Maryland, where she noticed "that there was mistrust of the health care system in the surrounding community."[129]

"Patients would use vernacular language and I was expected to interpret, because most of the people delivering the care were not minorities," Cooper recalled. "The differences between the patients and the staff were pronounced, and, while I think most people were well meaning, there were misunderstandings and a lot of assumptions being made."[130]

She came to Hopkins as a research fellow, knowing that she wanted to help minority patients on a broader level than just a one-to-one basis.[131]

Completing a research fellowship in Hopkins' School of Medicine and obtaining a master's degree in public health from the School of Public Health in 1993, Cooper joined the faculty in 1994 and continued her inquiries into what now is called "disparities research." In 1999, she published a landmark study in the *Journal of the American Medical Association*, reporting that minority patients felt they were less involved in their physicians' decisions on treatment than non-minorities. She also uncovered a direct link between the effort of physicians to involve patients in treatment decisions and the success of the patients' health care. These findings led Cooper to develop programs designed to overcome racial and ethnic disparities in medical care and research.[132]

The MacArthur arbiters also apparently were in agreement with the editors of *Time* magazine, which named Hopkins anesthesiologist and critical care specialist Peter Pronovost as one of the world's 100 "most influential people" of 2008. The magazine said Pronovost's research on preventing hospital-caused infections and improving communication between caregivers "may have saved more lives than any laboratory scientist in the past decade by relying on a wonderfully simple tool: a checklist." The MacArthur Foundation decided that Pronovost's championing

Anesthesiologist and critical care medicine specialist Peter Pronovost, perhaps the world's most prominent patient safety expert, received a $500,000, no-strings-attached 2008 MacArthur Foundation fellowship in recognition of his ongoing research into and efforts to prevent hospital-caused infections and errors while improving communication between caregivers.

of scientifically precise, common-sense approaches to improving patient safety also merited a no-strings-attached, $500,000 "genius" grant, awarded in September 2008.[133]

A Connecticut native whose own father died as a result of a medical error, Pronovost is a 1991 graduate of the School of Medicine and recipient of a 1999 Ph.D. from the School of Public Health. He's a professor of surgery as well as anesthesiology and critical care medicine. The praise for his patient safety work clearly has had an impact on the health care community nationwide. In what he called "the first quantitative outcome goal for quality and safety that the country has ever made," federal health officials in the summer of 2010 called for cutting bloodstream infections in half by 2013—and 30 states agreed to institute Pronovost's protocols in an effort to achieve that. In Maryland, which had one of the highest growth rate in such infections, 44 hospitals said they would adopt the Pronovost program.[134]

Winning a Nobel While Folding the Laundry

Unlike her colleague biochemist Peter Agre six years earlier, molecular biologist Carol Greider wasn't still in her pajamas when a 5 a.m. telephone call from the Royal Swedish Academy of Sciences in Stockholm informed her that she had won the 2009 Nobel Prize in Physiology or Medicine. She was folding laundry.

"I don't usually do the laundry so early in the morning," she told a reporter for the *New York Times*, "but I was already up, and there was all this laundry staring at me.

"I was supposed to later meet two women friends to take our morning spin class," she added. "After I got the call, I sent my friend an email: 'I'm sorry I can't spin right now. I've won the Nobel Prize.'"[135]

■ Donna Vogel, the physician who directs Hopkins Medicine's Professional Development Office, completed a four-day winning streak on *Jeopardy!* in July 2008 by walking away with $89,299 in winnings—all of which she donated to charity. In March 2009, she returned to the show as a contestant in its "Tournament of Champions." She was eliminated in the semi-final round but still received $10,000, which she again gave to charity. "Being on *Jeopardy!* was so much fun, but I felt I didn't 'earn' the money—I just received it for playing a game," she explained. "Need is enormous now, and contributions are down all over. I have a pretty comfortable life and so many people need this more than I do." One aspect of appearing on *Jeopardy!* that Vogel could not give away was a new-found fame. "It's a very special kind of status," she told an interviewer for the program's Web site. "One colleague of mine who works in New York said to me, 'People are more impressed that I know a *Jeopardy!* champion than that I know two Nobel Prize winners.' "

Hopkins molecular biologist Carol
Greider (below) receives the 2009 Nobel
Prize in Physiology or Medicine at the
awards ceremony held on Dec. 10, 2009
in the Stockholm Concert Hall before
an audience of hundreds of dignitaries,
scholars, and members of the Swedish
royal family.

Opposite page: Greider, director of
the Department of Molecular Biology
and Genetics, celebrates with her
children, Gwendolyn and Charles (right)
and her laboratory colleagues after
receiving word of her Nobel Prize.
Shared with her former colleagues,
University of California, San Francisco
biologist and physiologist Elizabeth
Blackburn and Harvard geneticist Jack
Szostak, Greider's prize recognized her
discovery of telomerase, an enzyme
that helps preserve the ends of a cell's
chromosomes when it divides. The
finding led to studies showing that
telomerase plays a major role in cancer
growth and conditions related to aging,
which could result in therapies to affect
the enzyme's action.

She also went upstairs to awake her children, Gwendolyn, then 10, and
Charles, then 13, to tell them that she had won the Nobel Prize and that,
no, they didn't have to go to school that day. They were coming with her to
Hopkins for a celebration.[136]

Greider, director of the Department of Molecular Biology and Genetics,
received the Nobel for her 1984 discovery of telomerase, an enzyme that
preserves telomeres. Telomeres, often likened to the protective plastic tips on
shoelaces, are the specialized regions of DNA that stabilize the ends of a cell's
chromosomes. Telomerase prevents telomeres from degrading when cells divide.
The finding established the groundwork for subsequent studies showing that
the enzyme plays a major role in cancer growth and conditions related to aging.
The scientists' ultimate goal is to see if they can develop therapies that could
affect the action of telomerase.[137]

For Greider, the Nobel Prize that she shared with her mentor, biologist
and physiologist Elizabeth Blackburn of the University of California, San
Francisco, and geneticist Jack Szostak of Harvard, served as vital confirmation
of the importance of basic, curiosity-driven science. All three also had shared
the 2006 Lasker Award for Basic Medical Research—frequently called the
"American Nobel"—for Greider's discovery and the groundbreaking research
that Blackburn and Szostak did on telomeres.[138]

The Nobel and Lasker are, Greider says, "a testament to
investing in people who have clever ideas that might not seem
directly applicable to diseases but become apparent later on." Not
until 1990 did researchers discover that telomerase run amok plays
a significant part in the growth of cancer cells—and its link to
degenerative diseases related to aging was uncovered long after
that.[139]

Raised in the Northern California city of Davis, not far
from the state capital of Sacramento, Greider was the daughter of
scientist parents—a physicist father and a biologist mother, who
died when Greider was six. She had dyslexia as a child, struggled in
school and was placed in remedial classes. "I thought I was stupid,"
she has said.[140]

Having difficulty reading and spelling, she honed her already
exceptional ability to memorize. This skill helped her excel in
chemistry and anatomy, which she enjoyed. At the University
of California in Santa Barbara, she earned a bachelor's degree in
biology in 1983, then obtained a Ph.D. in molecular biology in
1987 from the University of California at Berkeley. It was there
that she became a protégé of Blackburn, whose enthusiasm for
research into telomeres was contagious.[141]

Researchers knew that a small section of telomeres wasn't
copied as cells divided and chromosomes were replicated. This
should have left the resulting cells with shorter telomeres, but
it didn't. Blackburn wanted to know why. She suspected it was

because of a yet-undiscovered enzyme that protected them.[142]

Working in Blackburn's lab, Greider began searching for the elusive enzyme by studying a single-celled, pond-dwelling organism called *Tetrahymena thermophila*—which has more than 40,000 chromosomes (compared to a human's 23 pairs) and consequently far more chromosome ends to examine. Nine months of tinkering with *Tetrahymena* went by without success. On Christmas Day 1984, working alone in the lab, she pulled X-ray film from the developer and felt a rush of adrenalin. There in the images was the exact pattern researchers had predicted would be seen if the unknown enzyme was found. Confident she had cracked the mystery, Greider went home, put on Bruce Springsteen's just-released "Born in the USA," and "danced and danced and danced." [143]

Greider and Blackburn conducted countless additional tests to make sure they hadn't made a mistake about the Christmas Day finding. Dubbing the newly discovered enzyme "telomerase," they published their results—sparking an explosion in telomere and telomerase research that continues to this day.[144]

Coming to Hopkins in 1997, Greider founded a laboratory that remains in the forefront of that research boom. Although known as a determined, rigorously focused and thorough researcher, Greider has a quick, self-effacing sense of humor and natural exuberance. When the Associated Press asked if she hadn't expected to win the Nobel Prize in 2009, since others had predicted she would, Greider replied that no one ever expects to win a Nobel, even if others anticipate it. "It's like the Monty Python sketch, 'Nobody expects the Spanish Inquisition!'"[145]

With a mischievous sense of fun, Greider has been known to enjoy snowball fights, practical jokes, and occasionally urging those working in her lab to form a human pyramid. She'll then climb to the peak of the stack of students, technicians and postdoctoral fellows, strike a pose reminiscent of her high school days, when she would stand erect atop running horses, and remind them in all seriousness that she wouldn't be where she is without their support.[146]

Greider told a press conference on the day she won the Nobel that a key aspect about working at Hopkins "is it is such a collegial group of people." Being in a "setting where you have so many people thinking about clinical questions really changed my research into a much more clinical kind of direction," she said, noting happily that interdisciplinary collaborations could be arranged easily by a simple telephone call. "That has really been wonderful for me."[147]

■ Everyone has experienced how the power of odors can stir memories and emotions. Yet as recently as 1991, no one could explain scientifically how our olfactory ability plays such conjuring tricks. That was before molecular biologist Richard Axel, a 1971 School of Medicine graduate who zoomed through Hopkins in three years, began investigating the odor-detecting cells that line the nose. In his Columbia University lab, he discovered a pool of more than 1,000 genes that encode olfactory receptors, which allow us to distinguish more than 10,000 smells. His work, conducted with then-postdoctoral student Linda Buck, won them the 2004 Nobel Prize in Physiology or Medicine.

After accepting his Nobel, Axel wrote: "I was a terrible medical student, pained by constant exposure to the suffering of the ill and thwarted in my desire to do experiments. My clinical incompetence was immediately recognized by the faculty and deans. I could rarely, if ever, hear a heart murmur, never saw the retina, my glasses fell into an abdominal incision and finally, I sewed a surgeon's finger to a patient upon suturing an incision.... I was allowed to graduate medical school early with an M.D. if I promised never to practice medicine on live patients. I returned to Columbia as an intern in Pathology where I kept this promise by performing autopsies. After a year in Pathology, I was asked...never to practice on dead patients. Finally, I was afforded the opportunity to pursue molecular biology [research] in earnest..."

Physicians at Hopkins continually receive unusual public recognition for their skill and caring. From the true-life dramas of ABC News' 2008 documentary series *Hopkins* to the 2009 TNT dramatization of brain surgeon Benjamin Carson's 1996 autobiography, *Gifted Hands: The Ben Carson Story,* Hopkins remains in the spotlight. Starring Academy Award-winning actor Cuba Gooding, Jr. as Carson, the TV biopic traced the remarkable life journey he made from Detroit's inner city to the pinnacle of American medicine.[148]

2005
The U.S. Department of Homeland Security gives a $15 million grant to Hopkins' Office of Critical Event Preparedness and Response (CEPAR) to lead a Center of Excellence that will undertake in-depth research on issues related to emergency preparedness and response. The 21-member consortium's Center for the Study of Preparedness and Catastrophic Event Response (PACER) is launched in December.

2006
January: Hopkins' second Cancer Research Building (CBRII) opens. The $87 million building encourages researchers across departments to work together on solving some of the disease's most daunting problems. In December it is dedicated and named for philanthropist and university trustee David Koch, who donates $20 million in support of the structure.

ABC's *Hopkins*

Hopkins has a way of inspiring even non-physicians and non-research scientists to create their best work. Seven years after the award-winning miniseries *Hopkins 24/7* aired in 2000, its ABC News creators returned to Baltimore to once again plumb the depths of the human dramas continuously unfolding at Hopkins Hospital. They wanted to film another multipart series, this time entitled simply *Hopkins.* The riveting programs that emerged not only captured the nation's attention but won the prestigious George Foster Peabody Award for distinguished achievement in broadcasting from the University of Georgia, just as its *Hopkins 24/7* had garnered the equally coveted Alfred I. duPont Award for Excellence in Broadcast Journalism from Columbia University.[149]

When shooting of the new miniseries began in February 2007, ABC News executive producer Terry Wrong noted that many of the crew who worked on *Hopkins 24/7* were returning to work on its successor and that the prospect of telling more of the Hopkins story was exciting. "We felt that we had never captured people quite as smart and operating at such a high level. It was most rewarding and enriching."[150]

From Hopkins' point of view, having the news team of 16 producers and videographers back to tape hundreds of hours of material for what would be a seven-part documentary (compared with the six-part *Hopkins 24/7*) offered yet another opportunity to show a vast audience what really takes place inside a major academic medical center.[151]

From mid-February to mid-June, the news crews taped up to 20 to 30 hours daily. They were looking for compelling cases and cast a wide net. For example, in just one week in March, they covered brain surgery and two liver transplants, as well as witnessed the birth and successful treatment of a baby born with a rare birth defect requiring special care for which Hopkins is among the foremost centers.[152]

In between the clinical cases, the documentarians wanted stories about interns and residents—how they handled the difficult task of combining their extraordinarily intense training and clinical care duties with their private lives. Showing the public how some of the best physicians in the world are trained was the goal of the TV crews.[153] (Appendix F contains ABC's description of the key points in each episode's storyline.)

When *Hopkins* aired in June-July 2008, critics and viewers were captivated. *The Washington Post* called the series "fascinating." *Variety* said it was "fraught with emotion, [and] humor," and the *Philadelphia Daily News* proclaimed it "one of the most watchable dramas on television this summer."[154]

Just prior to the series' broadcast, Terry Wrong said he was impressed by Hopkins' continued resilience in facing major challenges, such as immense technological changes, the shifting expectations for residents, a more assertive patient population, and the growing financial perils involved in operating a large academic medical center. Hopkins, he said, is "a national treasure."[155]

Building a New Hopkins Hospital and School of Medicine Curriculum

When Jim Kaufman, then Hopkins Medicine's director of government relations, was working hard during this period to persuade Maryland legislators and budget officials to approve requests for state funds to revitalize Hopkins Hospital, perhaps his most effective tools were several antiquated elevators in the Blalock Building, opened in 1963, and the Carnegie Building, opened in 1927.[156]

Taking state officials on hospital tours, Kaufman initially walked them through the stunning new Weinberg Cancer Center. Then he'd guide them over to Blalock, where the crammed quarters were a far cry from Weinberg's well-lit, wide-open spaces. Although Blalock was home to Hopkins' preeminent cardiovascular care program, featuring some of the most modern clinical technology and finest patient care, just getting to the sixth floor was a challenge on its rickety, slow-moving elevator.[157]

Kaufman then might take the visitors to Carnegie, which housed clinics, labs, 10 general operating rooms—and an elevator in which riders sometimes had to jump up and down to get it started. If it broke down, parts no longer were available. They had to be built from scratch at a cost of thousands. Yet the Hospital's blood bank, one of the nation's busiest, was located on Carnegie's sixth floor; and the Moore Clinic, which was Maryland's largest HIV care program, serving hundreds of patients each week, was on the third floor.[158]

The conventional wisdom in the Maryland General Assembly in Annapolis was that Hopkins really didn't need state help. A tour such as Kaufman provided proved otherwise.

Raising the Funds

State grants represented only a fraction, however, of what was going to be needed to build 1.3 million square feet of new Hospital space on the 52-acre East Baltimore medical campus. When Hopkins' comprehensive 10-year master plan for rebuilding was unveiled in 2002, its total cost initially was estimated at $1.2 billion. Maryland would be asked to provide $50 million in grants, but Hopkins Hospital itself planned to raise $400 million by going into debt through the sale of revenue bonds. Another $200 million in charitable gifts would be sought, while some $125 million would come from projects sponsored by the Maryland Health and Higher Education Facilities Authority, an agency that helps sell hospital revenue bonds. Another $35 million would come from Hospital earnings—making the Hospital's projected total portion of the 10-year master plan's price tag a whopping $810 million.[159]

Looming over the entire rebuilding enterprise were a host of uncontrollable factors, such as potentially rising interest rates, unpredictable steel prices, a shortage of skilled construction workers, and the difficulty of persuading potential donors that bricks-and-mortar projects were essential to fostering the clinical programs and research to which many of them preferred to give their money.[160]

■ In 2008, Hopkins Medicine, known throughout the world, literally went out of this world with a unique contribution to a Space Shuttle Endeavor mission. At the request of astronaut Rick Linneman, who in the 1980s trained in Hopkins Division of Comparative Medicine, now the Department of Molecular and Comparative Pathobiology, a 18-by-6-inch poster of the animal species that had been sent into space was prepared by the Department of Art as Applied to Medicine. Depicting everything from the fruit flies first sent into orbit in 1947 to the harvester ants that flew in 2003 and the monkeys, mice, dogs, amphibians and other space-traveling animals in between, the poster was rolled into a packet the size of cigar and went aloft with Linneman. It orbited six million miles in space. It now is on display in the molecular and comparative pathobiology department in the Broadway Research Building.

Early on a June morning in 2010, medical photographer Jon Christofersen of the Department of Pathology climbed to the top of the Dome of the original Hopkins Hospital, now the Billings Administration Building (its shadow can be seen on the lower right center of the picture above), and took 16 separate photos. He then stitched them together digitally to compose this 360-degree panoramic view of the East Baltimore medical campus. Construction of the new, campus-revolutionizing clinical buildings was well advanced—with the Sheikh Zayed cardiovascular and critical care tower looming just left to center.

Johns Hopkins Medicine, Health System and Hospital board chairman C. Michael Armstrong and his wife, Anne, with David Nichols, vice dean for education, at the October 2009 dedication of the School of Medicine's new Mike and Anne Armstrong Education Building.

Despite all the difficulties, the great rebuilding proceeded. Two of its signature structures, the Broadway Research Building and the Koch Cancer Research Building, were well under way as Hopkins launched its campaign for state regulatory approval of its new clinical towers. The $140 million Broadway Research Building, begun in 2000 and quickly dubbed simply the BRB, was filled with researchers and administrators by December 2003 but had a grand opening ceremony in May 2004. The $87 million Koch Center, begun in 2003, was completed in 2006.[161]

Also completed in 2006 was a new, four-story pediatric medical office building that became the new home for the Harriet Lane Clinic, which moved from its old, by-then cramped quarters in the Park Building, opened in 1973. The new building was officially dedicated in 2007 and named for David M. Rubenstein, a member of the University and Johns Hopkins Medicine boards of trustees and co-founder and managing director of Carlyle Group, a large private equity firm.[162]

By 2006, the estimated cost of the overall campus redevelopment had risen to $2 billion, with the combined price tag for the cardiovascular and critical care tower and the children's tower projected to be $725 million, some $190 million more than initially estimated.[163] The project's budgeters had tracked the rate of construction inflation at 2.7 percent over 10 years and figured they'd hedged their bets by planning for a three percent rise in costs. Instead, building inflation between 2003 and 2006 spiked to eight percent annually, partly due to construction demands following Hurricane Katrina in 2005.[164]

Groundbreakings and Ribbon-Cuttings

Even as construction cost estimates rose, groundbreakings and ribbon-cuttings moved forward over the next few years. In June 2006, ceremonial shovels dug into the soil from which would spring the two, 12-story clinical towers. By September, shiny metal spades again were wielded to break ground for the $45 million, four-story, 100,000-square-foot education building to be named for Johns Hopkins Medicine, Health System and Hospital board chairman C. Michael Armstrong and his wife, Anne, whose generosity would cover nearly half of its cost.[165]

In 2007, construction began on a six-story, 200,000-square-foot research and patient care building for the Wilmer Eye Institute, paid for completely by philanthropists, led by the late Robert H. Smith, one of the nation's leading real estate developers, and his wife Clarice. Across the street, work started on a new four-story residence for long-term oncology patients and their families, fully funded by philanthropists Willard Hackerman and Sidney Kimmel.[166]

The John G. Rangos Sr. Building and the Science + Technology Park at Johns Hopkins

In April 2005, five top science students from the nearby, inner-city Dunbar High School were among those wielding shovels to give a symbolic start to construction of what by April 2008 would become the first major building in the East Baltimore Biopark—formally called the Science + Technology Park at Johns Hopkins—just north of Hopkins Hospital. The Biopark was slated to be the centerpiece of an 88-acre, $800 million urban redevelopment project that aimed to make the "New East Side" of Baltimore a hub of the biomedical revolution while revitalizing the community surrounding Hopkins.[167]

The subsequent recession and other unanticipated difficulties hampered much of that city-envisioned, urban redevelopment effort, but construction of the $54 million, seven-story research facility for which the ground had been broken was facilitated by a $10 million donation from the John G. Rangos Sr. Family Charitable Foundation, created by a Pittsburgh businessman with long ties to Hopkins. Targeting the foundation's gift to Hopkins' Institute for Basic Biomedical Sciences, which would occupy a third of the building once it opened, Rangos noted that his family's foundation had a history of supporting a wide range of health care and educational efforts. He chose Hopkins for his latest multimillion-dollar support of basic science research, Rangos said, because "this institution and its programs are a beacon of light in medicine and science."[168]

Approximately 200 Hopkins researchers now are using the laboratories and equipment in the Rangos Building to study cancer, kidney disease, HIV and hepatitis C, among other afflictions.[169]

Mike Weisfeldt (right), director of the Department of Medicine and physician in chief of Hopkins Hospital, shows Pittsburgh businessman and philanthropist John G. Rangos Sr. an architectural model of the Science + Technology Park just north of the Hospital, where the $54 million John G. Rangos Sr. building would open in 2008 and house Hopkins' Institute for Basic Biomedical Science—to which Rangos had directed his $10 million gift toward the building's construction.

The 42,000-square-foot, 39-suite Hackerman-Patz Patient and Family Pavilion, adjacent to the Kimmel Cancer Center, opened in December 2008 and offers a welcoming, homelike environment to long-term care cancer patients and their families. Its $10 million cost was covered by donations from philanthropists Willard Hackerman and Sidney Kimmel.

The Hackerman-Patz Patient and Family Pavilion

Improving the clinical experience was behind construction of a new home away from home for cancer patients and their families at the Sidney Kimmel Comprehensive Cancer Center. Opened in December 2008, the four-story, 42,600-square-foot Hackerman-Patz House, located directly across Orleans Street from the Weinberg Building, doubled the long-term residential space previously available for patients and their families in the original Hackerman-Patz and the aging Joanne Rockwell Memorial houses.[170]

The pavilion features 39 suites—each painted in soothing colors and containing living space, separate sleeping areas, a bath, a kitchenette, a flat-screen TV and decorations that include framed photographs, mixed media and paintings. It aims to give patients a homelike, welcoming environment that provides a restful setting that doesn't feel like a clinic.[171]

"Transformational" Gifts for the New Clinical Buildings

More major philanthropic gifts spurred the immensely complex, ongoing construction of the children's and adult care patient towers, as well as the completion of a new research and patient care building for the Wilmer Eye Institute.

In April 2007, Hopkins Medicine leaders (left to right) Ronald Peterson, president of Hopkins Hospital and Health System and executive vice president of Hopkins Medicine, Dean/CEO Edward Miller, and Johns Hopkins University President William Brody went to the United Arab Emirates' capital of Abu Dhabi and presented an architectural rendering of the Sheikh Zayed Tower to the late UAE ruler's son, Crown Prince Sheik Mohammed bin Zayed Al Nahyan, center. The "transformational" gift from the Crown Prince's older brother, Sheikh Khalifa bin Zayed Al Nahyan, the UAE's president and ruler of the Emirate of Abu Dhabi, paved the way for the new clinical tower, named in honor of their father.

The Sheikh Zayed Tower

In late April 2007, Ed Miller, Ron Peterson and Bill Brody flew to Abu Dhabi, capital of the United Arab Emirates (UAE), for a whirlwind, 17-hour visit. They went to meet with emirate's leaders, Sheikh Khalifa bin Zayed Al Nahyan, the UAE's president and ruler of the Emirate of Abu Dhabi, and his younger brother, Crown Prince Sheikh Mohammed bin Zayed Al Nahyan. They are the sons of the UAE's founder, Sheikh Zayed bin Sultan Al Nahyan, who led the federation from its formation in 1971 until his death in 2004. The journey of the Hopkins emissaries concluded with announcement of what was called a "transformational gift" from Sheikh Khalifa to support construction of Hopkins Hospital's new cardiovascular and critical care tower, as well as provide funds for cardiovascular and AIDS research. The specific amount of the gift was not disclosed at the donor's request, but the Hospital tower would be named for Sheikh Khalifa's late father.[172]

Hopkins had been developing a relationship with the UAE for some 20 years. Since at least 1988, Johns Hopkins International's team of representatives had been coordinating tailored care by Hopkins physicians for patients arriving from the UAE or by dispatching doctors to make intercontinental house calls.

Johns Hopkins Medicine's ties to the UAE extended far beyond treating patients. A business relationship developed over a number of years with Abu Dhabi's government health authority. It led to a 10-year contract, signed in 2006, for Hopkins to manage Tawam Hospital, the UAE's most prestigious health care facility. Harris Benny, then-CEO of Johns Hopkins Medicine International, observed that the "whole relationship has grown from strength

to strength." The gift from Sheikh Khalifa was made solely to honor his father and came with no political or cultural strings attached.[173]

The aim of the tower's design was to combine all the features necessary to address the needs of a modern teaching hospital while providing the finest medical care technology available—and to do so in a way that was flexible enough to make the building readily adaptable to future medical developments.[174] The 913,000-square-foot tower would become the new home of the Johns Hopkins Heart and Vascular Institute, state-of-the-art intensive care units for Medicine, Surgery, Neurosurgery, a new home for Gyn/Ob and the neonatal ICU, a new adult emergency department and a cluster of high-tech operating rooms, together with catheterization labs and imaging suites. The tower boasts 355 private patient rooms and the latest in communication and information technology, including wireless voice communications, wireless and high-speed internet access. All-digital medical imaging is also available to every laboratory, operating room and patient room in the facility.

In announcing Sheikh Khalifa's gift, Ron Peterson likened the UAE to the "Switzerland of the Middle East," saying it was open, tolerant of all religions, and maintains good relations with its neighbors and the United States. "It's a smaller world now," Peterson told Hopkins Medicine, and the "transformational gift" was the royal family's "way of giving back."[175]

The Charlotte R. Bloomberg Children's Center

Giving back to Hopkins—and honoring a parent—also was behind yet another substantial donation to Johns Hopkins Medicine by the university's most generous alumnus, New York City Mayor Michael Bloomberg. In keeping with his long association with Johns Hopkins, Bloomberg bestowed a mammoth gift for building of the children's inpatient tower. It would be named for his then-99-year-old mother, who still was active in Medford, Massachusetts, where she raised the future 1964 Hopkins electrical engineering graduate, business media mogul and later board of trustees chairman.[176] (Mrs. Bloomberg died in 2011 at the age of 102.)[177]

In this September 2011 photo, Hopkins' new clinical buildings, the Sheikh Zayed Tower on the left and The Charlotte R. Bloomberg Children's Center on the right, were nearing completion. The scope of the two buildings is staggering. Michael Iati, Hopkins Hospital's director of architecture, says the entire original Hopkins Hospital Billings Building with its iconic Dome could fit neatly within the new towers' front entry circle "and it wouldn't touch anything."

New York City Mayor Michael Bloomberg, a 1964 electrical engineering graduate of Hopkins, has become the university's most generous benefactor. The new, 12-story children's facility will bear his mother's name. "It's a rare privilege to be able to thank simultaneously two of the most important forces in my life: my mother and my alma mater," he said.

Patrice Brylske (left), director of Child Life, and Rebecka Carlson, Child Life specialist, check out the new two-story playroom during construction. The dimensions will accommodate a climbing wall and other activity areas. Brylske and Carlson say their job is to reassure children that The Charlotte R. Bloomberg Children's Center is a safe and playful place to heal. The new two-story playroom, as well as the expanded Children's and Family Resource Library, will certainly help.

Coming from a modest, middle-class background (his father was the book-keeper at a local dairy), Bloomberg helped support himself as an undergraduate in the Krieger School of Arts and Sciences by serving as a parking attendant at the Johns Hopkins Club. While he has jokingly admitted that his academic achievements were not particularly impressive, Bloomberg has been a tireless supporter of Hopkins for decades, from his chairmanship of the university's board of trustees to his dedication to what now is known as the Bloomberg School of Public Health.[178]

Because hospitals often are accessed only during times of crisis, creating a calming environment through the integration of art and architecture in the Bloomberg Children's Center was integral to its mission. In keeping with his longstanding commitment to public health and the arts, Michael Bloomberg played a major role in overseeing both the exterior design and interior artwork in the Children's Center. Its exterior is enveloped by patterned clear glass featuring 26 different hues and surrounded by gardens; its interior is enlivened by more than 300 works of original art. The result is a state-of-the-art medical facility that is not only a place of healing but one that elevates the hospital experience to match the quality of medicine provided at Johns Hopkins.

The Charlotte R. Bloomberg Children's Center is equipped with advanced medical technology, including ten new operating suites capable of handling the most complex procedures. It boasts 205 private inpatient rooms with extensive accommodations and amenities to help families support their children throughout hospitalization; the innovative, 45-bed neonatal intensive care unit and 40-bed pediatric intensive care unit offer private beds to enhance parent-child bonding and reduce infections.

Other features include a dedicated pediatric radiology unit and telecommunications facilities for consultations. The Bloomberg Children's Center also incorporates inpatient and outpatient areas dedicated to research, and close physical connections between areas dedicated to pediatric and adult care create opportunities for collaboration throughout the building. Expansion of pediatric subspecialties and fellowship programs is planned.[180]

At the height of the new clinical buildings' construction, close to 1,000 workers were on the site. In this photo, a group of them gather for safety orientation—given in Spanish as well as English.

Staggering Scope

The scope of the overall clinical towers project truly was staggering. The two 12-story buildings, connected on the first five floors and rising into separate but closely aligned facilities, encompass 1.6 million square feet—enough to house four Weinberg cancer centers, noted Michael Iati, Hopkins Hospital's director of architecture. The entire, original Billings Building with its fabled Dome could be fit neatly within the new towers' front entry circle "and it wouldn't touch anything," Iati said, with a hint of awe.[181] He likened the entire project to "building a 60-ton watch," with constant, synchronized coordination required between architects, engineers, contractors, the construction manager and suppliers.[182]

At the height of construction between 2009 and 2011, close to 1,000 workers were on the Orleans and Wolfe streets job site. "It takes an army," said Sally MacConnell, the Hospital's vice president of facilities. In many ways, every aspect of the towers was being custom-made—and all of its mechanical systems had to work properly from day one. "That's where it departs from being like your kitchen," MacConnell observed. "If the stove doesn't work on day one, you call the guy back; if the medical gas outlets don't work, that's a crisis."[183]

For six years prior to the groundbreaking, an in-house team composed of Iati, MacConnell, mechanical systems expert Anatoly Gimburg, architect Marge Siegmeister, construction manager Howard Reel, and support personnel from Hopkins Hospital's Facilities Design and Construction department worked with faculty, staff and outside architects to prepare preliminary design drawings. They asked the future "users" of the towers—nurse managers, nurses, physicians and others—what they needed and wanted.[184]

Driving the design was the need not just for more space but for space that was flexible, capable of expanding and contracting depending on the demands imposed by the patient population or evolving medical technology in the years ahead.[185]

What the planners called "adjacencies"—such as placing the adult emergency department adjacent to the pediatric ED, with imaging suites and trauma rooms that could be shared between them—also was a design priority. So was placing almost all of the operating rooms adjacent to their corresponding intensive care units. In addition, setting aside sufficient space for medical education was deemed crucial, making certain that medical students and residents doing their clinical rotations had rooms in which to sleep, work, study and gather.[186]

Completed late in 2011 and scheduled to be occupied in April 2012, the Sheikh Zayed Tower and The Charlotte R. Bloomberg Children's Center will change the face of Hopkins Hospital and alter not only the East Baltimore landscape but the entire city skyline. Situated on the same hill that Johns Hopkins himself had selected for the location of his namesake Hospital, they will be visible from the downtown and waterfront; lit at night, they will serve as a glowing beacon overlooking the city.[187]

2006

February 20: Johns Hopkins Medicine International signs 10-year agreement to manage the 469-bed Tawam Hospital in the United Arab Emirates in Abu Dhabi. In addition, the Clemenceau Medical Center, the first Middle East clinical affiliate of Johns Hopkins Medicine International, opens in Beirut, Lebanon. The affiliation began in 2002 as plans for the hospital were being drawn up.

May 15: Hopkins establishes a Center for Global Health to coordinate and focus its efforts against HIV/AIDS, malaria, tuberculosis, hepatitis, flu and other worldwide health threats, especially in developing countries.

Hopkins is involved in the controversy over embryonic stem cell research. Dean/CEO Ed Miller writes a column for *Hopkins Medicine* magazine, urging Congress to loosen current restrictions on stem cell research, prompting letters to the editor arguing against stem cell research. Hopkins scientists John Gearhart and Curt Civin are in the forefront of stem cell research; Civin lobbies Congress to permit it.

Hopkins is in the forefront of community-based efforts at cancer prevention. The Sidney Kimmel Comprehensive Cancer Center strives to reach out to patients, families and the community, initiating such programs as a Boost for Colon Cancer Screening program.

The five-story triangular window at the center of the façade of the Robert H. and Clarice Smith Building of the Wilmer Eye Institute, opened in October 2009, was designed to capture the reflection of the original Wilmer building's octagonal dome across the street.

2006

June 5: Despite rising construction costs, groundbreakings abound for new Hopkins Medicine structures, including the June 5 ceremonies marking the launch of work on the Hopkins Hospital's new clinical towers, the core of the medical campus's $1.1 billion redevelopment plan.

The School of Medicine's fabled Turtle Derby, which raises money for local charities, marks its 75th anniversary.

September 17: Carol Greider shares the Lasker Award, often called "America's Nobel," for her discovery of telomerace, a crucial enzyme within telomeres, which help protect the ends of chromosomes. The research of Greider and her colleagues will win them the Nobel Prize itself in 2009.

October 4: Hopkins' Institute for Basic Biomedical Sciences formally establishes eight new research centers to collectively tackle such complicated questions in biology as the genetic roots of obesity and the relationships among the five senses. The centers focus the combined talents of about 30 laboratories and new technologies on how living things receive and process information from all five senses; how genetic information outside the chromosomes is controlled and passed through generations; and how cells coordinate their shape and structure to either move or stay put, a critical element of everything from normal development to cancer metastasis.

The Robert H. and Clarice Smith Building

With a triangular-shaped, five-story window above its entrance that reflected—literally and figuratively—Hopkins' past and future, the Robert H. and Clarice Smith Building of the Wilmer Eye Institute was dedicated on October 16, 2009—80 years to the day after the opening of the original Wilmer Institute. The first Wilmer building's octagonal dome can be seen in the new building's soaring, mirror-like façade, which was designed to create that effect.[188]

Located at the corner of Broadway and Orleans Street, the $100 million, 207,000-square-foot building more than doubled the space devoted to what already is the country's largest eye-related research program. The top five floors, all housing laboratories, established what Wilmer director Peter McDonnell calls "research neighborhoods," where scientists can interact far more freely than they could in the old Wilmer's "disjointed patchwork of labs and offices in six different buildings."

With some 30 scientists already engaged just in tackling age-related macular degeneration (AMD), the leading cause of blindness in individuals over 60, the new research floors quadrupled the space not just for these researchers alone, but for those in such emerging fields as nano-technology and tissue engineering.[189]

The Smith building also dramatically improves Wilmer's clinical setting. Its Maurice Bendann Surgical Pavilion contains six of the most modern ophthalmic operating rooms in the world. These new ORs enable Wilmer's surgeons to perform 50 percent more procedures a day, adding to a clinical effort that already had reached nearly 9,000 sight-saving operations at Hopkins Hospital annually, as well as another 5,000 surgeries yearly at Wilmer's Hopkins Bayview and suburban centers.

The surgical pavilion's namesake, Maurice Bendann, belonged to a venerable, long-philanthropic Baltimore family, and he donated most of his estate to Wilmer when he died in 1969. His family's association with Hopkins goes back to before the University, Hospital and School of Medicine were founded. His uncle, the renowned Civil War-era photographer Daniel Bendann, took the most famous photo of Mr. Johns Hopkins in 1871. Maurice Bendann's father, David Bendann, began a Baltimore art gallery that remains in business today.[190]

Another major benefactor to the building was James Gills Jr., a Wilmer resident from 1962 to 1965 who went on to found St. Luke's Cataract and Laser Institute in Tarpon Springs, Florida, and perform more cataract and lens implant surgeries than any other eye surgeon in the world. (A onetime triathlon athlete, he also once owned the World Triathlon Corporation.) After donating $5 million to the Smith Building and the Bendann pavilion, as well as endowing two professorships, Gills said that having been part of the Wilmer Institute "is like being a Marine or a Green Beret, in that they represent excellence, just as Wilmer represents excellence."[191]

Education: A New Home for a New Curriculum

The Anne and Mike Armstrong Medical Education Building

The Armstrong Building, which opened in October 2009, was designed specifically to accommodate the School of Medicine's new "Genes to Society" curriculum, launched that fall. "To have a new facility that would complement and serve the new curriculum is really what I want this building to do," Armstrong said when he and his wife made their pledge. "I also hope that the people who are going to live in it—the faculty and the students—not only enjoy teaching and learning in the environment we create, but in the end just come to love the building."[192]

By all accounts, they have. Right before the Armstrong Building was dedicated, a first-year medical student from California told the *Baltimore Sun* that he had enrolled at Hopkins precisely because of the new curriculum and its custom-made home.[193]

"The one thing that tipped the scale was the Genes to Society curriculum and all the things we're going to be introduced to," said Jacob Rusevick. "Having a building that was specifically designed to make this happen was just incredible. I'm extremely happy."[194]

The Armstrong Building is packed with the latest digital communications technology. It features everything from virtual microscopy tools composed of high-resolution monitors and displays showing multiple images from centralized servers to virtual-reality simulations and dissection table digital reference tools that make information instantly available, literally, at students' fingertips. First-year medical students still will dissect real cadavers, but the computers at their worktables will also enable them to discover how regions of the human body appear in high-resolution radiologic images. The building's location adjacent to the Outpatient Center also gives students easy access to the $5 million, 15-room Simulation Center that opened on its top floor in March 2008. There they can hone their key clinical skills on computerized plastic mannequins and robots before performing them on patients.[195]

The Genes to Society curriculum emphasizes the importance of communication and teamwork among physicians and other health care providers. To foster such collegiality between students and faculty, one floor of the Armstrong Building is devoted to the School of Medicine's four advisory colleges. Once just known as "A," "B," "C," and "D," each was named in 2006 for a Hopkins legend: Daniel Nathans, Florence Sabin, Helen Taussig and Vivien Thomas. Every student is assigned to one of the colleges and each college's suite of rooms fosters a blend of intimate teaching, learning, camaraderie and community. Small focus rooms are perfect for one-on-one mentoring sessions, while meeting rooms and learning studios accommodate larger gatherings.[196]

"These new approaches to mentoring and collaborative learning play an essential role in turning our visionary curriculum into reality," said David Nichols, vice dean for education, when the Armstrong Building opened.[197]

The October 2009 dedication of the Anne and Mike Armstrong Building, the School of Medicine's new education center named for C. Michael Armstrong, then-chair of the boards of Johns Hopkins Medicine, The Johns Hopkins Health System Corporation and The Johns Hopkins Hospital, and his wife, ushered in the official, full implementation of Hopkins' new Genes to Society curriculum, for which the Armstrong Building specifically was designed.

2006
November 14: Hopkins transplant surgeons make medical history by performing the first five-way donor kidney swap among 10 individuals. All five organ recipients—three men and two women—are operated on during the marathon 10-hour surgeries that occupied six operating rooms staffed by twelve surgeons, eleven anesthesiologists and eighteen nurses at The Johns Hopkins Hospital.

John Singer Sargent's iconic quadruple portrait of Hopkins Medicine's founding physicians, "The Four Doctors," one of America's most famous medical masterpieces, marks its centennial. The emblematic painting even makes a cameo appearance in *Indiana Jones and the Kingdom of the Crystal Skull*.

Geriatrician Crystal Simpson (second from right) has long been eager to get minority students interested in medicine—especially geriatrics. Launching a Geriatric Summer Scholars program at Johns Hopkins Bayview, she has drawn minority medical students from around the country to Hopkins, not only helping to shape them into top-notch candidates for residency and well-rounded future physicians, but also increasing Hopkins' diversity by encouraging the program's participants—such as (left to right) Ashleigh Hicks, Luis Amaury Castro and Delvin Yazzie—to consider building their own careers at Hopkins.

Pulmonologist Charles Wiener (second from right) headed the curriculum reform committee that spent six years developing the innovative "Genes to Society" curriculum that went into effect in 2009. "It was if we were starting a brand new school," said Wiener.

He later did precisely that, spearheading the unique public-private partnership between Johns Hopkins, a private Malaysian company and the Malaysian government to create the brand new, Hopkins-inspired Perdana University Graduate School of Medicine in Kuala Lumpur, which opened in September 2011. Wiener implemented the Genes to Society curriculum there, too.

Genes to Society

Developed during six years of debate and planning, "Genes to Society" (GTS) likely will "usher in a new era of medical education that raises the bar even higher and that will be as transformative for this century as the original Hopkins model was for the last," Ed Miller said at the October 2009 dedication of the Armstrong Building.[198]

The lengthy period required to devise the new curriculum, a process that involved hundreds of faculty as well as students, administrators, and even former patients, turned out to be a blessing in many ways.

Not only did the new curriculum's long gestation allow time for the Armstrong Building to be funded and erected, it also showed that the revisers' initial 2003 expectations of how the practice of medicine could evolve over the next decade proved to be much as they anticipated.

For example, the federal government was poised to enact health care insurance reform, which likely will impose significant changes in the methods used to pay for health care and may put an emphasis on preventive medicine. The explosion of scientific knowledge continued unabated—with new genetic variations and their relationships to disease being discovered every day. The patient population was increasingly diversified, even international. Preparing students to take care of dengue fever or malaria was becoming just as important as learning how to treat influenza or the common cold, David Nichols said.[199]

Formulated around organ systems—often referred to as systems biology—the medical school curriculum teaches students to understand each patient's health through various levels of biological hierarchy, starting at the genetic level, moving to the cellular and organ levels, and beyond that to family, community and the environment.[200]

Essentially, the courses consider how each organ system is affected by genetic inheritance, biology, social and cultural factors. Interdepartmental collaboration between faculty instructors is a key aspect of Genes to Society. Such a comprehensive overview "moves medicine into the realm where each patient's disease becomes unique," Nichols said. "Establishing a framework for understanding that individuality is the challenge for modern medical education."[201]

Pulmonologist Charles Wiener, head of the curriculum reform committee, noted that Hopkins' top administrators let the faculty draw up GTS "from a blank slate, rather than ask us to manipulate the existing one. It was as if we were starting a brand new school."[202]

Nichols did suggest that the foundation for the new curriculum should be a farsighted 1999 book, *Genetic Medicine: A Logic of Disease*, by acclaimed Hopkins pediatric geneticist Barton Childs, by then a professor emeritus. Wiener said the book reinforced the faculty's growing sense that the basic structure of traditional medical school curricula no longer reflected the character and pace of scientific and social change.[203]

For example, under GTS, clinically-oriented instruction, previously a focus of second- and third-year courses, now is introduced during the first two years,

with students seeing patients and being taught how to take their first patient histories almost immediately after entering the Medical School. Basic science material, formerly the main concentration in the first two years of instruction, will be reintroduced during the third and fourth years, when students can better absorb—and understand—its clinical relevancy.[204]

Although the concept of integrating basic and clinical sciences throughout the entire four-year curriculum was not an entirely new idea, having been proposed during Hopkins' 1992 curriculum revision, it had never been completely realized. Now it is a core tenet of GTS.[205]

Another key aspect of Genes to Society is its focus on the future of medical care. In years to come, "it may be much more common to identify genetic variance with a blood test," Dave Nichols says, then use that information to draw conclusions about what is causing a patient's illness and design a personalized plan of treatment for it.

"A Hopkins graduate is being prepared not just for the 90 percent [of patients] but for the 10 percent—the outliers, the people who do not fit the classic case. The Hopkins graduate is also being prepared not just to practice now but to have a framework for thinking about the practice of the future. And that's part of what Genes to Society is attempting to accomplish."[206]

Another central tenet of the new curriculum is "doctor-patient communication," Nichols adds. "We argue that it's impossible to understand the patient as an individual unless you can communicate with the patient."[207] To ensure this, efforts to enhance communication between faculty and students, as well as researchers and clinicians, also are key aspects of the new curriculum.[208]

The seismic changes in Hopkins' curriculum quickly were felt throughout the medical education community. "We've already been asked by many leading institutions to consult as they consider curricular change," Nichols said only a month before Genes to Society went into effect.[209] Interest even has come from overseas, Nichols added. With the implementation of GTS, Hopkins had a revolutionary guide to offer on the future of medical instruction.[210]

The School of Medicine Class of 2013 was the first to begin its education under GTS—but fourth-year students in 2009 who took several new courses that were slated to become part of it found them to be among their most valuable experiences at Hopkins.[211]

"I don't feel like I ever left. I think that may be true for a lot of people [even if they go elsewhere]. Time at Hopkins is such an important part of our professional identity. Maybe I was a prodigal son for three years out at Case Western and three years at UNC Chapel Hill ... and three years at Duke. But people knew where I was from. There's a little bit of a missionary spirit. No other place can really be like Hopkins. We're unique."

—Peter Agre, 1974 graduate of the School of Medicine, co-recipient of the 2003 Nobel Prize for Chemistry for his discovery of aquaporins, the proteins that form the "water pore" channels through which water enters and exits cells, upon returning to Hopkins after a stint as Duke University's vice chancellor for science and technology

2007
February: ABC-TV documentarians return to Hopkins Hospital to begin filming *Hopkins,* another multi-part series. Of his experiences filming *Hopkins 24/7* seven years earlier, producer Terry Wrong says: "We felt that we had never captured people quite as smart and operating at such a high level. It was most rewarding and enriching." Wrong calls *Hopkins* "a national treasure." Hopkins later airs in the summer of 2008 and focuses on how young physicians-in-training struggle to keep balance between their professional and personal lives. In April 2009, it wins a George Foster Peabody Award for excellence. The Peabody Award is the oldest prize in broadcasting and considered among the most prestigious and selective accolades in electronic media.

John Singer Sargent's 10-foot, 9-inch painting of the founders of the Johns Hopkins School of Medicine, "The Four Doctors," marked the 100th anniversary of its Baltimore unveiling in 2007, looking every bit as grand in its mammoth gilded frame and current location, the Welch Medical Library. Its enduring impact also was a fitting symbol for the immense achievements of Johns Hopkins Medicine, which marked its 10th anniversary in 2007.

2007

Two Hands: The Leon Fleisher Story, a documentary about the crippling affliction to the right hand of famed pianist and Peabody professor Leon Fleisher, is nominated for an Academy Award. It is produced by the daughter of Solomon Snyder and features Daniel Drachman, professor of neurology.

Johns Hopkins Medicine International develops an array of programs designed to help patients with limited English skills. Its latest experiment is a Spanish-speaking robot.

March 22: Hopkins Medicine expands its reach by signing an agreement with Anne Arundel Health System, the parent organization of the 265-bed Anne Arundel Medical Center community hospital in Annapolis, where it will collaborate on developing a satellite health care center and conduct research. Another affiliation agreement later is signed with the Greater Baltimore Medical Center.

Top leaders of Hopkins Medicine, at right, willingly "cast aside their dignity" to appear in a tongue-in-cheek, in-house 2007 video "roast" of Dean/CEO Ed Miller (sixth from the left) celebrating his stewardship of Hopkins Medicine over its first decade. The good humor reflected a sincere appreciation of all he had done to ensure Hopkins Medicine's immense accomplishments over the preceding 10 years.

Johns Hopkins Medicine's First Decade

Early in 2007, "The Four Doctors," John Singer Sargent's iconic portrait of School of Medicine founders William Welch, William Halsted, Sir William Osler and Howard Kelly, marked its 100th birthday. Thanks to a $38,000 restoration completed a few years before, it looked as solemnly splendid as it had when it was unveiled.[212]

At the Jan. 19, 1907 ceremony, Welch described sitting for Singer at the acclaimed artist's London studio two years earlier. Sargent was frustrated while trying to arrange his four subjects. Observing the various poses in which he had placed them, he grumbled, "It won't do. It isn't a picture." At the next sitting, he added a massive Venetian globe to the composition, quickly sketched it in and stood back. "Now," he said, "we have got our picture."[213]

The addition of the huge globe might almost be seen as a prescient metaphor for the world-wide impact of the institution that was given such an extraordinary beginning by those four physicians. The 2007 centenary of the mammoth painting—still powerful in all its 10-foot, 9-inch glory—also was a fitting symbol for the immense achievements of Johns Hopkins Medicine, which marked its 10th anniversary that year. Its success ensured that the guiding philosophy and mission of the Hopkins Hospital and School of Medicine's founders would endure.

A dinner honoring Dean/CEO Edward Miller marked the Hopkins Medicine's decade milestone. The highlight of the affair was a video "roasting" Miller.[214] It featured many of the top leaders of Hopkins Medicine, from University President Bill Brody and Hopkins Medicine board chairman Mike Armstrong, to Hospital and Health System President and Hopkins Medicine

Chi Dang   David Nichols   David Hellmann   Ron Peterson   Bill Baumgartner   Ed Miller

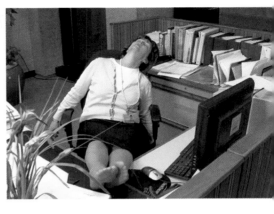

In this scene from the 2007 "roast" video for Dean/CEO Ed Miller, Kathy Long, a top assistant to Miller, sleeps at her desk, oblivious to his intercom pleas for her—for anyone—to come to his office and help him.

Executive Vice President Ron Peterson, senior vice president Steve Thompson, and vice deans Dave Nichols, Chi Dang, Bill Baumgartner, Janice Clements, Dan Ford and David Hellmann. To cap off the video, producer Jay Corey wrote a humorous disclaimer on behalf of all those who willingly "cast aside their dignity and, let's be honest, perhaps job security, for the making of this production." The DVD brought down the house, which included a number of Hopkins Medicine trustees, School of Medicine leadership, and a few select guests.[215]

Even Miller himself went along with Corey's script—although he had no idea how a recording of his voice would be used. On his office's intercom, he taped an increasingly urgent request for his secretaries or vice president and chief financial officer Rich Grossi to come see him.[216] This plaintive plea was heard over scenes of Miller's staff sleeping behind their desks, eating candy, reading paperback books, filing their nails, practicing their golf puts, making paper airplanes, drinking beer—and in Grossi's case, playing solitaire on his computer.[217]

The rousing good humor reflected an appropriate feeling of immense accomplishment—and sincere appreciation for all that Miller had done to help achieve it. The challenges of a dramatically changing, increasingly complex medical marketplace were being met successfully; some of the most respected names in American medicine had been recruited to head nationally acclaimed departments and divisions; substantial federal and state funding, along with remarkable philanthropic gifts, were fostering research and underwriting construction of landmark clinical buildings.[218]

Behind all of those accomplishments were the dedicated employees of Johns Hopkins Medicine. From the facilities' staff to the CEO's office; from the operating rooms to the research labs, their skills were recognized as the best. That July, *U.S. News & World Report* named The Johns Hopkins Hospital the finest in the United States for the 17th consecutive year.[219]

2007

March 26; November 26: The Johns Hopkins Hospital installs a 256-slice computed tomography (CT) scanner, believed to be the world's most advanced CT imaging software and machinery; and then eight months later, the first 320-slice computed tomography (CT) scanner in North America. The most powerful imaging machine in its class, the 4,400-pound (two-ton) device provides startling images of the entire heart or brain, down to tiny blood vessels.

April 30: The Russell H. Morgan Department of Radiology and Radiological Science and one of its faculty, Elliot Fishman, both are ranked the nation's best by 600 readers of *Medical Imaging Magazine*. Readers of the magazine made the departmental choices based on reputation and information related to equipment, patient care and outcomes, clinical research and staff. Individual radiologists were judged by their current research, patients care and outcomes, and industry interaction.

May 1: Announcement is made of the receipt of a "transformational gift" from His Highness Sheikh Khalifa bin Zayed Al Nahyan, president of the United Arab Emirates (UAE). Most of the gift, made in honor of Sheikh Khalifa's late father, Sheikh Zayed bin Sultan Al Nahyan, who created the UAE in 1971 and ruled it until his death in 2004, will support construction of The Johns Hopkins Hospital's new cardiovascular and critical care tower, currently under construction on the East Baltimore campus. In addition, some funds will go to the School of Medicine Dean's Discretionary Fund to be directed to cardiovascular research and some also have been earmarked for AIDS research at the Johns Hopkins University-Makerere University Collaborative Care Center, at Mulago Hospital in Kampala, Uganda.

Steve Thompson   Steve Rum   Joanne Pollak   Rich Grossi   Janice Clements   Dan Ford

2007

Johns Hopkins Medicine International exports the century-old Hopkins Medicine mission to the United Arab Emirates' Tawam Hospital, the most prestigious in that nation.

Special scholarships have enabled some of the nation's top minority medical students to attend Hopkins.

June 6: Groundbreaking for a new Wilmer Eye Institute building, a $100 million, 200,000-square foot structure to house additional research and outpatient eye surgery facilities across the street from The Institute's landmark 1929 facility adjacent to the original Johns Hopkins dome. The new Wilmer building opens in 2009.

Johns Hopkins at White Marsh opens a surgery center to relieve some of the pressure on the operating rooms at The Johns Hopkins Hospital and Johns Hopkins Bayview Medical Center.

The Johns Hopkins Bayview Medical Center's 134-acre campus grew by another 3.15 acres in 2009, when Baltimore businessman Jim Crystal and his family donated land adjacent to it so the hospital could "expand even further and continue to do great things." The $3.125 million gift was one of the largest ever received by Hopkins Bayview.

Hopkins Grows Locally, Nationally and Internationally

Building and expansion projects were not confined to Hopkins Medicine's East Baltimore campus and its immediate neighborhood between 2004 and 2011. Major expansion took place not only regionally but nationally and internationally.

Johns Hopkins Bayview

By 2007, the National Institute of Health's biggest-ever Baltimore building was ready for occupancy on the Johns Hopkins Bayview campus, where the NIH had been conducting research for six decades. Having taken three years to erect, the $250,000-million, 573,000-square foot Biomedical Research Center, a two-tower structure occupying an 11-acre site, was the new home for two of the campus's longtime NIH residents, the National Institute on Aging (NIA) and the National Institute on Drug Abuse (NIDA). Its features included the latest in laboratories and clinical space; a library; a vivarium (housing for animals used in research), and office space for some 800 scientists.[220]

Another new building, a 100,000-square-foot medical office complex providing space for the urology, neurology, neurosurgery, cardiology, renal and endocrinology divisions, as well as the Women's Center for Pelvic Health, also was completed in 2010. Named simply the 301 Building after its location at 301 Mason Lord Drive on the campus, the new facility not only provided room for expanded outpatient services but was home to the Geriatric Clinical Research Unit, the Clinical Research Unit, the Center in Sleep Research and Education, and Heart Health, a cardiology research program focusing on exercise physiology.[221]

Additional expansion at Hopkins Bayview was assured in 2009 by one of the largest private gifts ever received there, 3.15 acres of land adjacent to the campus, worth $3.125 million. Donated by Baltimore businessman Jim Crystal and his family, the land at 5400 E. Lombard Street would "allow the hospital to expand even further and continue doing great things," Crystal said in bestowing the property.[222]

Howard County General Hospital

Adapting to a growing—and aging—population, as well as already overcrowded facilities, Howard County General Hospital opened a new, four-story Patient Pavilion in July 2009. Housing a large outpatient facility and three floors of all-private inpatient rooms, the $73 million, 100,000-square-foot building almost doubled the hospital's outpatient space and

is linked to the original HCGH by a distinctive, curved glass "knuckle" of corridors with floor-to-ceiling windows. It also includes space for such support departments as pharmacy and information technology.[223]

The outpatient clinic is named for the family of Howard Hospital Foundation head Evelyn T. Bolduc and her husband, J.P. The Bolduc family's generosity was matched by that of HCGH's board vice chair, Harry L. "Chip" Lundy and his wife, Cathy. Together, the Bolduc and Lundy families donated $2 million to the hospital's campus development plan.

The new building brought together all of HCGH's outpatient services into one convenient location.[224] When the entire construction and renovation project was completed in late 2011, 43 new beds were added to the existing 227, as well as two new operating rooms.[225]

Howard County General Hospital, part of the Johns Hopkins Medicine family since 1998, opened a $73 million, four-story, 100-000-square-foot Patient Pavilion in 2009, with a large outpatient facility and three floors of all-private inpatient rooms, in response to its region's growing—and aging—population.

Health Care and Surgery Centers

Celebrating its 10th anniversary in 2004, Johns Hopkins at Green Spring Station could look back on an enviable record of success. Growing from a relatively small enterprise in 1994, with several dozen physicians based in a single pavilion, it had rapidly attracted additional practices and support services that required a second pavilion. By 2004, it had more than 235 part-time and full-time faculty physicians handling some 300,000 patient visits annually—representing approximately $100 million in business for Hopkins Medicine. By 2009, the number of patient visits each year had jumped to 350,000.[226]

Green Spring's success was mirrored at Hopkins' other suburban health care and surgery centers. For example, just four years after its July 2000 debut, the Johns Hopkins at White Marsh center opened a second, $5.8 million pavilion in October 2004. In June 2007, it added a $1.5 million ambulatory surgery center.[227]

By reallocating many outpatient surgical procedures to White Marsh, both Hopkins Hospital and Bayview were able to free up operating room time for additional, more complicated inpatient surgeries. By 2009, Johns Hopkins at White Marsh was handling more than 140,000 patient visits each year, up from some 115,000 patient visits just two years earlier.[228]

Johns Hopkins at Green Spring Station celebrated its 10th anniversary in 2004. That year, its 235 part-time and full-time faculty physicians handled some 300,000 patient visits.

In 2007, Johns Hopkins Medicine signed a strategic alliance with the Anne Arundel Health System, parent organization of the 265-bed Anne Arundel Medical Center (above) to develop satellite health care centers and collaborate on research.

2007
September 24: Howard County General Hospital holds a groundbreaking ceremony for expanded facilities.

September 28: The Department of Social Work marks its 100th anniversary. Its workers dig deep to provide patients with supportive resources.

Surgeon Patrick Walsh, M.D., professor of urology, performs his 4,000th nerve-sparing radical prostatectomy, an operation he created and perfected, revolutionizing treatment of prostate cancer.

Priority Partners, Hopkins' Medicaid care organization, marks its 10th anniversary. Serving some 113,000 members – nearly a quarter of the state's Medicaid population – it steers families through the healthcare maze. Its cadre of case managers transcend traditional medical care for low-income families.

September 25: Lisa Cooper, M.D., M.P.H., epidemiologist and professor of internal medicine who conducts landmark studies designed to understand and overcome racial and ethnic disparities in medical care and research, receives a MacArthur Foundation "genius award."

Affiliations Forged in Anne Arundel and Baltimore Counties

The Anne Arundel Health System

The success of Hopkins Medicine's suburban centers prompted an interest in developing more of them south of the city. In March 2007, an agreement was reached with the Annapolis-based Anne Arundel Health System (AAHS), the parent organization of Anne Arundel Medical Center (AAMC), a 265-bed community hospital, to form a strategic alliance to develop satellite health care centers, as well as collaborate on clinical research.[229]

Even before the broader alliance was inked, Johns Hopkins Community Physicians had entered into an agreement with AAHS to provide primary and urgent care for residents of Maryland's Eastern Shore at a new, 55,000-square-foot facility that AAHS planned to open early in 2008 on Kent Island. [230]

Greater Baltimore Medical Center

Just four months after the agreement with the Anne Arundel Medical Center, another affiliation was formed with the Greater Baltimore Medical Center (GBMC), a 292-bed acute care hospital in Baltimore County, north of the city. The alliance with GBMC, which remains an independent entity that also comprises two hospice centers and satellite facilities throughout Baltimore County, aimed at enhancing the access of patients in North Central Maryland to a wide range of clinical specialty programs.[231]

In part, the five-year agreement signed in July 2007 put full-time Hopkins cardiologists in charge of a GBMC cardiac care program, which was named Johns Hopkins Cardiology at GBMC.[232] Prior to the agreement, Hopkins and GBMC had a longstanding relationship in a number of specialties. The Wilmer Eye Institute had forged an affiliation with the Towson-based hospital in the 1960s, and both GBMC and Hopkins had integrated residency programs in ophthalmology, obstetrics and gynecology, and otolaryngology-head and neck surgery. Additionally, many GBMC physicians held teaching positions at Hopkins.[233]

Suburban, Sibley, All Children's Join the JHM Family

Suburban Hospital Healthcare System

Almost exactly two years after Johns Hopkins Medicine signed its affiliation agreement with GBMC, it swiftly concluded arrangements in June 2009 to assume control of the Suburban Hospital Healthcare System (SHHS) in Montgomery County, southwest of Baltimore and a key gateway to Washington, D.C. The health system's cornerstone is the 239-bed Suburban Hospital, a not-for-profit, highly regarded community acute care facility in Bethesda that has been serving Montgomery County since 1943 and long had ties to Hopkins.

In 1996, Hopkins had signed a business alliance with Suburban to help increase the clout of both institutions when they negotiated with managed care companies. A decade later, they both also worked closely with Suburban

Suburban Hospital Healthcare System, owner of the 239-bed Suburban Hospital (left) in Bethesda, Md., a Montgomery County suburb of Washington, D.C., joined the Johns Hopkins Medicine family in June 2009.

Hospital's neighbor across the street, the National Institutes of Health, to develop the NIH Heart Center at Suburban, which opened in 2006 and offers advanced cardiovascular specialty care, including heart surgery. Within three years, the prospect of a new era of tightening health care resources and the need to provide more efficient, integrated regional services to patients prompted SHHS officials to approach The Johns Hopkins Health System Corporation to ask about a possible formal integration. In the April 2009 announcement of the impending integration of Suburban with Hopkins, it was estimated that ironing out the details would take until the following fall. Instead, it took just two months. While Surburban became a wholly owned subsidiary of Johns Hopkins Medicine, no money was exchanged to complete the transaction, and Suburban retained both its name and current leadership.[234]

Suburban "is strong financially, very highly regarded in the community and located virtually on the doorstep of the nation's capital," Ron Peterson said when the agreement was announced.[235]

Sibley Memorial Hospital

Less than a year after the Suburban Hospital alliance, Hopkins Medicine moved across the doorstep of the nation's capital and right into it. Again citing the need to develop a comprehensive, regional approach to health care, Ed Miller and Ron Peterson announced in May 2010 that Sibley Memorial Hospital, a financially sound and widely respected 328-bed acute care facility in northwest Washington, D.C., would become another member of The Johns Hopkins Health System Corporation family. Once again, the integration did not involve any financial exchange.[236]

As was the case with Suburban, the newest member of the Johns Hopkins Medicine group of hospitals would retain its name and leadership team. It was expected also to maintain its voluntary medical staff and physician organization, which had developed highly regarded programs in oncology, orthopedics, obstetrics and geriatric services, as well as a high-end, assisted living facility on its campus.[237]

2008

January: In an effort to make medical care safer for patients and health care workers, The Johns Hopkins Hospital is the first major medical institution to become "latex safe" by ending all use of latex gloves and almost all medical latex products. William Stewart Halsted, Hopkins' first surgeon in chief, is widely credited as the first to popularize the use of rubber surgical gloves in the United States. That was in 1894, five years after the institution opened.

March 10: William Brody, president of The Johns Hopkins University for the past 12 years, announces his decision to retire within a year. On November 11, Ronald J. Daniels, former dean of the University of Toronto Law School and current provost of the University of Pennsylvania, is chosen as his successor and assumes office on March 2, 2009.

Sibley Memorial Hospital (below), a highly regarded 328-bed acute care facility in northwest Washington, D.C., became another member of the Hopkins Medicine family in November 2010.

All Children's Hospital (right) in St. Petersburg, Florida, became the first Hopkins Medicine family member outside of the Baltimore-Washington area in April 2011, shortly after its new $400 million, 259-bed medical center opened. Founded in 1926 as a center for crippled children, All Children's is one of only 45 stand-alone children's hospitals in the country.

2008

February 28: Cardiologist Myron "Mike" Weisfeldt, director of the Department of Medicine and Hopkins Hospital's physician in chief, receives the 2008 Diversity Award from the Association of Professors of Medicine, an organization whose members come from across the United States and Canada. The award recognizes Weisfeldt's achievements in improving diversity within his department and the School of Medicine as a whole. He considers this one of the most important accolades he—and Hopkins—has received.

April 8; April 23: Hopkins surgical teams perform what is believed to be the first six-way donor kidney swap among 12 patients. The 10-hour surgeries required six operating room and employed nine surgical teams at the Hopkins Hospital. One of the patients was Randall Bolten, brother of White House Chief of Staff, Joshua Bolten. President George W. Bush later meets with the patients, physicians and nurses who participated in the historic surgery.

April 11: The Rangos Building, the first of five proposed life sciences structures to be completed at the Science and Technology Park at Johns Hopkins, opens.

All Children's Hospital & Health System

The pace of Hopkins Medicine's expansion soon increased and even went out of state. Just five months after Sibley joined the burgeoning JHM family, All Children's Hospital, a 259-bed freestanding children's hospital in St. Petersburg, Florida, became the first Hopkins subsidiary outside of the Baltimore-Washington area.[238]

Executives from All Children's, one of only 45 stand-alone children's hospitals in the country, were interested in enhancing both its research and educational endeavors. They actually had contacted Hopkins initially about a possible merger three years earlier, Ed Miller told *The Baltimore Sun*. "This was really a desire by All Children's Hospital to become one of the top pediatric hospitals in the country. An opportunity came along that fit into our strategic plan. We weren't out there looking for opportunities."[239]

A financially secure institution that began as a center for crippled children in 1926, All Children's has a new $400 million, 1 million-square-foot facility, opened in 2010, trains pediatric residents and fellows, serves eight counties in west Florida, and draws patients from all over the nation, as well as 36 countries.[240]

As in the Suburban and Sibley alliances, no money was exchanged when All Children's joined Hopkins Medicine. It also kept its name, respected leadership, medical staff and physician affiliations, including with University of South Florida doctors who practice there. Hopkins' Jonathan Ellen, director of the Hopkins Center for Child and Community Research, as well as head of pediatrics at Johns Hopkins Bayview, was named Vice Dean and Physician in Chief at All Children's. He will develop residency and fellowship programs for Hopkins' post-graduate students who will have an opportunity to train in St. Petersburg beginning in 2013. Ellen also will develop an expanded research program at All Children's.[241]

All Children's will foster the overall Hopkins mission, as well as provide Hopkins Medicine with a presence in Florida that not only will serve patients in that potentially important new market, but from the Caribbean, Central America and South America.[242]

Johns Hopkins Medicine International

To export Hopkins' philosophy and practice of superior care to more patients worldwide than ever before, Johns Hopkins Medicine International also expanded its reach exponentially during this period. Just between 2006 and 2009, it added two additional clinical affiliations with hospitals in Chile and Japan to its two previous affiliations in Turkey and Lebanon. It also assumed management of three more hospitals—two in the United Arab Emirates and one in Panama—adding them to a third UAE hospital it had managed since 2003 and its Singapore center, created in 2001. In addition, during the same three-year period, it formed academic and clinical collaborations with hospitals, institutes and clinics in Trinidad and Tobago, Portugal, Italy, Mexico and Canada to provide educational and consulting support to promote quality care and patient safety.[243]

Dave Nichols considers JHI's "increasing formalization of the international reach of the institution" to be one of the most important Hopkins Medicine innovations over the past 20 years.

"Of course, Hopkins has always had an international reach, so I don't want to pretend that this is entirely new, but it was more ad hoc; individuals had their connections with peers in other countries or with institutions in other countries. And this has been a wonderful thing about the institution," Nichols observed in 2009.[244]

Overseas Clinical Affiliates

In April 2006 and February 2007, respectively, JHI added Tokyo's outpatient Midtown Medical Center and Santiago, Chile's 240-bed Clínica Las Condes to a lineup of clinical affiliates that had already included Beirut's 106-bed Clemenceau Medical Center and Gebze, Turkey's 204-bed Anadolu Medical Center, both of which became affiliates in 2002.[245]

Overseas Management

Between 2005 and 2008, JHI initiated or expanded the application of its health care management expertise from Asia and the Middle East to Latin America. In 2005, Johns Hopkins Singapore International Medical Center, originally opened in 2000, moved to the Tan Tock Sen Hospital, which became Hopkins' only jointly owned and managed facility outside of the United States. In 2006, the United Arab Emirates' largest health care facility, the 469-bed Tawam Hospital, signed a 10-year administrative and operational oversight agreement with JHI. Then, in 2008, the UAE's 163-bed Al Rahba Hospital and 235-bed Al Corniche Hospital both signed management agreements with JHI, as did Panama City's 75-bed Hospital Punta Pacífica.[246]

Tawam Hospital (above), the largest hospital in the United Arab Emirates, is a 469-bed facility owned by the UAE's General Authority for Health Services. In 2006, Johns Hopkins Medicine International assumed administrative and operational oversight of Tawam, which serves as a regional referral center for specialized medical care and a national referral center for cancer treatment.

The 204-bed acute care Anadolu Medical Center in Gebze, Turkey, a suburb of Istanbul, has been a Johns Hopkins Medicine affiliate since 2002. It is a specialty-focused health care provider recognized as the best of its kind in Turkey, southern Eurasia and the Middle East.

Signed as Johns Hopkins Medicine International affiliates between 2006 and 2007 were the 240-bed Clínica Las Condes in Santiago, Chile (above), one of the nation's leading private hospitals, and Tokyo's Midtown Medical Center (below), located in a 54-story, luxury building in the city's fashionable Roppongi District. It offers a wide range of services in an outpatient center, a concept not widely available in Japan.

Johns Hopkins International's former director of medical services, otolaryngologist Charles Cummings, observed: "We have dozens of projects in dozens of countries—some big, some small, but each unique—because the cultures and needs are all so different. But what's so meaningful is that the learning always flows both ways. We learn something new from every experience. Still, there is one resounding theme that unites all our projects: We're raising the standard of care around the world."[247]

2008

June 9: The announcement is made that the state-of-the-art 12-story tower being built to house the Johns Hopkins Children's Center will be named The Charlotte R. Bloomberg Children's Center at Johns Hopkins, in honor of the mother of New York Mayor Michael R. Bloomberg, a Johns Hopkins graduate and former head of the board of trustees.

June 19: Benjamin S. Carson Sr., director of pediatric neurosurgery, receives the Presidential Medal of Freedom from George W. Bush at a White House ceremony. At a July 10 tribute banquet, a Hopkins professorship formally is named in his honor, as well as for benefactor, Dr. Evelyn Spiro, R.N. On August 29, it is announced that Academy Award-winning actor Cuba Gooding, Jr. is chosen to portray Carson in a made-for-TV TNT movie, *Gifted Hands: The Ben Carson Story*, which airs February 7, 2009.

July: Donna Vogel, head of Hopkins Medicine's professional development office, has a four-day winning streak on the TV show *Jeopardy!* and wins $85,299, which she donates to charity. She is asked to return to take part in the show's Mar. 18, 2009 "Tournament of Champions" program. She is eliminated in the second round—but still wins $10,000, which she again gives to charity.

Overseas Academic and Clinical Collaborations

Through videoconferencing, continuing medical education and customized training programs and observerships, JHI has developed academic and clinical collaborations since 2007 with the Trinidad and Tobago Health Sciences Institute; the Instituto Tecnológico y de Estudios Superiores de Monterrey, commonly known as Monterey Tec in Mexico; Group José de Mello Saúde, one of Portugal's premiere hospital groups; the San Matteo Hospital in Pavia, Italy; and Canada's Medcan Clinic in Toronto.[248]

Telehealth

Using technology to advance the clinical and educational aspects of health care also has become a major activity of Johns Hopkins Medicine International. Videoconferencing, Web sites and computer-based programs all link Hopkins faculty with JHI's affiliates around the world. More than 350 Hopkins faculty, administrative and clinical staff have taken part in consulting activities overseas—and also throughout the United States—via these multiple means of communication. More than 320 international on-site and distance education events, reaching an audience of some 22,000 health care officials and workers overseas, took place in 2009 alone.[249]

Partnerships from China to Malaysia to Saudi Arabia

In 2006, Johns Hopkins Medicine, in the person of geriatrician Sean Leng, re-established a relationship with Peking Union Medical College that began with its founding in 1917 but was interrupted by China's political upheavals following World War II. Leng obtained a $1 million grant to oversee establishment of a full-blown geriatrics department at the Beijing medical school, now considered China's top academic medical center.[250]

Leng, an associate professor of geriatric medicine and gerontology who was born in China, got his four-year educational grant from the China Medical Board of New York—the division of the Rockefeller Foundation that had enlisted William Welch and Simon Flexner nearly 90 years earlier to help plan the Peking Union Medical College. They modeled it—and the Peking Union Medical College Hospital (PUMCH)—on Hopkins. Leng, who had trained at PUMCH, Texas A&M, Yale and Columbia before coming to Hopkins, went to China to recruit exceptional internists at PUMCH, instructed them in geriatric medicine at Johns Hopkins Bayview, then provided on-site consultation in Beijing on developing an multidisciplinary geriatric team.[251]

After observing JHI's success at improving medical services, staff training and patient safety in the United Arab Emirates, the Saudi Arabian government signed an agreement in 2010 to have the Wilmer Eye Institute establish a research, education and patient care partnership with its King Kahled Eye Specialist Hospital (KKESH) in Riyadh.

The Middle East's largest, best-equipped and best-financed ophthalmology facility, KKESH wanted to elevate its scientific inquiries, physician education and patient care. For its part, Wilmer saw the association as an opportunity to accelerate its efforts to cure blinding diseases.[252]

With more than 100,000 outpatients, 26,000 emergency room visits and 10,600 surgeries annually, "KKESH has two, three, sometimes four times as many patients with eye disease than we have," Wilmer's director Peter McDonnell noted at the time the partnership was formalized. "We can do studies much more quickly than if we could only enroll patients from our own country in clinical trials."[253] The Hopkins link to KKESH actually went back to its founding in 1982. Its first medical director was David Paton, a 1956 graduate of the School of Medicine and a resident at Wilmer who became a leader in international ophthalmology. Under the 2010 partnership agreement, KKESH's new medical director will be Wilmer's Ashley Behrens, a Venezuela-born specialist in cornea, refractive surgery and external eye diseases. He is a surgical innovator and developer of novel therapeutic approaches that have helped patients worldwide regain their vision.[254]

Despite the seemingly rapid expansion of JHI's activities overseas during this period, Ed Miller stressed at the time that the partnership with KKESH was announced that such affiliations were not undertaken without long, careful consideration. While income from overseas operations helped support academic and research activities in East Baltimore and also gave medical staff, house officers and administrators important international experience, "we're selective in sorting through the offers we constantly receive," Miller said.[255]

"We want the right partners in developing nations who are serious about lifting their health care delivery standards. JHI will not take on these challenges unless we feel certain Hopkins can bring value to that nation's peoples—and to our own institution."[256]

2008
July 22: Victor McKusick, M.D., the "father of medical genetics," dies at 86. The National Library of Science creates a Web site to honor him in its "Profiles of Science" project.

September 9: Molecular biologist, geneticist and ophthalmologist Jeremy Nathans and neuroscientist and ophthalmologist King-Wai Yau receive the $1.4 million António Champalimaud Vision Award, sometimes called the "Nobel Prize for vision." The Lisbon, Portugal-based Champalimaud Foundation praises the two scientists for their "ground-breaking discoveries in the laboratory that enhance our knowledge and understanding" of sight.

September: Johns Hopkins Bayview marks the 50th anniversary of its intensive care unit, the nation's first.

September: Gregory Schaffer announces his impending retirement as president of Johns Hopkins Bayview Medical Center after an eventful 10-year tenure at its helm.

The "Great Recession" economic downturn has an adverse impact on Hopkins Medicine, as it does everywhere else.

December 5: The Hackerman-Patz Patient and Family Pavilion opens at the corner of Orleans Street and North Broadway, directly opposite the Weinberg Cancer Building. The four-story, 42,500-square-foot brick building, twice the size of its predecessor, provides a residence for oncology patients undergoing prolonged treatment and their families. It begins housing new residents on December 17, as 16 patients and their families move in.

The King Kahled Eye Specialist Hospital (KKESH) in Riyadh, Saudi Arabia, has been partnering with the Wilmer Eye Institute in research, education and patient care since 2010. The Middle East's largest, best-equipped and best-financed ophthalmology facility, KKESH sometimes treats up to four times as many eye patients annually as Wilmer. This patient flow gives Hopkins researchers and their Saudi colleagues opportunities to undertake studies and initiate clinical trials for better treatments more quickly than could be accomplished at Wilmer alone.

U.S. Secretary of State Hillary Rodham Clinton (center) and Malaysia's Deputy Prime Minister Tan Sri Dato' Haji Muhyiddin Bin Mohd Yassin (left) applaud as Hopkins Dean/CEO Edward Miller shakes hands with Tan Sri Datuk Dr. Mohan Swami, a private Malaysian developer, after signing an agreement in November 2010 to create a U.S.-style medical school and hospital in Kuala Lumpur. Negotiated by Mohan Chellappa, vice president of global ventures for Johns Hopkins Medicine International (JHI), the public-private partnership agreement among the Malaysian government, JHI and JHSOM led to the September 2011 opening of the Perdana University Graduate School of Medicine. Modeled on Hopkins, it is Malaysia's first four-year graduate school in medicine and uses Hopkins' new Genes to Society curriculum.

2009
As the "Great Recession" drags on, Hopkins Medicine leaders Ed Miller and Ron Peterson hold a series of open forums in January, March and April to keep the staff informed about the impact of the downturn on the Hopkins Hospital and School of Medicine. A mandate to trim operating budgets by 5 percent prompts cost-cutting efforts throughout the enterprise and inspires a wide variety of money-saving initiatives.

Perdana University Graduate School of Medicine

In November 2010, the University, School of Medicine and JHI launched what might become its most comprehensive, far-reaching overseas endeavor ever, an agreement with Malaysia to help that Southeast Asian nation of 27.1 million develop its first fully integrated, private four-year graduate medical school and teaching hospital—based specifically on the Johns Hopkins model.[257]

During a ceremony in Kuala Lumpur, the country's capital, Ed Miller signed an agreement with Malaysia-based Academy Medical Centre and an associate company of Turiya to assist in creating the Perdana University Graduate School of Medicine. U.S. Secretary of State Hillary Rodham Clinton and Malaysia's Deputy Prime Minister Tan Sri Dato' Haji Muhyiddin bin Mohd Yassin looked on as the agreement was formalized.[258]

Under the agreement, Hopkins will assist in developing every major aspect of the tropical nation's new medical school and 600-bed teaching hospital. The medical school opened in September 2011 with its inaugural class in temporary quarters. Hopkins' new Genes to Society curriculum will form the basis for the medical education programs and Hopkins will assist in the campus' design, facilities planning and clinical affairs. The hospital will have a full compliment of ambulatory care facilities, diagnostic capabilities and ancillary support services.[259]

In a third major component of the agreement, Hopkins will advise its Malaysian colleagues on the development and integration of research programs across the entire medical enterprise. Johns Hopkins Medicine's organizational and operational processes will serve as the model for all of the new facilities' education, patient care and research activities. Hopkins faculty will provide the initial leadership team for the new school, including its founding dean and CEO, pulmonologist Charlie Wiener.[260]

Patient Care Advances at Home

The breadth of clinical experiences to which Hopkins medical students are exposed mirrors the nature of the frequently complex cases handled routinely by their physician-teachers. Out of some 200 operations performed at Hopkins Hospital by one young gastrointestinal surgeon late in 2004, only five were simple appendectomies. The rest were complicated, high-risk procedures involving the liver, pancreas and gastrointestinal tract. Such a variety of cases goes a long way toward revealing what makes Hopkins different from other hospitals—perhaps even different from most academic medical centers.[261]

The patients are referred from across the country and overseas. A Georgetown University medical school graduate who later trained in pediatrics at Yale told *Dome* in 2005 that she chose Hopkins for a fellowship because she knew it would provide the most comprehensive training available anywhere.

That proved to be true. Patients came from as far away as North Carolina and Michigan, with problems more complicated than she had ever seen. "It's been a very steep learning curve," she said. Another fellow, on the heart failure service, was awed by the number of transfers Hopkins agreed to take from "other hospitals where treating physicians simply cannot manage the patients any longer, given their complex medical issues."[262]

Patient Safety—and Satisfaction

Whether the clinical challenges are exceptional or everyday, the physicians, nurses, pharmacists, administrators and support personnel throughout Johns Hopkins Medicine concentrate intensely on enhancing patient safety and improving patient satisfaction with the services they receive. Having ultimate responsibility for the quality of patient care at The Johns Hopkins Hospital, its board of trustees long had maintained a committee on patient safety. Following the tragic and well-publicized deaths of Josie King, Ellen Roche and Briana Cohen between 2001 and 2003, Beryl Rosenstein, vice president for medical affairs at Hopkins Hospital, formed a patient safety committee which became a standing committee of the medical board. A pediatrician and former director of Hopkins' cystic fibrosis clinic, Rosenstein knew Hopkins' culture—its strengths and weaknesses—as well as anyone. He arrived at the Hospital in 1961 as an intern in pediatrics, earning 40 cents an hour. Except for a two-year tour at the United States Public Health Service, he had spent his entire career at Hopkins. As chairman of the new patient safety committee, he became a key force in shaping the comprehensive series of institution-wide patient safety initiatives that followed. Working with Karen Haller, the hospital's vice president for nursing and patient care services, Rosenstein saw to it that all the patient safety policies were listed in a single manual.[263]

Safety Initiatives

Among the initiatives devised and adopted with the cooperation of Bob Feroli, the pharmacy's medication safety officer, and Stephanie Poe, program coordinator in the Department of Nursing, was a "medication reconciliation" procedure, implemented in 2006. By creating an individualized "home medication" list that was attached to every patient's medical record and required to be signed by that patient's physician, communication was improved between patient care teams, as well as between patients and providers inside and outside of the Hospital.[264]

A stunning drop in catheter-related bloodstream infections was achieved by the strict adoption of Peter Pronovost's protocol for reducing these potentially deadly complications, which can be caused by poor procedures in the insertion of so-called central line fluid-and-medication catheters. By late 2004, central line bloodstream infections had been reduced to zero in the intensive care units of both the Weinberg Cancer Center and the surgery department; while the medical department's ICU reported only one.[265]

In the ongoing, unrelenting effort to improve patient safety, Hopkins dispensed with some time-honored—even revered—medical standbys, including latex surgical gloves.

Hopkins long had been recognized as the hospital where original surgeon in chief William Halsted popularized a surgeon's use of rubber gloves in the

L. Randol Barker (right), a 1966 graduate of the School of Medicine and a 1973 graduate of the School of Public Health, is a master of the doctor-patient relationship. For more than 30 years he has directed the residency practice at Johns Hopkins Bayview, mentoring generations of young physicians on how to become practicing internists. A professor of medicine, Randy Barker has the Hopkins way of patient care in his blood: His grandfather was Lewellys Barker, successor to William Osler as physician in chief of Hopkins Hospital and director of the Department of Medicine. (See page 40.)

Pediatrician Beryl Rosenstein, a 50-year Hopkins veteran, was vice president for medical affairs at Hopkins Hospital for 15 years. He was instrumental in shaping its patient safety program, working to improve medication safety and helping Hopkins to be compliant with myriad regulations without sacrificing excellence. Yet Hospital leaders and colleagues say his most significant contributions may have been as a force for culture change by improving civility and communication among disciplines and encouraging a work environment that is more transparent about its shortcomings.

1890s. By late 2007, however, Hopkins jettisoned what once was among its most symbolic claims to fame. It became the first major hospital to end all use of latex gloves and nearly all medical products using latex, citing latex's potential for causing a severe allergic reaction in approximately 6 percent of the general population and up to 15 percent of health care workers. Instead of latex, Hopkins adopted surgical gloves made of such synthetic materials as sterile neoprene and polyisoprene.[266]

Rheumatologist David Hellmann (left), vice dean and director of the Department of Medicine at Johns Hopkins Bayview, and internist Scott Wright, insist on fostering an often-forgotten and much-missed aspect of medical care: good bedside manner. Hellmann, Wright and internist Roy Ziegelstein collaborated on a 2005 article for the *American Journal of Medicine* that provided insights and advice on how to enhance the personal interaction of physicians with patients.

At the time he ended his 15-year tenure as medical affairs vice president in 2009, Rosenstein was lauded for being a central figure in changing the Hopkins culture by improving civility and communication among disciplines and encouraging a work environment that was more transparent about its shortcomings.[267] At about the same time, a Johns Hopkins Medicine Quality, Safety and Service Executive Council was created by Hopkins Hospital's chief operating officer Judy Reitz to maintain the patient safety momentum. It continues to be a forum at which leaders from the major entities of Johns Hopkins Medicine gather to share information on best demonstrated practices for patient safety, many of which emanate from The Johns Hopkins Hospital.

Patient Satisfaction

The April 2005 issue of the *American Journal of Medicine* featured a Hopkins-originated article, "52 Precepts That Medical Trainees and Physicians Should Consider Regularly." Its aim was to foster an often-forgotten and much-missed aspect of medical care: good bedside manner. Compiled by Johns Hopkins Bayview's Department of Medicine director David Hellmann, internist Scott Wright, and internal medicine head Roy Ziegelstein, the insights were meant to remind even brilliant physicians that they risk providing insufficient care if they ignore patients' worries or fears and don't comprehend how a disease takes a toll on daily life.[272]

Hopkins Bayview's co-director of internal medicine, David Kern, earlier promoted the same concept when he and other clinicians created the Osler Center for Clinical Excellence three years before the *American Journal of Medicine* article. Its guiding principle was summarized in one of Sir William Osler's most famous dictums: "Care more particularly for the individual patient than for the special features of the disease."[273]

Osler's interest in learning about and responding to an individual patient's specialized needs—apart from those that were medical-related—had been among the reasons behind creation of Hopkins Hospital's social work program in 1907. By its 100th anniversary in 2007, the Department of Social Work had

some 80 social workers at Hopkins Hospital, with additional workers at both Hopkins Bayview and Howard County General Hospital. They see thousands of patients annually. Its five divisions deal with patients in a broad spectrum of specialties, including medicine/surgery, the AIDS service, oncology, pediatrics/obstetrics and gynecology, and psychiatry. Its workers explore every avenue for helping patients deal with frequently complex issues before, during and after hospitalization or outpatient care.[274]

Also instrumental to Hopkins Medicine's response to patient needs has been an ongoing initiative to exceed expectations and provide "service equal to our science," says Judy Reitz.

For example, should an incoming patient's scheduled surgery have to be cancelled for whatever reason, such simple gestures as offering a sincere apology along with a "service recovery kit"—a box containing such tokens of regret as vouchers for parking, cafeteria meals, gas and the gift shop—can have an impact. "There is a science to this," says Rebecca Zuccarelli, senior director of service excellence of the Johns Hopkins Health System.[275]

Hopkins' science of patient satisfaction appears to be succeeding. According to Press Ganey, a national health care consultant that regularly conducts surveys of consumer response to health care service performance, Hopkins Hospital's overall patient satisfaction score has soared in recent years. Between 2007 and 2009, it moved up from the 20th to the 70th percentile of satisfaction among all of the hospitals surveyed. Among academic medical centers, Hopkins ranked in the 80th percentile.[276] In 2010, the American Alliance of Healthcare Providers, an Alexandria, Va.-based organization, awarded Hopkins its "Hospital of Choice" award for the seventh consecutive year, recognizing its "customer-friendly" service; and in 2011, the National Research Corporation gave Hopkins Hospital its Baltimore region "Consumer Choice Award" for the 16th consecutive year.[277]

In May 2011, less than two months before stepping down as chairman of the board of trustees of Johns Hopkins Medicine, C. Michael Armstrong made yet another visionary gift to the institution, pledging $10 million to create the Armstrong Institute for Patient Safety and Quality. Named to head the new institute was Hopkins' acclaimed patient safety expert, anesthesiologist Peter Pronovost, a driving force behind patient safety initiatives not just at Hopkins but worldwide.[268]

"We have been making excellent progress on patient safety and quality," Armstrong said, "but we can do better. We must take our patient safety research and results to the next level to be the best."[269]

The Armstrong Institute will rigorously apply scientific principles to the study of safety for all patients, not just those at Hopkins. Its focus will be on eliminating preventable harm for patients, removing health disparities, ensuring clinical excellence, and creating a culture that values collaboration, accountability and organizational learning. Hopkins itself will serve as a learning laboratory to test the best that the institute's researches have to offer in patient safety and quality improvement.[270]

Following this latest flourish of Armstrong's beneficence and foresight, he was succeeded as head of the Hopkins Medicine Board by Baltimore attorney and businessman Francis B. Burch Jr.[271]

Holders of membership cards for the mid-Atlantic region's Uniformed Services Family Health Plan (USFHP), the Hopkins-affiliated managed care service for dependents of the military and armed forces retirees and their dependents, repeatedly have expressed satisfaction with the care they receive. Their approbation has been echoed by national consumer advocates. In 2011, *Consumer Reports* magazine named Hopkins' USFHP as the highest rated health insurance plan among health maintenance organizations and preferred provider organizations in Maryland. USFHP also received the National Committee for Quality Assurance's highest rating of excellence in 2011. Operated by Hopkins on behalf of the Department of Defense, USFHP provides comprehensive health care benefits to more than 35,000 members living in Maryland, Washington, D.C., and surrounding areas of Virginia, West Virginia, Pennsylvania and Delaware.

A Boom in Outpatient Services

Hopkins' wide range of outpatient services experienced considerable growth during this period. All saw patient numbers and patient satisfaction steadily increase—from Johns Hopkins Community Physicians (JHCP) and Johns Hopkins Home Care Group (JHHCG) to Hopkins' managed care programs. These include the Medicaid care organization, Priority Partners, and Uniformed Services Family Health Plan (USFHP), the service for members of the military, their loved ones and armed forces retirees.

Uniformed Services Family Health Plan

Between 2005 and 2011, USFHP's coverage of military dependents, retirees and their dependents in Maryland, Washington, D.C., and surrounding areas of Virginia, West Virginia, Pennsylvania and Delaware rose from 24,000 to 35,000, while patient satisfaction with the Hopkins USFHP care, provided by Johns Hopkins Community Physicians since 1997, was nearly off the charts. Independently conducted surveys of the plan's members in 2008 and 2009 found that it rated in the 98th percentile for health plan satisfaction both years, compared to commercial adult health care plans. Its overall satisfaction rate of 84 percent was 20 points above the national benchmark.[278]

Priority Partners

Hopkins Medicine's other managed care entity, Priority Partners, serves some 200,000 members—about a quarter of all of Maryland's Medicaid recipients. It is one of seven organizations authorized by the state to provide health care services for low-income families on Medicaid or the Maryland Children's Health Insurance Program.[279]

Jointly owned by Johns Hopkins HealthCare and the Maryland Community Health System, a group of eight federally qualified health centers, Priority Partners cares for patients in every one of Maryland's 23 counties and Baltimore City.[280]

Priority Partners used to lose money—lots of it. In the first eight years since its founding in 1997, it recorded $57.2 million in losses. In large part, this is because Priority Partners operates under what's known as a "capitated" system—meaning the amount it receives from the state for serving each Medicaid patient is "capped" at a set fee, regardless of the costs incurred.[281]

As Ed Miller has noted, Hopkins Medicine tapped into its extensive health care delivery system to devise methods for managing the care of Priority Partners' patients in a cost-effective, quality way. Seventy percent of Priority Partners' patients say they are pleased with the care they receive—and between 2006 and 2007, the organization actually managed to reverse its losses and report a $28 million profit.[282]

Ashley Asilis (left) suffered from a metabolic disorder that causes frequent, disabling seizures. Yet by the time she was five, in 2007, she had become almost seizure-free, gained weight and began talking, thanks to the comprehensive care coordinated for her by nurse case manager Kathy Just (right) from Hopkins' Medicaid managed care organization, Priority Partners.

In 2008, however, Maryland more than doubled the eligibility level for the program from 40 to 116 percent of the federal poverty level, and the ensuing flood of an additional 30,000 patients pushed Priority Partners back into the red. It lost a $15 million in just nine months.[283]

Although Ed Miller and Johns Hopkins Medicine supported efforts to reform the nation's health care system—believing that its goal of providing millions more of Americans with health care coverage is in keeping with Hopkins' 120-year-old mission to serve the needy—Miller also warned of potential problems. Increasing Medicaid eligibility without reforming the capitated Medicaid system "to adjust payments based on risk" could lead to disaster, he said.[284]

Johns Hopkins Home Care Group

Financial woes also had troubled Johns Hopkins Home Care Group. It is the service that provides home visits by nurses, therapists, health aides and social workers—as well as medical and respiratory supplies, home infusion therapy, pharmaceuticals and hospice care—to thousands of adults and children in Central Maryland.[285]

By 2001, Home Care was staggering under $5.5 million in losses. That year, Dan Smith, an 18-year veteran of Hopkins Medicine and senior director of finance for the Health System since 1995, took over as Home Care's chief financial officer. He quickly helped turn its business around. By 2003, it was generating a $1.6 million profit.[286]

Smith was named Home Care's president in April 2005, when it faced the prospect of further financial bumps down the road in the form of reduced Medicare payment rates. Yet handling tough assignments was nothing new to Smith. He had begun his Hopkins career in 1983 as part of the team that helped transform the then-financially moribund Baltimore City Hospitals into today's thriving Johns Hopkins Bayview Medical Center.[287]

Under Smith's leadership, Home Care not only has significantly enhanced both its quality of care and developed its now-nationally praised safety procedures, it has expanded its services. JHHCG employees currently make more than 55,400 visits to patients annually—and the service continues operating solidly in the black.[288]

Johns Hopkins Community Physicians

The leadership of Johns Hopkins Community Physicians—the largest primary care physician group in Maryland—also changed during this period. Barbara Cook, a former National Health Services Corps physician who had become head of the practice in 2000, retired in January 2009 after nearly a decade of significant achievements. Her successor was Steven Kravet, a Johns Hopkins Bayview physician with a passion for patient-centered primary care.[289]

Through a network of 31 locations throughout the state and in Washington, D.C., JHCP currently handles more than 660,000 patient appointments annually.[290] Kravet, a third-generation pharmacist as well as a physician, joined the Hopkins Bayview faculty in 1995. He had served as medical director of the center's ambulatory care services, chief medical officer for quality and patient safety, and as deputy director for clinical activity in Bayview's Department of Medicine. As did Cook, he continues to see patients even as he heads JHCP.[291]

Employees of Johns Hopkins Home Care Group (JHHCG) such as Carla Cook and Dorothy (Dottie) Hall, make more than 55,400 visits to patients annually, providing home care by nurses, therapists, health aides and social workers, as well as medical and respiratory supplies, home infusion therapy, pharmaceuticals and hospice care. JHHCG has received national praise for its patient safety initiatives.

2009
January 29: In what is believed to be a first-ever procedure, surgeons at Johns Hopkins successfully remove a healthy donor kidney through a small incision in the back of the donor's vagina. The transvaginal donor kidney extraction, performed on a 48-year-old woman from Lexington Park, Md., eliminates the need for a 5-to-6-inch abdominal incision and left only three pea-size scars on her abdomen, one of which is hidden in her navel.

With efforts under way to enhance diversity at Hopkins, ensuring that multi-ethnic issues are addressed is a goal of employees such as (left to right) Arden Bongco, Wanda Smith, Willie Ferrer and Rhonda Cole, who foster such awareness in the Pathology Department's Core Lab in Hopkins Hospital. "Conflict comes when you don't know others' culture or traditions," says Ferrer, a Philippine native who employs a touch of diplomacy while supervising more than 40 medical technologists from multiple backgrounds. The Core Lab's rich employee mix includes not only other Filipinos but Chinese, Indian, Hispanic and African Americans. Ferrer, Bongco, Smith and Cole have organized pot-luck dinners for lab workers to bring dishes from their native lands; others have encouraged adoption of days when employees don the garb of their home countries. Pathology also hired a consultant to train lead personnel in communication skills and cross-cultural interaction.

2009
February 4: Hopkins emergency medicine experts offer a free, Web-based tool, developed with the Johns Hopkins Applied Physics Laboratory, that calculates and predicts in advance the impact on individual hospitals of a flu epidemic, bioterrorist attack, flood or plane crash, accounting for such elements as numbers of victims, germ-carrying wind patterns, available medical resources, bacterial incubation periods and bomb size. Called EMCAPS (Electronic Mass Casualty Assessment & Planning Scenarios), the software program lets users put their own information into the modeling software, customize it to their needs, and predict what they will need to handle a surge in casualties. It is believed to be the first that generates the anticipated outcomes of disaster planning scenarios developed by the Department of Homeland Security. The scenarios include patient estimates by injury type, estimated level of care required, and the need for decontamination facilities. The program is available for download free of cost from Johns Hopkins' CEPAR Web site.
February 7: Academy Award-winning actor Cuba Gooding, Jr., portrays Benjamin S. Carson, Sr., head of pediatric neurosurgery, in a made-for-TV movie, *Gifted Hands: The Ben Carson Story*. Based on Carson's memoir, *Gifted Hands*.

Enhancing Hopkins' Diversity

For an institution long deemed slow-moving and stuck in its ways, Hopkins Medicine undertook remarkable initiatives during this period to expand its diversity. Efforts to enhance the role of women, increase the number of under-represented minorities in the student body and staff, and improve opportunities for advancement were implemented or strengthened. The diversity campaign continues. It has attained notable successes—some of them receiving national recognition—but Hopkins' leadership believe much remains to be done.

"Today, the talent pool is worldwide, so we have to recruit worldwide talent, whether it's in Baltimore or Bangladesh, Dundalk or Dubai," Ed Miller said in 2006, soon after the Johns Hopkins Medicine Diversity Committee was formed. Headed by George Dover, director of the Children's Center, and Pamela Paulk, Health System vice president for Human Resources, the group seeks to ensure that Hopkins maintains its leadership role in recruiting, retaining and promoting more women, underrepresented minorities and people of otherwise diverse backgrounds.[292]

Enlarging the Role of Women

Although the School of Medicine had been committed for nearly a decade to improving the status of women on the faculty, progress was proving slow. Full equity remains frustratingly elusive, notes pediatric nephrologist Barbara Fivush.[293] Salary equity has been achieved, but progression up the leadership ladder remains problematic. It now is a major focus.[294] In 2008, Fivush was appointed the first director of the school's Office of Women in Science and Medicine. Among its chief aims is an increase in the number of women in leadership throughout Hopkins.[295]

Women Professors

Despite the sometimes frustrating pace of progress, the School of Medicine proudly held a daylong symposium and gala dinner in November 2005 to mark the promotion of the first 100 women to the rank of full professor since the school's founding in 1893. (The actual number of women full professors by then was 115.) It was a substantial milestone, considering the fact that less than 20 years earlier, in 1979, only seven women had been promoted to professor in the three-quarters of a century since the school opened its doors.[296]

By 2010, the number of women who had attained full professorships in the School of Medicine had jumped to 155. At the celebration of that achievement, Ed Miller noted that 100 of Hopkins' 155 women professors had been promoted in the preceding 10 years. "Twenty-one percent of our professors now are women," Miller said. "We'll get to 40 percent pretty soon, I think, and that's exactly where we should be."[297]

Janice Clements, vice dean for faculty affairs, notes that with women making up about 20 percent of the professors in the School of Medicine, Hopkins is ahead of the curve nationally. Only about 15 percent of the professors at academic medical centers nationwide are women. The proportion of female assistant professors is about 45 percent. Medical school classes are 50-50. "We're building the pipeline," Clements says.[298]

"A Woman's Journey"

Hopkins' focus on the medical concerns of women goes back to its founding, with Howard Kelly's pioneering work in gynecology and William Halsted's development of modern surgical techniques for breast cancer. Since 1995, however, Hopkins has held an annual, day-long conference specifically focusing on women's health. Entitled "A Woman's Journey," it features more than 30 seminars during which Hopkins faculty physicians and scientists describe the latest advances in medical specialties of particular concern to women—from combating memory loss to protecting the heart, avoiding chronic digestive disorders and surviving cancer. In 2008, a second "Women's Journey" health conference also was held in West Palm Beach, Florida. A third now is held in Naples, Florida, as well. At the 15th annual "A Women's Journey" in Baltimore, more than 1,050 women attended seminars, setting an attendance record.[299]

Opening Doors for All

The commitment to increasing diversity at Hopkins has not by any means been limited to advancing the careers of women on the faculty. Initiatives also have been undertaken to foster diversity in the student body, the house staff and the workforce.

For example, the 2006 formation of the Medicine Diversity Committee aimed to accelerate improvements already under way.[300] Gains indeed had been made. In 2002, underrepresented minorities made up just 3 percent of the residents in the Department of Medicine's Osler house staff. By 2008, 20 percent of Osler residents were minorities—compared to a national average of 13.5 percent—thanks in part to changes in how the department advertised and interviewed for positions. At Johns Hopkins Bayview, the results were even better. Twenty-five percent of its house staff were minorities in 2008, meaning that if it were a standalone institution, Bayview would eclipse its East Baltimore sibling.[301]

Yet there was room to do better. On average, nearly 15 percent of Hopkins medical students in 2006 were underrepresented minorities, but the School of Medicine's 20-member board of advisors was concerned that Hopkins still was lagging far behind its peers in attracting students from diverse backgrounds.

A day-long symposium and gala dinner were held on Nov. 1, 2005 to celebrate the appointment of "100 Women Professors" in the School of Medicine. In fact, from the inception in 2001 of the drive to reach the goal of "100 Women Professors" to the 2005 celebration of that milestone, nearly three dozen more women on the School of Medicine faculty were promoted to full professor—bringing the actual grand total then to 115. Among those at the celebration were Janice Clements (left), vice dean for the faculty, and dinner keynote speaker Kathleen Sander (right), author of *Mary Elizabeth Garrett: Society and Philanthropy in the Gilded Age*, a biography of the woman whose generosity made it possible for the Johns Hopkins School of Medicine to open—and whose insistence on high admission standards and the equal acceptance of women students made it a model for all medical schools to follow.

February 14: Surgical teams at The Johns Hopkins Hospital, Barnes-Jewish Hospital in St. Louis and INTEGRIS Baptist Medical Center in Oklahoma City successfully complete the first six-way, multihospital domino kidney transplant under the supervision of Robert Montgomery, Hopkins' chief transplant surgeon. All six donors— one man and five women, and six organ recipients—four men and two woman—are in good condition.

In 2007, Baltimore's Reginald F. Lewis Museum of Maryland African American History and Culture opened an exhibit with 7-foot-high panels celebrating 17 African American academic surgeons. Five of the honorees were from Hopkins—the largest group from a single medical school: Malcolm Brock, a thoracic oncology surgeon; Ben Carson, director of pediatric neurosurgery; Edward Cornwell, then chief of adult trauma surgery; Claudia Thomas, former assistant professor of orthopedic surgery and the first African American female orthopedic surgeon in the country; and Levi Watkins, associate dean for postdoctoral programs, professor of cardiac surgery, and the surgeon who performed the first implantation in a human of an automatic heart defibrillator.

C. Michael Armstrong, then head of the board of advisors, said he feared Hopkins was "losing some of the best and brightest students, irrespective of their background, because we didn't have competitive scholarships to offer them."[302]

The board sprang into action, pledging to fund 12 full scholarships—quickly moving Hopkins into the higher ranks of medical schools offering such assistance. By 2009, four of the 126 graduating medical students held a special place in their class: They were the first Board of Advisors Scholars to become newly minted physicians.[303]

To ensure that many more will follow them, the advisory board's initiative now is named the Johns Hopkins Medicine Scholars program, opening it up to a wider pool of sponsors. "The program represents the institution's commitment to diversity," says Daniel Teraguchi, assistant dean of student affairs and director of the Office of Student Diversity.[304]

Helping Hopkins employees bolster their educational credentials and upgrade their job positions has been the goal of several programs launched in recent years.

Project REACH (Resources and Education for the Advancement of Careers at Hopkins) has opened the door for hundreds of employees to enter

training programs for allied health care positions that chronically face critical shortages. Begun in 2004 with supplemental funds initially provided by the U.S. Department of Labor—but now entirely paid for by the Health System—Project REACH enables employees with at least a year of satisfactory full-time service to attend classes during work hours (up to 16 hours a week) while being paid, so long as they commit to repaying Hopkins in years of service.[305]

Under this and other programs, it has been common for entry-level employees who may have begun by performing maintenance work to advance routinely to such positions as respiratory therapist, pharmacy technologist, or medical form coder.[306]

The School of Medicine added to its diversity drive late in 2008 by appointing Brian Gibbs as its first associate dean for diversity and cultural competence. Gibbs formerly headed the Program to Eliminate Health Disparities in the Harvard School of Public Health's Department of Health Policy and Management. He is putting his social-change expertise to work on implementing the Johns Hopkins Medicine Diversity and Inclusion Vision 2020 Plan.[307] Beverly White-Seals, a member of the boards of trustees of Johns Hopkins Medicine and Howard County General Hospital, also was named director for Hopkins Hospital's Office of Workforce Diversity.[308]

All of these efforts have not gone unnoticed. In 2008, the U.S. Department of Labor gave its "Opportunity Award" to the Johns Hopkins Health System

Johns Hopkins Medicine's mission to build a more diverse, inclusive culture got a boost with the arrival in January 2009 of Brian Gibbs as the first associate dean for diversity and cultural competence. Gibbs had served as director of Harvard's Program to Eliminate Health Disparities in its school of public health. Now he is directing his attention to realizing the Johns Hopkins Medicine Diversity and Inclusion Vision 2020 Plan, which sets concrete goals for recruitment and retention of underrepresented minority employees, cultural competency in patient care, and eliminating potential disparities in quality of care and outcomes.

in recognition of the equal opportunity employment efforts at Johns Hopkins Bayview—which became the first hospital ever to receive the honor. That same year, the Association of Professors in Medicine bestowed its annual Diversity Award on Myron "Mike" Weisfeldt, director of the Department of Medicine, for his achievements in improving diversity within his department and the School of Medicine as a whole. Since Weisfeldt returned to Hopkins in 2001 to head Medicine, his diversity initiatives had become a departmental hallmark—and led to a tripling of underrepresented minority residents and fellows, as well as a near-doubling of assistant professors of medicine who are underrepresented minorities. Additional recognition came from the National Institutes of Health, which honored Hopkins for excellence in minority recruiting.[309]

Pioneers and Role Models

That excellence will remain the ultimate guide in Hopkins Medicine has been emphasized repeatedly in recent years by the success of many who, in previous decades, might have been denied the opportunities to shine.

In 1984, when Michael Ain graduated from Brown University, where he had built a solid academic and extracurricular record, he was rejected by 30 medical schools—including Hopkins. He believed that some or all of those rejections were due to one thing: His height. He is a 4-foot, 3-inch achondroplastic dwarf, a small person with an average-size trunk but shorter limbs.[310]

Finally, Albany Medical College in upstate New York admitted Ain, and by the time he had obtained his medical degree and finished pediatric and surgical residencies, Hopkins had changed. In 1995, Ain was offered an orthopedic surgery fellowship by Paul Sponseller, now chief of pediatric orthopaedics. Sponseller was eager to establish a program for treating skeletal dysplasias, the large group of disorders that result in the abnormal shape and size of bones frequently linked to dwarfism.[311]

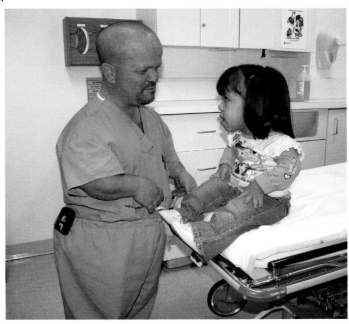

Pediatric orthopedic surgeon Michael Ain with achondroplasia patient Carmen Chavez, age 4.

Ain has been at Hopkins ever since. Today he may be the only dwarf in the world who is an orthopedic surgeon. He treats patients with a wide variety of orthopedic illnesses and injuries but specializes in the orthopedic problems of dwarfism and related disorders. Standing on a stool in the operating room, wearing a specially tailored, shortened surgical gown, he fuses painfully curved spines, reduces bones that compress the spinal cord, and performs other procedures for complications that can result from skeletal dysplasia disorders.[312] He is among those specialists whose work makes Hopkins' Greenberg Center for Skeletal Dysplasia, named for philanthropists Alan C. and Kathryn Greenberg, a place to which dwarfs from all over the world come for treatment. "I like being at Hopkins," he says. "It's a center of excellence. I'm able to foster caring for people." He has arrived, he says, where he wanted to be.[313]

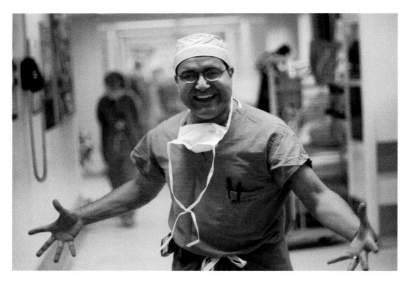

The astonishing odyssey of Alfredo Quiñones-Hinojosa from being an illegal immigrant picking weeds in Southern California cotton and tomato fields to a renowned brain surgeon at Hopkins can only be described as an extraordinary American success story—based on incredible determination, hard work, compassion and a preternatural ebullience that is irrepressible and inspiring. (It almost inevitably became the focus of one of the episodes of ABC-TV's *Hopkins*, just as Michael Ain's story had been part of *Hopkins 24/7*.)

"The challenge in what we do is not the surgery—it's the emotional connection you form with the patients."

—Alfredo Quiñones-Hinojosa, associate professor of Neurosurgery, Oncology Neuroscience and Cellular and Molecular Medicine, and the director of the Brain Tumor Surgery Program

When Alfredo Quiñones-Hinojosa clambered over a chain link fence between Mexicali, Mexico and Calexico, California in 1987, he simply wanted to be somewhere other than his economically depressed native land.[314]

At 18, his original plan just was to enter the United States, make a lot of money, and return to Mexico to support his parents and siblings. Like many illegal immigrants in Southern California, he began working in the fields, pulling weeds at cotton and tomato farms near Fresno. After a year, he decided he had to learn English and go to school.[315]

What happened after that epiphany can only be described as an extraordinary American success story—based on incredible determination, hard work, compassion and a preternatural ebullience that is irrepressible and inspiring. He won a scholarship to attend the University of California at Berkeley. He excelled at calculus, physics and chemistry.[316]

Inspired by the memory of a grandmother who had been a respected *curandera*, or village healer, he decided to become a physician. He went to Harvard, where he distinguished himself in neurobiology. He became a U.S. citizen in 1997 and graduated *cum laude* from Harvard in 1999, giving the commencement address for his medical school class.

Quiñones came to Hopkins in 2005 as an assistant professor of neurosurgery and oncology and director of the brain surgery program at Johns Hopkins Bayview. An associate professor since 2008 and now also director of the Neurosurgical Pituitary Center, he performs between 230 to 240 brain tumor operations a year, conducts extensive research, and teaches. He has garnered such accolades as the $150,000 Howard Hughes Medical Institute Physician-Scientist Early Career Award and the $15,000 Nickens Faculty Fellowship from the Association of American Medical Colleges. He uses the $15,000 fellowship award to give research stipends to minority students who work in his lab—one of the most racially and ethnically diverse on campus.[317]

Called "Dr. Q" by all, Quiñones' intensity is matched by compassion for his patients.[318] Quiñones is focusing his research on finding non-invasive ways to treat the often-intractable brain tumors he confronts almost daily, as well as on uncovering whether the molecular switches that turn normal cells into cancer cells can be reset. He also is determined to be a mentor to those who train under him.[319]

From left, the gas station in Mexicali, Mexico, where Alfredo Quiñones worked for his father starting at age 5; the fence he jumped to cross the border from Mexicali to Calexio, Calif., as seen from the United States side; in the San Joaquin Valley, his first home in the United States.

Careers in medicine can have as many origins as there are physicians, but few had so stark a beginning as that of nephrologist Hamid Rabb.

Now vice director of the Department of Medicine and head of kidney transplantation, Rabb was only eight in 1971 when his family fled the war-ravaged region between India and Burma that ultimately became Bangladesh—and he began his long journey to Johns Hopkins.[320]

When a brutal firefight erupted on March 25, 1971, around the faculty compound where his family lived, young Hamid saw men in army uniforms operating bulldozers, scooping up dead people, then dumping them into earthen pits. Part of what Hamid didn't know was that all members of his ethnic group, the Bengalis, were next on the soldiers' list for elimination.

Rabb's father, philosophy professor at the University of Dhaka, made "a 10-second decision" to lead his family on a desperate escape. Grabbing only a few possessions, they jammed into their Volkswagen Beetle. Avoiding military checkpoints, they headed to the home of Rabb's uncle, a successful government official who lived in a compound of Kashmiris, who were not targeted by either side in the fratricidal conflict between Pakistanis and Bengalis.

They joined a flood of 8 million other refugees who spread around the globe. They ultimately landed in Montreal, where his family previously had lived while his father was obtaining his doctorate in theology and mysticism.

As a member of a minority group struggling to integrate into a new homeland, Rabb was encouraged by his father to develop skills that no one could every take away from him. Medicine, his father said, was such a profession. "At least if you're a doctor," he said, "it doesn't matter where you go. You will always have employment."

Hamid lost no time racing through the Canadian school system, then rapidly advancing into medicine with a scholarship at McGill University, Sir William Osler's alma mater. After a series of subsequent training stops that took him to UCLA and Harvard, Rabb was persuaded to speak at Hopkins in 2000 by former nephrology chief Joseph Handler and clinical director Paul Scheel.

During that visit, Handler reminded Rabb that he'd once spent time in India, tending to Bengali refugees fleeing the Bangladesh atrocities. "I know your background," Handler told Rabb. "We could use your sort of temperament and skills here. Would you be interested in becoming the medical director of our kidney transplant program?"

The fit made sense on many levels to Rabb, who also enjoys Hopkins' high-level access to international communities, which allows him quietly to address medical inequities far from American shores.

Yet even as he hones his own leadership skills, Rabb has no appetite for combating the old wrongs that uprooted his life four decades ago.

He also does not consider himself a war refugee—but rather a child of good fortune who now, at Hopkins, can help improve the lives of others.

2009

March 9: Johns Hopkins Bayview announces it has received one of its largest-ever private gift, 3.15 acres of land at 5400 E. Lombard Street, valued at $3.1 million, donated by Baltimore businessman James Crystal and his family. Crystal says he hopes the gift will enable Hopkins Bayview "to expand even further and continue doing great things."

April 9: Richard Bennett, professor of geriatric medicine and a key participant in the transformation of the once-beleaguered Baltimore City Hospitals into today's vibrant Johns Hopkins Bayview Medical Center, is appointed president of the 600-bed facility. He assumes office on July 1.

June 10: A formal ribbon-cutting takes place for the Wilmer Institute's new Robert H. and Clarice Smith Building, ushering in a new era for the world-acclaimed eye care institution. A formal dedication is scheduled for October 16.

Though the horrors of 1971 still haunt him, Hamid Rabb refuses to see himself as a war refugee. From left: Wife Nausheen, Hamid, Samy, Adam, and Neil.

The process of increasing diversity at Hopkins—perhaps slow in the beginning—is unlikely ever to end. As far back as 1998, the University-wide Johns Hopkins Diversity Council, with the support of the Medical School Council and other institutional organizations, played catch-up with such peer institutions as Yale, Duke and Stanford by finally implementing the policy of extending University benefits to same-sex domestic partners.[321]

A lack of such inclusion and tolerance several decades earlier would likely have prevented Hopkins medical students from launching the grassroots campaign they undertook in 2004 to urge the appointment of psychiatrist Tom Koenig as associate dean for student affairs. Clearly they were unconcerned that Koenig and his partner, Ciro Martins, a Brazilian-born dermatologist, had been a couple since 1989 and never hidden their relationship.[322]

The only things that mattered to the students who wanted Koenig to be their dean were his exceptional teaching skills, empathy as a former Hopkins medical student himself (Class of '89), and commitment to them as a counselor. When the news broke in 2004 that hepatologist Frank Herlong, the much-admired associate dean of students for the preceding 15 years, intended to relinquish that post, a group of fourth-year students began an e-mail campaign on behalf of Koenig.[323]

Psychiatrist Tom Koenig , Associate Dean for Student Affairs (right), and his partner, Associate Professor of Dermatology Ciro Martins, have been together since 1989. "We're totally out with everybody: deans, students, department directors, faculty," Martins said in 2004. "We attend faculty functions together, and nobody bats an eye. We're treated as a couple." Both men credit Hopkins with helping them overcome some difficulties, with the University offering domestic partner benefits and helping Martins, a native Brazilian, obtain a semi-permanent visa.

He got the job—and quite a job it is. In addition to his regular work as a physician, teacher, and researcher, Koenig now heads an office that advises 4,000 medical students, routinely providing the kind of support that can make or break any student's year and significantly affect his or her future.[324]

Perhaps the most crucial—certainly the highest-profile—part of Koenig's job is overseeing the process of preparing the "match letters" of recommendation that fourth-year students need to have sent to the residency programs of their choice.[325] "If the students are to have faith in my role as an advocate, they have to know we'll be unbiased to all," Koenig says.[326]

Striving to ensure a complete lack of bias throughout Hopkins Medicine has been the mission of cardiac surgeon Levi Watkins Jr. ever since he became the first black surgical resident at Hopkins Hospital in 1970. His decades-long role in this ongoing endeavor was celebrated in 2006 with a mammoth reception honoring his 35 years at Hopkins and creation of The Levi Watkins Jr. Professorship of Cardiac Surgery, funded by his family.[327]

Hopkins Hospital and Health System President and Hopkins Medicine Executive Vice President Ron Peterson praised Watkins as a "medical pioneer,

2009

April 14: The 1,000th kidney transplant is performed since the merger of the Hopkins Hospital and Hopkins Bayview transplant programs in 1996.

April 24: An agreement is signed to integrate Suburban Hospital Healthcare System into the Johns Hopkins Health System (JHHS), bringing into the fold a highly regarded Montgomery County community hospital "virtually on the doorstep of the nation's capital," in the words of Ron Peterson. The move seeks to build on longstanding ties and to address growing regional interest in more efficient, integrated regional health care services for patients, say officials of Suburban Hospital Healthcare System (SHHS) and The Johns Hopkins Health System Corporation The 238-bed Suburban Hospital has annual admissions of nearly 15,000 and is a cornerstone of Montgomery County medical care.

April 30: *Time* magazine names anesthesiologist and critical care specialist Peter Pronovost as among the publication's 100 "most influential people." Medical director of Hopkins' Center for Innovation in Quality Patient Care, Pronovost is a champion of patient safety whose creation of a checklist to reduce physician errors is credited with a significant worldwide impact. The previous September 23, Pronovost was awarded a MacArthur Foundation "genius grant."

role model extraordinaire and all-around humanitarian." In addition to being the first African American medical student admitted to Vanderbilt in 1966; his groundbreaking residency at Hopkins, and his 1980 achievement as the first surgeon to implant an automatic heart defibrillator in a human, Watkins launched a highly effective, one-man minority recruitment drive when he joined the School of Medicine's admissions committee in 1979.[328]

In 1982, Watkins also founded the annual Martin Luther King Jr. Commemoration at Hopkins Hospital. In the ensuing decades, he has brought an extraordinary array of speakers to the campus for the event. Among them have been civil rights leaders Coretta Scott King (whom he had known since he was a teenager), Rosa Parks and Jesse Jackson, as well as celebrities such as poetess Maya Angelou, actors Harry Belafonte, James Earl Jones, Danny Glover, Lou Gossett Jr. and Cicely Tyson, and musician Stevie Wonder.[329] Although he has set down his surgical scalpel, Watkins remains fully active as associate dean for postdoctoral affairs and a driving force behind the ongoing diversity initiatives. At the celebration honoring Watkins' three-and-a-half decades at Hopkins, he recalled: "I came up when color was everything." Now, gazing at the diverse crowd that had gathered to honor him—black, white, Asian, male, female, young, old, gay and straight— he said, "But looking out at all of you today, I don't see color at all."[330]

Pioneering cardiac surgeon Levi Watkins (left) has spent a lifetime putting into practice the lessons he learned as a teenager from Dr. Martin Luther King Jr., whom he met when King was pastor of the Dexter Avenue Baptist Church in Montgomery, Ala. Each January for three decades, Watkins has shared his mentor's wisdom with Hopkins and the community. Founding the annual Martin Luther King Jr. Commemoration at Hopkins Hospital, he has brought such civil rights leaders as former Rep. Kweisi Mfumi (above, left) and the Rev. Jesse Jackson (above, right) to speak to overflow audiences in the medical campus' 500-seat Turner Auditorium.

In Hopkins Hospital's annual Turtle Derby, a group of the phlegmatic reptiles are challenged to lumber out of a circle on hot pavement and be the first to seek shade—with all of the race's proceeds being donated to community causes. Begun in 1931, the Turtle Derby is going strong, even if the competitors remain slow, as the 2005 photo above suggests.

2009
June 18: The Johns Hopkins Children's Center is among the top 10 children's hospitals in the nation by *U.S. News & World Report* in its annual rankings of children's hospitals. The magazine's "honor roll" of the 10 top pediatric hospitals is in no particular order but singles out those that ranked in all key specialties.

■ The members of the Hopkins School of Medicine Class of 2009 were an eclectic and accomplished group—as have been their predecessors and their successors. Among those in the Class of 2009 were a West Point graduate and Army Ranger School leader; a breeder of pythons, boa constrictors and rattlesnakes; and a one-time resident of a Tibetan monastery in Nova Scotia.

Hopkins' Impact Continues on Community

Although the feisty, combative Blue Jay is the emblem of Johns Hopkins athletics, it is a group of plodding, lumbering turtles that long have been associated with Hopkins Medicine's commitment to—and impact on—its community.

Turtles?

They're steady. They're imperturbable. They are determined to reach their goal. And they're tough.

For 80 years, all of Hopkins Medicine—students, faculty, interns, residents, administrators, employees and their families—has been captivated by a unique spring sporting event that pits a phlegmatic group of reptilian competitors in a race. It's called the Turtle Derby, held on the third Friday in May, the day before each year's running of the Preakness Stakes, Baltimore's leg of the storied Triple Crown of thoroughbred racing.[331]

The unusual tradition began in 1931. Forty turtles, kept as pets in a pen on the Broadway side of the Hospital by "Colonel" Benjamin Frisby, the Hospital's doorman from 1889 to 1933, were borrowed by gynecologist Edward Kelly. He thought it would be fun to create a "race" by putting the heat-averse turtles in a sun-splashed circle and seeing which one hustled the fastest to get to a nearby shady spot. Kelly trumpeted the event as a way to create camaraderie among the house staff and faculty. It worked. Witty—sometimes salacious—names were appended to the turtle contestants. The winner of the first Turtle Derby was dubbed "Sir Walter," in honor of brain surgeon Walter Dandy; one year, the contestants included a turtle named "Desperation out of Unhappy Marriage by Philandering Husband." Over time the race became a major spring event, as the crowd of cheering spectators grew to include not only physicians and medical students but nurses, patients and even neighborhood families.[332]

Since 1980, the Turtle Derby has served as a fund-raiser for community causes. All of its proceeds go to supporting local charities and other worthwhile endeavors, including the Child Life Program at the Johns Hopkins Children's Center, and the Hospital-sponsored Perkins Day Care Center in East Baltimore.[333]

The Turtle Derby's contributions to the community are symbolic of Hopkins' long-standing and ongoing commitment to improving the lives of those who live in its neighborhood and city at large. Over the past half-dozen years, Hopkins' initiatives have run the gamut from maintaining its century-and-a-quarter record of providing free care for the poor, to examining a remarkable variety of species—be they elephants, snakes, tarantulas, dolphins or whales—in the city zoo or the National Aquarium downtown, to encouraging inner city schoolchildren to become health care professionals and opening employment opportunities to the homeless.[334]

Between Hopkins Hospital, Johns Hopkins Bayview and Howard County General Hospital, Hopkins Medicine gave away $67 million in charity care in fiscal year 2009. It considers this a commitment to following the instructions in Johns Hopkins' 1873 will to care for the "indigent sick of this city and its environs without regard to sex, age, or color… without charge."[335]

With the dramatic influx of Central and South American immigrants to East Baltimore, the neighborhoods surrounding Johns Hopkins Bayview, and Howard County, Hopkins' efforts to serve the region's needy Hispanics have ranged from holding Latino health fairs and increasing the number and availability of Spanish-speaking interpreters to providing free Spanish classes to faculty, students and staff, as well as free English classes for Hispanic employees.[336]

In 2009, the faculty's billing arm, the Clinical Practice Association, began to make offering free care less complicated, at least at Hopkins Hospital, by creating The Access Partnership (TAP). Its physicians—now led by Barbara Cook, former president of Johns Hopkins Community Physicians—provide free specialty care such as radiology, cardiology and ophthalmology services to patients in five ZIP code areas near Hopkins and Hopkins Bayview.[337]

Elsewhere, readily acknowledging his share of the Hopkins DNA is Rick Hodes. He has spent more than two decades caring for indigent people in Ethiopia—adopting five children in the process, touching the lives of thousands of patients, and inspiring many young physicians.[339] It was during Hodes' 1982–1985 internal medicine residency at what now is Johns Hopkins Bayview that he first went to Ethiopia as a relief worker during the 1984 famine there. He returned on a Fulbright Fellowship to teach internal medicine. In 1990, he was hired by the American Jewish Joint Distribution Committee as a medical advisor for Ethiopian Jews who were hoping to immigrate to Israel. Ultimately, 60,000 of them did. Hodes chose to remain in Ethiopia to care for the critically ill, especially youngsters suffering from grotesquely deformed, crippled backs due to often-fatal spinal tuberculosis.[340] "I just like to help people," Hodes says modestly—as if those simple words could explain such a dramatic life at the extreme edge of modern medicine. He ministers to people in sprawling refugee camps and also serves in a string of clinics he founded in Ethiopia's capital, Addis Ababa, and Gondar, a city of 140,000. He patrols small Ethiopian villages, administering medicine on the fly. He has set up a self-styled global network that brings him cheap medicines from India and enables him to refer patients for surgery in neighboring Ghana or the U.S. He is the sole Western physician at the Mother Teresa Mission in Addis Ababa, where—despite his open observance of his Orthodox Jewish faith—he sometimes is called "Father Teresa."[341]

Hodes' work was lauded on CNN's *Heroes* program in 2007; been the subject of a 2010 HBO documentary, *Making the Crooked Straight* (a reference to his focus on fixing the backs of children with spinal tuberculosis); and a book, *This is a Soul: The Mission of Rick Hodes*. Just as Hodes acknowledges the importance of his early experiences at Hopkins, those at Hopkins are honored to celebrate his work. He is, says Hopkins Hospital trustee Shale Stiller, "one of the most selfless men I've ever met."[342]

Maurice Pretto, above, once penniless and homeless, living at Baltimore's Helping Up Mission, has been a Hopkins Hospital biomedical engineering technician since 2007, having moved up from an Environmental Services job he got in 2003. He is one of dozens of former shelter residents who have been hired by Hopkins Medicine since it inaugurated a program to do so in 2001.[339]

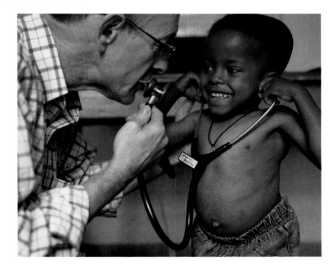

Rick Hodes, an internal medicine resident at Hopkins in the early 1980s, has spent more than two decades caring for poor people in Ethiopia. "I just like to help people," he says modestly. His work—which has earned him an accolade as the region's "Father Teresa"— has been chronicled in an HBO documentary, *Making the Crooked Straight*, and a book, *This is a Soul: The Mission of Rick Hodes*, written by veteran *Washington Post* and NBC News reporter Marilyn Berger.

2009
July 7: Johns Hopkins kidney transplant experts lead the first eight-way, 16-patient, multi-medical center "domino" kidney transplant procedure as surgical teams at The Johns Hopkins Hospital, Barnes-Jewish Hospital in St. Louis, INTEGRIS Baptist Medical Center in Oklahoma City and Henry Ford Hospital in Detroit successfully complete multiple operations. The transplants involved eight donors—3 men and 5 women along with eight organ recipients— 3 men and 5 women. One of the donors is Pamela Paulk, vice president of human resources at The Johns Hopkins Hospital and Johns Hopkins Health System, who volunteered to donated a kidney to a needy recipient. The ultimate beneficiary of Paulk's generosity is Robert Imes, a painter on the Hopkins Hospital's facilities department who received someone else's kidney because of her donation. Paulk got to know Imes nine years earlier during contract negotiations with Imes' local union. Paulk says Imes' kidney had failed three years ago but she only learned about it recently when he had to take 10 months off from work.

Richard (Rich) Grossi (above right), senior vice president of Johns Hopkins Medicine and its chief financial officer since 1996; and Ronald Werthman (lower right), vice president/chief financial officer of the Johns Hopkins Health System and Hopkins Hospital, are the widely praised "financial wizards" who for three decades have managed to keep Hopkins Medicine in the black during lean as well as flush times.

Financial Management to Maintain the Highest Quality

As the Hopkins Medicine enterprise expanded exponentially during this period, maintaining its superior standards in patient care, research and teaching required constant attention to management and money. Such an effort may be less exciting, perhaps, than clinical triumphs, laboratory breakthroughs, revolutionary curricula, or community service—but without it, none of those would be possible.

Richard (Rich) Grossi

Financial Wizards

Regularly credited by Ed Miller and Ron Peterson for the "strokes of genius" that have kept Hopkins Medicine's financial ship on an even keel are two longtime management specialists and numbers-crunchers extraordinaire, Richard (Rich) Grossi and Ronald Werthman.[343]

Grossi has been associate dean for finance and administration of the School of Medicine and Hopkins Hospital since 1978. He became vice president and chief financial officer of Johns Hopkins Medicine in 1996 and was promoted to senior vice president in 2011. Werthman, another three-decade veteran in health care and higher education, became the vice president/chief financial officer of the Johns Hopkins Health System and Hopkins Hospital in 1993.

Despite nationwide economic turmoil, Hopkins Medicine has maintained positive operating margins under Grossi and Werthman's leadership—yet the space between black and red ink on the ledger often has been slim. Hopkins may now be a $6 billion enterprise, but because "we've never had large margins, our organization is required to manage every operational and financial issue at a detailed level," Grossi observed late in 2008, just as the Great Recession was getting under way.[344]

Spending money on patient safety and efficiency is a key to curbing expenses and

Ronald Werthman

remaining on budget, Hopkins' top administrators note. Richard Bennett, a professor of geriatric medicine who became president of Johns Hopkins Bayview in May 2009, proudly observed that the elimination of central-line bloodstream infections throughout the medical center's intensive care units not only is good for the patients but "saves hospital resources and helps the hospital financially."[345]

Leadership Changes

The Dean's Office, Hopkins Hospital, Hopkins Bayview—and The University

While recruitment of Hopkins Medicine luminaries by other institutions is common, occasionally it also happens within The Johns Hopkins University itself. A number of key leadership changes during this period occurred just in this fashion. Michael Klag, an eminent epidemiologist and cardiovascular and kidney disease expert, had been the School of Medicine's vice dean for clinical investigation since that post was created in 2001. In 2005, however, he was named dean of the University's Bloomberg School of Public Health. As Klag's successor, Ed Miller chose internist Daniel Ford, who long had specialized in pioneering studies of the relationship between mental disorders and chronic medical conditions. He also had become an early leader in the use of the Internet for clinical research.[346] Ford had arrived at Hopkins in 1982 as an Osler resident. He subsequently joined the internal medicine faculty and received joint appointments in psychiatry in the School of Medicine, as well as in epidemiology and health policy and management in the Bloomberg School. He earned an international reputation for groundbreaking clinical studies documenting depression as an independent risk factor in heart disease.[347] Under Ford, Hopkins continues receiving more in federal research grants from the National Institutes of Health alone than any other academic medical center. Several thousand research projects are under way at any given time, with well over half of them involving human subjects. Ford ensures that the clinical researchers have all the tools they need to conduct safe, ethical studies that meet all federal regulations.[348]

When pediatrician Beryl Rosenstein decided to step down as Hopkins Hospital's vice president for medical affairs in 2009, he was succeeded by Redonda Miller, a 1992 graduate of the School of Medicine who had spent nearly half of her life—and all of her career—at Hopkins.[349]

With a typically engaging laugh, Miller wonders aloud how and why she could have followed such a professional path. "Isn't that crazy? What am I thinking?" Noting that she came to Hopkins as a medical student in 1988 and has been here ever since, she readily answers her own question: "Why would you leave when you're already at the best place?"[350]

Miller says that keeping Hopkins "the best place"—and making it even better—is a formidable but obtainable goal in an era of increasingly strict oversight; payer and consumer demand for quality; and uncertainty about the future impact of health care reform.[351]

An associate professor of medicine and previously the Department of Medicine's vice chair for clinical operations, as well as an assistant dean for student affairs, Miller now oversees medical staff affairs; Hospital epidemiology and infection control; medical records; the pharmacy and a host of other institutional activities.[352]

Cardiac surgeon Bill Baumgartner, already vice dean for clinical affairs, director of the Surgical Research Laboratory, president of Hopkins' 1,700-member Clinical Practice Association, and executive director of the Chicago-based American Board of Thoracic Surgery, accepted another assignment in 2011. He became senior vice president of Johns Hopkins Medicine and head of its new Office of Johns Hopkins Physicians. Given Hopkins Medicine's rapid expansion through affiliations and integrations with community hospitals and their physician groups, Dean/CEO Edward Miller said the new office ensures that "all physicians who practice under our JHM umbrella… are all working in concert to fulfill our tripartite mission" of patient care, research and medical education.

A 1992 graduate of the School of Medicine, Redonda Miller, associate professor of medicine and vice president for medical affairs in The Johns Hopkins Hospital, has spent nearly half of her life at Hopkins. She says: "Why would you leave when you're already at the best place?"

A few years after he left Hopkins Medicine's flagship, The Johns Hopkins Hospital, in 2000 to become head of the Department of Medicine at Johns Hopkins Bayview, rheumatologist David Hellmann said, "When I got to Bayview, I'd hear about the disadvantages of the place: We're not Hopkins Hospital. But I said, What are the first two words of our name? Johns Hopkins is known around the world, it's one of the most famous names in medicine. Bayview is Johns Hopkins with more convenient parking. This is Johns Hopkins with *grass.*" In 2005, he became vice dean of the School of Medicine at Bayview. Here he holds a pyramid that symbolizes the three key parts of Johns Hopkins Medicine's mission: patient care, teaching and research.

2009
July: Howard County General Hospital opens its new, four-story patient pavilion in July. The building houses a large outpatient facility and three floors of all-private inpatient rooms, plus additional space for several support departments, including Information Technology and Pharmacy.

Johns Hopkins Bayview

When anesthesiologist L. Reuven Pasternak, named the first vice dean for Johns Hopkins Bayview in 2001, was recruited to become chief medical officer at the six-hospital Health Alliance of Greater Cincinnati in 2005, Dean/CEO Ed Miller said the School of Medicine was fortunate to have so deep a reservoir of talent that acclaimed rheumatologist David Hellmann was immediately available to succeed Pasternak. Hellmann had been head of the Department of Medicine on the Eastern Avenue campus since 2000.[353] Hellmann's arrival at Johns Hopkins Bayview had brought even greater energy to the already dynamic development of the campus. He had recruited accomplished division directors and attracted other highly regarded clinicians and administrators to join the medical center's departments. In particular, he had been a strong supporter of programs that reinforced the bonds between Hopkins Hospital and Bayview. Miller said that Hellmann's appointment as vice dean emphasized Hopkins Medicine's "commitment to the same goal."[354]

A significant change for Johns Hopkins Bayview occurred in 2009, when Gregory Schaffer retired as president of the 700-bed medical center following an eventful decade at its helm.[355]

Schaffer had joined Hopkins Bayview in 1995 as vice president of support services. He succeeded Ron Peterson as head of the medical center in 1999. During his 10-year tenure, Schaffer developed new clinical programs, expanded others, and established centers of excellence in bariatric surgery, wound care, pelvic health and stroke. Patient volumes roles approximately 83 percent under his leadership, going from 17,705 in 1999 to 21,359 by the fall of 2008, when he announced his intention to retire. (Schaffer's retirement proved to be short-lived. Early in 2010, he joined Johns Hopkins Medicine International and became the chief executive officer of Hopkins-managed Tawam Hospital in Abu Dhabi.)[356]

Schaffer's successor as Hopkins Bayview president was geriatrician Richard Bennett, a 1982 graduate of the School of Medicine and 22-year veteran of the faculty. Two years earlier, when Bennett had been appointed executive vice president and chief operating officer for Hopkins Bayview, Schaffer had praised him as "an outstanding physician with exceptional

Gregory Schaffer

interpersonal skills who also understands the business of medicine."[357] Shortly after his appointment as Johns Hopkins Bayview's president, Bennett said that the medical center's trustees recognized the need to move as rapidly as possible toward creation of all-private patient care rooms and to expand the emergency department and obstetrics facilities to meet the growth in demand for both services. By late 2009, four new operating rooms had been opened, including one which was among the first in the country to have intraoperative computed tomography (CT) scanners, permitting complex neurosurgical procedures to be performed more safely. The 301 Building for clinical offices and research, also was completed.[358]

Marketing and Communications

Not all of the changes in Hopkins Medicine leadership involved physicians. One veteran, non-physician official who had done much to make the work of Hopkins' clinicians, researchers and educators known not only to their colleagues but the nation decided to retire in 2005. Elaine Freeman, by then vice president of corporate communications and head of Hopkins Medicine's award-winning public affairs and marketing office, chose to step down after 23 years of tireless advocacy on Hopkins' behalf.[359]

A former newspaper reporter and freelance journalist, Freeman recalled that when she was put in charge of communications for Hopkins Hospital and the School of Medicine in 1982, then-dean Richard Ross "told me my job was to keep the tarnish off the Dome" of the emblematic Billings Building, "with the corollary that I should help make it shine."[360]

Geriatrician Richard Bennett, a 22-year Hopkins veteran, became president of Johns Hopkins Bayview Medical Center in 2009. Born into a family that long had owned and operated a small Northwest Baltimore home for the elderly, Bennett was raised on its campus and learned from an early age not only how to deliver compassionate care to older adults but how to do so within a budget. Bennett had been a key participant in the transformation of the debt-laden old Baltimore City Hospitals into the vibrant Hopkins Bayview. He had trained at City Hospitals and witnessed its phenomenal growth first-hand. He aims to maintain its momentum.

Elaine Freeman

For the next two decades, she did just that. Under her direction, medicine at Hopkins became an increasingly hot topic in the press, television, even the movies. Hopkins' long-standing reputation for research and clinical excellence—upon which so many present-day rankings are based—was never allowed to lag. No matter how important a scientific or clinical advance may be, its significance might go unnoticed if few are aware of it.

Consequently, the Hopkins Medicine accomplishments that continuously—and justifiably—were trumpeted by the communications and marketing team ensured

From 1982 until 2005, Elaine Freeman (left) kept "the tarnish off the Dome" and helped make it shine as head of Hopkins Medicine's public affairs and marketing office.

Dalal Haldeman, named Hopkins Medicine's vice president of marketing and communications in late 2005, is an experienced health care industry specialist in marketing, public relations, business development and operations.

2009
The Center for Clinical and Translational Research is launched by the Brain Science Institute to channel expertise from Hopkins' various imaging-dedicated centers into creating a university-wide surge in the understanding and use of new imaging technologies in neuroscience research.

August 21: Lloyd Minor, the director of the Department of Otolaryngology—Head and Neck Surgery, is named provost and senior vice president of academic affairs, the second-in-command position for the University.

September 13: Ronald J. Daniels is installed formally as the 14th president of The Johns Hopkins University. He had assumed the presidency the previous March 2.

that Hopkins' Dome sparkled brightly. This certainly helped both the Hospital and medical school to remain atop national surveys for excellence year after year. In response to the tremendous growth in Hopkins' research efforts, faculty and staff, the public affairs and marketing office itself tripled in size and expanded dramatically in scope, producing publications and promotional materials that repeatedly garnered awards from professional groups.[361]

Freeman's successor is Dalal Haldeman, who was named vice president of marketing and communications in December 2005. An experienced health care industry specialist in marketing, public relations, business development and operations, Haldeman had compiled a substantial record in these fields for various institutions, including the MetroHealth System in Cleveland and the Cleveland Clinic.[362] She also served as an assistant professor at several universities, including the University of Pittsburgh.

In announcing Haldeman's appointment, Ed Miller, Ron Peterson and Steve Thompson all cited her impressive credentials in quality improvement, strategic planning, market research, Web site development and branding as being vital during a time of immense growth of the Johns Hopkins Medicine enterprise and intense competition among academic medical centers. [363]

Haldeman brings a lifelong interest in science and a penchant for data-driven decision-making to her role.

Describing herself as a "connector" whose mantra is "together, we're better," Haldeman has initiated and marshaled the results of specialized research on market changes and consumer behavior to both shore up and expand Hopkins' already high national name recognition.

Haldeman has been melding her attention to detail with her panoramic outlook on other fronts as well. Under her leadership, Marketing and Communications has widened awareness of what Johns Hopkins Medicine does by creating a more cohesive Web presence, taking advantage of new media channels, expanding the reach of print publications and producing videos such as *The DNA of Johns Hopkins Medicine* to capture the essence of Hopkins and its influence on research, medical discovery, education and clinical care.

At the same time, never losing sight of the needs of the neighborhoods surrounding the East Baltimore campus, Haldeman has helped produce health education information for the local community.

"My goal," she says, "has always been to ask the right questions, cross-pollinate our talent to exchange and share ideas, gather and understand the data so we can act on it and focus our efforts to make sure that Johns Hopkins Medicine is known as the nation's premier research, teaching and patient care destination."[364]

A New President at Hopkins' Helm

After a dozen dynamic years as president of The Johns Hopkins University, Bill Brody announced in March 2008 that he would retire after the official, December 2008 conclusion of the then-ongoing $3.2 billion fundraising campaign—the second successful, billion-dollar drive he had overseen.[365]

As he had with every aspect of the large, diverse Johns Hopkins enterprise, the 64-year-old Brody created an enduring legacy within Johns Hopkins Medicine—an entity he knew well as a former professor and director of the radiology department and a professor of biomedical engineering and electrical and computer engineering.

Brody's fundraising efforts had helped underwrite the major transformations then taking place on the East Baltimore campus, as well as enhancements at Johns Hopkins Bayview and Howard County General Hospital. He also had been involved in creation of the multiple research institutes, centers and offices that had blossomed throughout Hopkins Medicine, as well as helped foster the creation and expansion of Johns Hopkins Medicine International, exporting Hopkins' mission of teaching, research and patient care worldwide.[366]

Under Brody's direction, Hopkins Hospital and Health System, along with the University, also collaborated to solve problems much closer to home. He established the Urban Health Institute to focus Hopkins' resources on East Baltimore's health problems and gave financial support to the New EastSide Project, the urban revitalization initiative north of the medical campus.[367]

Following a seven-month, international search that included the review of some 300 candidates and dozens of discreet interviews in New York City conference rooms, the Hopkins board of trustees announced in November 2008 that Ronald J. Daniels, the 49-year-old provost of the University of Pennsylvania, had been chosen to succeed Brody.[368]

A Canadian-born attorney, Daniels had served 10 years as dean of the University of Toronto's law school before becoming Penn's provost in 2005. In Toronto, he had cut the faculty-student ratio in half, increased the faculty's endowment from $1 million to $57 million, and improved diversity among the student body.[369]

Such a record was of particular interest to Hopkins Medicine, which confronted the increasing financial needs of medical science and research in the face of a faltering economy and mostly stagnant government funding for research. Shortly after his appointment as Hopkins' president, Daniels told *The Chronicle of Higher Education*, "You can have the loftiest of dreams and the most daring of ambitions for your institution, but at the end of the day, if you don't have the resources, you will fail in terms of advancing your mission."[370]

Ronald J. Daniels, a Canadian-born attorney and provost of the University of Pennsylvania, was named The Johns Hopkins University's 14th president in November 2008 and took office in March 2009. Within a month of his official installation in September 2009, he had to undergo unexpected emergency surgery at Hopkins Hospital for an abdominal tumor. Following his successful treatment and swift recovery, he wrote, "Not that I needed it, but I now have even more reason to be proud and thankful to be at Johns Hopkins."

Lloyd Minor (right), a gifted researcher, scientist and surgeon, was an internationally renowned otolaryngologist and director of the Department of Otolaryngology–Head and Neck Surgery when Hopkins' President Ron Daniels chose him to become provost and senior vice president for academic affairs—the second-in-command at the University—in August 2009.

2009

October 5: Carol Greider, Ph.D., professor and director of Molecular Biology, shares the Nobel Prize in Physiology or Medicine for her 1984 discovery of telomerase, an enzyme that maintains the length and integrity of chromosome ends and is critical for the health and survival of all living cells and organisms. Greider shares the prize with Elizabeth Blackburn, a professor of biochemistry and biophysics at the University of California, San Francisco, and Jack Szostak, Ph.D., of Harvard Medical School, who discovered that telomeres are made up of simple, repeating blocks of DNA and that they are found in all organisms. Greider, Blackburn and Szostak also shared the 2006 Albert Lasker Award for Basic Medical Research for this work.

October 12: Hopkins President Ron Daniels undergoes surgery at Hopkins Hospital for removal of a small mass discovered in his abdomen, behind his pancreas. It turns out to be a rare, gastrointestinal stromal tumor in his intestine (duodenum). In a letter to students, faculty and staff, he writes 10 days later: "Not that I needed it, but I now have even more reason to be proud and thankful to be at Johns Hopkins." He recovers fully.

To emphasize that its suburban centers are not administrative or research facilities but provide health care, Johns Hopkins at Green Spring Station, White Marsh and Odenton are renamed "Health Care and Surgery Centers."

One resource that Daniels tapped almost immediately was the administrative and medical expertise of Hopkins Medicine—in one instance, intentionally; in the other, unexpectedly.

In August 2009, Daniels appointed Lloyd Minor, director of the Department of Otolaryngology–Head and Neck Surgery, as provost and senior vice president of academic affairs, the second-in-command position for the University.[371]

Minor, who had been on the School of Medicine faculty since 1993 and head of his department since 2003, was renowned for treating Meniere's disease, a syndrome involving dizziness, hearing loss and ringing or pressure in the ear. As head of his department, he had recruited and retained an outstanding and diverse faculty, expanded annual research funding by more than half, increased clinical activity by more than 30 percent, and strengthened teaching and student training.[372]

In becoming provost, Minor succeeded Steven Knapp, who had left Hopkins in 2007 to become president of The George Washington University in Washington, D.C. As provost, Minor oversees all the university's academic divisions, scattered across 11 campuses and locations in Maryland, the District of Columbia, and overseas. His responsibilities include accreditation, compliance with federal regulations, and research collaborations between the University's schools, all of whose deans report to him.[373]

When making the appointment, Daniels praised Minor's well-known "driving passion to make Johns Hopkins stronger in all its crucial dimensions: research, education and service."

What Daniels nor anyone else knew at the time was how soon he himself would have to use Hopkins' medical services.

Officially installed as the University's president on September 13, 2009, Daniels was on an operating table at Hopkins Hospital almost precisely a month later, on Oct. 12. Boyish-looking and lean, Daniels had radiated energy and health from the moment he'd stepped on the university's flagship Homewood campus.[374]

Yet just weeks after his installation, Daniels underwent surgery to remove a "small mass in my abdomen," as he explained in a university-wide email distributed immediately following his operation. "The growth was located behind my pancreas and surgeons got access to it by performing a Whipple procedure, during which part of my pancreas was removed."[375]

Fortunately for Daniels, the tumor was small, not on his pancreas, hadn't spread to any other organs, and was removed completely during the Whipple procedure—for which Hopkins, under former head of surgery John Cameron, had become the world's leading hospital.[376]

Within 10 days of the surgery, Daniels dispatched another university-wide email. In it, he conveyed his prognosis for a complete recovery and thanks to physicians Richard Schulick, John Flynn, John Ulatowski, Chris Saudek, Marcia Canto and their superb support teams. "Not that I needed it," he wrote, "but I now have even more reason to be proud and thankful to be at Johns Hopkins."[377]

Twenty-One Years on Top

As of this writing, Hopkins Hospital's tenure at the top of *U.S. News & World Report*'s ranking of 4,852 U.S. hospitals continues unabated. In July 2010, Hopkins earned the No. 1 spot for the 20th year in a row—an astonishing achievement in any such survey of professionals. In July 2011, the streak was extended to 21 consecutive years. It is an accomplishment of which everyone in Hopkins Medicine is proud, but which they also have always sought to put in perspective.[378]

In 1992, when Hopkins Hospital happily marked just its second consecutive year as No. 1, Edward Wallach, then-director of the Department of Gynecology and Obstetrics, observed that the accolade, while appreciated, was not something the Hospital consciously tried to achieve.

"Everybody here works very hard and strives for excellence," Wallach told *Dome*. "We do it for our patients, for ourselves, for the people we train and develop, to give them the highest of standards. We don't set out to be number one; it's just a wonderful and unexpected reward."[379]

In 1997, when the Hospital topped the rankings for the sixth year in a row, Charles Cummings, then head of the Department of Otolaryngology–Head and Neck Surgery, wrote a commentary about the *U.S. News & World Report* survey for *Change*, a biweekly Hopkins Medicine newsletter. He entitled it, "Don't Get Carried Away by the Rankings."[380]

Cummings wrote that "as we bask in the distinction of having been ranked #1 for six years in a row, I, as well as many other members of the Hopkins family, feel a certain uneasiness about whether we'll do it again."

A cross-section of Hopkins Hospital employees celebrate its designation as the nation's top hospital for two decades in a row.

Harry Koffenberger, a 24-year-veteran of the Baltimore City Police Department, became vice president of security for Johns Hopkins Medicine in 2006, succeeding Hopkins' first top security officer, former Secret Service official Joseph Coppola, who had held the post since its creation in 1994. Koffenberger also had joined Hopkins in 1994 as director of external security for Hopkins' East Baltimore campus. He became senior director of corporate security in 2002, overseeing parking, transportation and security not just at Hopkins Hospital but Hopkins Bayview, Howard County General Hospital and Suburban Hospital. As vice president of security, Koffenberger oversaw development of a sophisticated communications program, including an emergency text-message system, a robust security intranet site, and detailed "worst case scenario" planning. He also maintained a close relationship between Hopkins security and federal, state and local law enforcement agencies. These efforts were instrumental in the swift, coordinated and effective response to a September 16, 2010 shooting incident at Hopkins Hospital that left one physician wounded and a patient and her son dead. A 50-year-old man grew distraught while being updated by an orthopedic surgeon on his 84-year-old mother's condition. He pulled a small, semiautomatic handgun from his waistband and shot the physician in the abdomen. The man then ran into his mother's hospital room and slammed the door shut. An ensuing three-hour standoff, involving not only Hopkins Medicine security but a huge contingent of Baltimore City police, ended when it was determined that the man had shot both his mother and himself to death. The surgeon, David Cohen, was seriously injured but recovered following emergency surgery. Hopkins' security efforts have included not only close collaboration with city police but with neighborhood groups as well, reflecting a commitment to community involvement that had been a hallmark of Koffenberger's entire law enforcement career. In January 2012, Koffenberger will retire and be succeeded by John Bergbower, a 27-year veteran of the Baltimore City Police Department who has been a central figure in Hopkins' security system since 2003.

2009

October 20: the School of Medicine inaugurates a new curriculum, "Genes to Society," following six years of debate and planning. The incoming Class of 2013 is the first to be taught it – and the first to begin their studies in the new $54 million Anne and Mike Armstrong Medical Education Building. It is named for the chair of the boards of Johns Hopkins Medicine, The Johns Hopkins Health System Corporation, and The Johns Hopkins Hospital and his wife and is dedicated on October 24. Dean/CEO Ed Miller says the new curriculum and the new building "usher in a new era of medical education that raises the bar even higher and that will be transformative for this century as the original Hopkins model was for the last."

November 14: "A Woman's Journey," a Hopkins-sponsored, day-long conference on women's health issues, marks its 15th year. The full-day health conference includes 32 sessions presented by Hopkins experts on topics ranging from ways to maintain a sharp memory to why women ignore their risk for heart disease, to the latest cancer treatments. A dozen new sessions join the lineup this year, including "Top Ten Infections" with tips on how to prevent them.

"I'd like to suggest that this newfound nervousness over whether Hopkins will once more come out on top doesn't really make much sense…. The truth is, there are several very distinguished institutions that justifiably could be ranked #1…. A healthy outlook for us at Hopkins, in fact, would be to acknowledge that being in the top 10 is our just due and being #1 may truly be a roll of the dice."[381]

In 2010, when Hopkins extended its reign in the top spot to 20 consecutive years, it did so by placing first in five medical specialties—two more than in 2009—and among the top five in 10 others. In 2011, it again placed first in five medical specialties and in the top five in 10 others.

In 2010, repeating the outcome from 2009, Hopkins accumulated 30 points in 15 of the 16 specialties ranked to earn the No. 1 position, edging out Mayo Clinic's 28 points. In 2011, Hopkins also accumulated 30 points, surpassing Massachusetts General's 29 points. In 2011, Hopkins placed first neurology and neurosurgery, urology, psychiatry, rheumatology, and ear, nose and throat. It ranked second in gynecology and ophthalmology; third in nephrology, diabetes and endocrinology, gastroenterology, geriatrics, cardiology and heart surgery, and cancer; fourth in pulmonology; fifth in orthopedics; and 15th in rehabilitation medicine. Of the 4,852 hospitals for which data was accumulated and assessed, only 146 ranked in even one specialty, and of that number, only 17 made the top honor roll.[382]

"We know that rankings and ratings all have their weaknesses and strengths," Miller and Peterson observed in a "Dear Colleagues" letter issued after the 2010 *U.S. News* hospital rankings were released, "but as health care choices become increasingly important, independent evaluations continue to interest our patients, the public, referring physicians and insurers. The elements of what makes us rank high are firmly in place, and we will, as always, strive to make Hopkins Hospital the best choice year after year."[383]

As the annual ritual of waiting for the *U.S. News & World Report* hospital rankings continues, Hopkins Medicine also keeps in mind the caveats about them expressed by Charles Cummings in 1997. The School of Medicine can empathize with the Mayo Clinic, long the perennial No. 2 in the hospital survey until being displaced by Massachusetts General in 2011. In 1991, the second year of the *U.S. News* survey and the one in which Hopkins Hospital dethroned Mayo as top hospital, Hopkins' School of Medicine was ranked second only to Harvard—but first as a graduate medical school.[384] The School of Medicine retained that No. 2 ranking for the next 13 years.

In the 2004 *U.S. News* survey of 125 medical schools, however, Washington University in St. Louis edged Hopkins out of second place and into third by one point (96 to 95). In 2005, Hopkins regained the second spot and kept it for another five years before being bumped down to third again in 2010, this time by the University of Pennsylvania, as Harvard remained No. 1. The bottom line: *All* of these medical schools are superior, and often a year's difference in the rankings are due—as Charles Cummings observed about the hospital surveys— to "a roll of the dice."[385]

Legends Lost

As the first decade of the 21st century wound to a close, Hopkins Medicine lost a foursome of figures within 18 months whose stature was almost as legendary as that of the original Four Doctors: Victor McKusick, the father of genetic medicine; Lockard Conley, a hematologist and superb teacher whose influence on his field and future physicians was incalculable; pediatrician and geneticist Barton Childs, whose pioneering work formed the foundation for the School of Medicine's new "Genes to Society" curriculum; and pediatrician Henry Seidel, a beloved dean of students whose legacy will endure for decades.

Genetic Medicine's Father

The first of the legendary figures to depart was Victor McKusick, who died at his Baltimore home of cancer on July 22, 2008 at the age of 86. Ever intellectually active and curious, he spent part of his last day viewing a computer screening of the latest session of his landmark Short Course in Medical and Experimental Mammalian Genetics, then under way for the 49th consecutive year at Bar Harbor, Maine.[386]

McKusick's extraordinary 65-year tenure at Hopkins, from medical student to an ever-active, iconic figure, was chronicled in Chapter 2. He created a "legacy to medicine…so pervasive, even fundamental, that it will be difficult to pinpoint but impossible to avoid," said Hal Dietz III, the first Victor A. McKusick Professor of Genetics and Medicine. (The chair was created in 2004 with $2.07 million donated by more than 450 of McKusick's colleagues and friends, including many former protégés and several Nobel laureates.)[387] Even before undertaking his pioneering work in genetics, McKusick's "lifelong creative genius" led to early advances in cardiovascular medicine, observed cardiologist Mike Weisfeldt, who now is the William Osler Professor of Medicine and heads Hopkins' Department of Medicine, as McKusick once did.[388]

Lean and laconic, McKusick leavened his New England reserve with a sly sense of humor. Never without his small notepad and pocket camera, he constantly scribbled observations and snapped pictures of events, places and people. Perhaps his most memorable extracurricular exploit was a tour of the original Hopkins Hospital, now the Billings administration building, which he conducted for medical students and residents while giving a lecture on Hopkins history. For decades, he concluded the tour with a physically challenging (for others) climb up the narrow, winding stairs that lead to the top of the building's towering dome, with its spectacular view of Baltimore.[389]

McKusick's honors were abundant. Among the most notable were the 1997 Albert Lasker Award for Special Achievement in Medical Science, often called the "American Nobel"; the 2001 National Medal of Science, the country's top scientific accolade; and the 2008 Japan Prize in Medical Genomics and Genetics. Three months before he died, he went to Tokyo to receive that medal—and a 50 million yen ($470,000) award—at a formal presentation attended by the Japanese Emperor and Empress. He called the Japan Prize, considered that country's equivalent of the Nobel, an honor that also lauded "the contributions and support of Johns Hopkins, and of my colleagues and students over many decades."[390]

Victor McKusick's continuous 65-year tenure at Hopkins Hospital was the longest in its history. As the father of genetic medicine, his legacy is "so pervasive, even fundamental, that it will be difficult to pinpoint but impossible to avoid," observed his protégé Hal Dietz III, who became the first Victor A. McKusick Professor of Genetics and Medicine.

Pioneering hematologist C. Lockard Conley (1916–2010) was readily recognized by many of his colleagues and students as "the smartest doctor at Hopkins" for some 30 years. Modest and soft-spoken, he was internationally acclaimed as a researcher, teacher and clinician. He was the first Hopkins physician who was not the head of a department to be appointed to a full professorship.

2009

December 15: Hopkins transplant surgeons successfully complete their 100th kidney swap — a procedure popularized at Hopkins to enlarge the pool of kidneys available for donation and provide organs to patients who might have died waiting for them. One form of kidney swap relies on a so-called "domino donor" effect, made possible by altruistic donors willing to donate a kidney to any needy person and other willing donors who are not a match for their loved ones.

The National Science Foundation reports that The Johns Hopkins University spent $1.68 billion on medical, science and engineering research in fiscal year 2008 – putting it at the top of all research universities in the nation. JHU also ranked at the top on the foundation's separate list of federally funded research and development, spending $1.42 billion in research and development in FY08.

Hopkins Medicine gives away at least $67 million in charity care annually – a fact not formally reported before but which will become mandatory in 2010 for Hopkins to maintain its tax-exempt status.

Modest Giant

As famed for his soft-spoken manner and modesty as for his brilliance in the laboratory, the classroom and at his patients' bedsides, C. Lockard Conley persistently deflected praise. He insisted that his extraordinary career as an internationally acclaimed hematology researcher, clinician and teacher was "wholly unplanned" and the result of "unexpected circumstances."[391]

Yet for four decades, Conley—known as "Lock" to colleagues and friends—was considered by generations of Hopkins medical students, house staff and fellows to have been the finest physician they ever knew. He died in January 2010 at 94 of Parkinson's disease.

Former School of Medicine Dean Richard S. Ross liked to recall that as a young resident in 1947, he encountered a patient, an Italian tomato farmer from New Jersey, whose blood was baffling. When a sample was sent to the chemistry lab, the red cells settled out rapidly and the plasma clotted almost instantaneously. Ross took the blood to the one man he thought could untangle the mystery—Conley, then the newly appointed head of the nascent hematology division.

Conley's subsequent research showed that the patient's blood contained a potent anticoagulant—and when it later became clear the patient had lupus, Conley's findings defined the existence of lupus anticoagulants, antibodies linked to thromboses, miscarriages and other conditions. This was just the beginning of Conley's landmark inquiries into blood coagulation, blood platelets, hemorrhagic diseases, hemoglobins and sickle cell anemia. He also made crucial contributions to developing therapy for vitamin B-12 deficiency.

A Baltimore native and 1935 graduate of Hopkins' School of Arts and Sciences, Conley obtained his medical degree from Columbia and served in the U.S. Army Air Corps during World War II. He joined the Johns Hopkins faculty in 1946, and by 1956, he became the first Hopkins physician who was not head of a School of Medicine department to be made a full professor. He was named a University Distinguished Professor of Medicine in 1976.

Many medical students, residents and fellows considered Conley to be their most important teacher and mentor, attributing much of their subsequent achievements directly to his impact on them. Of the 70 fellows Conley supervised during 33 years as head of the Hematology Division, 10 were elected to the American Society for Clinical Investigation, and a dozen became heads of hematology divisions or chairmen of departments of medicine.

Among those whom Conley mentored was David Hellmann, now vice dean on the Johns Hopkins Bayview campus and head of the Department of Medicine there. Hellmann says he repeatedly saw Conley rapidly make a correct diagnosis in a patient whose problems had stumped physicians elsewhere. Once, a man "who had been seen in multiple hospitals by multiple doctors" came to Hellman, plagued by painful, blue fingers. The previous diagnosis: vasculitis, or inflammation of the blood vessels; the ineffective remedy: massive doses of morphine.

Hellmann consulted Conley, who examined the patient's peripheral blood smear and swiftly diagnosed a rare condition: polycythemia, not vasculitis. The effective remedy: a daily aspirin.

"I have a framed picture of Dr. Conley in my office to remind me every day of what an impact a great doctor and teacher can have on the lives of patients and students," Hellmann says. Those skills have been permanently enshrined via the American College of Physicians' C. Lockard Conley, M.D. Award for Excellence in Medical Resident Education.

In response to such praise, Conley would smile slightly and characteristically turn the encomiums aside, saying he didn't understand why so many superb physicians wanted to study with him.

For Hellmann, that was no mystery: "When I was a medical student, one of the senior residents I most admired told me, 'Dr. Conley is the smartest doctor at Hopkins.'"[392]

Pioneering Pediatric Geneticist

As an adopted child, world-renowned pediatric geneticist Barton Childs liked to joke that he had "nothing to worry about all these years … no concerns of the kind that geneticists are likely to arouse in people when they tell them they have a hereditary disease."[393]

Childs' anonymous genes, combined with the wonderfully supportive upbringing he received from the Chicago couple who adopted him, produced a man of exceptional, wide-ranging brilliance, foresight and energy. Known as a genuine Renaissance man of medicine, the bow-tie-and-red-sock-wearing Childs was equally at home in philosophy, the history of science, and education.

He was among the first to urge integration of genetics into all of medicine and to directly analyze the impact of genetic testing and counseling on patients and families. He published articles on topics ranging from cerebral palsy to human development and evolution; and his insatiable curiosity about human nature compelled him to study ethical and philosophical questions unrelated to medicine.

A 1942 graduate of the School of Medicine, Childs died in February 2010 following a brief illness at The Johns Hopkins Hospital—where he had spent virtually his entire seven-decade career. Active until just a few weeks before he died, Childs was 93. (Although, in another irony, Childs was born on Feb. 29, 1916—a leap year—so he actually celebrated only his 23rd "birthday" in 2008.)[394]

As the Department of Pediatrics' first director of genetics, Childs made critical contributions to the understanding of the genetic underpinnings of many diseases, including adrenal hyperplasia, Addison's disease and hypoparathyroidism. He formulated the now classic study definitively proving that one of the two X-chromosomes in human females is inactivated during

Barton Childs (1916–2010) was the Department of Pediatrics' first director of genetics and made landmark contributions to the understanding of the genetic origins of many diseases. His farsighted 1999 book, *Genetic Medicine: A Logic of Disease*, became the inspiration for the School of Medicine's new "Genes to Society" curriculum, implemented only a few months before Childs died.

2010
July 15: For the 20th consecutive year, The Johns Hopkins Hospital earns the top spot in *U.S. News & World Report*'s annual rankings of more than 4,800 American hospitals, placing first in five medical specialties and in the top five in 10 others. The first place rankings are in ear, nose and throat; neurology and neurosurgery; gynecology; rheumatology; and urology. The other top 10 rankings came in cancer; diabetes and endocrinology; gastroenterology; geriatrics, heart and heart surgery; kidney disorders; ophthalmology; orthopedics; psychiatry; and pulmonology. The hospital ranked 24th in rehabilitation. In 2011, Hopkins received top ranking for the 21st consecutive year.

■ Two roads on the Johns Hopkins Bayview Medical Center campus are named for famous doctors: Mason Lord Drive and Nathan Shock Drive. One of the main entrances to the campus is called Cassell Drive—but no one really knows for whom it is named. During a 1953 renovation of the original, 19th century main hospital building now named for Mason Lord, a construction worker found a pair of boots buried in a portion of one of its walls. Inside one of the boots was a note that read "buried in October 1868" and bearing the signature "Cassell." The first name was illegible, but research showed that a man named Cassell was one of the founders of the local bricklayer's union in 1866. The road leading from Eastern Avenue to the building was named Cassell Drive in his honor—even though his full identity was never uncovered.

early development, a fundamental biological mechanism. He encouraged many Hopkins colleagues to consider the human diseases they studied in the context of genetics, including prostate cancer, inflammatory bowel disease and dyslexia. Additionally, he collaborated with his wife, psychiatric epidemiologist Ann Pulver, also of Hopkins, in seminal studies of the genetic basis of schizophrenia and other neuropsychiatric diseases.

Childs' numerous accolades included the John Howland Award, the American Pediatric Society's highest honor; the William Allan Award, the highest honor of the American Society of Human Genetics; and the Research Career Award of the National Institutes of Health. From 1972 to 1975, Childs was chairman of the National Academy of Sciences' national research committee on inborn errors of metabolism, which laid the groundwork for the current policies on genetic-disease screening in newborns.

Beloved Dean

As Assistant Dean for Student Affairs at the School of Medicine from 1968 to 1970 and Associate Dean from 1977 to 1990, pediatrician Henry Seidel guided future physicians not only through the labyrinth of their medical training, but also through personal crises. Former students recalled that in counseling them, Seidel would sometimes use the same techniques he normally reserved for his pediatric patients.[395]

As "the pediatrician for all of us medical students, he cared for all of us and cared about all of us," said Robert Chessin, a Connecticut-based pediatrician and 1973 School of Medicine graduate. "I always hoped I could live up to the type of physician and human being he was."[396]

Seidel, who died in March 2010 of lymphoma at the age of 87, had a profound, enduring influence not just on medical students but virtually everyone with whom he came in contact at Hopkins, including fellow faculty. John Freeman, now professor emeritus of pediatrics and neurology and former chief of pediatric neurology and pediatric epilepsy, said Seidel "taught me to listen, to care more deeply and to be more compassionate. When in need, he was the one I could turn to."

"As if it were no more complicated than breathing, Henry Seidel allowed everyone he came in contact with to feel that he cared for them as an individual," recalled Dave Nichols. "It is for that reason that generations of pediatricians and students learned the ideal embodiment of the physician, because they had seen it in Henry Seidel."

Seidel endeared himself to class after class of students during the angst-ridden process of national "Match Day," on which all graduating medical students in the United States learn where they will receive residency training. Historically, matches had been posted for all to see, leaving a few students at risk of being embarrassed at not receiving a match. Instead, Seidel ordered individual envelopes for each student, affording more privacy. In the days leading up to Match Day, Seidel consulted hospitals around the country to find places for students who had not been matched.

Born in Passaic, N.J., Seidel said his interest in medicine was ignited by the family doctor he saw as a child. "After poking me a bit and looking into my ears and throat, he would pull a leather case from his bag," Seidel wrote in a 2002 essay

Pediatrician Henry Seidel (1923–2010) often employed the same techniques he used for dealing with pediatric patients when gaining insights on and counseling the thousands of medical students he helped during his terms as Assistant Dean for Student Affairs from 1968 to 1970 and Associate Dean from 1977 to 1990—earning their enduring gratitude for his wisdom and compassion.

published in the *Journal of the American Medical Association*. "It was in my child's mind's eye huge. When opened, it presented a double row of small bottles filled with variegated pills." Only years later did Seidel realize that his doctor had been the renowned poet William Carlos Williams—and that what he had been offering his young patients were sugar pills.

Seidel first arrived at Hopkins in 1938 as an undergraduate, going on to earn his bachelor's degree in 1943 and his medical degree in 1946. He did postdoctoral pediatric training at the Harriet Lane Home for Invalid Children, precursor to the present-day Johns Hopkins Children's Center, and was one of the founding editors of the *Harriet Lane Handbook*, the now-multi-edition pediatrician's reference book still in wide use today.

Years later, he was the co-author of the now-classic textbook, *Principles of Pediatrics: Health Care of the Young*, and served as lead editor of the acclaimed *Mosby's Guide to Physical Examination* for more than two decades. The seventh edition was published shortly before his death. He also was a co-author of a history of Hopkins pediatrics entitled *The Harriet Lane Home: A Model and a Gem*.

He joined the School of Medicine faculty in 1950. He spent his entire professional career at Hopkins but also split his time between academic interests and private practice. He was an attending physician for Sinai Hospital of Baltimore between 1953 and 1968 and also a part-time instructor in pediatrics at the University of Maryland. Acclaimed as a pediatrician's pediatrician, he continued to see patients and teach students after he retired formally in 1990 at the age of 68.

Like all good clinicians, Seidel knew that medicine is not an exact science. He considered empathy to be at least as important a skill for a doctor as clinical expertise and scientific knowledge.

As dean of students, Seidel pushed to develop a socially diverse student body, arguing for and encouraging careful and controlled distribution of scholarship funds. He also urged fellow physicians to educate lawyers, judges and legislators about children's rights.

"Henry was the best that Hopkins could be," said Ed Miller after Seidel's death. "All you had to do was to look into his eyes, listen to his voice and know that you were in the presence of a wonderful person who just happened to be a doctor."[397]

2010

November 1: In a move to address growing interest in more efficient, integrated regional health care services for patients, officials of Washington, D.C.-based Sibley Memorial Hospital and The Johns Hopkins Health System Corporation sign documents integrating Sibley Hospital into the Johns Hopkins Health System (JHHS). The transaction does not involve financial exchanges. In becoming a member of Johns Hopkins Medicine, Sibley retains its name, leadership, current medical staff, and commitment to its community.

2011

A major restructuring of Johns Hopkins Medicine's strategic planning processes is begun. A broadly based Johns Hopkins Medicine Operating Committee is established to create an inclusive, transparent process for addressing and deciding long-range strategic planning decisions that arise within Hopkins Medicine concerning such key areas as expansion of international business, addressing health care reform and cost-effectiveness, greater system-wide clinical integration, and enhanced patient safety, quality outcomes and performance measures.

Still Leading the Way into the Future

Health Care Reform

Hopkins Medicine had a major stake in the lengthy health care reform debate that embroiled Congress and the nation between 2009 and 2010. Should the new law, being challenged in court at this writing, fully go into effect in 2014, millions of Americans will be gaining health coverage, private insurers will be offering policies to individuals with preexisting conditions, and care providers are slated to see reduced reimbursements from Medicare and Medicaid. Hopkins Medicine leaders believe the changes will mean fewer dollars for the institution as it delivers quality care to many more people.[398]

"We have to adjust to that reality," Ron Peterson told the in-house publication, *Change.* "We will have to do a much better job managing costs and figuring out ways to provide more thoughtful and efficient delivery of health care services."[399]

Despite the immense challenges imposed by health care reform, Hopkins Medicine endorsed the effort as it moved through Congress—and had an impact on what ultimately became the law. Ed Miller joined with his predecessor as Hopkins School of Medicine dean, Michael Johns, now at Emory, to form a coalition of leaders from top academic medical centers. Their aim was to talk to as many key congressional leaders as possible to urge adoption of proposals that would make the legislation more effective—and realistic.[400]

To accomplish this, Miller and Johns recruited to their group the heads of nine more major academic medical centers: Cornell-affiliated New York Presbyterian Hospital; New York's Mount Sinai Medical Center and School of Medicine; Boston's Partners HealthCare, founded by Harvard-affiliated Brigham and Women's Hospital and Massachusetts General; the University of Chicago Medical Center; the University of Michigan Health System; the University of Florida's Shands Hospital in Gainesville; Washington University's Barnes-Jewish Hospital in St. Louis; the University of California Medical Center in San Francisco; and the University of Washington Medical Center in Seattle.

Joining Miller and Johns, these leaders, along with the American Association of Medical Colleges president Darrell Kirch, spent considerable time on Capitol Hill, working toward creation of a health care reform bill that would recognize academic medical centers as models for how to meet the needs of those with little or no insurance without destroying state and federal budgets. [401]

To foster that concept, Miller urged—and Congress enacted—a provision in the new law that would create demonstration projects, known as Healthcare Innovation Zones. In these regions, academic medical centers would assume responsibility for the medical needs of most individuals in a broad geographic areas.[402]

Miller also pointed out that Hopkins does something that few academic medical centers do: It runs managed care plans efficiently and cost-effectively. These include its employee health plan, which serves 51,000 members; the U.S. Family Health Plan (USFHP), which provides services for more than 32,000 military dependents and retirees; and perhaps most important, given the mammoth increase in Medicaid recipients likely under the new law, Hopkins' Priority Partners, the managed-care program that serves 175,000 Medicaid recipients in Maryland. All emphasize lower-cost

care solutions, preventive health programs and case management for individuals with chronic conditions.[403]

In an article written for *The Wall Street Journal* and in a speech before the National Press Club, Miller cited Priority Partners in particular as a model for addressing the health care reform law's three central themes: coverage, quality and cost.

Miller noted Priority Partners' success at achieving above-national benchmarks on all clinical quality measures for its dialysis patients, reducing the monthly costs for patients with substance abuse and highly complex medical needs, and improving prenatal care compliance—all while instituting rigorous cost-management strategies and attaining a 70 percent satisfaction rate among patients. Priority Partners, Miller suggested, offers "a system of care that can be duplicated around the nation."[404]

To prepare Hopkins for the changes being mandated by Washington, Miller and Peterson have appointed a 17-member, in-house committee. It is headed by Patty Brown, president of Johns Hopkins HealthCare, which runs USFHP and Priority Partners, and Bill Baumgartner, vice dean for clinical affairs and president of the Clinical Practice Association. It will have to determine how Hopkins' physicians can best respond to the anticipated crush of new patients.[405]

"We'll meet the demands placed on us because serving poor and disadvantaged populations is part of our century-old mission," Miller wrote in *The Wall Street Journal* prior to final enactment of the health care reform law. "But without an understanding by policy makers of what a large Medicaid expansion actually means, and without delivery-system reform and adequate risk-adjusted reimbursement, the current health-care legislation will have catastrophic effects on those of us who provide society's health-care safety-net. In time, those effects will be felt by all of us."[406]

Restructuring Required

By the end of 2010, the extraordinary expansion of Johns Hopkins Medicine in the preceding three years compelled its leadership to review its organization and structure. What had been a compact institution with a highly successful, relatively straightforward operation overseeing three hospitals, small home care and managed care entities, and one suburban outpatient center had grown to include three additional hospitals, a multibillion-dollar managed care business, triple the number of ambulatory care locations, and the largest primary care practice in the state. The institution needed a significant administrative restructuring. "In short, we've outgrown ourselves," Ed Miller explained in a column in *Dome*.[407]

Adding to the challenges facing Hopkins Medicine were uncertainties attendant to federal health care reform, potentially reduced Medicare, Medicaid and NIH funding, and possible changes in the hospital reimbursement rates by the Maryland Health Services Cost Review Commission.

Hopkins Medicine's past and future are linked in this view of the 1889 Billings Building dome as seen through the rainbow-colored, paneled windows in the new Charlotte R. Bloomberg Children's Center. Each window pane is playfully marked by countless white waves and brush strokes in the glass, a technique known as ceramic frit.

A key architect of the significant, ongoing administrative restructuring of Johns Hopkins Medicine has been Joanne Pollak, senior vice president and general counsel for Johns Hopkins Medicine. Pollak has been named chief of staff for the Office of Johns Hopkins Medicine and Chair of the new Johns Hopkins Medicine Operating Committee, the group of vice deans, clinical chiefs, enterprise-wide division heads, community hospital leaders, faculty and community physician representatives. With the JHM 3.0 process, the Office strives to ensure that its decision-making for future expansion initiatives and long-range strategic planning is more transparent and inclusive. Pollak, the head attorney for Hopkins Medicine since 1994, was involved in the formation of Johns Hopkins Medicine in 1996 and long has worked closely with Hopkins leadership.

"You are going to be a role model whether you want to or not. Young people are going to look up to you because of your educational background, your degree from Hopkins and your accomplishments, so be a good role model. Role modeling is particularly important in medicine and science. If we have really done our job, 20 years from now there will be graduates, from universities all over the country, who will be imbued with the Hopkins spirit and culture, because you have been their role model."

—John Cameron, (above, center) 1962 graduate of the School of Medicine, Professor and former Director of the Department of Surgery (1984–2003), who has operated on more patients with pancreatic cancer and performed more Whipple surgeries than any other surgeon in the world, addressing the Class of 2004 commencement.

Miller initiated the restructuring, called JHM 3.0, using a consultant with experience in evaluating the operating processes of health care organizations. Early in 2011, Miller and Peterson began sending out a series of institution-wide emails to all employees, explaining how a restructuring would be undertaken to ensure that Hopkins Medicine was positioned positively for its future advancement in four key areas: international business and revenue; maintaining cost-effectiveness and long-range planning to address health care reform's likely changes; greater system-wide clinical integration, specifically with Hopkins Medicine's community based hospitals and relations with community physicians; and enhancing patient safety, quality outcomes and performance measures.[408]

To enhance strategic planning, streamline decision-making and allow the institution to respond more quickly to pressures, the architects of the restructuring created a Johns Hopkins Medicine Operating Committee. The committee would strive to make its decision processes regarding future expansion initiatives more transparent and inclusive, with its membership including the vice deans, clinical chiefs, enterprise-wide division heads, community hospital leaders, faculty and community physician representatives.

Johns Hopkins Medicine "remains the most well-known health care institution," Miller wrote, "[and] we must protect and enhance the value of our brand.[409]

"You could say that the new structure looks pretty corporate, and quite honestly, it is. But my response is, we can't afford not to be corporate when we're a $6 billion business. It's not done on the back of a matchbook any longer."[410]

The Recruitment, Retention and Replenishment of Faculty

Throughout its history, Hopkins has been a key breeding ground and major exporter of medical talent. Developing and then dispatching elsewhere its superior clinicians and researchers always was among Hopkins' key goals. William Welch once was caricatured by Hopkins' founding medical artist, Max Broedel, as the progenitor of a large brood of "Welsh Rabbits" whom he then let loose on the medical world to disseminate and spread the Hopkins philosophy. Welch himself observed: "We have become so accustomed at The Johns Hopkins University to academic suitors for…[our faculty] that our position has been compared to that of the benignant father of a large family of girls…who replied to the young man who graciously requested the privilege of marrying one of his daughters: 'Take her, young man, take her. God bless you. Do you know who wants another?'"[411]

After the Flexner Report of 1910 proclaimed Hopkins the gold standard for all U.S. hospitals and medical schools, the recruitment of Hopkins faculty accelerated. Although what Ed Miller calls the "brain drain" at Hopkins has waxed and waned occasionally since then, it really has never let up. In recent years, it has reached another peak. A concerted effort is under way to retain—and recruit—the bright stars who will ensure that Hopkins remains Hopkins.[412]

Usually those who leave Hopkins do not do so for more money, Miller notes—even though, compared with its peer academic medical institutions in the Association of American Medical Colleges (AAMC), most Hopkins' salaries fall well below the 50th percentile. Faculty come to Hopkins and stay here "because you want to make a difference," says urology legend Patrick Walsh. "To me, it's the reason this place attracts and binds."[413]

While a higher salary frequently isn't a deciding factor for faculty who leave, Miller believes that boosting faculty salaries to the AAMC's 50th percentile should be a major priority. Hopkins typically puts its financial resources into "this new program, that new idea, this building, that whatever, and oh, yeah, by the way, the salary," Miller says. "Let's put the salary at the top of the list as a priority," he suggests, particularly for faculty who concentrate on clinical care and teaching.[414] He wants to raise money to create an endowment to "build human capital" and support teaching and scholarship, as well as intensify the effort to increase salaries for Hopkins' incredibly hard-working clinicians.

In 2009, "we took the proceeds that we made out of our profits from the Hospital and we directed it to faculty salaries," Miller says. "First time ever." A bonus, based on a carefully compiled measurement of the amount of time a physician spends in the clinic, was given to increase the salaries of faculty whose income was below the AAMC's 50th percentile. Miller is intent on continuing this effort. "This is a major issue and it can't continue," he says.[415]

Hopkins also has not been shy about turning the tables and recruiting some impressive faculty from other institutions—as it did in 2010 by hiring away University of Pittsburgh star plastic surgeon and hand-transplant pioneer W. P. Andrew Lee to head a new Department of Plastic Surgery at Hopkins. What made that coup even sweeter, perhaps, was the fact that Lee is a 1983 graduate of the School of Medicine and completed his internship and residency here, so his return is a kind of homecoming.[416]

The New Buildings' Bright Clinical Future

With the construction of Hopkins Hospital's new clinical towers completed in late 2011 and the transfer of services and patients into them in early 2012, Hopkins Medicine will have "among the most spectacular facilities anywhere in the world," Ron Peterson says.[417]

"A lot of thought has gone into making what we're building as flexible as we possibly can make it for the future—not knowing where medicine may be five, 10 years from now, let alone 20 years from now," Peterson says. A great deal of what the planners call "interventional space" was created to allow for installation of whatever emerging medical technology may arise.

"None of us has a crystal ball, but the advances in medicine are occurring so rapidly that the practical applications that we deal with today may be totally modified in the next five to 10 years, and that will have a bearing on how we care for patients," Peterson explains.[418]

As far as pediatrics director George Dover is concerned, the foresight employed by the buildings' designers guarantees that The Charlotte R. Bloomberg Children's Center alone will be "the outstanding children's hospital in America, bar none."

"Because it will be so closely linked to both the adult tower and to our research facilities it will be more comprehensive and more innovative than any free-standing children's hospital is now, or ever could be."[419]

2011
April: All Children's Hospital, a 259-bed freestanding children's hospital in St. Petersburg, Fla., and one of only 45 stand-alone children's hospitals in the country, becomes the first Hopkins Medicine member hospital outside of the Baltimore-Washington area. A financially secure institution that began as a center for crippled children in 1926, All Children's has a new $400 million, 1 million-square-foot facility, opened in 2010, trains pediatric residents and fellows, serves eight counties in west Florida, and draws patients from all over the nation, as well as 36 countries. All Children's keeps its name, its medical staff and physician affiliations, including with University of South Florida doctors who practice there, and its leadership. Hopkins' Jonathan Ellen, director of the Hopkins Center for Child and Community Research as well as head of pediatrics at Johns Hopkins Bayview, is named head of a new Johns Hopkins pediatrics department at All Children's.

Neurodegenerative disease experts Ted Dawson (left), head of the Institute for Cell Engineering, and his wife, Valina (right), are among those whose vitally important cellular and molecular research could be hampered by political restrictions and funding cutbacks.

2011

May 4: Hopkins President Ronald J. Daniels announces that Johns Hopkins Medicine Dean/CEO Edward Miller will retire on June 30, 2012 following an "extraordinary" 16-year tenure. Appointed Dean/CEO and university vice president for medicine in 1997 after serving for a year on an interim basis, Miller guided Hopkins Medicine through an exponential growth in its mission, launching some $2 billion in building projects to support advanced medical care and research, expanding its family of hospitals and international reach, adopting a revolutionary new "Genes to Society" medical curriculum, and increasing its diversity. Miller oversaw the undertaking of perhaps the largest hospital building and renovation project in the nation's history, the $1.1 billion construction of the entirely new Johns Hopkins Hospital patient care buildings, the Sheikh Zayed Cardiovascular and Critical Care Tower and The Charlotte R. Bloomberg Children's Center at Johns Hopkins, as well as some $688 million in other major buildings; appointed 31 of the 33 department directors in the School of Medicine, including the heads of three new departments; added four new hospitals to the Johns Hopkins Medicine family; and forged partnerships with medical entities throughout the world. During his tenure between 1998 and 2010, more than a dozen multidisciplinary research and clinical institutes and centers were founded, heralding a significant trend in medical advances; Hopkins Medicine's two clinical practice groups were consolidated, creating one of the largest academic group practices in the nation; widely influential patient safety innovations were fostered and adopted; diversity initiatives were undertaken, especially concerted drives to increase the percentage of women who are full professors and the number of underrepresented minority medical students.

The Promise—and Funding Challenges—of Research

Both Stephen Desiderio, head of the Institute of Basic Biomedical Sciences, and Nobel laureate Peter Agre cite the curiosity-driven, Nobel Prize-winning research of telomerase discover Carol Greider as emblematic of the promises of and challenges to the future of basic science research.

Agre says Greider is a role model for the next generation of researchers. "Young scientists have to realize Carol was in graduate school. On her first rotation she did an experiment that changed everything. So science is for young people."[420]

Desiderio, noting that the National Institutes of Health decides how its grants will be used—and sometimes shies away from the kind of basic research Greider pursues—says Hopkins has to find "pilot funding" from other sources "in order to attract young faculty who are going to be the Carol Greiders of the future and to support their laboratories or to support risky projects or to support scientists when the NIH doesn't deem their work fundable according to their tenets."[421]

Although Hopkins scientists have spent decades atop the NIH's list of research fund recipients (one of whom actually is Carol Greider), the level of support has been flat or unsteady—going up, remaining the same and even going down, Ed Miller observed in 2009. Such stagnation or unpredictability in funding hampers the progress of research that could be vital to the nation's future health.[422]

Miller said he has tried to persuade NIH officials that what Hopkins and other research institutions need is assured, "sustained funding—whether it's one percent above inflation…true inflation, medical inflation."

"It takes years for us to build the facilities; it takes years to recruit the faculty; and they have to know there's going to be funding at the end of the tunnel," Miller explained. He is worried that he and other medical leaders have not made a strong enough case for such sustained, long-range funding—and fears that Congress no longer has the strong advocates for NIH support that it once did.[423]

While researchers foresee dazzling advances in medical care as a result of their inquiries, the complexities of deciphering what's wrong in the human body and how to cure its afflictions are complicated even more by the body politic, notes Ted Dawson, director of the Institute for Cell Engineering (ICE). He fears the continuing battle over stem cell research could hobble the efforts of ICE researchers and others elsewhere to help those suffering from cruel diseases.

"The real obstacles to stem cell therapies aren't scientific, they're political and financial. We may lose a whole generation of scientists to other countries if political restrictions and funding constraints don't ease soon. People like me will stay here and continue to make contributions, but young researchers already are looking overseas.

"It would be a shame to lose them—a shame for Hopkins and a shame for this country. Think of all the people with diabetes and Parkinson's and heart disease who could be helped by breakthroughs in stem cell research."[424]

Despite such frustrations, however, years of discovery at the molecular and cell level during the late 20th century are enabling researchers now to plot steps to therapies in a way that didn't exist a decade ago. In this second century of molecular and cell research, "if" is beginning to give way in discussions to "when." Discoveries continue being made. The hope for their clinical use is real.

Enhancing the Resources for Medical Education

Hopkins remains a premier place for the education of future physicians. As described earlier, with adoption of the "Genes to Society" curriculum in 2009, the School of Medicine refashioned its approach to educating tomorrow's doctors. With a new building that has every technological bell and whistle to implement and enliven this innovative curriculum, Hopkins is committed to maintaining its distinction as an exemplar of medical education.[425]

Ed Miller has been a staunch advocate for enhancing the resources for—and recognizing the importance of—the faculty's teaching role. Resources previously committed elsewhere are now being directed to improving the financial rewards for teaching. Faculty who devote more time to teaching are advancing through the academic ranks as they had not in the past.[426]

Fundraising campaigns have been focused on bricks and mortar in recent years, Miller notes, but "now we're going to build human capital, and part of that has got to be an endowment for teaching and scholarship. There's no question of where we need to go."[427]

Research Grant Money

Under Michael Amey, a 35-year veteran in research administration for the School of Medicine, Hopkins' record as the No. 1 recipient of federal research grants has been maintained for more than two decades.[428]

Amey, a 1969 graduate of Hopkins' School of Arts and Sciences, joined the Department of Medicine as its financial manager for university affairs in 1975. He became the School of Medicine's Director of Research Administration in 1980, Assistant Dean for Research Administration in 1986, and Associate Dean for Research Administration in 2003.[429]

Amey's office reviews and negotiates grant award contracts and works closely with academic research teams in every discipline to ensure their compliance with the policies established by both the university and the grant-giver for acceptance and use of the funds. During his tenure, the School of Medicine's annual research budget has grown from $90 million to more than $400 million.[430]

Despite Hopkins Medicine's astounding, quarter-century reign as the top recipient of NIH research funding, the trend of such grants has drifted down a little in recent years. In fiscal year 2005, the total grant amount had been $449 million; in fiscal year 2006, it had slumped to $414 million; by 2009, it had picked up again to $435 million.[431]

Alternative sources of funding must be pursued. The creation of multidisciplinary research institutes and centers begun a decade earlier is having a profound effect on Hopkins' ability to undertake scientific research on a large scale. By pooling their resources in these institutes, researchers across a broad spectrum of departments have been able to purchase advanced equipment no single department could have afforded. By having such unparalleled technology available, these equipment investments, totaling some $11 million, has led to a stunning $135 million in research awards. In addition, the enhanced proximity of researchers in different fields—with their offices now abutting each other in new laboratory buildings—is sparking new collaborations and accomplishments.[432]

Moreover, the exciting collaborative atmosphere in the institutes and centers, coupled with the technology at their disposal, are vitally important to maintaining Hopkins' ability to attract top scientific recruits.[433]

The Johns Hopkins Technology Transfer Office

Grants and philanthropy are not the only sources of the funds essential to foster medical research.

Within months of becoming Dean/CEO of Johns Hopkins Medicine in 1997, Ed Miller wrote of the need to use the "entrepreneurial road to save research."[434]

In his first column for the School of Medicine's alumni magazine, Miller said that the leader of an academic medical center needed to have well-honed entrepreneurial skills. "The fact is, unless we can devise new methods for bringing in revenue, Hopkins' best hope for the future—its research enterprise—is at risk," he wrote.[435]

Among the changes Miller envisioned was development of a far more effective, aggressive licensing effort to turn Hopkins' research work and discoveries into a steady revenue source. Doing so took some time. Nearly a decade later, in 2006, Miller wrote that Hopkins still was "way behind its peers" in transforming its medical insights into marketable products.[436]

To address the problem, Miller undertook a complete revamping of the Johns Hopkins Technology Transfer office (JHTT), the licensing arm for technologies developed by Hopkins faculty and staff in such fields as molecular genetics, proteomics, chemistry, biomedical engineering and material sciences. In late 2006, he asked Wesley Blakeslee, an attorney, one-time corporate manager, and former computer software designer for the National Aeronautic and Space Administration (NASA), to serve as head of JHTT in the School of Medicine. (Blakeslee just had been named acting director of technology transfer for the university, and Miller wanted to be sure the School of Medicine was an important part of his portfolio.)[437]

Despite the bleak economic times, in fiscal year 2008, JHTT oversaw the creation of 12 startup companies based in whole or in part on technology developed at the School of Medicine—triple the number of startups launched by his predecessors. Together, those firms received $76 million in private funding to underwrite their research—and Hopkins' part in it.[438]

In addition to the jump in startups, JHTT also had strong growth between fiscal years 2006 and 2008, with inventions up 25 percent; licenses and options up 60 percent, and licensing revenues up 44 percent. Also rising steadily have been the office's material transfer agreements, which are contracts detailing how Hopkins exchanges its research materials, such as cell lines, cultures, bacteria and proteins, with private companies, other academic entities or the government.[439]

"Only about 10 percent of the faculty probably is truly interested in this, but those 10 percent—if we get one, big Gatorade hit, it would be important," Miller said. Gatorade was invented in 1965 by researchers at the University of Florida and named for the school's football team, the Florida Gators. Players on the team were struggling with dehydration problems in the Florida heat and humidity and Gatorade was formulated to help them. It now commands 80 to 90 percent of the North American sports drink market and reportedly has earned more than $80 million in royalties for the university since 1973.[440]

JHTT is maintaining its aggressive outreach to potential business prospects. In mid-2010, it created what was believed to be the first free app (mobile application), providing instant access to the office for investors and entrepreneurs using their smartphones or similar devices.[441]

2011

September: The Perdana University Graduate School of Medicine opens in Malaysia. That nation's first four-year graduate medical program, it was created in collaboration with The Johns Hopkins University's School of Medicine, uses Hopkins' new Genes to Society curriculum, and follows the American system of medical education instead of the British system traditionally used in Malaysia. Twenty-five trailblazing medical students, three-quarters of whom are women, begin their studies in the school's temporary headquarters near Kuala Lumpur. Most of their teachers are Hopkins faculty members and the school's dean and CEO is Charlie Wiener, former director of the Osler residency service at Hopkins. In only nine months, Wiener succeeded in overseeing the recruitment of the students, the faculty and the development of a high-tech teaching center for the 2011–2012 academic year. Johns Hopkins Medicine also will help design the new Perdana University Hospital, a 600-bed facility, and a research institute, scheduled to be completed in 2014.

Future Leadership

When the dust had settled from the turmoil that led to the creation of Johns Hopkins Medicine in 1997, and the institution proceeded not only to continue excelling in patient care, research and education but to achieve even greater heights, a sort of chicken-or-egg question arose: Was it the unified leadership structure of Johns Hopkins Medicine that made it tick along so well, or was it the remarkable chemistry between Ed Miller and Ron Peterson?

"I think the real answer is it's some of each," Peterson says. "I think the structure is the right one for these institutions, but it certainly helps when the chemistry between two human beings at the top of an organization works—and it has worked well."[442]

Both Miller and Peterson say that they complement each other. "We have built on each other's strengths—and weaknesses," Miller says. "Ron is very detail-oriented. He knows the numbers to their decimal points, and he'll give the history of [any aspect of Hopkins]...almost like a recording." Peterson describes Miller as "more of a broad-brush, broad-stroke kind of guy"—or as Miller characterizes himself, "I'm much more up in the clouds. So that works very well for us."[443]

"I have a lot of medical knowledge that he doesn't have. He has a lot of administrative knowledge that I don't have. So we can actually put those together and kind of push clinical programs and research programs together," Miller says.[444]

Peterson "loves this place and he works his heart and soul out for it," Miller says. In turn, Peterson praises Miller as "a wonderful boss and colleague" who is "a natural at meeting people; he has great interpersonal skills." Yet Peterson's own interpersonal skills are formidable. "He has intense loyalty to people that he's worked with over the years," Miller notes. Not well known is that Peterson makes a concerted effort to visit any colleague who becomes an inpatient at Hopkins Hospital.[445]

What Miller and Peterson also have in common is their tendency to be "operationally oriented," Peterson says. Observes Miller: "I think we both understand the importance of Johns Hopkins Medicine and the responsibility we've been given to not only do the day-to-day management work but make sure that we're looking ahead enough at positioning Johns Hopkins Medicine for the future."[446]

Miller has announced his intention to retire effective June 30, 2012 and his replacement, Dr. Paul B. Rothman, currently dean of the Carver College of Medicine at the University of Iowa and leader of that university's clinical practice plan, has been named. Ron Peterson has said he hopes to remain president of Hopkins Hospital and Health System and executive vice president of Johns Hopkins Medicine for awhile longer after that. How will Johns Hopkins Medicine work without the Miller-Peterson chemistry?

"I think Ron and I have kind of come to the conclusion that it was successful at the beginning because of the personalities," Miller says, "and I think that we have cemented so many things that are so tightly together it would be hard to pull this thing asunder now."[447]

If the two top members of a future leadership team approach questions "just as a School of Medicine issue or just as a Hospital or a Health System issue, you're dead in the water. It just doesn't work. It has to be a balance," Miller says.[448]

The chemistry between Ronald Peterson and Edward Miller did much to make Hopkins Medicine work once it was formed in 1997, but they both believe the structure they've put in place is firmly enough established to flourish under new leadership—provided their successors maintain the essential balanced focus between the School of Medicine, the Hospital and the Health System.

2011
October 19: The installation process for Epic, a new system-wide, electronic medical record system, officially is launched in Johns Hopkins Medicine's ambulatory care centers by Dean/CEO Edward Miller and Ronald R. Peterson, president of The Johns Hopkins Hospital and Health System. The ambulatory care centers of The Johns Hopkins Hospital, Johns Hopkins Bayview Medical Center, Howard County General Hospital, and Johns Hopkins Community Physicians have had different electronic health record systems. With the addition of Suburban Hospital, Sibley Memorial Hospital and All Children's Hospital to the Johns Hopkins Medicine family, and a much larger primary care network, more coordination and integration of care was needed, and the Epic system can supply providers with the information necessary to implement that. It also will facilitate the use of electronic health records to measure quality and quantity of care, as now required by federal rules, as well as assist in the mandatory, enterprise-wide conversion by October 2013 to new billing codes. The cost of installing Epic in Hopkins Medicine's ambulatory centers is estimated to be $100 million; it will be phased in for inpatient services within a few years. It will help fulfill Hopkins' goals of better coordinating patient care and becoming more patient- and family-centered in care delivery.

December 19: Paul B. Rothman, dean of the Carver College of Medicine at the University of Iowa, is named to succeed Edward Miller as dean/CEO of Johns Hopkins Medicine.

It is not uncommon for Ed Miller or Ron Peterson to be asked how it is that Hopkins Hospital has maintained its No. 1 ranking for two decades or how Hopkins Medicine as a whole has remained the top recipient of federal research money for years. "I think that it largely has to do with the fact that we've been blessed with the opportunity, generation after generation, to recruit the best of the best in terms of those folks who are leading the various disciplines that we have here in medicine," Peterson says.[449]

"It's all about human capital," Peterson says. Despite the emphasis in recent years on rebuilding the physical structure of Hopkins, "the truth is that the strength resides in the human beings who make it work."

"It's a very special place in that regard. This is an institution that enjoys a tremendous amount of collegiality, and that's not a trivial statement because it's my observation that because people work so well across departmental lines, that facilitates the kind of environment that fosters innovation and discovery. That may be at its core the key to the success that is Johns Hopkins Medicine."[450]

"The Spirit of the Place"

Even as the excitement over the opening of the new clinical towers mounted in 2011 with the buildings' impending completion; and as enormous planning efforts were being undertaken to ensure the swift and smooth transition of departments and patients into the glistening new structures in April 2012, the Hopkins "environment" of which Ron Peterson spoke remains one that Hopkins' "Big Four" would recognize immediately.

As it has since the founding of The Johns Hopkins Hospital in 1889 and the Johns Hopkins School of Medicine in 1893, Hopkins has sought, nurtured and been a magnet for men and women who have been drawn to what Sir William Osler, nearly a century ago, called "the spirit of the place."[451]

On the Hospital's 25th anniversary in 1914, Osler wrote that Hopkins' unique atmosphere bound together everyone connected with the Hospital and medical school. It was a "sweet influence," Osler wrote, "whence we knew not, but the teacher and taught alike felt the presence and subtle domination. Comradeship, sympathy with one another, devotion to work, were its fruits...."[452]

William Welch echoed Osler's sentiment. At the same 25th anniversary celebration, he looked to Hopkins' future and said, "All who are here today from those early years feel that there was an environment, an atmosphere and ideals which will always be cherished and will continue to be an abiding influence."[453]

It is clear that Osler's and Welch's evocation of the atmosphere and spirit that they and others fostered in The Johns Hopkins Hospital and the School of Medicine remains valid—and abiding—as Johns Hopkins Medicine continues to lead the way in every aspect of patient care, medical education and scientific research.

Appendix A

1999–2011: Leadership Changes Across the Board

New Directors and Chiefs, New Departments and Divisions

Both immediately prior to and during the tumultuous period that led to Hopkins' dramatic enhancements in patient safety and research oversight, substantial changes were under way in its Medical School and Hospital leadership. Within just the five-year period between 1999 and 2003, some 22 new department and division heads were named; two new departments were created; and Hopkins Medicine continued to produce—and often export—exceptional talent, as it had since its founding.

Dean/CEO Edward Miller says that apart from overseeing the transformation of Johns Hopkins Medicine from "an idea to a reality," his most significant accomplishment has been "the recruitment of talented leaders," both in the dean's office itself and in departments. He has appointed the heads of 31 of the 33 departments in the School of Medicine.[1]

Following the appointments of pediatrics head George Dover and GYN/OB chief Harold Fox in 1996, when Miller was interim dean; Gerald Hart as head of biological chemistry in 1997; and then-neurology chief John Griffin in 1998, a far swifter pace of change began in 1999 as more department heads began to retire or were recruited away by outside institutions, appropriately impressed by their Hopkins credentials. Miller's appointments furthered his goal of making Hopkins Medicine a more interdisciplinary, more diverse institution.[2]

Miller's Men, Women and Departments

Department of Anesthesiology and Critical Care Medicine

After two years of serving as both Dean/CEO of Hopkins Medicine and director of the Department of Anesthesiology and Critical Care Medicine—acting as an exemplar of the ongoing effort to conserve funds—Miller finally relinquished the directorship in May 1999. As his successor, he appointed University of Virginia anesthesiologist Roger A. Johns—the brother of Michael Johns, the former School of Medicine dean whom Miller himself had replaced.[3]

Johns stepped down as department director in 2003 to concentrate on his research, as well as pursue his interests in national health care policies. He served as a health and science policy advisor to the U.S. Senate from 2005 to 2007 and continues studies of a broad range of health care issues.[4]

John Ulatowski, an expert on cerebral blood flow and oxygen delivery in the brain, was appointed interim director of the department in 2003. In 2004, he became the Mark C. Rogers Professor and director of the department and anesthesiologist in chief for Hopkins Hospital.[5]

Anesthesiology and critical care medicine (ACCM) is one Hopkins' largest departments. It services all of the operating rooms at Hopkins Hospital and Johns Hopkins Bayview, as well as out-of-operating-room locations such as radiation areas, the cardiovascular diagnostic labs, MRI and gastrointenstinal (GI) endoscopy.[6]

Department of Pharmacology and Molecular Science

In 1999, Miller also named Philip Cole, head of Rockefeller University's bioorganic chemistry laboratory, as director of the Department of Pharmacology and Molecular Science. Cole, who had obtained his M.D. and Ph.D. degrees from Hopkins in 1991, succeeded J. Thomas August, who had headed the pharmacology department, one of Hopkins Medicine's oldest, since 1975—and had emphasized its devotion to molecular research by adding the words "molecular science" to its name.[7]

Under Cole, the pharmacology faculty has grown to 62 members—23 of them focusing exclusively on its work, 22 of them associated with other Hopkins Medicine departments. A dozen more are part-time faculty and two have inter- or intradivisional appointments. Cole believes the multidisciplinary nature of the faculty is among its greatest strengths.[8]

Remaking and Creating Departments

The pace of departmental, division and administrative leadership changes picked up dramatically between 2000 and 2003 as more than two dozen major appointments were made.

Department of Medicine

In July 2000, the largest department in the School of Medicine learned that hematologist Edward Benz, Jr., its director since 1994, physician in chief for Hopkins Hospital, and holder of Hopkins' immensely prestigious Sir William Osler Professorship of Medicine, had agreed to become the next president of Harvard's Dana-Farber Cancer Institute.[9]

A year-long search ensued, resulting in the August 2001 decision to bring home one of Hopkins' own to head the mammoth department—cardiologist Myron "Mike" Weisfeldt, head of the Department of Medicine at Columbia University, a 1965 graduate of the School of Medicine, as well as a 1962 graduate of the School of Arts and Sciences.[10]

Weisfeldt is a former president of the American Heart Association, internationally known for his research in cardiopulmonary resuscitation (CPR)—including his forceful call for public access to portable cardiac defibrillators in such places as airport terminals.[11]

The Department of Medicine's 13 divisions cover an enormous range of medical specialties in addition to cardiology, including clinical immunology, clinical pharmacology, endocrinology, gastroenterology, general internal medicine, geriatrics, hematology, infectious diseases, nephrology, pulmonary medicine, rheumatology, and immunogenetics. It continues to obtain more research grant money from the National Institutes of Health than any other department of medicine in the nation, with the overall, outside research funds received in recent years totaling more than $200 million. Its more than 450 faculty members also lead other departments in direct contact with medical students and treat about a third of the patients discharged from Hopkins Hospital and Johns Hopkins Bayview. In addition, Weisfeldt's concerted efforts to increase diversity among the department's residents, fellows and faculty earned him the 2008 Diversity Award from the Association of Professors of Medicine, an organization whose members span the U.S. and Canada.[12]

From East Baltimore to Eastern Avenue

In 2000, the same year that Benz chose to leave Hopkins for Harvard, renowned rheumatologist David Hellmann, then executive vice chair of the Department of Medicine and head of its residency program, decided to make a far shorter—but in some respects, equally dramatic—move from Hopkins Hospital's inner city East Baltimore campus to the 130-acre Johns Hopkins Bayview campus, three and a half miles to the east in the city's Highlandtown area. Succeeding the retiring Philip Zieve as director of the Department of Medicine at Hopkins Bayview, Hellmann set out to end the perception of the Eastern Avenue facility as somehow a smaller planet in the Hopkins Medicine universe.[13]

Hellmann, a 1977 graduate of the School of Medicine, had spent all but six years of the preceding three decades engaged in clinical and educational work at Hopkins Hospital. He was not one to be fazed by the challenge of improving Bayview's image. "Bayview is Johns Hopkins with more convenient parking. This is Johns Hopkins with grass," he said.[14]

Hellmann's move to Hopkins Bayview was considered by many insiders as a major change in the dynamics between "downtown" (Hopkins Hospital) and its Eastern Avenue sibling.[15] Hellmann did much to achieve that goal. He was named the School of Medicine's vice dean for the Hopkins Bayview campus in 2005, succeeding anesthesiologist L. Reuven Pasternak, another long-time Hopkins faculty member and administrative leader.[16]

Department of Cell Biology

In May 2000, Peter Devreotes, who had earned his Ph.D. in biophysics at Hopkins and gone on to become a world authority on the chemical signaling that takes place between cells, was chosen from among 70 applicants to become director of what then was known as the Department of Cell Biology and Anatomy. His appointment ended a long interregnum in the department's leadership, begun when his predecessor, Thomas Pollard, had left Hopkins in 1996 to become president of the Salk Institute for Biological Studies in La Jolla, Calif.[17] (The Salk Institute subsequently developed a taste for top Hopkins scientists and leaders: in 2008, it picked Bill Brody, who was retiring as Hopkins' president, to become its new president and CEO.)

Within a year of assuming the directorship of the department, Devreotes decided its focus should be entirely on advancing its significant work in cellular research. In 2001, he renamed it the Department of Cell Biology, and Hopkins' anatomical and anthropological studies were concentrated in a new Center for Functional Anatomy and Evolution under the direction of Christopher Ruff, a Ph.D. in biological anthropology. In the Center, medical students still learn basic anatomy and study human development and evolution by dissecting cadavers, as well as employing the latest computer programs and other tools—including pull-out shelves lined with the bones of paws and jaws, arms and legs, and umpteen teeth, unearthed by paleontologists in the jungles of Costa Rica, the savannas of Africa, and the Badlands of Wyoming.[18]

Department of Orthopaedic Surgery

With orthopedics poised to experience revolutionary advances in its research and practice, Frank Frassica was named head of the Department of Orthopaedic Surgery (which retains the British spelling for orthopedics in its name) and orthopaedic surgeon in chief of Hopkins Hospital in October 2000. He had served as interim vice chair of the department following the death of his predecessor and mentor, Richard Stauffer, in 1998.[19]

After joining the Hopkins faculty in 1992, Frassica established a multidisciplinary musculoskeletal cancer program that quickly became one of the largest in the region seving patients with bone and soft-tissue tumors. In the early years of Frassica's directorship, one of the brightest stars of Hopkins orthopedics was James Wenz, head of the Department of Orthopaedic Surgery at Johns Hopkins Bayview. Wenz became one of the first physicians in the country to use a "mini-hip" procedure, totally replacing a patient's hip through a four-inch incision.

Wenz had performed hundreds of these minimally invasive total hip replacements and earned worldwide fame for his surgical skills prior to a tragic auto accident in 2004 that claimed both his life and that of his wife, Lidia Wenz, a child psychiatrist at Hopkins. A new orthopedic unit at Hopkins Bayview was named in their honor in 2006.[20]

Shortly before his death, Wenz also had been working on a procedure known as autologous chondrocyte implantation (ACI), a new approach for treating damage to cartilage, the critical tissue that cushions or protects the joints in knees, shoulders, elbows, ankles and hips. It involves the extraction of some 300,000 chondrocyte cells, the key cellular component of cartilage, from a patient's own body, then culturing them in a laboratory to grow millions more. The surgeon then injects the newly grown cartilage cells into the patient's injured joint—with knees being the most common one benefiting from this procedure.[21]

Within four years of Wenz's death, a multi-location Johns Hopkins Cartilage Restoration Center was providing this and other new treatments for repairing, regenerating or replacing cartilage.[22]

Department of Neurosurgery

Late in 2000, Donlin Long, director of the Department of Neurosurgery for 27 years, was succeeded by Henry Brem, a 16-year veteran of the faculty and one of the nation's most acclaimed brain tumor researchers and surgeons.[23]

As head of the department and neurosurgeon in chief of Hopkins Hospital, Brem—also a professor of oncology and ophthalmology—has been instrumental in building a research and clinical center widely credited with changing the outlook for brain tumor patients. He treats patients referred to him from around the globe, and his neurosurgical practice—plus that of 19 other full-time neurosurgeons in the nation's largest neuroscience critical care unit, opened in 2002—has made Hopkins one of the highest clinical volume brain tumor centers in the U.S.[24]

Along with its affiliate department at Johns Hopkins Bayview, directed by neurosurgeon Alessandro Olivi; and in collaboration with the Department of Neurology, headed by neurologist Justin McArthur, and its Hopkins Bayview affiliate, headed by neurologist Richard O'Brien, neurosurgery at Hopkins has literally every area of the field represented abundantly through the services of more than 100 specialists. They perform approximately 4,000 brain, tumor, vascular, and peripheral nerve surgeries annually.[25]

In addition to its groundbreaking work on brain tumors, the department's major areas for clinical treatment and research include hydrocephalus, commonly called "water on the brain," meaning the accumulation of fluid surrounding the brain; spinal tumors; chronic pain; cerebrovascular disease; and the functional organization of the thalamus, the portion of the brain that handles sensory perception and movement.[26] Its major areas for basic research include the mechanisms of pain production, transmission and perception; genetic alterations that control the growth of brain tumors; the molecular basis of cerebral vasospasm (a constriction or narrowing of the blood vessels in the brain); vascular malformations; applications of technology to neurosurgery; and new drug delivery systems for brain and spinal tumors.[27]

Department of Dermatology

In 2001, Daniel Sauder, former head of dermatology at the University of Toronto, succeeded Thomas Provost as head of the Department of Dermatology. Provost, who headed the department for 15 years before retiring in 1996, had recruited what became an internationally acclaimed research faculty, prompting the Baltimore-based Noxell Company, manufacturer of Noxzema skin cream and Cover Girl cosmetics, to create what was then one of only four endowed dermatology professorships in the country. Provost became the first Noxell Professor in Dermatology, the chair since held by every head of Hopkins dermatology.[28]

Sauder left Hopkins in 2005 to enter private practice. He was succeeded on an interim basis by Bernard (Buddy) Cohen, a 1977 graduate of the School of Medicine and director of pediatric dermatology. In 2008, Sewon Kang, head of dermatologic translational research at the University of Michigan, was appointed the new Noxell Professor in Dermatology, head of the 20-member dermatology department and dermatologist in chief at Hopkins Hospital.[29]

A mentor to dozens of U.S. and foreign medical students and post-graduate research fellows prior to and since coming to Hopkins, Kang also owns nine patents on dermatologic medications and procedures. He is directing an exciting period of growth and expansion for the department, which offers medical, surgical and cosmetic dermatology not just as Hopkins Hospital and Johns Hopkins Bayview, where Gerald Lazarus is director of dermatology and head of the wound center, but Hopkins' suburban centers as well.[30]

Department of Pathology

In the fall of 2001, the School of Medicine's oldest department got its ninth director when J. Brooks Jackson, who had been interim director for a year, was named the new head of Pathology, assuming the post originally held by William H. Welch.[31]

A clinical pathologist who joined the Hopkins faculty in 1996, Jackson had revolutionized HIV-prevention efforts in developing countries. He succeeded Fred Sanfilippo, a transplant immunology expert who had been a dynamic leader of the department for seven years. Sanfilippo had succeeded John H. Yardley and John K. Boitnott, who had undertaken a rare joint leadership of the department from 1988 until Sanfilippo's appointment in 1993.

One of the largest and most active departments of its kind in the country, Pathology has continued to grow under Jackson. Its more than 1,300 employees include 130 full-time faculty, 32 residents, 35 graduate students, more than 100 fellows, and approximately 1,000 staff. It has full-time employees not just in Baltimore but in Brazil, Peru, Uganda, Zimbabwe, Malawi, Zambia, India and China, as well as 21 divisions. These include autopsy pathology; cardiovascular pathology; the continuous quality improvement programs office; cytopathology (the study and treatment of changes in cells caused by disease); gastrointestinal and liver pathology; gynecologic pathology; medical microbiology; neoplastic hematopathology and flow cytometry (the study and treatment of malignancies in the blood and examination of cells); neuropathology; phlebotomy services; pediatric pathology; surgical neuropathology; surgical pathology; transfusion medicine; urologic pathology; genitourinary pathology; hematology pathology; informatics; kidney urologic pathology; molecular pathology; and pathology at Hopkins Bayview.[32]

Department of the History of Medicine

Randall M. Packard, a world-renowned expert in the study of international health and non-Western medicine, was picked to be director of the Department of the History of Medicine and the Institute of the History of Medicine in November 2001. He succeeded Gert H. Brieger, who had headed them since 1984.[33]

Packard has been a pioneer in the study of how social factors such as history and economics affect the appearance and spread of epidemic diseases, including malaria, tuberculosis and AIDS. He had headed the Department of History at Emory University from 1995 to 1998, as well as been director of the Center for the Study of Health, Culture and Society.[34]

The Department of the History of Medicine's 12 faculty members offer courses to medical, nursing and public health students, as well as collaborate with the Department of the History of Science and Technology in the Krieger School of Arts and Sciences to administer the Program in the History of Science, Medicine and Technology, which trains historians in all those fields at the undergraduate, master's and doctoral levels. As founder of the *Bulletin of the History of Medicine*, the official publication of the American Association for the History of Medicine (and produced by The Johns Hopkins University Press); as well as holder of one of the finest collections of books and journals in the field at its Library of the Institute of the History of Medicine, the Hopkins Department of the History of Medicine continues to have an unusually strong impact in its discipline.[35]

Department of Psychiatry and Behavioral Sciences

In June 2001, Paul McHugh, the cherubic, charismatic—and sometimes controversial—head of the Department of Psychiatry and Behavioral Sciences stepped down after 25 exceptional years at its helm. The following February, J. Raymond DePaulo, Jr., a 1972 graduate of the School of Medicine; an internationally acclaimed expert in the study and treatment of mood disorders; and himself a quarter-century veteran of the psychiatry department's faculty, was named as the new Henry Phipps Professor, director of the department, and psychiatrist in chief for Hopkins Hospital.[36]

An active clinician, teacher and researcher, DePaulo is one of the world's foremost experts on the genetic basis of such disorders as manic depression, panic disorder and general depression.[37]

Within five years of DePaulo's appointment, 15 Hopkins researchers led by Akira Sawa, associate professor of psychiatry and neuroscience and director of the department's program in molecular psychiatry, collaborated with two Japanese scientists to take advantage of the recent discovery of a gene, DISC1 (the first three initials of which stand for "disrupted in schizophrenia"), to engineer the first mouse that modeled both the anatomical and behavioral defects of schizophrenia, a complex and debilitating brain disorder that

affects more than 2 million Americans. (The magazine *Wired* dubbed this finding one of the "Top 10 New Organisms of 2007.") [38]

By 2009, the Department of Psychiatry and Behavioral Sciences was doing a great deal to help the mentally ill, treating 2,411 adult and 387 child and adolescent inpatients that year; handling more than 48,000 outpatient visits; conducting the research about which DePaulo is so passionate; and teaching the next generation of psychiatric practitioners, researchers and educators.[39]

Department of Molecular and Comparative Pathobiology

In 2002, another quarter-century veteran of the faculty, Janice Clements, a professor of comparative medicine, neurology and pathology—as well as vice dean for faculty affairs—was named the first director of the new Department of Comparative Medicine, which already had compiled a storied history at Hopkins long before it finally received departmental status.[40]

Clements, who directed the Retrovirus Laboratory in what had been the Division of Comparative Medicine in the Department of Medicine and had been the division's interim director since 1999, successfully urged that it be named a department. The elevation in status recognized the increasing importance to the School of Medicine's research mission of the studies performed by the biochemists, molecular biologists and other researchers who work with animals. By 2002, more than 500 Hopkins scientists, almost all of them in the School of Medicine, were engaged in more than 1,400 research projects involving some 116,500 animals (110,557 of which were rats and mice).[41]

A few years after the department's creation, Clements decided that its academic and research mission would be described more accurately if it was renamed the Department of Molecular and Comparative Pathobiology. In 2008, while retaining her posts in the dean's office, at the Retrovirus Laboratory, and the women's council, Clements chose to relinquish the department's directorship, which soon was filled by a two-decade veteran of the School of Medicine, M. Christine Zink, a Canadian-born doctor of veterinary medicine and a Ph.D. in macrophage biology (the study of cells that absorb harmful substances in the bloodstream or tissues).[42] Internationally known for her research on the effects of HIV in the brain, Zink is credited with discovering the potential HIV-suppressant properties of the common acne antibiotic, minocycline.[43]

Department of Surgery

The efforts spearheaded by Catherine DeAngelis, Janice Clements and others to encourage the advancement of women in Hopkins Medicine—enhanced by Ed Miller's commitment to achieving that goal—helped lead to an important, historic milestone in January 2003, when Julie Freischlag, chief of vascular surgery and director of the University of California at Los Angeles' Gonda Vascular Center, was named the William Stewart Halsted Professor and Director of the Department of Surgery. She was the first woman and only the sixth person to hold that position in the School of Medicine's 110-year history, and the first woman to head a clinical department of such size.[44] (Barbara de Lateur had become the first woman to head a clinical department when she was named director of Physical Medicine and Rehabilitation in 1994.) Freischlag succeeded John Cameron, master of the Whipple operation for pancreatic cancer, who had headed the department for 29 years.[45]

The 11-division, 73-surgeon department provides everything from acute care and adult trauma surgery to vascular surgery. Other specialties include burn surgery, cardiac and transplant surgery, surgical oncology and endocrinology, pediatric surgery, plastic and reconstructive surgery, thoracic surgery, and general surgery. It maintains operating rooms not just at Hopkins and Johns Hopkins Bayview but at the suburban centers where outpatient surgical procedures are performed.

The department's faculty members also engage in extensive research—involving many medical and postgraduate students, research fellows, and even some Hopkins undergraduates. They delve into subjects ranging from the fundamental mechanisms of disease at the cellular and molecular levels to clinical evaluations of surgical strategies and techniques. [46]

Department of Molecular Biology and Genetics

Seven months after Julie Freischlag became the head of surgery, another woman was tapped to be director of one of Hopkins' basic sciences departments. Carol Greider, a Ph.D. in molecular biology, was appointed in July 2003 to become the Daniel Nathans Professor and head the Department of Molecular Biology and Genetics. She had been a member of the faculty there since 1997, conducting the kind of research that would win her a Lasker Award in 2006 and a Nobel Prize in 2009, as described in Chapter 5. She later would say she hopes such achievements will encourage and enable more women to enter the field of basic, curiosity-driven science.[47]

Greider has focused her career on just that, basic biological questions—and coming up with answers that have dramatic implications for research into cancer, aging and other conditions. When the department's then-director, Thomas Kelly, invited Greider to join the Hopkins faculty, it already was the top-ranked department in her field, having been the home to Daniel Nathans and Hamilton Smith when they made their Nobel Prize-winning discoveries about restriction enzymes and how they could be used to launch what became the entire field of gene-splicing.

Kelly, who had succeeded Nathans as department head, expanded its size from six laboratories to 14 and forged a strong relationship with the Howard Hughes Medical Institute (HHMI), begun when Nathans was appointed as a senior HHMI investigator and recruited young Hopkins faculty who became HHMI investigators.[48]

The department remains tops under Greider, with nearly 200 personnel, including 33 faculty members, working on studying such simple but immensely useful organisms as yeast, fruit flies and worms, mice and rats, and human cells. Their projects aim to unravel mysteries in areas ranging from the fundamental molecular processes inside cells to the development of smell and vision. More than 40 postdoctoral fellows from all over the world are pursuing

advanced training in the department, preparing for careers in academic or industrial research.[49]

The Wilmer Eye Institute

After 14 dynamic years as director of the Wilmer Eye Institute, Morton Goldberg chose to relinquish his administrative responsibilities yet remain on the faculty, conducting research and treating patients. Selecting Goldberg's successor as director in May 2003, Ed Miller chose another distinguished Wilmer alumnus, Peter McDonnell III, a 1982 graduate of the School of Medicine and former member of the Wilmer house staff who had gone on to become an award-winning expert in refractive surgery and diseases of the cornea, as well as head of ophthalmology at the University of California, Irvine.[50]

McDonnell's credentials certainly reflected the impact of his own training at Wilmer two decades earlier. He had led the clinical trials for excimer laser technology and been the first physician to use its high-powered, ultraviolet pulses to treat astigmatism. He had received substantial grants from the National Eye Institute for research studies of cataract surgery, corneal transplantation, viral diseases of the eye and laser surgeries. *Ophthalmology Times* ranked him as one of the best ophthalmologists in the United States. [51]

The Wilmer Institute that McDonnell leads is by far the largest department of ophthalmology in the nation, with 120 full-time faculty. Indeed, some divisions at Wilmer are much larger than the average department of ophthalmology elsewhere.[52]

Clinical services at Wilmer now comprise comprehensive eye care; contact lens medical and surgical evaluation and management; cornea, cataract, and external diseases service, including cataract removal, corneal transplants and excimer laser therapy; and, as Maryland's designated Eye Trauma Service, emergency care, 24/7.[53]

In addition, Wilmer offers comprehensive consultations and long-term care for all forms of glaucoma; low vision and visual rehabilitation service; and a neuro-ophthalmology/orbit unit, which treats patients with blurred vision caused by optic disorders, double vision caused by strokes and brain tumors, defects in the visual fields, and involuntary spasms of the face and eyelids.[54]

Other specialized divisions include ocular immunology; ocular oncology; an oculoplastic service, which treats the eyelids and structures around the eyes that protect them; ophthalmic echography, which is an imaging technique useful in the evaluation of structures within or around the eye that cannot be discerned when an ophthalmoscopic evaluation is limited or totally obscured; pediatric ophthalmology and adult strabismus (crossed eyes); retina service; and visual physiology, which performs special tests to diagnose night blindness, color vision disorders, hereditary retinal disorders and other diseases.[55]

Department of Otolaryngology–Head and Neck Surgery

Seven months after Pete McDonnell succeeded the hard-to-follow Mort Goldberg at the Wilmer Institute, another much-admired, hard-to-follow department director, otolaryngologist Charles Cummings, stepped down as director of the Department of Otolaryngology–Head and

Neck Surgery following an immensely successful 12 years at its helm. This time, the search committee for his successor didn't have to go across the country to find an outstanding candidate: They unanimously recommended and Ed Miller heartily approved the appointment in December 2003 of Lloyd Minor, an Arkansas native, 10-year veteran of the faculty, and an outstanding clinician, researcher and teacher, at the new Andelot Professor and director of the department.[56]

Cummings, who succeeded Michael Johns as head of otolaryngology when Johns became dean of the School of Medicine, had done much during his directorship to recruit and mentor an extremely talented young faculty, including Minor. Under Cummings' leadership, Hopkins otolaryngology became one of the most comprehensive departments of its kind in the country and consistently was ranked #1 by *U.S. News & World Report* in the magazine's annual survey. His four-volume *Cummings Otolaryngology: Head and Neck Surgery* is acknowledged universally as the most comprehensive, authoritative reference book in the field.[57]

In choosing Lloyd Minor as Cummings' successor, Miller selected a world-renowned expert in hearing and balance disorders who also held faculty appointments in the departments of biomedical engineering and neuroscience while heading the otology and neurotology fellowship training program for physicians who want to focus on both the clinical and basic science aspects of chronic ear diseases.[58]

The 10-division otolaryngology department provides services in audiology; dentistry and oral surgery; facial plastic and reconstructive surgery; head and neck surgery; cancer; snoring disorders and sleep apnea; swallowing disorders; thyroid and parathyroid surgery, and voice disorders. Its listening center's services include speech language pathology and rehabilitation therapy, pediatric otolaryngology, and balance disorders treatment. Its cochlear implant program is the largest in the nation.[59]

Although the School of Medicine is used to having some of its top physicians hired away by other universities, Minor was persuaded by Hopkins' new president, Ronald Daniels, to leave the school in August 2009 to become The Johns Hopkins University's provost and senior vice president of academic affairs—the second-in-command post at the university. Surgeon John Niparko, among the nation's leaders in cochlear implants, has served as an exceptional interim head of the department following Minor's departure, maintaining its status as the best in the country. In the fall of 2011, it was announced that David Eisele, for 13 years an award-winning professor in the department before being recruited away in 2001 to become head of otolaryngology–head and neck surgery at the University of California, San Francisco, had been recruited to return to Hopkins as the new director of the department here, effective early in 2012. Niparko will remain as chief of the department's Division of Otology, Audiology, Neurology and Skull Based Surgery, as well as director of the Listening Center at Johns Hopkins, internationally renowned for its cochlear implant program.[60]

Eisele specializes in treating malignant and benign tumors of the head and neck, with special interest in the salivary gland and thyroid. His research interests include head and neck cancer, upper airway physiology, and dysphagia, or swallowing difficulties. Named by *U.S. News & World Report* as one of the "best doctors" in the country, Eisele leads a department that itself still remains the premiere ear, nose and throat program in the nation, according to *U.S. News & World Report*.[61]

Department of Radiation Oncology and Molecular Radiation Services

Extraordinary advances in the use of radiation for the treatment of cancer patients—including breakthroughs attained by researchers in the Kimmel Center's Division of Radiation Oncology—led to the creation of a new, separate Department of Radiation Oncology and Molecular Radiation Services in July 2003. As its first director, Ed Miller named long-time Hopkins radiation oncologist Theodore DeWeese, recognized internationally as a leader in the study of urologic malignancies, for his use of subtly calibrated doses of radiation to treat prostate tumors, and for developing gene therapy concepts to combat the disease.[62]

As part of the Johns Hopkins Kimmel Cancer Center, DeWeese's department concentrates on accelerating research in and delivery of novel cancer treatments.[63] The department's 32 faculty members use the latest electromagnetic energy—intense, shortwave beams—to treat cancer, as well as conduct extensive research to develop innovative new cancer treatments. The department's division of molecular radiation sciences focuses on deciphering how human cells respond to genetically based lesions; how those lesions on a cell's DNA are sensed and repaired; and how those pathways to repair are disrupted in cancer. The scientists' research has two goals: understanding how these cellular mechanisms have evolved to protect us from cancer; and harnessing that understanding to diagnose a change in cellular makeup early, predict a cell's future development, and launch new therapies based on these discoveries.[64]

New Leaders for Divisions and Centers

New leaders also were named by various department directors to head several major departmental divisions during this period—prompted, in part, by the venerable Hopkins tradition of having top faculty lured away by other major institutions. Indeed, several of the newly named divisional chiefs would themselves be recruited to go elsewhere within only a few years of their appointments.

Division of Rheumatology

Early in 2001, then-chief of the rheumatology division Fredrick Wigley took the unprecedented step of requesting that Antony Rosen, a creative and effective researcher, clinician and teacher, be named his co-chief. The unusual request, approved by Michael Klag, then-interim director of the Department of Medicine, made rheumatology the only departmental division with two leaders—one, Wigley, a basic scientist; the other, Rosen, a clinician.[65]

Wigley and Rosen both had done much in the preceding few years to expand and build the division into a model of translational research, moving discoveries in the laboratory to the bedside of patients for treatment. While working together on a grant application to establish a new research center, Wigley said he realized how important Rosen's leadership had been to the division's success and wanted him to take on a greater role in planning its future.[66]

By 2002, Wigley chose to focus on his work as director of the division's Scleroderma Center, which he had founded in 1992 and has developed into one of the largest of its kind in the world. Rosen then became the division's sole chief.[67]

Consistently ranked by *U.S. News & World Report* as the top division of its kind in the country, it is home to subspecialty clinics intensely concentrating on developing the latest treatments for such common or uncommon disorders as arthritis, lupus, myositis, scleroderma, Sjögren's syndrome, and vasculitis, as well as providing consultative rheumatology at Baltimore's Good Samaritan Hospital.[68]

Division of Cardiology

In 2002, Kenneth Baughman, the much-admired head of the Department of Medicine's Division of Cardiology since 1992, and a 24-year veteran of the faculty, left to become director of the advanced heart disease section at Brigham and Women's Hospital in Boston and a professor of medicine at Harvard. In 2003, he was succeeded as cardiology chief by Eduardo Marbán, a 20-year Hopkins veteran.[69]

In May 2007, Marbán left Hopkins to become director of the Cedars-Sinai Heart Institute in Los Angeles.[70] Gordon Tomaselli, a 23-year Hopkins veteran and conductor of landmark research on irregular heartbeats (known as heart arrhythmias) and cellular electrophysiology (the electrical activity within cells), was named the new head of the cardiology division and co-director of the Johns Hopkins Heart and Vascular Institute.[71]

Under Tomaselli, who became president of the American Heart Association in 2011, the division treats more than 4,000 inpatients and nearly 30,000 outpatients annually.[72]

Division of Surgical Oncology and Endocrinology

A short time elapsed between the 2003 appointment of Charles Yeo, a 1979 graduate of the School of Medicine and the John L. Cameron Professor for Alimentary Tract Diseases, as chief of what then was the Division of General and Gastrointestinal Surgery and his 2005 acceptance of the chairmanship of the Department of Surgery at Thomas Jefferson University's medical school in Philadelphia.[73]

After Yeo's departure, the Division of Gastrointestinal Surgery was split into three divisions. One became the Cameron Division of Surgical Oncology, headed by Richard Schulick, a 1989 graduate of the School of Medicine who was named the new holder of the Cameron professorship. Another was the Handelsman Division of Surgical Oncology and Endocrinology, named for Jacob Handelsman, a 1943 graduate of the School of Medicine who trained under Alfred Blalock and spent six decades devoting himself to Hopkins surgery. The head of this division—and holder of the newly created Jacob C. Handelsman, M.D. Professorship in Abdominal Surgery—was Michael Choti. A graduate of the Yale medical school who joined the Hopkins faculty

in 1992, he heads both the Colon Cancer Center and Liver Tumor Center.[74]

The third division created was the Ravitch Division of Gastrointestinal Surgery, named for Mark Ravitch, another Blalock protégé once dubbed the master's "alter ego." Its first head was Frederic Eckhauser, a Hopkins intern and resident in the 1970s who now is director of clinical operations in surgery and a member of the Sol Goldman Pancreatic Cancer Research Center. The current Ravitch Division leader—and holder of the Ravitch Professorship in Surgery—is Jonathan Efron, a University of Maryland medical school graduate who was an associate professor of surgery at the Mayo Clinic before coming to Hopkins.[75]

In July 2009, the Cameron and Handelsman divisions were merged into a single Division of Surgical Oncology and Endocrinology, headed by Schulick. Choti became vice chair for finance and operations of the division.[76]

The John M. Freeman Pediatric Epilepsy Center

Retirement, not departure for another institution, prompted the 2003 appointment of neurologist and pediatrician Eileen P.G. Vining to succeed John Freeman as head of the Pediatric Epilepsy Center. Freeman's immense impact on the field of children's epilepsy and seizures led him to create the center that Hopkins later named for him, and which his one-time protégé Vining now would head.[77]

Freeman, a 1958 graduate of the School of Medicine, had spent most of his career at Hopkins. A professor of neurology and pediatrics, he created and was the long-time head of the Child Neurology Program. His championing of a long-neglected treatment—the high-fat, meticulously managed ketogenic diet—helped return it to prominence as an effective way to significantly reduce or eliminate seizures in children and adolescents.[78] Along with pediatric neurosurgeon Benjamin Carson, he also emphasized the role of hemispherectomy, the removal of one-half of the brain of seizure-plagued children, to control and cure such a child of these intractable, incapacitating incidents.

Under Vining, a 1972 graduate of the School of Medicine and former fellow in developmental pediatrics, the Freeman Center provides a comprehensive, "whole child" treatment program that incorporates medications, vagal (cranial) nerve stimulation, diet and surgeries to address the impact of epilepsy on a child's mental function, language skills, school performance, behavior and family life.[79]

Division of Geriatric Medicine and Clinical Gerontology

Change in the leadership of the pediatric epilepsy center, one of Hopkins' key divisions for treating children, was mirrored in the division that focuses on the elderly, its geriatric and gerontology division, home to an internationally prominent program. John Burton, director since 1982 of the division he had created at the Johns Hopkins Bayview Medical Center, retired in 2003 and also was succeeded by one of his protégés, Linda Fried, a geriatrician and epidemiologist.[80]

Acknowledged as one of the founding fathers of modern geriatrics, Burton led Hopkins' division to top ranking in the nation. Fried emulated Burton's multifaceted approach to geriatric care and research. She produced studies that changed nationwide preventive approaches to heart disease and stroke in men and women over 65, as well as led to a clinical definition of frailty as a distinct syndrome.[81]

In January 2008, Fried became dean of Columbia University's Mailman School of Public Health. Her successor as division director is Samuel (Chris) Durso, whose research interests include "training the trainer" education to prepare faculty to teach geriatric medicine and how to establish priorities in the health care of older adults with many complicating conditions.[82]

Appointment Pace Continues Unabated: 2004–2011

The pace of departmental and divisional change did not diminish, with six new directors and chiefs being named in 2004 alone—in one instance, twice in a single month.

2004

Department of Urology and
The Brady Urological Institute

As described above, Johns Ulatowski was appointed director of the Department of Anesthesiology and Critical Care Medicine on November 2, 2004. Ten days later, Alan Partin, a Hopkins-trained, world-renowned expert on prostate cancer, was chosen to succeed his even more famous mentor, Patrick Walsh, who had headed the department at the Brady Urological Institute for three decades. Under Walsh, the Brady Institute had been the acknowledged international leader in urology and topped the *U.S. News & World Report* rankings as the No. 1 urology department in the nation. It continues that record under Partin.[83]

As Walsh's protégé, Partin has followed the same practice of maintaining a detailed database on all of the cancer patients on whom he operates, keeping track of every piece of clinical information about them. While Walsh's name has become synonymous with his prostate removal surgery, Partin's name perhaps has become just as familiar as creator of the Partin tables, the diagnostic prostate cancer computer models (called nomograms) that he developed in 1993. These tables use a mathematical formula to predict the probability of a prostate cancer's aggressiveness and potential for spreading.[84]

Because Partin and Hopkins recognized the immediate clinical importance of Partin's work, the tables were promptly made available to clinicians everywhere—for free. Their swift adoption by other physicians led one oncologist to call them the "prognostic paradigm of the 1990s."[85]

An Ohio native, Partin received his Ph.D. and M.D. degrees from Hopkins in 1988 and 1989, respectively.[86] He says that Donald Coffey, the charismatic, then-director of research at Brady "brainwashed me into becoming a urologist."[87]

Among Partin's current goals is to solidify the reputation of Hopkins' 54-member urology department not only as a place that meets all the urological needs of men, but of women, too.[88] Given the scope of the department's clinical and research activities, that goal already is well met. In addition to prostate cancer, its specialists treat patients with cancers of the bladder, kidney and testis. Brady also has acclaimed specialists in female and pediatric urology as well as experts in treating incontinence, stone disease, ureteropelvic junction obstruction (a condition where blockage occurs where the ureter, the tube that carries urine from the kidney to the bladder, attaches to the kidney), and Peyronie's Disease (not a "disease," actually, but a connective tissue disorder of the penis that can be likened to arthritis). Treatment also is provided for erectile dysfunction and male infertility.[89]

Under urological research director Robert Gertzenberg, who succeeded Donald Coffey at the same time Partin succeeded Walsh, seven research laboratories are delving into everything from advancing the use of robotic technology for urological procedures to discovering and developing biomarkers—specific physical traits that are used to measure the effect or progress of a disease or condition—to improve the early detection, analysis, monitoring and prognosis of prostate and other urological cancers.[90]

The Russell H. Morgan Department of Radiology

When School of Medicine executive vice dean and radiology department director Elias Zerhouni left Hopkins in mid-2002, following his appointment as head of the National Institutes of Health, Bob Gayler, a pillar of the department for more than 20 years, was named interim director.

Following a nearly year-and-a-half, nationwide search, Jonathan Lewin, a leading expert in magnetic resonance imaging (MRI), neuroradiology and biomedical engineering was chosen to become the new Martin W. Donner Professor and director of the Russell H. Morgan Department of Radiology. With that appointment came the post as radiologist in chief of The Johns Hopkins Hospital.[91]

The holder of numerous patents related to MRI applications and a researcher who has headed inquiries underwritten by more than $10 million in federal and state grants, Lewin is an acknowledged pioneer in interventional magnetic resonance imaging. This is a new technology that uses MRI to guide surgery and minimally invasive procedures for the treatment of cancer, vascular disease and other disorders. He also has received a number of grants from the MRI industry to develop computer software and hardware for this emerging discipline.[92]

The department has 11 divisions at Hopkins Hospital and a similarly broad lineup of services provided by eight divisions at Johns Hopkins Bayview. These include: cardiothoracic and abdominal computed tomography (CT); body MRI; positron emission temography (PET); the breast center; general and diagnostic; general interventional; musculoskeletal; neuro interventional; nuclear medicine; pediatrics; and ultrasound.[93] The department also provides radiology services at Hopkins' suburban centers at Green Spring Station and White Marsh, as well as at Howard County General Hospital.[94]

The Division of General Internal Medicine

As was the case when the directorship of radiology was left open with Elias Zerhouni's 2002 elevation to the top of the National Institutes of Health, the appointment of Michael Klag as the School of Medicine's vice dean for clinical investigation left the division of general internal medicine looking for a new leader in 2004.

Hopkins' always-deep bench of in-house talent filled the vacancy this time, when Department of Medicine head Mike Weisfeld chose Frederick Brancati, a 15-year veteran of the staff with exceptional skills as a teacher, mentor and highly respected obesity and diabetes researcher, to become the new division chief.[95] Brancati had arrived at Hopkins in 1989 as a post-doctoral fellow in general internal medicine and joined the faculty in 1992.[96]

The general internal medicine (GIM) division that Brancati heads likes to describe itself as practicing "medicine without limits" since its founding in 1977. Historically, all internists were generalists who provided comprehensive, continuous care to patients with a variety of health problems. Some also specialized in particular areas of medicine that interested them.[97]

Today GIM has more than 65 primary, full-time faculty, 16 fellows and more than 130 staff. Placement in its fellowship program is one of the most sought-after in the country, with the vast majority of its graduates becoming leaders in research, medical education, clinical practice and administration in American medicine and public health.[98]

Brancati's goals for the division include increasing the size of Hopkins Hospital's program for hospitalists, or hospital-based general physicians who devote most of their time to the treatment of hospitalized patients; and enhancing the division's research efforts on cancer, women's health, and medical informatics, with a special emphasis on the technologies now emerging for the management of chronic diseases.[99]

Under the direction of internist John Flynn, the division's clinical centers include not only Hopkins Hospital but its Outpatient Center, where comprehensive outpatient and internal medicine services are provided, and the Green Spring Station Health Care and Surgery Center. Its services also include the Executive Health Program, a comprehensive preventive medicine program designed for the convenience of people with busy schedules; and a Comprehensive Consultation Service, which provides advice to hospitalized patients on surgical, psychiatric and other services that are not part of general medicine's traditional portfolio. GIM faculty also serve as doctors to Hopkins doctors, overseeing the University Health Service that provides acute and ongoing care for Hopkins students, residents, trainees and their dependents.[100]

The Department of Physical Medicine and Rehabilitation

In June 2004, Barbara de Lateur, head of the Department of Physical Medicine and Rehabilitation since 1994, decided to step down.[101] As de Lateur's successor, Miller chose Jeffrey Palmer, a 21-year veteran of the faculty who not only is an expert on swallowing disorders but an accomplished pianist who studied at New York's Juilliard School of Music.[102]

As Lawrence Cardinal Shehan Professor and director of the department, as well as physiatrist in chief of Hopkins Hospital, Palmer heads a department devoted to restoring the health and functional abilities of people after acute illness or injury, such as stroke, spinal disorders, heart surgery,

amputation and joint replacement or spinal cord injuries and sports injuries. He also is chairman of the Department of Physical Medicine and Rehabilitation at Good Samaritan. [103]

In addition, Palmer is director of the department's research and heads its oral functional and swallowing laboratory. Along with his wife, psychologist Sara Palmer, an assistant professor in the department, and Bethesda, Md.-based psychologist Kay Harris Kriegsman, Palmer also has co-authored *Spinal Cord Injury: A Guide for Living*, the definitive book for people who have sustained damage their spinal cords, as well as for their families.

The department offers a full spectrum of services for individuals with a wide range of disabilities and musculoskeletal problems. Among the services is a comprehensive evaluation by a physiatrist who deals not only with the patient's physical illness but its impact on the person's family and home life, as well as interactions with society. The physiatrist then works with the referring physician and the rehabilitation team to design and implement an individualized plan of care. This level of individual attention is necessary to ensure the best possible treatment for patients with a broad range of disabilities.[104]

Other services provided by the department include physical and occupational therapy, speech and language pathology, rehabilitation nursing, comprehensive inpatient rehabilitation, and an outpatient neurorehabilitation program. Departmental specialties include amputation rehabilitation, burn injury rehabilitation, and treatment of cancer patients' therapy-related pain and other physical problems. Additional rehabilitation service specialties include carpal tunnel syndrome; cardiac and pulmonary disorders; cognitive and neurobehavioral disabilities; neuromuscular disorders; gait and mobility problems and sports injuries. The department's physicians also specialize in serving patients who have suffered strokes, have vestibular (balance) disorders; require wheelchair and seating assessments and evaluations of work conditioning and functional capacity. Women's health problems as related to rehabilitation issues also are a specialty.[105]

2005

The Baltimore Regional Burn Center

A branch the Department of Surgery's burn division and headquartered at Johns Hopkins Bayview since its founding in 1968, the center admits about 600 patients annually and cares for an additional 1,600 outpatients, serving burn and wound victims from not just Maryland but also Pennsylvania, Delaware and West Virginia.[106]

The center's team concept of care was developed by Andrew Munster, its director from 1976 until 2000.[107] Under Munster, patients were helped to achieve recovery not only by surgeons but by psychologists, sociologists, vocational rehabilitation counselors, clergy, home health nurses, physical and occupational therapists, nutritionists, and the Burn Victim's Aid Society, made up of former burn patients. The center became one of the few of its kind in the country that linked such support efforts to an acute burn facility, and its philosophy of all-encompassing care has led to a 98 percent patient survival rate, making the Hopkins Bayview program one of the nation's premier burn centers.[108]

A month after Munster's death in 2003, his long-time co-director, plastic surgeon Robert Spence, was appointed director of the center. A 1972 graduate of the School of Medicine, Spence was and remains nationally known for his work in burn reconstruction and a rehabilitation ethic that reflects Munster's belief that caring for burn victims goes far beyond doing everything possible to restore a patient's physical appearance. It is through the combination of counseling, rehabilitation and reconstructive surgery that the center seeks to give burn patients their lives back.[109]

In July 2005, Spence stepped down as director of the center but remained as head of burn reconstruction there until February 2008, when he founded the National Burn Reconstruction Center at Good Samaritan Hospital.[110]

Department of Surgery director Julie Freischlag appointed Stephen Milner to be Spence's successor as the chief of the burn division and the center's head in 2005. A British-born surgeon who honed his skills in treating burn victims as a lieutenant colonel in the Royal Army Medical Corps during the 1990-91 Gulf War (Operation Desert Storm), Milner had been a professor of surgery and director of Southern Illinois University's regional burn center prior to coming to Hopkins.[111]

The Burn Center maintains two locations, its adult burn and wound unit at Hopkins Bayview, which admits patients from the age of 15 and up, and a pediatric burn unit at Hopkins Hospital's Children's Center. Its 19-member staff treats all types of burns, including flame, scald, electrical, chemical and radiation injuries, as well as repair complex wounds.[112]

The Burn Center also is the only civilian facility in the country that provides burn, trauma and critical care training for United States military personnel. Its physicians, caregivers and counselors have trained more than 1,000 U.S. Air Force medical personnel—including physicians, nurse practitioners, nurses, and independent medical technicians. In addition, the center has provided Advanced Burn Life Support Provider Certification Courses for both the Army and Navy.[113]

The Division of Hematology

As one of the oldest research branches of the Department of Medicine, the Division of Hematology—which explores the causes and treatments of blood disorders—has a distinguished lineage of former leaders. Excellence in hematology, it appears, runs in Hopkins' blood.[114]

It certainly seems to be in the genes of Robert Brodsky, who in 2005 became head of the division, as well as chief of the medical oncology and hematologic malignancies program at the Kimmel Cancer Center. He is a second-generation specialist in those fields. At the time of his appointment, his late father, Isadore, directed parallel services at Hahnemann University Hospital in Philadelphia. It is associated with Drexel University's College of Medicine, where the elder Brodsky headed the Isadore Brodsky Institute for Blood Diseases and Cancer and became the first physician in the Philadelphia area to perform a stem-cell transplant.[115]

The Brodskys, pére et fils, even collaborated on a Hopkins-Drexel study. Patients identified by the father in Philadelphia volunteered to participate in Hopkins experiments conducted by the son that led to a revolutionary treatment for the frequently fatal autoimmune condition called severe aplastic anemia (SAA).[116]

Hopkins' leadership in hematology began with its first head, Maxwell Wintrobe (1901–1986), who actually had become a hematologist before hematology even existed as a discipline. He established the modern methods for measuring a blood sample's volume of red and white blood cells and platelets; invented the most commonly used instrument for performing these measurements; and single-handedly wrote *Clinical Hematology*, which quickly became recognized internationally as the most authoritative work in the field.[117]

Hematology at Hopkins became a formal division of the Department of Medicine in 1947, when C. Lockard Conley—then at the beginning of a career that would also win him world-wide renown—was named to be its chief. He headed the division for 33 years.[118]

Conley was succeeded in 1980 by one of his former fellows, Jerry Spivak, who went on to be a mentor himself to the young Robert Brodsky. Spivak remains at Hopkins as head of the Center for the Chronic Myeloproliferative Disorders. These are a unique set of blood afflictions that share in common an overproduction of one or more of the cellular elements of the blood: the red cells, white cells or platelets. They can cause significant damage to the bone marrow, liver and spleen if not properly diagnosed and treated.[119]

In addition to Spivak's center, the Division of Hematology also collaborates with Hopkins' Department of Pharmacy to operate a separate Anticoagulation Management Service, which helps Hopkins cardiologists, oncologists and other physicians monitor anticoagulation therapy for patients at risk of deep vein thrombosis, pulmonary embolisms and other hypercoagulable states. In addition, the division provides special services for patients with Sickle Cell Disease in its Sickle Cell Center for Adults (see Chapter 5).[120]

Division of Gastroenterology and Hepatology

Two other divisions in the Department of Medicine were given new directors in 2005, with Anthony Kalloo taking over as head of the Division of Gastroenterology and Hepatology and Paul Scheel Jr. becoming chief of the Division of Nephrology. Both had more in common than their desire to ensure that their patients' digestive and renal systems properly discharge what their bodies take in. Each is a technology geek, determined to harness the power of computers and other high-tech devices to improve medical care at Hopkins.[121]

Kalloo has been a pioneer in natural orifice surgery, a technique that involves entering the abdomen through such an opening as the throat to enable abdominal surgery to be done without the use of incisions. He also has patented a number of medical devices and procedures. These include endoscopic cryotherapy, or the use of gastrointestinal endoscopes to freeze diseased tissue, and the winged biliary/pancreatic stent, used to treat blockages in the bile ducts or pancreas due to cancer or other conditions.[122]

Kalloo, a Trinidad native and graduate of the University of West Indies Medical School, joined the Hopkins faculty in 1988—just in time to witness the computer era being born and eager to make the most of it. He became the founder and medical director of the Hopkins Gastroenterology and Hepatology Resources Center, a 3,000-page, multilingual Web resource for patients and physicians. He also is the impresario of the division's remarkably interactive Web site, created with the help of the Department of Art as Applied to Medicine.[123]

The multilingual GI Web site has more than 1,000 animations, original anatomical drawings, still photographs, videos, podcasts and other depictions that give those who log onto it easy-to-comprehend images of the 27-foot-long digestive system—replete with sloshing sound effects—and its maladies.[124]

With faculty at both Hopkins Hospital and Johns Hopkins Bayview, the division provides comprehensive gastroenterology care and conducts extensive research into the field. The areas of care and research covered by its experts include the esophagus and stomach, the pancreas and biliary tract, the liver, and the small and large intestine.[125]

The Division of Nephrology

Like Kalloo, Paul Scheel Jr. has set out to use technology—even package-delivery technology—to improve patient care at Hopkins, and not just in his own Division of Nephrology.[126]

Scheel arrived at Hopkins as an internal medicine intern in 1987.[127] His research interest is retroperitoneal fibrosis, a rare condition of mysterious origin that results in persistent inflammation of the area between the abdominal wall and the peritoneum, the membrane that lines the abdominal cavity and covers most of the abdominal organs, including the kidneys, adrenal glands, ureters, part of the small intestine, pancreas and colon. The unrelenting inflammation leads to the development of scar tissue (fibrosis), which can seriously affect the function of these organs.[128]

Other research concentrations being pursued by the division's 22 full- and part-time faculty and 15 fellows at Hopkins Hospital and Johns Hopkins Bayview include studies of glomerulonephritis, a group of kidney diseases characterized by inflammation of the filtering units of the kidney, called glomeruli; and polycystic kidney disease (PKD). This is a group of diseases that cause the kidney's tubules to become impaired. The kidney tubules process the immense amount of fluid that is filtered daily by the glomeruli, and when the tubules cannot perform this function properly, the result can be fluid retention, high blood pressure, kidney failure requiring dialysis—and kidney transplantation.[129]

As a holder of a master's degree in business administration as well as his medical credentials, Scheel was in the forefront of a multidisciplinary collaboration to create "MedBed," a paperless, Web-based, single-page patient assignment system. It employs the technology perfected by package delivery companies to streamline the process of finding available beds for patients.[130]

"Federal Express can tell you exactly where a package is anywhere in the world," Scheel told *Hopkins Medicine*

magazine in 2005. "We couldn't tell you whether we had an empty bed or not. That's ridiculous. I figured if they could do it, we could do it."

Coordinated with a computerized admission/discharge/transfer system known as the electronic bedboard, MedBed now provides a powerful tool for speeding up the admissions and bed-assignment process for patients.[131]

2006

The Solomon H. Snyder Department of Neuroscience

When Richard Huganir was chosen in February 2006 to become director of what just had been renamed the Solomon H. Snyder Department of Neuroscience in honor of its founder and his predecessor, he thought it appropriate to paraphrase a famous remark by Thomas Jefferson. Upon arriving in Paris to become the United States' second ambassador to France after Benjamin Franklin, Jefferson heard someone call him Franklin's "replacement." As Huganir noted, Jefferson replied, "You don't replace Ben Franklin. You succeed him."[132]

As an 18-year veteran of the 25-year-old department, Huganir is an internationally acknowledged authority on how molecular signals created in the neurons of the brain bring about human learning and the construction of memories.[133]

Huganir came to Hopkins as an assistant professor in 1988.[134] A professor of neuroscience since 1993, a member of the National Academy of Sciences, and a Howard Hughes Medical Institute investigator, Huganir is widely acclaimed for the novel experiments he has conducted to shed light on the makeup and activity of the proteins and other chemicals that are involved in the communication between nerve cells in the brain.[135]

The wide-ranging, collaborative research undertaken in the department is focused on fundamental issues surrounding synaptic plasticity, or the constantly changing communication between the trillion neurons in the brain. This is critical for understanding such higher brain functions as perception, learning, memory and decision-making and also has profound implications for treatment of an array of neurological diseases and movement disorders, including Lou Gehrig's Disease, stroke and dementia, as well as schizophrenia, depression, drug addiction, mental retardation and autism. The goal, Huganir has said, is to make "a bridge between molecules and the mind."[136]

Under Huganir, the department is searching for answers to the most difficult questions about the human mind. Its studies are revealing how our eyes use light to set an internal clock, how neurons grow and establish appropriate connections, and how memories are encoded by changes in the strength of communication between neurons.[137]

The department's researchers seek to understand the causes of neurological disease. They are working on determining how the devastating consequences of strokes can be ameliorated; how neurons might be replaced and stimulated to regrow their connections following degeneration or injury; how environmental factors cause neurons to die.[138] By unraveling the mysteries surrounding the inner workings of the human mind, they aim to find the causes of and treatments for some of the most devastating and disabling of human disorders.[139]

The Department of Physiology

For William Guggino, a childhood question led to a distinguished, quarter-century career as a Hopkins scientist and, in May 2006, directorship of the Department of Physiology, where studies of the mechanical, physical and biochemical functions of human organs and cells have been a cornerstone of the medical school since 1893.[140]

As a boy, Bill Guggino loved the ocean and wondered "how do fish live in the sea?" As a Ph.D. student in zoology at Yale, seeking a research topic, he remembered that question and began studying the chloride-secreting cells that enable fish to survive in a salty environment.

He continued those inquiries and others when he joined the Hopkins faculty in 1982. Over the next two and a half decades—which included service as course director in organ systems physiology and histology (the anatomy of cells and tissues)—he became an authority on chloride transport and a key figure in developing a gene therapy for cystic fibrosis (CF), the chronic, inherited disease that affects the lungs and digestive systems of about 30,000 children and adults in the United States and some 70,000 victims worldwide.[141]

"The protein involved in secreting chloride in fish is like the one that's defective in CF," Guggino says, so he still is studying the basic science of the chloride channel and its malfunctions in that disease. He became director of the Cystic Fibrosis Research Development Program in 1989 and vice chairman of research in the Department of Pediatrics in 1996. In 1992, he co-authored with Peter Agre the paper detailing discovery of aquaporin1, the first water channel protein—the research that won Agre the Nobel Prize for Chemistry in 2003. Guggino's own work has won him top awards for cystic fibrosis research and training clinician scientists who are dedicated to unraveling CF's mysteries and developing new treatments for it.[142]

Guggino's immediate predecessors as head of physiology were interim director Peter Maloney, associate dean of graduate student affairs and a researcher into the mechanisms by which molecules move through cell membranes; and William Agnew, who directed the department until 2004. He then stepped down to concentrate on his research into the molecular mechanisms that regulate the structure and function of the human nervous system's ion channels—the pathways through which electrical impulses travel.[143]

The overall research mission of the 28 faculty, 12 postdoctoral fellows and 23 graduate students in physiology is to explore how the complex phenotypes—or observable physical or biochemical characteristics—of the cells that underlie our tissue and organ systems emerge from the genetic code. To do this, their studies encompass everything from bacteria and yeast to zebrafish, mice—and humans.[144]

As Guggino notes, it is becoming clear that mutations in the genes that influence the formation of the proteins within cell membranes are associated with such inherited disorders as cystic fibrosis and polycystic kidney disease, which causes numerous, potentially dangerous cysts to form in the kidneys and sometimes the liver.[145]

For patients suffering CF, the advantage of being treated at the Hopkins Cystic Fibrosis Center is that for more than 60 years, Hopkins basic scientists and clinical investigators have achieved significant CF breakthroughs, from developing the standard "sweat test" to diagnose CF, defining the course of the disease, identifying mutations in the CF gene, and developing counseling programs and genetic and drug therapies used to keep its symptoms under control.[146] The result of these multifaceted efforts has been a three-fold increase in the average lifespan for cystic fibrosis patients, from as little as 8 or 10 years to an average lifespan of 36, with many living into their 40s and beyond.[147]

The Department of Biophysics and Biophysical Chemistry

In appointing physical chemist L. Mario Amzel as the new head of the Department of Biophysics and Biophysical Chemistry in July 2006, Ed Miller called him "an extraordinary ambassador for Hopkins science and medical education."[148]

No wonder: in the preceding year alone, Amzel had led seminars, appeared on panels and lectured from New Jersey and North Carolina to Mexico City, Bogota, Columbia, Mar del Plata, Argentina and Zaragoza, Spain.

The son of a Buenos Aires shoe store owner, Amzel obtained his Ph.D. in physical chemistry at the Universidad de Buenos Aires in 1968. He came to Hopkins on a postdoctoral fellowship in 1969. He has been here ever since, becoming a full professor in 1984.

Amzel's early work focused on understanding how proteins take on their final shape, which can shed light on what they do and how they work in cells. He was the first to define the structure of part of an antibody, the molecule that helps the human immune system fight off infection. He also was part a Hopkins research team that produced the first high-resolution pictures of how antibodies interact with antigens, the foreign molecules that invade the body.[149]

Amzel since has determined the structure of many proteins and how proteins act together. He also has studied how proteins fold, take on their 3-dimensional shapes, and serve as the catalysts for biochemical reactions within cells.[150]

The department Amzel heads had an unusual origin in the late 1940s as a joint venture spanning both the schools of medicine and arts and sciences. A separate School of Medicine Department of Biophysics was created in 1961 and headed by Howard Dintzis, a mentor to future Nobel Prize-winner Richard Axel and cancer researcher nonpareil Bert Vogelstein. Dintzis remains an emeritus member of its faculty, all of whom specialize in applying fundamental principles from biology, chemistry and physics to gain insights into the mechanisms of medically relevant biological processes.[151]

For example, to improve the design and enhance the effectiveness of drugs, they study metabolic and cell-signaling processes. To do this, they use a wide variety of techniques, including X-ray crystallography, which analyzes the structure and function of hemoglobin, the protein that transports oxygen from the lungs to the tissues via red blood cells, and myoglobin, which stores and distributes oxygen in muscle cells.[152] They also employ microcalorimetry, which measures the heat or energy produced during chemical reactions, useful for assessing the actions of drugs[153]; and fluorescence spectroscopy, for determining the elements within, structure, diffusion and interaction of molecules.[154]

The Department of Psychiatry at Johns Hopkins Bayview

In 2006, Chester Schmidt Jr. decided to step down as director of the psychiatry department at Johns Hopkins Bayview after an astonishing 34 years at its helm. Named as his successor was Constantine (Kostas) Lyketsos, one of the world's top experts on the treatment and long-term care of patients with dementia.[155]

Relinquishing the psychiatry directorship didn't mean Schmidt, a 1960 graduate of the School of Medicine, was retiring. He simply moved on to become chief medical officer for Johns Hopkins Healthcare, the organization that oversees Hopkins' three managed care programs—Priority Partners, the Uniformed Services Family Health Plan, and the Johns Hopkins Employer Health Programs. Combined, they cover nearly 300,000 individuals.[156]

Lyketsos, the son of an academic psychiatrist, obtained his medical degree from Washington University in 1988, then completed his psychiatry internship, residency and fellowships in psychiatric epidemiology and neuropsychiatry at Hopkins. He also obtained a master's degree in clinical epidemiology.[157]

For more than 20 years, Lyketsos has worked to understand Alzheimer's disease and ensure enlightened care of its sufferers. His research has focused on the range of psychiatric problems, such as depression and delusions, that patients suffer in addition to the decline in their comprehension of the world around them. He has formed a broadly-skilled team of researchers to probe for Alzheimer's biomarkers, the biological signatures of the disease that not only would quicken its diagnosis but could also lead to therapies.[158]

The Hopkins Bayview psychiatry department's extensive clinical services include treatment not only for Alzheimer's disease and dementia but chronic mental illness, substance abuse disorders, traumatic brain injury, sleep disorders and behavioral medicine—which connects psychiatry with other areas of medicine. These include HIV/AIDS, burn treatment, cardiovascular disease and chronic pain. Inpatient, partial hospitalization and outpatient programs treat adults, adolescents and children in all these areas.[159]

The department's research concentrations include substance abuse, memory disorders, mind-body interaction, sleep disorders and behavioral biology. The two divisions of the National Institutes of Health that are located on the Hopkins Bayview campus—the National Institute on Aging and the National Institute on Drug Abuse—are active collaborators in the department's research, as well as provide training opportunities for interns and residents there.[160]

The Division of Infectious Diseases

The Department of Medicine's Division of Infectious Diseases of which David Thomas became chief in mid-2006 had become a colossus under his predecessor, John Bartlett.[161]

In the 26 years that Bartlett led the division, it grew from just three full-time staff and a budget of $200,000 in 1980 to a roster of 55 faculty and a staff of 177 in 2006. It treats more than 5,100 patients annually and has a research budget of $40 million—one of Hopkins' largest. Bartlett, whose clinical and research achievements were detailed in Chapter 4, remains active on the faculty.[162]

Thomas, who arrived at Hopkins as a post-doctoral fellow in 1990, has become a world-renowned expert on hepatitis C, the leading cause of liver disease in the United States and an infection that kills an estimated 10,000 to 12,000 Americans annually. He also has investigated how co-infections with hepatic C viruses and HIV grow in intravenous drug users with weakened immune systems. Because hepatitis C is transmitted via infected blood and possibly other body fluids, drug-users' self-injections are its leading cause.[163]

In addition to the division's studies on hepatitis C, other major research initiatives focus on HIV/AIDS, tuberculosis, influenza, and methicillin-resistant *Staphylococcus aureus* (MRSA), a type of bacteria that is resistant to antibiotics such as methicillin and hence is much more difficult to treat. It can infect the skin, surgical wounds, the bloodstream, lungs or urinary tract and is particularly common among patients with weak immune systems, as well as those in hospitals and nursing homes. Although some antibiotics still can combat MRSA infections, it is a bacteria that constantly is adapting. MRSA now causes more than 60 percent of the staph infections in U.S. hospitals and research is intense on developing new antibiotics to fight it.[164]

The Department of Neurology

When John Griffin stepped down as head of the Department of Neurology in 2006, Justin McArthur was named interim director—an appointment that became permanent in 2008.[165]

A 1979 graduate of Guys Hospital Medical School in London, McArthur began his Hopkins career in 1980 with an internal medicine internship and residency. He then stayed on to complete a residency in neurology and obtain a masters degree in public health. He has devoted much of his career to investigating the neurological manifestations of AIDS—first advancing the development of methods for managing it, then treating it.[166] In addition, McArthur has become internationally known for his research on multiple sclerosis and other neurological infections and immune-mediated neurological disorders.[167]

A professor of pathology and epidemiology as well as neurology, McArthur is also the Director the of the Johns Hopkins/National Institute of Mental Health Research Center for Novel Therapeutics of HIV-associated Cognitive Disorders. The center's experienced, interdisciplinary research team pools their talents to study the nature of HIV-associated cognitive disorders. Their aim is to develop novel therapies for HIV-associated dementia (HIV-D).[168]

Among McArthur's own innovative efforts has been creation of his Cutaneous Nerve Laboratory in 1993. It has developed a better, simpler and less invasive method of performing skin biopsies in search of the small nerve fibers that are subject to sensory neuropathy, a disorder associated with diabetes and chemotherapy as well as HIV. When these small nerves are disrupted, such unusual sensations as pins-and-needles, pricks, tingling and numbness can afflict patients, in some cases causing disabling pain. The more McArthur's lab learns about these nerves and what affects them, the better will be the chances of developing treatments for such disorders, especially as patients live longer with AIDS.[169]

The 113-member faculty of Hopkins' neurology department contains some of the most productive neuroscientists in the world, McArthur notes, with many of them publishing research findings that are among the most widely cited by others in the field.[170]

At the bedside, physicians in the department's Division of Neuroscience Critical Care strive to preserve the brain function of stroke and tumor victims; in the Division of Cognitive Neuroscience, they are dementia detectives, searching for clues on Alzheimer's causes and how to treat or cure it; and in the Multiple Sclerosis Center, they are pursuing nerve-restoring therapies both for it and its more quickly acting first cousin, transverse myelitis. In other department branches, they're decoding the causes of autism; stalking the brain and spinal cord tumors caused by neurofibromatosis, a genetic disease that is 10 times more common than cystic fibrosis; and unraveling the mysteries of the more than 100 types of peripheral neuropathies, many of which cause pain that won't respond to current medications.[171]

2007

The McKusick/Nathans Institute of Genetic Medicine
The McKusick/Nathans Institute of Genetic Medicine, described in earlier chapters, and sometimes known for short as the Institute of Genetic Medicine (IGM), got a new director in March 2007, when Aravinda Chakravarti stepped down following seven years as its leader. He wished to concentrate on his research and became the founding director of the new Center for Complex Disease Genomics. Named to succeed him was David Valle, a nearly three-decade veteran of Hopkins who heads the Center for Inherited Disease Research and also treats patients with genetic diseases.[172]

Valle came to Hopkins in 1975 as a resident in pediatrics. His research interests have focused on inborn errors of the metabolism, retinal degeneration, and peroxisomal disorders. These are congenital diseases caused by the absence of normal peroxisomes—special parts within a cell that contain enzymes necessary for critical cellular processes. Children born with these disorders suffer mental disabilities, blindness, hearing loss, physical deformities and often die during infancy. Valle first made a name for himself with his 1978 work on an inherited form of blindness called gyrate atrophy, and he ran the pediatric genetics clinic during his early years on the faculty.[173] In recent years, Valle has focused his research on identifying genes that increase the risk for developing such neuropsychiatric disorders as schizophrenia and bipolar disease.[174]

As head of the IGM, Valle wants to enable every member of the Hopkins Medicine faculty to benefit from its research.

He doesn't necessarily mean that every person on the faculty should participate in genetic research, but he encourages them to avail themselves of everything the institute has to offer—including its host of exotic genetic research tools and the genetic advice available from its 13 genetic counselors and 69 full-time and part-time faculty. Citing the host of dramatic advances in genetic medicine discoveries in recent years, Valle envisions a future in which IGM's impact can only grow.[175]

The Department of Biomedical Engineering
To hear Elliot McVeigh tell it, happenstance has played a major part in his career.[176]

Named as the new director of the Department of Biomedical Engineering in September 2007, McVeigh says he didn't set out to pursue MRI technology at all.[177]

As an undergraduate at the University of Toronto, he majored in physics. Then he just happened to pick up an issue of *Scientific American* that featured an article on some of the first commercially available MRI machines. "I was hooked," he says. After obtaining a Ph.D. in medical biophysics in 1988, he came to Hopkins. Instead of visiting the person with whom he initially was supposed to interview, he just happened to drop by to see a young radiologist in body MRI named Elias Zerhouni. Zerhouni showed him "some very early images of myocardial tagging—and, again, I was hooked."[178]

McVeigh went to work with Zerhouni in the Department of Radiology. Instead of following his plan to spend just a year here and then return to Toronto, where a faculty appointment awaited him at his alma mater, he settled contentedly in Baltimore.[179]

McVeigh joined the biomedical engineering (BME) department in 1991. Over the next eight years, he mentored 13 Ph.D. students, oversaw the work of master's candidates, and taught undergraduates.[180] In 1999, offered an opportunity to become a senior investigator in the Laboratory of Cardiac Energetics at the NIH, he relocated to its Bethesda headquarters while maintaining a part-time faculty appointment at Hopkins. Then came the opportunity to return to Hopkins and succeed Murray Sachs as head of biomedical engineering—with its exceptional record of accomplishments and ongoing achievements, as described earlier.

McVeigh is maintaining the growth and excellence of the department's current 89-member faculty, many of whom have joint appointments in other departments and divisions of the university. Spanning both the School of Medicine and Homewood campuses, BME now encompasses an undergraduate training program with more than 500 students, a graduate program with more than 200 master's degree and doctoral candidates, and more than 30 postdoctoral fellows engaged in research that fills laboratories in the schools of medicine, engineering and arts and sciences. BME also maintains close ties with Hopkins' Applied Physics Laboratory, the Center for Hearing and Balance, the Mind Brain Institute, the Center for Cardiovascular Bioinformatics and Modeling, the Institute for Computational Medicine, the Center for Imaging Science and the Center for Magnetic Resonance Micro-imaging.[181]

2008

The Sidney Kimmel Comprehensive Cancer Center

In an elegant downtown Baltimore hotel banquet hall, the Kimmel Cancer Center's annual faculty dinner on May 3, 2007 featured an enormous outpouring of praise and affection for Martin Abeloff. As the center's head for 15 years, he had doubled its faculty, increased its research funding sixfold, presided over construction of three separate cancer facilities with nearly 1 million square feet of treatment and research space, and established a world-wide reputation for Hopkins' oncology.[182]

Abeloff, a 1963 graduate of the School of Arts and Sciences and 1966 graduate of the School of Medicine, was relinquishing the center's directorship, and the accolades for him were appropriately expansive, often good-humored, and in a way bittersweet. They were marking the end of what many were sure would be known as "The Abeloff Era" in cancer treatment and research.[183] As much as the speakers and a moving video tribute lauded Abeloff as a clinician and researcher—an internationally acclaimed expert on lung and breast cancer who planned to remain actively involved in the center—even more so they heralded "Marty the man," a preternaturally humble person whose key purpose was the relief of suffering.[184]

What most of the attendees in the packed banquet hall did not know was that Abeloff had been fighting leukemia for a year. He died of it four months later at the age of 65—immensely saddening the entire Hopkins Medicine community.[185]

Selecting a successor to Abeloff took more than a year, but in November 2008, prostate cancer and genetics expert William Nelson, a 1987 graduate of the School of Medicine with both an M.D. and Ph.D., and an oncology faculty member since 1992, was named head the Kimmel Cancer Center.[186]

Nelson, a national leader in transferring research findings to clinical use, had been part of the Hopkins team that discovered the most common genome alteration in prostate cancer. That discovery led to new diagnostic tests for the disease and fueled interest in developing innovative drugs and other treatment options for it.[187] He served as the cancer center's associate director for translational research and co-director of the Prostate Cancer Program. He also was a leader of the center's partnership program with Howard University, an effort aimed at building research capabilities at that predominantly African American institution in Washington, D.C. and enhancing minority participation in cancer research.[188]

Nelson considers the advances made in his particular specialty, prostate cancer, to be a model for the type of progress that needs to be made in battling all types of malignancies. "When I started in [prostate] oncology, men were commonly diagnosed at an advanced stage and death rates were far too high," he says. Since then, development of the PSA blood test has made earlier diagnosis and treatment—in some cases, innovative treatment—possible.[189]

"Our mission, our brand, if you will, is to have better than the state-of-the-art cancer therapy," Nelson says. His vision for the cancer center is to make it a leader in translational research—ensuring that what is learned in the lab gets into the hands of clinicians as rapidly as possible.[190]

2009

A number of the key divisional leadership positions filled in 2009 had to do with matters of the heart.

The Division of Cardiology

In March 2009, little more than a month after Gordon Tomaselli was named chief of the Division of Cardiology, he tapped Edward Kasper, an expert in chronic heart failure and the heart transplants that such an illness often requires, to be the new clinical director of the division and co-director of the Heart and Vascular Institute.[191]

Kasper had spent nearly his entire academic medical career at Hopkins. A member of the class of 1979 in the School of Arts and Sciences and then a graduate of the University of Connecticut medical school, he served his internship, residency and a cardiology fellowship at Hopkins Hospital. Joining the faculty in 1987, he left briefly to head the heart transplant program at Vanderbilt University before returning to Hopkins in 1993 to take over the leadership of the transplant service here. He rose to become head of cardiology at Johns Hopkins Bayview while treating thousands of patients and establishing an impressive record as a teacher of medical students, clinical residents and research fellows.[192]

As a researcher, Kasper is an expert on the biological origins of heart failure and the underlying reasons why the body rejects some transplanted hearts and not others. He has been in the forefront of the research on the possible use of special blood tests to predict the earliest signs of heart failure and an organ recipient's risk of having a transplanted heart rejected.[193]

As clinical director of the division at Hopkins Hospital, he oversees more than 700 people, including 86 cardiologists among the 102 faculty and 87 fellows who treat more than 4,000 inpatients and nearly 30,000 outpatients annually.[194]

The Division of Pediatric Cardiac Surgery

Treating the littlest of human hearts is the specialty of Luca Vricella, an ebullient native of Italy who was named head of Pediatric Cardiac Surgery in April 2009.[195]

Born and raised in Rome, Vricella comes from a family of physicians, including his father, two uncles and his twin brother, who is a plastic surgeon. His own decision to become a pediatric cardiac surgeon occurred while he sat watching an open heart operation on a small black and white television. "I thought, This is just absolutely fantastic." He was 10 at the time and never looked back.[196]

Graduating from the Catholic University School of Medicine in Rome, Vricella completed various residencies and fellowships in the U.S. and England before coming to Hopkins in 2001 for further training. He joined the faculty in 2003 as an assistant professor of surgery.[197]

When Duke Cameron became acting chief of the Division of Cardiac Surgery and cardiac surgeon in chief at Hopkins Hospital in 2009—an appointment that would become permanent in August 2010—he picked Vricella to succeed him as head of pediatric cardiac surgery.[198]

Vricella, whose work featured prominently in ABC-TV's award-winning 2008 documentary series, *Hopkins*, says the thrill of his specialty never has diminished.[199] In March 2009, he and Cameron travelled to Pavia, Italy, to sign a three-year collaboration agreement between Johns Hopkins Medicine International (JHI) and San Matteo Hospital. The collaboration was JHI's first cardiac surgery project of its kind in Europe. Hopkins cardiac surgeons, perfusionists, nurses and anesthesiologists agreed to join their Italian colleagues at San Matteo to offer their expertise and conduct up to 20 surgeries a year with them. In turn, three observers from San Matteo would visit Hopkins Hospital each year to learn about its latest clinical research and innovations.[200]

The Division of Cardiac Surgery

After 17 years as chief of the Division of Cardiac Surgery and cardiac surgeon in chief of Hopkins Hospital, Bill Baumgartner decided to step down from those positions in 2009. Retaining his jobs as vice dean for clinical affairs, director of the cardiac surgery research laboratory, president of the Clinical Practice Association—and accepting a new post as executive director of the Chicago-based American Board of Thoracic Surgery—he had plenty on his plate.[201]

Succeeding Baumgartner was Duke Cameron, a 25-year Hopkins veteran who has earned international acclaim for his work in preventing catastrophic internal bleeding by surgically repairing the aorta, the heart's main blood vessel. He also became director of the division's Dana and Albert "Cubby" Broccoli Center for Aortic Diseases, named for the late Hollywood producer of the James Bond movies, a long-time Hopkins heart patient, and his wife. The center conducts research into the causes and treatments of disorders that affect the aorta, such as Marfan's and Loeys-Dietz syndrome, described in Chapter 6.[202]

Cameron began his Hopkins career as a resident in cardiac surgery in 1984. An award-winning researcher and leader of professional organizations, Cameron also has undertaken humanitarian efforts that have spanned the globe. He has performed life-saving heart surgeries free of charge for both children and adults in Cuba, Egypt, and Saudi Arabia.[203]

As head of the cardiac surgery division, he leads a group of nearly 250 faculty, trainees and staff, including surgical residents, perfusionists, physician assistants and operating room and intensive care nurses. He oversees an annual research budget of $1.3 million, of which $1 million comes from the National Institutes of Health. This research delves into the study and evaluation of new methods of surgical treatments and the improvement of surgical results, including clinical and laboratory inquiries into such areas as protection of the brain and neurological systems during bypass operations and the future use of stem cells for treating patients with congestive heart failure.[204]

The division offers a complete range of cardiac surgeries, from coronary artery bypass operations to valve replacement, correction of congenital heart defects, minimally invasive cardiac surgery, and correction of atrial fibrillation, or irregular heart beats that can lead to blood clots and stroke. Comprehensive treatment also is provided for congestive heart failure and severe lung disease through surgical remodeling of the heart's ventricles (chambers), implantation of devices to help the ventricles work better, or heart and lung transplantations.[205]

The division's surgeons—some of the most experienced and skilled in the world—also can give patients the advantage of an on-site, multi-disciplinary team of colleagues in other divisions and departments, including cardiologists, electrophysiologists, interventional radiologists, vascular surgeons, advanced practice nurses, nurse practitioners, dieticians, physical therapists and social workers. This many-faceted approach and expertise have given Hopkins' cardiac surgery division the reputation as among the best-prepared practices for handling procedures that many be complex and/or complicated by a patient's other underlying illnesses or age.[206]

The Division of Pediatric Oncology

It isn't every day that a gene is cloned, let alone one that, when it mutates, plays a paramount role in a common form of childhood leukemia. Yet in 1992, that's what Donald Small and his research team did in his laboratory: discover and clone a mutated gene they dubbed FLT3. It is linked to acute myelogenous leukemia (AML), one of the most common blood cancers afflicting both children and adults.[207]

Once Small and his team identified the gene responsible for AML growth, they set about to stop it in its destructive tracks. They have identified small molecules able to turn off the FLT3 gene receptor, target and kill the cancer cells, and leave the normal blood cells unharmed. FLT3 inhibitor drugs now are being tested in clinical trials to examine their effectiveness in children and adults with FLT3-mutant AML, as well as in infants with acute lymphocytic leukemia (ALL).[208]

"It's an exciting period in research because we're finding mutant genes that cause cancer and have existed since the beginning of time," Small says. "Once we find them, we can develop drugs specific to their mutations…. It will enable us to improve cure rates and decrease treatment side effects. It's really the molecular era."[209]

It's also the Donald Small era in the Kimmel Comprehensive Cancer Center's pediatric oncology division. Seven months after his own appointment as director of the Kimmel Comprehensive Cancer Center, Bill Nelson chose Small to be chief of the division, which he had served as interim director since 2006.[210]

A 1979 graduate of the Krieger School of Arts and Sciences, a 1985 M.D. and Ph.D. graduate of the School of Medicine, and an intern, resident and fellow at Hopkins Hospital, Small had spent 32 years at the university, 19 of them as on the faculty.

Along with Small, clinical director Kenneth Cohen, and 11 other specialists, the division's staff tackles a potentially deadly baker's dozen of malignancies. In addition to leukemia, these include brain and spinal tumors; Ewing's sarcoma, a bone tumor that often occurs in the arms, legs, pelvis or chest; germ cell tumors that arise in children's reproductive organs and other locations, such as the lower back abdomen, chest and brain; histiocytosis, which are disorders of the immune system; and lymphoma, or cancers of the lymphoid system, which is part of the immune system.[211]

The researchers and physicians also study and treat neuroblastoma, which are solid tumors that are second only to brain tumors as malignancies affecting children; osteogenic sarcoma, another form of bone cancer; retinoglastoma, tumors of the eye that begin in the retina; rhabdomyosarcoma, which are muscle cancers; and Wilms' tumor, a kidney cancer that mostly afflicts children.[212]

2010

The Department of Plastic Surgery

Plastic surgery has a long and distinguished history at Hopkins—in spite of the fact that for decades the leadership of the Department of Surgery gave it surprisingly short shrift.[213]

That short-sightedness long has been reversed. With 13 superb plastic surgeons on the faculty, expert treatment in a wide variety of cases is provided. These include adult and pediatric craniofacial surgery, bariatric surgery, breast reconstruction following cancer treatment, microsurgery, head and neck cancer and reconstruction surgery, liposuction, wound healing, facial paralysis, nerve repair, cosmetic surgery and hand surgery.[214]

The importance to Hopkins of hand surgery and all the other specialties within the plastic surgery field was emphasized forcefully in the autumn of 2010, when W.P. Andrew Lee, a Hopkins-trained surgeon who had gone on to perform pioneering hand-transplant surgery at the University of Pittsburgh's Medical Center (UMPC), was recruited to become director of a new Department of Plastic Surgery.[215]

The elevation of plastic surgery to the departmental status from the divisional classification it had held since the early 1940s would have pleased John Staige Davis (1872–1946), an 1899 graduate of the School of Medicine who became perhaps the first physician in the nation to concentrate his entire career on plastic surgery.[216]

Working under surgeon in chief William Halsted following his graduation, Davis produced the first-ever textbook on reconstructive surgery, *Plastic Surgery, Its Principles and Practice*, in 1919. He sent copies to Halsted, Welch and Whitridge Williams, then dean of the medical school. All of them failed to acknowledged receipt of it.[217]

Davis pressed on. While engaged in private medical practice in Baltimore, he continued performing plastic surgery research, surgeries and outpatient treatment at Hopkins, becoming so renowned that in 1925 he earned an honorary degree from Yale, which cited him as "a pioneer in plastic surgery." He became the first chairman of the American Board of Plastic Surgeons, president of the Southern Surgical Association, and ultimately president of the American Association of Plastic Surgeons.[218] When Alfred Blalock created a plastic surgery division in 1942, he asked Davis, then 70, to head it. He did so until shortly before his death in 1946. Today, Maryland's professional organization for his specialty is called the John Staige Davis Society of the Plastic Surgeons of Maryland.[219]

A number of distinguished physicians followed in Davis' path as head of the plastic surgery division, including Paul Manson, who became its head in 1990. Internationally known for his work in facial and jaw injury repair, he is the author of hundreds of research papers and recipient of more than $10 million in research grants, awards and endowments. He has served as president of the American Association of Plastic Surgeons, the Association of Program Directors in Plastic Surgery, the American Society of Academic Chairman in Plastic Surgery—and the Maryland society of plastic surgeons named for his illustrious predecessor.[220]

In recruiting Andrew Lee to head the new Department of Plastic Surgery, Hopkins was bringing home a 1983 graduate of the School of Medicine who had spent a decade here as a medical student, intern and resident, then gone to become renowned for performing breakthrough hand-transplant surgeries that set new standards for the specialty.[221]

On Mar. 14, 2009, Lee led a team of UPMC surgeons who performed an historic, 11-hour surgery on a 25-year-old Marine from Pennsylvania who had lost his right hand during a military accident, giving him the hand of a West Virginia man who had died. The operation was only the fifth one of its kind performed in the United States. Lee followed it up with a second, even more historic operation on May 4, 2009, when he performed the first two-hand transplant in the U.S. on a Georgia man who had lost both of his hands to a bacterial infection. Lee performed a second double hand transplant on a Pennsylvania man on Feb. 5, 2010. All three operations involved the use of just one anti-rejection drug instead of the usual three in a procedure that came to be known as "the Pittsburgh Protocol."[222]

Julie Freischlag, director of the Department of Surgery and a member of the team that recruited Lee, said the hand surgeon "brings to Hopkins not only his surgical credentials but a demonstrated ability to envision the future and build world-class programs."[223]

In other words, he epitomizes what every other Hopkins Medicine clinician, researcher and teacher strives to do.

2011

The Division of Surgery at Johns Hopkins Bayview

In the summer of 2011, Thomas Magnuson, who had arrived at Hopkins as a surgical resident in 1992 and went on to found the Johns Hopkins Center for Bariatric Surgery, one of the nation's top obesity surgery programs, was named chief of the Division of Surgery at Johns Hopkins Bayview.[224]

Along with Magnuson, Mark Duncan, chief of surgical oncology at Hopkins Bayview, was named vice chair of the Department of Surgery in the School of Medicine. A native of Sydney, Australia, Duncan has been at Hopkins since 1996 and headed surgical oncology at Bayview since 1998.[225]

The 20 surgeons on the Hopkins Bayview staff have at their command the latest in surgical technology, including videoscopic and minimally invasive approaches to disorders requiring surgery. In addition to general surgery, its specialties include bariatric surgery, providing a weight-loss option for morbidly obese patients; burn surgery; plastic and reconstructive surgery; thoracic surgery for diseases of the lung, esophagus, chest wall and pleural cavity; and vascular surgery in the Johns Hopkins Vein Center located at Bayview.[226]

Appendix B

Nobel Prize Recipients

Hopkins alumni, faculty and fellows who have received the Nobel Prize for their contributions to basic science or clinical research.

Thomas Hunt Morgan
Ph.D. 1890 (Zoology); LL.D. 1915
Nobel Prize in Medicine, 1933
Awarded for his discoveries concerning the role played by the chromosome in heredity.

George R. Minot
Assistant in Medicine, 1914–1915
Nobel Prize in Medicine, 1934

George Hoyt Whipple
M.D. 1905; Associate Professor in Pathology, 1910–1914
Nobel Prize in Medicine, 1934
Awarded jointly for their discoveries concerning liver therapy in cases of anemia.

Harold C. Urey
Associate in Chemistry, 1924–1928
Nobel Prize in Chemistry, 1934
Awarded for his discovery of heavy hydrogen.

Joseph Erlanger
M.D. 1899; Assistant in Physiology, 1900–1901; Instructor, 1901–1903; Associate, 1903–1094; Associate Professor, 1904–1906; L.L.D. 1947
Nobel Prize in Physiology, 1944

Herbert Spencer Gasser
M.D. 1944
Nobel Prize in Physiology, 1944
Awarded jointly for their discoveries relating to the highly differentiated functions of single nerve fibers.

Vincent du Vigneaud
National Research Fellow, Pharmacology, 1927–1928
Nobel Prize in Chemistry, 1955
Awarded for his work on biochemically important sulphur compounds, especially for the first synthesis of a polypeptide hormone.

Peyton Rous
M.D. 1905
Nobel Prize in Medicine, 1966
Awarded for his discovery of tumor-inducing viruses.

Haldan Keffer Hartline
M.D. 1927; Professor of Biophysics, 1949–1954; L.L.D. 1969
Nobel Prize in Physiology, 1967
Awarded for discoveries concerning the primary physiological and chemical visual processes in the eye.

Christian Anfinsen
Professor of Biology, 1982–1985
Nobel Prize in Chemistry, 1972
Awarded for his work on ribonuclease, especially concerning the connection between the amino acid sequence and the biologically active conformation.

Hamilton O. Smith
M.D. 1956; Assistant Professor of Microbiology, 1967–1969; Associate Professor, 1969–1973; Professor, 1973–1998; Professor Emeritus, 1998–present
Nobel Prize in Medicine, 1978

Daniel Nathans
Assistant Professor, 1962–1965; Associate Professor, 1965–1967; Professor of Molecular Biology and Genetics, 1967–1999; Interim President, 1995–1996
Nobel Prize in Medicine, 1978
Awarded jointly for their discovery of restriction enzymes and the application of them to problems of molecular genetics.

David H. Hubel
Assistant Resident, Neurology, 1954–1955; Fellow, Neuroscience, 1958–1959
Nobel Prize in Medicine, 1981

Torsten N. Wiesel
Fellow, Ophthalmology, 1955–1958; Assistant Professor, 1958–1959
Nobel Prize in Medicine, 1981
Awarded jointly for discoveries concerning information processing in the visual system.

Martin Rodbell
B.A. Biology 1949
Nobel Prize in Medicine, 1994
Awarded for discovery of G-proteins and the role they play in signal transduction in cells.

Paul Greengard
Ph.D. Biophysics 1953
Nobel Prize in Medicine, 2000
Awarded for discoveries concerning signal transduction in the nervous system.

Peter Agre
M.D. 1974; Postdoctor fellow, Department of Pharmacology, 1974–1975; Research Associate/Instructor, Cell Biology and Anatomy, and Medicine, 1981–1983; Assistant Professor, 1984–1988; Associate Professor, 1988–1993; Professor of Biological Chemistry and Medicine, 1993–2005; director, Malaria Institute, 2008–present.
Nobel Prize in Chemistry, 2003
Awarded for the discovery of aquaporins, the water channels in cells.

Richard Axel
M.D. 1971
Nobel Prize in Medicine, 2004
Awarded for discoveries of odorant receptors and the organization of the olfactory system.

Andrew Fire
Adjunct Professor of Biology, 1989–present
Nobel Prize in Medicine, 2006
Awarded for discovery of the RNA interference-gene silencing by double-stranded RNA.

Carol Greider
Daniel Nathans Professor and Director of Molecular Biology and Genetics, Institute of Basic Biomedical Sciences, 1997–present.
Nobel Prize in Physiology or Medicine, 2009
Awarded for the discovery of how chromosomes are protected by telomeres and the enzyme telomerase.

Robert Edwards
Visiting Fellow, Crytogenetics Laboratory, Department of Gynecology and Obstetrics, 1965
Nobel Prize in Physiology or Medicine, 2010
Awarded for the development of in vitro fertilization.

Appendix C

Award Winners

Lasker Awards

Hopkins Medical School alumni and faculty who have received the Albert and Mary Lasker Foundation Medical Research Awards, the "American Nobel," or the Lasker Special Achievement or Public Service Awards, which recognize "the contributions of scientists, physicians, and public servants who have made major advances in the understanding, diagnosis, treatment, cure, and prevention of human disease" through basic medical or clinical research.

Sir David Weatherall (2010)

A fellow in medicine and hematology from 1963 to 1965, Weatherall received the 2010 Lasker-Koshland Special Achievement Award in Medical Science for 50 years of international statesmanship in biomedical science—exemplified by discoveries concerning genetic diseases of the blood and for leadership in improving clinical care for thousands of children with thalassemia throughout the developing world.

Charles Sawyers (2009)

A 1985 graduate of the School of Medicine, Sawyers received the 2009 Lasker-DeBakey Clinical Medical Research Award for development of molecularly-targeted treatments for chronic myeloid leukemia, converting a fatal cancer into a manageable chronic condition.

Carol Greider (2006)

A member of the faculty since 1997 and director of the Department of Molecular Biology and Genetics since 2003, Greider received the 2006 Award for Basic Medical Research jointly with Elizabeth Blackburn of the University of California, San Francisco, and Jack Szostak of Harvard, for her 1984 discovery of telomerase, an enzyme that maintains the length and integrity of chromosome ends known as telomeres and is critical for the health and survival of all living cells and organisms. In 2009, she shared the Nobel Prize in Physiology or Medicine with Blackburn and Szostak for their pioneering research on the structure of telomeres.

Alfred Sommer (1997)

A 1973–1976 resident and fellow in ophthalmology and Dean of the Bloomberg School of Public Health from 1990 to 2005, Sommer received the 1997 Award for Clinical Research for understanding and demonstrating that low-dose vitamin A supplementation in millions of third world children can prevent blindness, as well as death from infectious diseases.

Victor McKusick (1997)

A 1946 graduate of the School of Medicine, McKusick received a Special Achievement in Medical Science Award for his lifetime career as a founder of the discipline of medical genetics.

John Allen Clements (1994)

A member of the Hopkins Hospital house staff from 1956 to 1961, Clements received the 1994 Award for Clinical Medical Research for his brilliant studies defining and describing the role of pulmonary surfactant and in developing a life-saving artificial surfactant.

Maclyn McCarty (1994)

A 1937 graduate of the School of Medicine, McCarthy received a Special Achievement Award in 1994 for his seminal and historic investigation which revealed that DNA is the chemical substance of heredity and for ushering in a new era of contemporary genetics.

Leroy Hood (1987)

A 1964 graduate of the School of Medicine, Hood received the 1987 Award for Basic Medical Research for his prolific and imaginative studies of somatic recombinations in the immune system detailing in molecular terms the genetics of antibody diversity.

Vernon Mountcastle (1983)

A 1942 graduate of the School of Medicine and director of the Department of Physiology from 1964 to 1980, Mountcastle received the 1983 Award for Clinical Medical Research for discovering the cellular structure of the brain and illuminating its ability to perceive and organize information and translate sensory impulses into behavior.

Ronald Finn (1980)

A 1962 post-graduate fellow in medicine, Finn received the Award for Clinical Medical Research for his dedicated and painstaking studies revealing how the unborn child is protected from the immune system of the mother, and for documenting the sequence of events leading to Rh disease in the child.

Solomon Snyder (1978)

A member of the faculty since 1965 and founding director of what now is the Solomon H. Snyder Department of Neuroscience, leading it from 1980 until 2005, Snyder received the Basic Medical Research Award in 1978 for discovering the brain's opiate receptors.

Robert Austrian (1978)

A 1941 graduate of the School of Medicine and a member of the faculty from 1943 to 1952, Austrian received the Award for Clinical Medical Research for the development and clear demonstration of the efficacy of purified polysaccharide vaccine against pneumococcal diseases.

Henry G. Kunkle (1975)

A 1942 graduate of the School of Medicine, Kunkle received the 1975 Award for Basic Medical Research for his outstanding contribution to the creation of a new medical discipline, immunopathology.

C. Gordon Zubrod (1972)

A fellow in medicine from 1946 to 1949, Zubrod received a Special Award for Clinical Medical Research for his leadership in expanding the frontiers of cancer chemotherapy.

Lyman C. Craig (1963)

A 1933 post-doctoral fellow in chemistry, Craig received the 1963 Award for Basic Medical Research for his countercurrent distribution technique, used as a method for the separation of biologically significant compounds, and for the isolation and structure studies of important antibodies.

Arnall Patz (1956)

A member of the Wilmer Eye Institute faculty beginning in 1955 and its director from 1979 to 1989, Patz received the Award for Clinical Research jointly with V. Everett Kinsey for the discovery that giving excessive oxygen to premature babies is a cause of retrolental fibroplasias, a blinding disorder.

Francis Peyton Rous (1958)

A 1905 graduate of the School of Medicine, Rous received the 1958 Award for Basic Medical Research for his invaluable contributions to the essential understanding of the causes of cancers, the source of antibodies, and the mechanism of blood cell generation and destruction in humans.

Alfred Blalock and Helen Taussig (1954)

A 1922 graduate of the School of Medicine, Blalock received the Award for Clinical Research jointly with Helen Taussig, a 1927 School of Medicine graduate, and Robert E. Gross, a 1927 graduate of Harvard, for distinguished contributions to cardiovascular surgery and knowledge.

Edwin B. Astwood (1954)

A 1935-37 Fellow in Pathology, Astwood received the 1954 Basic Research Award for basic contributions to our knowledge of endocrine function, leading to the control of hyperthyroidism.

Frederic A. Gibbs (1951)

A 1929 graduate of the School of Medicine, Gibbs received the 1951 Award for Clinical Research jointly with William C. Lennox for their research in epilepsy.

Florence Sabin (1951)

A 1900 graduate of the School of Medicine and the first woman to be named a full professor in the School of Medicine, received the 1951 Public Service Award for outstanding accomplishments in public health administration as head of the Health Committee of the Governor of Colorado's Post-War Planning Committee.

William S. Tillet (1949)

A 1917 graduate of the School of Medicine, Tillet received the Basic Medical Research Award jointly with L.R. Christensen for the discovery and purification of the enzymes streptokinase and streptodonase.

Martha M. Eliot (1948)

A 1918 graduate of the School of Medicine, Eliot received the 1948 Public Service Award for administrative achievement in the organization and operation of the Emergency Maternal and Infant Care Program.

Alan F. Guttmacher (1947)

A 1919 graduate of the School of Medicine, an associate professor of obstetrics from 1926 to 1952, and founder of the famed Guttmacher Institute in New York, he received a special award bestowed by Planned Parenthood for his work in birth control.

W. Horsley Gantt (1946)

Protégé of legendary Russian physiologist Ivan Pavlov and founder of the Hopkins School of Medicine's Pavlovian Laboratory, which he headed from 1929 to 1980, Gantt received a special award from the National Committee Against Mental Illness.

Science Medal

Hopkins Medicine alumni and faculty who have received of the National Medal of Science, the nation's highest scientific honor.

Solomon Snyder (2003)

A member of the faculty since 1965 and founding director of what now is the Solomon H. Snyder Department of Neuroscience, leading it from 1980 until 2005, Snyder received the 2003 National Medal of Science during a March 2005 White House ceremony for his "major contributions to the understanding of neurotransmitters, their receptors in the nervous system, mechanisms of action of psychoactive drugs, and pathways of signal transduction in the brain."

Daniel Nathans (1993)

A School of Medicine faculty member from 1962 until his death in 1999; director of the Department of Molecular Biology and Genetics from 1981 to 1995; interim president of The Johns Hopkins University from June 1995 to August 1996; and co-winner of the 1978 Nobel Prize in Medicine or Physiology for his role in the discovery of restriction enzymes, he was honored with the National Medal of Science for "his seminal research in molecular genetics that formed the foundation for contemporary biotechnology."

Donald A. Henderson (1986)

A 1960 graduate of the School of Public Health and its dean from 1977 to 1990, he directed the smallpox eradication campaign of the World Health Organization (WHO) from 1966 to 1977, he also was instrumental in initiating WHO's global program of immunization, which has vaccinated 80 percent of the world's children against six major diseases and aims to eradicate polio. Now a Hopkins Distinguished Service Professor and a Resident Scholar at the Center for Biosecurity of the University of Pittsburgh, he was honored in a White House ceremony for "his leading role as chief architect and implementer of the World Health Organization's successful global eradication of smallpox."

Vernon Mountcastle (1986)

A 1942 graduate of the School of Medicine and director of the Department of Physiology from 1964 to 1980, Mountcastle received the medal in a White House ceremony for "his fundamental research on how the brain functions in processing and perceiving the information gathered through the somatic sensory system."

Presidential Medal

Hopkins Medicine recipients of the Presidential Medal of Freedom, the nation's highest civilian honor.

Benjamin S. Carson, Sr. (2008)

Beginning as a member of the surgical house staff in 1977, joining the faculty in 1984, and at the age of 32, becoming Hopkins' youngest-ever director of pediatric neurosurgery, Carson was honored at a White House ceremony for his innovative and groundbreaking neurosurgeries, which include separations of craniopagus (conjoined) twins joined at the head, and heispherectomies, surgeries in which a portion of a child's brain is removed to stem intractable seizures; as well as for work as a motivator of youth and philanthropist.

Arnall Patz (2004)

A member of the Hopkins faculty since 1955; head of the Wilmer Eye Institute from 1979 to 1989; and an active professor emeritus until his death in 2010, Patz was praised for establishing "the gold standard in the field of researching the causes and treatment of eye disease" during the White House ceremony in which he received the Presidential Medal of Freedom.

Helen Taussig (1964)

A 1927 School of Medicine graduate and pioneering pediatric cardiologist who joined with Alfred Blalock and Vivien Thomas to develop the landmark "blue baby" operation, which launched modern cardiac surgery, Taussig was honored for her distinguished contributions to cardiovascular research, treatment, and mentorship of the next generation of top pediatric leaders.

MacArthur Foundation Fellowship

Hopkins Medicine faculty who have been among the recipients of the John D. and Catherine T. MacArthur Foundation fellowships, awarded since 1981 and popularly known as the "genius award." These fellowships provide unrestricted grants (currently $500,000) to talented individuals in the arts, sciences, humanities, education, business and other fields who have shown extraordinary originality and dedication in their creative endeavors and a clear capacity for future achievements.

Alan Walker (1988)

A professor of cell biology and anatomy in the School of Medicine from 1978 to 1996 and a renowned paleoanthropologist, his research on fossils has yielded remarkable insights into the evolution of humans.

Kay Redfield Jamison (2001)

A professor of psychiatry in the School of Medicine, one of the world's leading experts on bipolar disorder, Jamison is the author of such acclaimed books as *Manic-Depressive Illness*, the classic text on bipolar disorder, and *An Unquiet Mind*, describing her own experiences with severe depression and mania. A member of the Hopkins faculty since 1987, she was recognized for her efforts to enhance mental health treatment, to improve patient support and advocacy, and to increase public awareness of psychiatric disorders. She has helped to increase understanding of suicide and serious mood disorders.

Geraldine Seydoux (2001)

A professor of molecular biology and genetics who joined the School of Medicine faculty in 1995, Seydoux is an innovative researcher whose pioneering work is helping to illuminate some of the most complex processes in biology. Her studies on the molecular genetics of the roundworm *C. elegans* focus on how cells develop into a fully formed adult animal. Although her work concentrates on the roundworm, many of the mechanisms she elucidates are evolutionarily conserved in other organisms and may well provide insights into the development of mammals and other animals.

Lisa Cooper (2007)

A native of Liberia, a 1993 graduate of the School of Public Health, and a professor of medicine in the School of Medicine, with joint appointments in the schools of Public Health and Nursing, Cooper has conducted landmark research that seeks to better define barriers to equitable care across ethnic groups and identify ways for medical science to address a growing awareness of racial and ethnic disparities in disease prevalence, disease risk and care delivery. Her scholarship on clinical communication is improving medical outcomes for minorities in the United States. Cooper has developed culturally tailored education programs designed to improve the diagnosis and treatment of hypertension and depression among African Americans.

Peter Pronovost (2008)

A 1991 graduate of the School of Medicine and a professor of anesthesiology, critical care medicine and surgery, with joint appointments in the schools of Public Health and Nursing, Pronovost is an internationally prominent critical care specialist, patient safety researcher and advocate whose seemingly simple tools for greatly improving patient safety and care—such as his "safety checklist" for reducing catheter-related bloodstream infections and methods for enhancing communication between caregivers—are changing the way the world thinks about medical care. By rigorously evaluating and skillfully implementing effective safety procedures, Pronovost is sparing countless lives from the often deadly consequences of human error and setting new standards of health care performance in the United States and internationally.

Appendix D

Members of the National Academy of Sciences

Name	Primary Institution	NAS Section	Election Year
Peter Agre	Johns Hopkins University (SOM and JHSPH)	Physiology and Pharmacology	2000
Nancy Craig	Johns Hopkins University (SOM)	Biochemistry	2010
Peter Devreotes	Johns Hopkins University (SOM)	Physiology and Pharmacology	2005
Harry C. "Hal" Dietz III	Johns Hopkins University (SOM)	Genetics and Pediatric Cardiology	2011
Carol Greider	Johns Hopkins University(SOM)	Biochemistry	2003
Diane Griffin	Johns Hopkins University (SOM and JHSPH)	Microbial Biology	2004
Richard Huganir	Johns Hopkins University (SOM)	Cellular and Molecular Neuroscience	2004
M. Daniel Lane	Johns Hopkins University (SOM)	Biochemistry	1987
John Littlefield	Johns Hopkins University (SOM)	Medical Genetics, Hematology and Oncology	1977
Vernon Mountcastle	Johns Hopkins University (SOM)	System Neuroscience	1966
Jeremy Nathans	Johns Hopkins University (SOM)	Cellular and Molecular Neuroscience	1996
Saul Roseman	Johns Hopkins University (SOM)	Biochemistry	1972
Gregg Semenza	Johns Hopkins University (SOM)	Medical Genetics, Hematology, and Oncology	2008
Solomon Snyder	Johns Hopkins University (SOM)	Cellular and Molecular Neuroscience	1980
Alfred Sommer	Johns Hopkins University (SOM and JHSPH)	Medical Physiology and Metabolism	2001
Paul Talalay	Johns Hopkins University (SOM)	Biochemistry	1987
Bert Vogelstein	Johns Hopkins University (SOM)	Medical Genetics, Hematology and Oncology	1992
King-Wai Yau	Johns Hopkins University (SOM)	Cellular and Molecular Neuroscience	2010

Appendix E

Members of the Institute of Medicine

Name	Affiliation	Department	Year Inducted
Peter Agre, M.D.	JHSPH/SOM	Molecular Microbiology and Immunology / Director of JH Malaria Research Institute	2005
Marion Ball, EdD	SON/SOM	SON professor / Affiliate Professor in the Division of Health Sciences Informatics (SOM)	1996
John Bartlett, M.D.	JHSPH/SOM	Founding Director, Center for Civilian Biodefense Strategies in Epidemiology / Division of Infectious Diseases	1999
Robert Black, M.D., M.P.H.	JHSPH	International Health	2002
Robert Blum, M.D., Ph.D., M.P.H.	JHSPH	Population, Family and Reproductive Health	2006
Henry Brem, M.D.	SOM	Neurosurgery / Professor in Oncology and Opthamology	1998
Gert Brieger, M.D., Ph.D.	SOM	History of Medicine	1985
Frederick Burkle, M.D., M.P.H.	JHSPH	Visiting Professor / Senior Scholar in International Emergency, Disaster and Refugee Studies	2007
Jacquelyn Campbell, PhD, RN, FAAN	SON	Community Public-Health	2000
Aravinda Chakravarti, P.H.D.	SOM/JHSPH	Medicine / Pediatrics / Molecular Biology / Genetics and JHSPH Biostatistics	2007
Barton Childs, M.D.	SOM	Genetics in the Department of Pediatrics	1978
Lisa Cooper, M.D., M.P.H.	JHMI/JHSPH	Epidemiology / Department of Medicine, Division of General Internal Medicine	2008
Chi Dang, M.D., Ph.D.	SOM	Vice Dean for Research / Professor of Medicine / Cell Biology / Oncology / Pathology	2006
Barbara de Lateur, M.D., M.S.	SOM	Physical Medicine and Rehabilitation / Joint Professor of Health Policy and Management, School of Hygiene and Public Health	1996
Kay Dickersin, Ph. D.	JHSPH	Center for Clinical Trials / Epidemiology	2007
Harry "Hal" Dietz, III M.D.	SOM	Institute of Genetic Medicine	2008
Ruth Faden, Ph.D., M.P.H.	JHSPH	Berman Institute of Bioethics / Department of Health Policy and Management	1994
Andrew Feinberg, M.D., M.P.H.	SOM	Medicine / Oncology / Molecular Biology and Genetics	2007
Manning Feinleib, M.D., Dr.P.H.	JHSPH	Epidemiology	1997
Charles Flagle, D.Eng.	JHSPH	Health Policy and Management	1978
Morton Goldberg, M.D.	SOM	Wilmer Eye Institute	1998

Name	Affiliation	Department	Year Inducted
Lynn Goldman, M.D., M.P.H.	JHSPH	Environmental Health Sciences (Health Policy and Management; Epidemiology)	2007
Leon Gordis, M.D., Dr.P.H.	JHSPH	Epidemiology	1986
Diane Griffin, M.D., Ph.D.	JHSPH/SOM	Molecular Microbiology and Immunology / Neurology, Medicine	2004
John Griffin, M.D.	JHMI/SOM	Neurology/ Neuroscience / Pathology	2004
Bernard Guyer, M.D., M.P.H.	JHSPH	Population, Family and Reproductive Health	1995
Martha Hill, Ph.D., R.N., FAAN	SON/SOM/JHSPH	Dean of the SON / Professor in Medicine, Nursing and Public Health	1998
Richard Huganir, Ph.D.	SOM	Professor of Neuroscience	2011
Richard Johns, M.D.	SOM	Department of Biomedical Engineering	1986
Richard Johnson, M.D.	SOM/JHSPH	Professor of Molecular Microbiology and Immunology / Neurology and Neuroscience	1987
Nancy Kass, Sc.D.	JHSPH	Professor of Bioethics and Public Health Professor in the Department of Health Policy and Management	2008
Gabor Kelen, M.D.	SOM	Professor and Chair, Department of Emergency Medicine and Director, Johns Hopkins Office of Critical Event Preparedness and Response, Director of the Center for the Study of Preparedness and Catastrophic Event Response	2005
Robert Lawrence, M.D.	JHSPH/SOM	Environmental Health Sciences, Health Policy and Management and Medicine	2005
Paul McHugh, M.D.	JHMI/SOM	Former Director of the Department of Psychiatry and Behavioral Sciences	1992
Guy McKhann, M.D.	SOM	Neurology	1991
Edward Miller, M.D.	SOM	Dean of the School of Medicine, Chief Executive Officer, Johns Hopkins Medicine	2000
Vernon Mountcastle, M.D.	SOM	Neurology	1975
Jeremy Nathans, M.D., Ph.D.	SOM	Professor of Biology and Genetics, Neuroscience, and Ophthalmology	2011
Donald Price, M.D.	SOM	Professor of Pathology, Neurology and Neuroscience	1998
Peter Pronovost, M.D., Ph.D	SOM	Professor of Anesthesiology and Critical Care Medicine and Surgery	2011

Name	Affiliation	Department	Year Inducted
Thomas Quinn, M.D.	SOM/JHSPH	Professor of Medicine, Director of the Division of Infectious Diseases / International Health, Molecular Microbiology and Immunology	2004
Richard Ross, M.D.	SOM	Dean emeritus of the Johns Hopkins School of Medicine	1977
Murray Sachs, Ph.D.	SOM	Department of Biomedical Engineering, Professor of Neuroscience and Otolaryngology–Head and Neck Surgery	1990
Solomon Synder, M.D., D.Sc., D.Phil. (Hon. Causa)	SOM	Neuroscience	1988
Alfred Sommer, M.D., M.H.S.	JHSPH/SOM	Former dean of JHSPH, professor of Epidemiology, International Health, and Ophthalmology	1992
Barbara Starfield, M.D., M.P.H.	JHSPH/SOM	Health Policy and Management, Population and Family Health Sciences, Pediatrics	1978
Donald Steinwachs, Ph.D.	JHSPH/SOM	Director, Health Services Research and Development Center in the Department of Health Policy and Management, Department of Mental Health	1993
David Valle, M.D., Ph.D.	SOM	Director of the Institute of Genetic Medicine, Professor, Departments of Pediatrics, Ophthalmology and Molecular Biology & Genetics, Director, Predoctoral Training Program in Human Genetics, Director, Center for Inherited Disease Research	2002
Bert Vogelstein, M.D.	JHH/SOM	Sydney Kimmel Comprehensive Cancer Center, Clayton Professor of Oncology and Pathology	2001
Henry Wagner, M.D.	JHSPH	Environmental Health Sciences	1986
Edward Wallach, M.D.	SOM	Director, Assisted Reproductive Technologies Program in Department of Gynecology and Obstetrics	1995
Patrick Walsh, M.D.	JHMI	Former Chair of Urology	1995
Myron Weisfeldt, M.D.	SOM	Chair, Department of Medicine	1996
Scott Zeger, Ph.D.	JHSPH	Professor of Biostatistics / Department of Epidemiology	2006
Elias Zerhouni, M.D.	SOM	Former Vice Dean of SOM	2000

Appendix F

Episodes of ABC-TV *Hopkins 24/7* and *Hopkins*

In promoting the broadcasts of both *Hopkins 24/7* and *Hopkins*, ABC provided press materials supplying brief descriptions of each episode.

Hopkins 24/7, Broadcast Aug. 31, 2000 – Sept. 28, 2000

Episode 1
Experience the fear in the emergency room when a 27-year-old doctor is exposed to HIV infected blood; the race against time to implant new lungs in 17-year-old Zach Chamberlin, who may die without a new organ; nervous medical students face their first days at Johns Hopkins.

Episode 2
Focuses on Dr. Edward Cornwell, Chief of Trauma Surgery, as he tries to save the wounded of East Baltimore's street wars, both in the operating room and through preventive outreach to local youngsters; unprecedented access to a "Mortality and Morbidity" doctors' conference, a self critical discussion of doctor's mistakes; a doctor's heroic battle to save the life of a young girl stricken with uterine cancer; on a lighter note—a 180-pound sea turtle from the National Aquarium in Baltimore arrives with a mysterious illness.

Episode 3
Provides a look into the world of Dr. Michael Ain, who has persevered to become a leading surgeon in spite of being a little person; the plight of the pediatric intensive care staff, who have formed an unusual bond with a one-year-old desperately needing a new heart; an intimate portrait of Dr. Rick Montz, whom some call eccentric and irreverent, but who ... [then was] also considered one of the best surgical oncologists in the business; feel the anger of the pediatric emergency staff when a battered child is brought in unconscious.

Episode 4
Probe the fear and frustration of parents who must decide whether or not to allow their eight-year-old to undergo a complicated and risky "Hemispherectomy" that will remove half of her brain; meet Dr. Christina Catlett, head of the ER, who tries to keep its doctors' bodies and souls together; follow a "lowly" first-year intern through his 140-hour week—sleep-deprived and "motivated by fear"—a rite of passage in his quest to succeed in becoming a surgeon

Episode 5
Go behind closed doors of the psychiatric unit to hear candid conversations between therapists and anorexic young women; witness the amazing moment when a profoundly deaf child hears for the very first time, made possible by a microchip implant.

Episode 6
Examine the emotional turmoil as a family comes to grips with mortality; third-year resident Dr. Risa Moriarty, overcome by the demands of her profession, abruptly quits her job after seven years in medical school (and residency) and thousands of dollars in loans to pay for it; meet Dr. Brett Christiansen on a bad night at the ER, as he loses two patients in a row and his composure falters.

Hopkins, Broadcast June 26, 2008–July 31, 2008

Episode 1
Recounts how Mexican-born brain surgeon Alfredo Quiñones-Hinjosa climbed a 20-foot border fence 21 years earlier to join other illegal immigrants picking fruit in California; became a citizen, and ultimately emerged as one of the country's finest—and most charismatic— neurosurgeons; how Karen Boyle became the first female urological surgeon at Hopkins; balances her family and job; and provides unabashedly candid counseling on sexual health and intimacy to her patients; and how cardiothoracic surgery resident Brian Bethea's demanding training has imperiled his marriage—a story that will be followed in subsequent episodes.

Episode 2
Provides a gripping account of the difficulties faced by Brenda Thompson, suffering from an obscure and always fatal lung disease while awaiting a transplant; and introduces Mustapha Saheed, a six-foot, seven-inch tall emergency medicine resident who cuts a striking figure as he rushes through the emergency room and plans on getting married, despite the advice of a colleague who tells him not to marry "the girlfriend who got you through residency."

Episode 3
Follows Ann Czarnik as she confronts the often-harrowing cases that cram Hopkins' emergency room, which serves the city's frequently violence-riddled East Baltimore neighborhoods; and shows how Earl Ingemann, an energetic, 19-year-old Bermudan with a damaged heart, both charms and frustrates physicians and nurses by roaming the hospital, going out for junk food, playing video games, and getting his hair braided instead of staying put and behaving like the other patients on the heart transplant waiting list.

Episode 4
Covers the initially distressing but ultimately upbeat story of an apparently healthy toddler who suddenly develops a serious illness that could lead to his death unless a risky heart transplant is performed—should a heart become available; and showcases irreverent, chatty and candid vascular surgeon Tom Reifsnyder, whose surgeries can include amusing moments during which he asks his residents about everything from their romantic lives to the kind of cars they drive.

Episode 5
Follows the activities of third-year medical students Sneha Desai, Herman Bagga and Gina Westhoff, in whose worldview "the gross stuff is cool," delivering a baby is a career-decider, and wearing a turban proves an icebreaker with patients; turns an unblinking eye on the agonizing decisions faced by the parents of a precocious son who fear surgery for a devastating brain tumor may harm him, and the family of a child who nearly drowned in a pool but is being kept technically alive only through life support; and profiles tough-talking plastic surgeon Anthony Tufaro, who thinks no surgical challenge is too great.

Episode 6

Shows how Oscar Serrano, a first-year surgery resident, commits a major rookie blunder by puncturing a patient's lung, but finds that both the patient and his mentors are understanding, aware that making mistakes is part of becoming a physician; how pediatric resident Carmen Coombs must learn how to separate herself from the emotions that buffet her during shifts in the neonatal intensive care unit; and how master transplant surgeon Robert Montgomery engineers a bicoastal, three-couple kidney swap, beginning with the wife of a patient who needs a kidney but for whom she is not a donor match.

Episode 7

Watches Hopkins heart specialists tackle the baffling case of an adopted child from China who had undergone a mysterious heart operation there that now requires repair; witnesses the ebullient Alfredo Quiñones-Hinjosa perform stunning brain surgery on a patient who remains wide awake during the operation; and sees how young transplant surgeon Andrew Cameron handles the challenge of working in the same hospital as his father, John, a former head of surgery at Hopkins who is a living legend in the field, renowned for his expertise in pancreatic cancer operations.

Appendix G

Officers and Trustees

Johns Hopkins Medicine

2011–2012 BOARD OF TRUSTEES

Francis B. Burch, Jr., Esquire, *Chair*
William C. Baker, *Vice Chair*
Francis X. Knott, *Vice Chair*

Jeffrey H. Aronson
Janie Elizabeth Bailey
Lenox D. Baker Jr., M.D.
Sherry F. Bellamy, Esquire
Richard O. Berndt, Esquire
Evelyn T. Bolduc, *ex officio*
Deidre A. Bosley, *ex officio*
George L. Bunting, Jr.
Philip M. Butterfield
Edward L. Cahill
N. Anthony Coles, M.D.
Ronald J. Daniels, J.D., LL.M., *ex officio*
James C. Davis
Christopher J. Doherty, *ex officio*
James T. Dresher, Jr.
R. Christian Banghart Evensen
Josh E. Fidler
Pamela P. Flaherty
Richard A. Forsythe
William Thomas Gerrard
Sanford D. Greenberg, Ph.D.
Michael D. Hankin
David C. Hodgson
Stuart S. Janney, III, Esquire
Christopher W. Kersey, M.D., M.B.A.
Jack W. Kirkland, Jr., *ex officio*
Robert D. Kunisch
Jeffrey A. Legum
Edward J. Miller, Jr., *ex officio*
J. Mario Molina, M.D.
Walter D. Pinkard, Jr.
Kevin A. Plank
Michael F. Price
Arnold I. Richman
Sharon Percy Rockefeller
Peter J. Rogers, Jr., *ex officio*
Theo C. Rodgers
Barry Rogstad, Ph.D.
David M. Rubenstein
Mark E. Rubenstein
Stephen J. Ryan, M.D.
Alton J. Scavo
Charles P. Scheeler, Esquire
Mayo A. Shattuck III
Donald J. Shepard
Rajendra Singh, Ph.D.
Barry S. Strauch, M.D.
Louis B. Thalheimer
Selwyn M. Vickers, M.D.

OFFICERS

Edward D. Miller, M.D., *Chief Executive Officer*
Ronald R. Peterson, Executive *Vice President*
William A. Baumgartner, M.D., Senior *Vice President, Office of Johns Hopkins Physicians*
John Bergbower, *Vice President, Corporate Security*
John M. Colmers, *Vice President, Health Care Transformation and Strategic Planning*
Richard O. Davis, Ph. D., *Vice President, Innovation and Patient Safety*
Richard A. Grossi, *Senior Vice President and Chief Financial Officer*
Dalal J. Haldeman, Ph.D., *Vice President, Marketing and Communications*
Thomas S. Lewis, *Vice President, Goverment Affairs and Community Relations*
Joanne E. Pollak, Esquire, *Senior Vice President and General Counsel, Chief of Staff, Office of Johns Hopkins Medicine*
Peter J. Pronovost, M.D., Ph.D., *Senior Vice President, Patient Safety and Quality*
Stephanie L. Reel, *Vice President, Management Systems and Information Services*
Judy A. Reitz, Sc.D., *Vice President, Quality Improvement*
Steven A. Rum, *Vice President, Development and Alumni Relations*
G. Daniel Shealer, Jr., Esquire, *General Counsel and Secretary*

EMERITUS TRUSTEES

C. Michael Armstrong
Robert C. Baker
H. Furlong Baldwin
David H. Bernstein
Andre W. Brewster
Constance R. Caplan,
A. James Clark
Charles W. Cummings, M.D.
Leslie B. Disharoon
Edward K. Dunn, Jr.
Manuel Dupkin, II
Robert D. H. Harvey
Alan P. Hoblitzell, Jr.
A. B. Krongard, Esquire
Raymond A. Mason
Harvey M. Meyerhoff
Frederick O. Mitchell
Morris W. Offit
Christian H. Poindexter
Francis G. Riggs
Henry A. Rosenberg
Richard S. Ross, M.D.
B. Francis Saul II
Huntington Sheldon, M.D.
R. Champlin Sheridan
Wendell A. Smith, Esquire
Shale D. Stiller, Esquire
Calman J. Zamoiski

HONORARY MEMBERS

C. Michael Armstrong
H. Furlong Baldwin
Edward K. Dunn, Jr.
Manuel Dupkin, II
Sidney Kimmel
Harvey M. Meyerhoff
Arthur Modell
Morris W. Offit
Huntington Sheldon, M.D.
Shale D. Stiller, Esquire

Johns Hopkins School of Medicine

DEANS

Edward D. Miller, M.D., *Dean of the Medical Faculty, CEO*
Jonathan Ellen, *Vice Dean for All Children's*
David B. Hellmann, M.D., *Vice Dean for Johns Hopkins Bayview*
William A. Baumgartner, M.D., *Vice Dean for Clinical Affairs*
Daniel E. Ford, M.D., *Vice Dean for Clinical Investigation*
David G. Nichols, M.D., *Vice Dean for Education*
Janice E. Clements, Ph.D., *Vice Dean for Faculty Affairs*
Landon King, M.D., M.D., Ph.D., *Vice Dean for Research*
Todd Dorman, M.D., *Senior Associate Dean for Education Coordination*
Richard A. Grossi, *Senior Associate Dean for Finance and Administration, CFO, JHM*
James L. Weiss, M.D., *Associate Dean for Admissions*
Patricia Thomas, M.D., *Associate Dean for Curriculum*
Brian Gibbs, Ph.D., *Associate Dean for Diversity*
Peter Greene, M.D., *Associate Dean for Emerging Technologies*
Julia McMillan, M.D., *Associate Dean for Graduate Medical Education*
Peter Maloney, Ph.D., *Associate Dean for Graduate Students*
Julie Gottlieb, *Associate Dean for Policy Coordination*
Levi Watkins, Jr., M.D., *Associate Dean for Postdoctoral Affairs*
Mary E. Foy, *Associate Dean, Registrar*
Michael B. Amey, *Associate Dean for Research Administration*
Thomas W. Koenig, M.D., *Associate Dean for Student Affairs*
Harry Goldberg, Ph.D., *Assistant Dean and Director of Academic Computing*
Melissa Helicke, *Assistant Dean for Johns Hopkins Bayview*
Linda Dillon Jones, *Interim Assistant Dean for Faculty, Development and Equity*
John Rybock, M.D., *Assistant Dean and Compliance Officer for Graduate Medical Education*
Judith Carrithers, *Assistant Dean for Human Subjects Research Compliance*
Christine H. White, *Assistant Dean for Medicine*
Michael Barone, M.D., *Assistant Dean for Student Affairs*
Sarah Clever, M.D., *Assistant Dean for Student Affairs*
Daniel Teraguchi, *Assistant Dean for Student Affairs and Director of the Office for Student Diversity*
Gloria J. Bryan, *Senior Director, Office of Human Resources*
John E. Grinnalds, *Senior Director, Office of Facilities Management*
Susanne U. Boeke, MBA, *Director, Office for Faculty Research Resources*
Wesley Blakeslee, *Executive Director for Johns Hopkins Technology Transfer*
James Erickson, *Executive Director, Office of Financial Affairs*

Johns Hopkins School of Medicine

DEPARTMENT DIRECTORS

L. Mario Amzel, Ph.D.
  Department of Biophysics and Biophysical Chemistry,
  Professor And Director
Henry Brem, M.D.
  Department of Neurosurgery, Director, Harvey Cushing Professor
Philip A. Cole, M.D., Ph.D.
  Department of Pharmacology and Molecular Science, Director,
  E. K. Marshall and Thomas H. Maren Professor
Charles W. Cummings, M.D.
  Department of Orthopaedic Surgery, Professor and
  Interim Director
Ted M. Dawson, M.D., Ph.D.
  Institute for Cell Engineering, Director, Leonard and Madlyn
  Abramson Professor of Neurodegenerative Diseases
J. Raymond DePaulo Jr., M.D.
  Department of Psychiatry and Behavioral Sciences, Director,
  Henry Phipps Professor
Stephen Desiderio, M.D., Ph.D.
  Institute for Basic Biomedical Sciences, Director, Molecular
  Biology and Genetics Professor
Peter Devreotes, Ph.D.
  Department of Cell Biology, Professor and Director
Theodore L. Deweese, M.D.
  Department of Radiation Oncology and Molecular Radiation
  Sciences, Professor and Director
George J. Dover, M.D.
  Department of Pediatrics, Director, Given Foundation Professor
David W. Eisele, M.D.
  Department of Otolaryngology—Head and Neck Surgery,
  Director, Andelot Professor of Laryngology and Otology
Harold E. Fox, M.D., M.Sc.
  Department of Gynecology and Obstetrics, Director,
  Dr. Dorothy Edwards Professor
Julie A. Freischlag, M.D.
  Department of Surgery, Director, William Stewart Halsted
  Professor
Carol W. Greider, Ph.D.
  Department of Molecular Biology and Genetics,
  Daniel Nathans Director and Professor
William B. Guggino, Ph.D.
  Department of Physiology, Professor and Director
Gerald W. Hart, Ph.D.
  Department of Biological Chemistry, Director, Delamar Professor
Richard L. Huganir, Ph.D.
  Department of Neuroscience, Professor and Director

J. Brooks Jackson, M.D.
  Department of Pathology, Director, Baxley Professor
Sewon Kang, M.D.
  Department of Dermatology, Director, Noxell Professor
Gabor D. Kelen, M.D.
  Department of Emergency Medicine, Professor and Director
W. P. Andrew Lee, M.D.
  Department of Plastic and Reconstructive Surgery, Director,
  Milton T. Edgerton Professor
Gary Lees, C.M.I., F.A.M.I.
  Department of Art as Applied to Medicine, Associate Professor
  and Director
Jonathan S. Lewin, M.D.
  Department of Radiology, Director, Martin W. Donner Professor
Justin McArthur, M.B.B.S., M.P.H.
  Department of Neurology, Professor and Director
Peter J. McDonnell, M.D.
  Department of Ophthalmology, Director, William Holland Wilmer
  Professor
Elliot McVeigh, Ph.D.
  Department of Biomedical Engineering, Director, Bessie Darling
  Massey Professor
William G. Nelson, M.D.
  Department of Oncology, Marion I. Knott Director and Professor
Randall M. Packard, Ph.D.
  Department of History of Medicine, Director, William H. Welch
  Professor
Jeffrey B. Palmer, M.D.
  Department of Physical Medicine and Rehabilitation, Director,
  Lawrence Cardinal Shehan Professor
Alan W. Partin, M.D., Ph.D.
  Department of Urology, Director, David Hall McConnell Professor
Jeffrey D. Rothstein, M.D., Ph.D.
  The Brain Sciences Institute, John W. Griffin Director and Professor
John A. Ulatowski, M.D., Ph.D., M.B.A.
  Department of Anesthesiology and Critical Care Medicine,
  Director, Mark C. Rodgers Professor
David Valle, Ph.D.
  McKusick Nathans Institute of Genetic Medicine,
  Professor and Director
Myron L. Weisfeldt, M.D.
  Department of Medicine, Director, William Osler Professor
M. Christine Zink, D.V.M., Ph.D.
  Department of Molecular and Comparative Pathobiology,
  Professor and Director

Johns Hopkins Medicine International

2011–2012 BOARD OF DIRECTORS

Christopher W. Kersey, M.D., M.B.A., *Chair*
Edward D. Miller, M.D., *Vice Chair, ex officio*

Janie Elizabeth (Liza) Bailey
David H. Bernstein
Francis B. Burch, Jr., Esquire
Daniel Ennis
R. Christian B. Evensen
Julie A. Freischlag, M.D.
Martha Hill, Ph.D., *ex officio*
David C. Hodgson
Michael L. Klag, M.D., M.P.H., *ex officio*
Lloyd Minor, M.D.
Morris W. Offit
Ronald R. Peterson, *ex officio*
Stephen J. Ryan, M.D.
Alfred Sommer, M.D., M.H.S.
Steven J. Thompson, *ex officio*
James L. Winter
Samuel H. Clark, Jr., Esquire, *Secretary*

OFFICERS

Steven J. Thompson, M.B.A., *Chief Executive Officer*
John Ulatowski, M.D., *Vice President and Executive Medical Director*
Guilherme (Gui) Valladares, M.B.A., *Chief Financial Officer*
Mohan Chellappa, M.D., *President, Global Ventures*
Salim Hasham, M.H.A., *Senior Vice President, Global Ventures*
Burak Malatyali, M.B.A., *Vice President, Operations and Administration*
Jane Shivnan, M.Sc.N., R.N., A.O.C.N., *Executive Director, Clinical Quality and Nursing*

Acknowledgments

The standard advice to all would-be authors is: "Write about what you know."

I knew a bit about the history of The Johns Hopkins Hospital and School of Medicine long before beginning this book. In part, that was because more than a century ago, my great grandparents, Millard and Rebecca Grauer, owned a row house just four blocks north of the Hospital. My grandfather Albert and his two younger brothers, Milton and Edwin, grew up within the shadow of Hopkins' fabled—and then brand-new—dome. What's more, as a medical student from 1908 to 1910, pioneering neurosurgeon Walter Dandy, one of Hopkins' future legends, lived as a boarder in my great grandparents' home at 1029 North Broadway. He became a close chum of my grandfather and great uncles, who were his age, and admired my great grandparents. A decade after he graduated from the medical school—at a time when he was beginning to garner world-wide acclaim—Dandy wrote to my great grandmother, "I often think of you and how good you were to me when I was a mere student. I will never forget you."

Hence, Hopkins Medicine's history always has been a source of personal interest—yet as this project progressed, the revelations of its vast depth, scope and impact have been astounding and often daunting.

Invaluable assistance in comprehending and chronicling this history has come from every branch of Hopkins' immense medical enterprise—beginning right at the top. Exceptionally generous, cooperative and encouraging supporters of this book have been Edward D. Miller, the first chief executive officer of Johns Hopkins Medicine, 13th dean of the School of Medicine, and vice president for medicine of The Johns Hopkins University; and Ronald R. Peterson, president of The Johns Hopkins Hospital and Health System and Executive Vice President of Johns Hopkins Medicine. They have given unstintingly of their time, expertise and insights.

Indeed, all of Hopkins Medicine's leadership have been remarkably helpful and supportive, including William Baumgartner, vice dean for clinical affairs; Chi V. Dang, former vice dean for research; David Nichols, vice dean for education; Stephanie Reel, vice president for information services for Hopkins Medicine, as well as vice provost for information technology and chief information officer for the university; and Steven Thompson, former senior vice president for Hopkins Medicine and now chief executive officer of Johns Hopkins Medicine International. My immediate boss, Dalal Haldeman, vice president of marketing and communications for Hopkins Medicine, has been kind and patient as I plowed my way through all the research and drafts needed to complete this work.

It also has been my good fortune—and honor—to interview a host of Hopkins Medicine luminaries in addition to its leadership. Despite being internationally renowned, each of them found time to be warm, welcoming, helpful—and candid. Special thanks are due to Peter Agre, Randol Barker, Richard Bennett, Henry Brem, Denton Cooley, Ted Dawson, Stephen Desiderio, Charles Flexner, William Greenough III, Vincent Gott, Alex Haller, Richard Johns, Howard W. Jones Jr., Edward McCarthy, Peter McDonnell, Elliot McVeigh, Vernon Mountcastle, Richard S. Ross, Chester W. Schmidt Jr., Solomon Snyder and Bert Vogelstein.

The Hopkins Medicine Office of Marketing and Communications in which I work has a remarkable cadre of talented writers, superb graphic designers, and (hard as it is for a writer to admit) good editors. When I joined the office a decade ago, it was led by Elaine Freeman, then-vice president for marketing and communications, and media relations chief Joann Rodgers, two skillful and mind-bogglingly knowledgeable public affairs advocates for Hopkins. Their continued interest in my efforts is greatly appreciated. Patrick Gilbert, director of editorial services, demonstrated stoicism, determination and sympathy in tackling the unenviable job of trying to edit the manuscript down to manageable size. Justin Kovalsky displayed patience and persistence in proof-reading—and, most vital—additionally trimming the manuscript. Support also has been shown by Amy Goodwin, director of institutional internal communications; and Gary Stephenson, current media relations and public affairs director.

The extraordinary artistry of David Dilworth, associate director of graphic design, has given this book its captivating appearance. Director of design Maxwell Boam, senior designer Abby Ferretti, and photographer Keith Weller also made important contributions. Eileen O'Brien did a splendid job of proof reading, as did indexer Sally Lawther. We also are indebted to Jennifer Fairman and David Rini of the Department of Art as Applied to Medicine for the magnificent cover.

Colleagues past and present, both in editorial services and media relations and public affairs, produced a substantial supply of exemplary articles and background materials on Hopkins Medicine's accomplishments over the past two decades. It was from this trove that I was able to glean the information needed to resume the narrative of Hopkins Medicine's history begun some seven decades ago by Alan Mason Chesney and continued, in succession, by Thomas B. Turner, A. McGehee Harvey, Gert H. Brieger, Victor A. McKusick, and Susan L. Abrams. A great deal of this book's early history of Hopkins Medicine owes much to their work.

I owe a tremendous debt not just to these formidable predecessors but to those current and erstwhile colleagues whose skillful accounts of Hopkins Medicine's achievements have been exceedingly helpful. These include editorial services writers and editors Mary Ann Ayd, Marjorie Centofanti, Sue De Pasquale, Marlene England, Ramsey Flynn, Elaine Freeman, Mark Guidera, Michael Levin-Epstein, Gary Logan, Mary Ellen Miller, Judith Minkove, Edith Nichols, Maggie Pedersen, Deborah Rudacille, Stephanie Shapiro, Linell Smith, Anne Bennett Swingle, and Janet Farrar Worthington. Also of considerable value was the work of media relations experts Christen Brownlee, Stephanie Desmon, Joanna Downer, Audrey Huang, John Lazarou, David March, Ellen Beth Levitt, Valerie Matthews Mehl, Amy Mone, Ekaterina Pesheva, Beth Simpkins, Eric Vohr, Vanessa Wasta, and Mary Alice Wensel-Yakutchik; as well as the efforts of videographer Jay Corey.

In addition, the works of numerous freelance writers supplied worthwhile background information, including stories by Maria Blackburn, Lydia Lewis Bloch, Sharon Bondross, Sandy Budd, J. Pat Carter, Lynn Crawford Cook, David Dudley, Mat Edelson, Mike Field, Karen L. Helsley, Randi Henderson, Kate Ledger, Jack Lessenberry, Carol Pearson, Barry Rascovar, Sandra Salmons, Rebecca Skloot and Merrill Witty. Intern Dana Davis swiftly compiled the lists of Hopkins Medicine's winners of prominent awards and honors, as well as its leadership rosters.

No hospital or medical school has a finer repository of history than Hopkins does in its Alan Mason Chesney Medical Archives. This book could not have been possible without the tireless assistance I received from archives director Nancy McCall and her staff, including Andrew Harrison, Marjorie Kehoe, Phoebe Evans Letocha, Kate Ugarte and Timothy Wisniewski; as well as volunteers Diane Abeloff, Elizabeth M. Peterson, Chris Ponticas, Betty Scher, Ann Snead and Jerriann Myers Wilson. In addition, James Stimpert, the archivist for The Johns Hopkins University, always was happy to be of help, as were Mame Warren, former head of the Sheridan Libraries' Hopkins History Enterprises program, and the staff of the Welch Medical Library.

Although, as noted, writers are advised to know the subjects they address, I really had not planned on becoming a *patient* at Hopkins during the time I worked on this book. Yet I did—and hence got to learn about Hopkins' medical care from that perspective. I am enormously grateful for the splendid treatment I received and wish to thank the physicians, nurses and technicians who provided it. These include John Fetting III, M.D.; Joel Fradin, M.D., Mehran Habibi, M.D.; Nagi Khouri, M.D.; Michael Phelps, M.D.; Richard Zellars, M.D.; Andrea Cox, B.S.N., C.H.P.N.; Melissa Shelby, B.S., N.P.; Lillie Shockney, R.N., B.S., M.A.S., CBCN, CBPN-C; and Pam Simpson, R.N. In addition, vital services were supplied by Tammy Atkins, R.T.T., Jennifer Jones, R.T.T., Dwayne Lewis, R.T.T., Kim McNeal, R.T.T., Gail Reese, R.T.T., and Lauren Rowe.

Finally, regardless of how solitary a writer's work may be, few authors can navigate the treacherous shoals of so lengthy a writing voyage as this has been without the support of family and friends. Naturally, such a list begins with my late father, Dr. William S. Grauer, my brother Tony, and his late wife, Helene; my nieces, their husbands and offspring, Myndi, Mike, Hannah and Max Weinraub and Robyn, Adam and Isabelle Freund. In addition, although only a partial listing, my warmest thanks go to John Bainbridge, Russell Baker, Robert Brugger, Dr. Roger Blumenthal, Jim Burger, Neil L. Buttner, Dr. Robert and Judy Chessin, the brothers Steve, John, Brent and Henry Ciccarone, Jr., and their mother, Sue, Tom and Connie Cole, Joe and Ozzie Cowan, Lou D'Angelo, Alice M. Davis, Caleb, Mary Jo, Emily and Zooey Deschanel, Daniel Mark Epstein, Robert Erlandson, John Fairhall and Elaine Eff, Charles S. Fax and Michele Weil, Michael Federico, Sr., Stan and Bailey Fine, Graeme and Ronda Fox, Robert and Ginny Green, Mark and Sarah Lee Greenberg, Henry Greenberg and Ann LoLordo, John Halperin, Mark and Pat Heaney, Richard and Ellen Hollander, David M. Howland, Jr., David and Nancy Huntley, Geoff and Sue Huntting, Rob and Sue Kasper, Mitch and Sue Kearney, Brendan and Tracy Kelly and their children, Mac, Maggie and Bridget, Dr. John and Carol Kelly, Kevin Kilner and Jordan Baker, Jim and Sue Ledley, Richard Macksey, Dr. Les and Julie Matthews and their sons, Nolan, Nathan and Scott, Alec and Rachel MacGillis, H. Downman and Helen McCarty, Emmett and Karen McGee, Steve and Meghan Mitchell, Michael and Mary Page Morrill, Kenneth and Nancy Niman, Pat and Gale O'Connor, Dave and Colleen Pietramala, Lawrence and Colleen Quinn, Patrick and Robin Russell, Dan and Lil Rodricks, Stephen and Sheila Sachs, Bob and Margo Scott, Mark and Jeanne Shriver and their children, Molly, Tommy and Emma, Dr. Gus and Katherine Slotman, Louise Sullivan and Rob Ferris, Glen Thomas, Brook E. White, and Jack and Byrd Wood. Profound thanks also are due the entire Johns Hopkins Lacrosse family.

Neil A. Grauer

Endnotes

CHAPTER 1

1 Thom, Helen Hopkins, *Johns Hopkins: A Silhouette*, Johns Hopkins University Press, 1929 (first edition), pgs. 69–70; hereafter referred to as "Thom."

2 *The Baltimore Sun*, Dec. 25, 1873, "Death of Johns Hopkins," reprinted in the *Johns Hopkins Gazette,* Jan. 4, 1999, Vol. 28, No. 16.

3 Ibid.

4 Ibid.; John Kastor, *Governance of Teaching Hospitals: Turmoil at Penn and Hopkins,* Johns Hopkins University Press, 2004, pg. 160; hereafter referred to as "Kastor."

5 A. McGehee Harvey, Gert H. Brieger, Victor McKusick, Susan L. Abrams, *A Model of Its Kind: A Centennial History of Medicine at Johns Hopkins*, Vol. 1, Johns Hopkins University Press, 1989, pg. 7; hereafter referred to as *"Model."*

6 *The Baltimore Sun*, Dec. 25, 1873, op. cit.

7 *The Johns Hopkins Magazine*, Jan. 1974, Vol. 25, No. 1, "Mr. Johns Hopkins," Kathryn A. Jacob; Thom, pgs. 1–4. WorldLingo Web site, "Johns Hopkins," http://www.worldlingo.com/ma/enwiki/en/Johns_Hopkins.

8 *The Johns Hopkins Magazine,* Jan. 1974, op. cit.

9 Ibid.

10 Ibid.; Thom, pg. 10.

11 Thom, pg. 14

12 *The Johns Hopkins Magazine*, Jan. 1974, op. cit.; Thom, pgs. 16–27.

13 Ibid.; Thom, pgs. 23–29.

14 *The Johns Hopkins Magazine*, Jan. 1974, op. cit.

15 Ibid.; "Who Was Johns Hopkins?" (prepared for press kit for Hopkins Hospital centennial, 1989). On file with the author.

16 Ibid.

17 *The Johns Hopkins Magazine*, Jan. 1974, op. cit.

18 Ibid.

19 Ibid.

20 News in History Web site, "Pro-Southern Mob in Baltimore Riots, Attacks Union Troops," http://www.newsinhistory.com/blog/pro-southern-mob-baltimore-riots-attacks-union-troops.

21 Clayton Colman Hall, *Baltimore: Its History and People*, Lewis Publishing Co., 1912, pgs 192–193; The Abraham Lincoln Papers at the Library of Congress, transcribed and annotated by the Lincoln Studies Center, Knox College, Galesburg, Ill., "Johns Hopkins to Abraham Lincoln, Thursday, October 30, 1862 (General Wool)," http://memory.loc.gov/cgi-bin/query/r?ammem/mal:@field(DOCID+@lit(d1927200)).

22 Fort Wiki Web site, "John E. Wool," http://fortwiki.com/John_E._Wool; Hall, *Baltimore: Its History and People*, op. cit.

23 Thom, op. cit.

24 *Johns Hopkins Magazine*, Jan. 1974, op. cit.; Thom, op. cit., pgs. 46–47.

25 *Johns Hopkins Magazine*, Jan. 1974, op. cit.; *The New York Times*, "On This Day" in history, Sept. 22, 1860, http://www.nytimes.com/learning/general/onthisday/harp/0922.html; *The Baltimore Sun*, Nov. 7, 2008, "Samuel Hopkins," http://articles.baltimoresun.com/2008-11-07/news/0811060105_1_samuel-hopkins-sam-hopkins-johns-hopkins; Thom, op. cit, pg. 102 (Johns Hopkins' will re: number of acres at Clifton).

26 *Johns Hopkins Magazine*, Jan. 1974, op. cit.

27 Ibid.; Peabody Library Web site, http://www.peabodyevents.library.jhu.edu/history.html ; Mame Warren, *Knowledge for the World: A History of Johns Hopkins, 1876–2001*, Chronology, pg. 252.

28 *Johns Hopkins Magzine*, Jan. 1974, op. cit.

29 Ibid.; *Model*, pg. 8; *Johns Hopkins Magazine*, June 1989, "Merchant With a Plan and a Vision," Sue DePasquale.

30 Ibid.; *Model*, pgs. 8–10.

31 Ibid.

32 *Model*, pg. 10

33 Ibid.; John C. Schmidt, *Johns Hopkins: Portrait of a University*, Office of University Publications, 1986, pg. 5.

34 *Model*, pg. 11; Schmidt, op. cit., pg. 5.

35 Ibid; *The New York Times*, May 21, 1910, "Daniel Coit Gilman A Biography"; Fabian Franklin, *the Life of Daniel Coit Gilman*, Dodd Mead & Co., 1910; James Stimpert, Hopkins Archivist, Email, Oct. 15, 2010.

36 *Model,* pg. 11.

37 Ibid.

38 Ibid., *Annals of Surgery*, Sept. 2001, Vol. 234, "Early Contributions to the Johns Hopkins Hospital by the 'Other' Surgeon: John Shaw Billings," John L. Cameron, M.D.

39 Ibid.; *Model.*, pgs. 14–16

40 Ibid., pgs. 13–14.

41 Ibid.; *Annals of Surgery*, Sept. 2001, op. cit.

42 *Medical History Journal*, Oct. 1957, Vol. 1, pgs. 367–368, "John Shaw Billings, Florence Nightingale and The Johns Hopkins Hospital, citing *American Medical Biography* (1920), http://www.ncbi.nlm.nih.gov/pmc/articles/PMC1034322/?page=1 Florence Nightingale Museum Web site, on her later years, http://www.florence-nightingale.co.uk/cms/index.php/florence-old-age; New York State Health Department (definition of brucellosis), http://www.health.state.ny.us/diseases/communicable/brucellosis/fact_sheet.htm.

43 *Medical History Journal*, Oct. 1957, op. cit.

44 *Model*, pg. 16; "What Hopkins Built, and Where," (1989 centennial media kit, op. cit.) On file with the author.; Cameron, op. cit.

45 Ibid.

46 Ibid.

47 "The Four Founding Physicians" (1989 centennial media kit, op. cit.). On file with the author.

48 Cameron, op. cit.; Simon Flexner and James Thomas Flexner, *William Henry Welch and the Heroic Age of American Medicine*, Johns Hopkins Press, 1941; 1968; 1993, pgs. 92–93; "The Four Founding Physicians" (1989 centennial media kit, op. cit.). On file with the author.

49 "Deans and Directors," Hopkins Hospital Centennial press kit, 1989, op. cit. On file with the author.

50 *Model*, pgs. 17–19.

51 Ibid.; "The Four Founding Physicians," op. cit.

52 Lewellys F. Barker, *Maryland Historical Magazine,,* March 1943, Vol. XXXVIII, No.1, pg. 6.

53 *Model*, pgs. 20–21.

54 Grover M. Hutchins, Brendan P. Lucey, Archives of Pathology and Laboratory Medicine, Oct. 2004, "William H. Welch, MD, and the Discovery of Bacillus welchii," http://findarticles.com/p/articles/mi_qa3725/is_200410/ai_n9454778/; *William Henry Welch and the Heroic Age of American Medicine*, Simon Flexner and James Thomas Flexner, Hopkins Press, 1993 edition, pgs. 179–180, 209, op. cit.

55 *Model,* pgs. 20–22; "The Four Founding Physicians," op. cit.

56 Ibid., pg. 62

57 "The Four Founding Physicians," op. cit.; Flexner, op. cit., pgs. 243–244; *Model*, pg. 26.

58 *The Great Influenza: The Epic Story of the Deadliest Plague in History*, John M. Barry, Penguin, pg. 56

59 Ibid., pg. 63

60 Flexner, *Welch*, op.cit., pg. 445; *The Baltimore Sun*, Oct. 24, 1974; *The Evening Sun*, July 29, 1976 (on club closing).

61 Flexner, *Welch*, op. cit. pg. 168.

62 Ibid.

63 Ibid., pgs. 173; 336

64 Ibid., pgs. 3–8.

65 Ibid., pgs. 7, 170; *Time* magazine, April 14, 1930, cover image, http://www.time.com/time/covers/0,16641,19300414m00.html

66 Ibid., pg. 126; "The Four Founding Physicians," op. cit.; "A Brief Sketch of the Medical Career of Dr. William Stewart Halsted," Alan Mason Chesney Medical Archives, Johns Hopkins, Web site, http://www.medicalarchives.jhmi.edu/halsted/hbio.htm.

67 Ibid.; *Model*, pg. 22.

68 "A Brief Sketch," Chesney Archives, op. cit.

69 *Hopkins Medical News*, Fall 1997, Vol. 21, No. 1, "Halsted Did *Not* Invent Rubber Gloves," letter to the editor, Joseph M. Miller, M.D., Assistant professor, emergency surgery, 1953–1980)

70 "The Four Founding Physicians," op. cit.; *Hopkins Medical News*, Fall 1997, op. cit.

71 "The Four Founding Physicians," op. cit.

72 Ibid.

73 Ibid.; "A Brief Sketch," Chesney Archives, op. cit.; *Annals of Surgery*, March 2006, Vol. 243, No. 3, "William Stewart Halsted: A Lecture by Dr. Peter D. Olch," Edited by J. Scott Rankin, M.D., pg. 421.

74 "The Four Founding Physicians," op. cit.

75 *The Baltimore Sun Magazine*, Jan. 30, 1994, "A Casualty of Cocaine," Scott Shane.

76 Ibid.

77 Olch lecure, op. cit.; *William Stewart Halsted: Surgeon*, W.G. MacCullum, Johns Hopkins Press, 1930, pgs. 107–108.

78 Ibid.

79 Flexner, *Welch,* op. cit., pg. 157; *Model*, pg. 22; "The Four Founding Physicians," op. cit.

80 *The Life of Sir William Osler*, Harvey Cushing, Oxford University Press (1940), pg. 297.

81 Ibid., pg. 297; "The Four Founding Physicians," op. cit., *Hopkins Medicine*, Winter 2008, Vol. 31, N. 2, "Circling the Dome: How a Bad Boy Fooled His Fellows," Kate Ledger.

82 "The Founding Four Physicians," op. cit.

83 Ibid.; *Model*, pr. 22; "The Founding Four Physicians," op. cit., Amazon.com link to 23rd edition of Osler's *Principles*, http://www.amazon.com/Principles-Practice-Medicine-John-Stobo/dp/0838579639#reader_0838579639.

84 Brainy Quotes Web site, Sir William Osler, http://www.brainyquote.com/quotes/authors/w/william_osler.html.

85 "The Founding Four Physicians," op. cit.

86 Ibid.

87 *Hopkins Medicine*, Winter 2010, Vol. 33, No. 1, "The Imperturbability Club," Janet Farrar Worthington.

88 *Model*, pgs. 22–23.

89 "The Founding Four Physicians," op. cit.

90 Ibid.; *Model* , op. cit, pg. 62; author's interview with Howard Jones, Aug. 5, 2008; Ranice Crosby; Jody Cody, *Max Brödel: The Man Who Put Art Into Medicine*, Springer-Verlag, N.Y. (1991), pgs. 2; 24.

91 *Hopkins Medicine*, Winter 2007, Vol. 30, No. 2, "On the Streets Where They Lived," Neil A. Grauer.

92 "The Founding Four Physicians," op. cit.; *Hopkins Medicine*, Winter 2007, op. cit.; Richard W. TeLinde, address to the American Gynecological Society meeting, May 1954.

93 "The Four Founding Physicians," op.cit.

94 *Dome*, Dec. 2003, vol. 54, No. 10, "A Provocative Icon," Lindsay Roylance; *Johns Hopkins Magazine*, June 2002, "All-Star Statues," Mary Mashburn; *The Johns Hopkins Medical Journal*, Nov. 11, 1982, "The Statue of the Christus Consolator at The Johns Hopkins Hospital: Its Acquisition and Historic Origins," Nancy McCall.

95 *The Johns Hopkins Medical Journal*, Nov. 11, 1982, op. cit.

96 Ibid.; *Dome*, Dec. 2003, op. cit.

97 *Model*, pgs. 23–24; *Governance of Teaching Hospitals: Turmoil at Penn and Hopkins*, John A. Kastor, Johns Hopkins University Press (2004), pg. 161.

98 Kastor, pg. 162; Cameron, op. cit.; Cushing, op. cit., pg. 314.

99 *Model*, frontispiece.

100 Kastor, op. cit., pgs. 177–178; Cushing, *Osler*, op. cit., pg. 314.

101 Flexner, *Welch* op.cit., pg. 225; *Model*, pgs. 28–29; Kastor, op. cit., pg. 179.

102 Department of Physiology Web site, http://physiology.bs.jhmi.edu/pages/about/index.aspx; AllAboutScience Web site on Huxley, http://www.allaboutscience.org/thomas-huxley-faq.htm; *Model*, pg 58.

103 *Model*, pg. 17.

104 Ibid., pgs. 18–19.

105 Ibid., pg. 19; *The British Medical Journal,* Nov. 7, 1896, pg. 1419, "H. Newell Martin, M.B., D.Sc. Lond., F.R.S.," http://www.ncbi.nlm.nih.gov/pmc/articles/PMC2510933/?page=1.

106 *Molecular Interventions*, April 2009, Vol. 9. No. 2, "John Jacob Abel: The Fifth Horseman," Rebecca J. Anderson.

107 *Model*, pgs. 29–30.

108 *Model*, pg. 29; Named professorships Web site, http://webapps.jhu.edu/namedprofessorships/professorshipdetail.cfm?professorshipID=102.

109 Medterm.com Web site, definition of epinephrine, http://www.medterms.com/script/main/art.asp?articlekey=3286.

110 *Model*, pgs. 29–30.

111 Ibid.; *Molecular Interventions*, April 2009, op. cit.

112 Ibid.

113 *Model*, pg. 29.

114 Ibid.

115 Ibid.; pgs. 273–274.

116 *Model*, pgs. 28–29, 218

117 Ibid., pg. 160; Turner, Thomas B., *Heritage of Exellence: The Johns Hopkins Medical Institutions*, Johns Hopkins Press, 1974, pg. 163.

118 *Model*, pgs. 161–162, 29.

119 Ibid., pg. 29.

120 Cameron, op. cit.

121 Kastor, op. cit., pg. 171 footnote (for title of Hospital head); *Model*, pg. 23.

122 *Model*, pgs. 25–26; "Deans and Directors," Hospital centennial press kit, 1989. On file with the author.

123 "Deans and Directors," op. cit.

124 Judith Robinson, *Tom Cullen of Baltimore*, Oxford University Press, 1949, pg. 89.

125 Ibid.

126 "Deans and Directors," op. cit.; *Model*, pg. 69.

127 *Model*, pgs. 27–19; 137–140; *Johns Hopkins Magazine*, Sept. 2008, "A Pleasure to Be Bought," Kathleen Waters Sander; *Johns Hopkins Gazette,* Feb. 12, 2001, Vol. 30, No. 21, "Hopkins History: Mary Elizabeth Garrett, Founding Benefactor of the School of Medicine," Nancy McCall.

128 Ibid.

129 *Johns Hopkins Magazine*, Sept. 2008, op. cit.

130 Ibid.; "Welch," Flexner, op. cit., pg. 220.

131 "Welch," Flexner, op. cit., pg. 220.

132 Ibid., pg. 227; *Model*, pg. 30.

133 "Welch," Flexner, op. cit, pg. 227; *Model*, pg. 31.

134 *Model*, pg 142; *Style* magazine, Nov. 1, 2008, "Baltimore Blues," Deborah Rudacille.

135 "Welch," Flexner, op. cit., pgs. 228–229.

136 *Model*, pgs. 31–32.

137 Ibid.

138 Ibid.

139 Ibid.; pgs. 32, 51–56.

140 Ibid., pg. 32.

141 Ibid.

142 *Model,* pg. 158.

143 *Model*, pgs. 36–38.

144 Ibid.

145 *Model*, pg. 34; Young, H., *Hugh Young, A Surgeon's Autobiography*, Harcourt, Brace and Co., N.Y., 1940, pg. 76.

146 Hopkins Urology Web site, "Hugh Hampton Young, Chairman, 1897–1941," http://urology.jhu.edu/about/young.php; Chesney Medical Archives Web site, Young papers, http://www.medicalarchives.jhmi.edu/sgml/young.html.

147 Hopkins Named Professorship Web site, http://webapps.jhu.edu/namedprofessorships/professorshipdetail.cfm?professorshipID=125; U.S. National Library of Medicine Web site, abstract of "Harvey Cushing at Johns Hopkins," D. M. Long, http://www.ncbi.nlm.nih.gov/pubmed/10549918; *Model*, op. cit, pgs. 34–35; *Neurosurgery*, Jan. 2000, "The Hunterian Neurosurgical Laboratory: the first 100 years of neurosurgical research," Sampath, P.; Long, D.M., Brem, http://www.ncbi.nlm.nih.gov/pubmed/10626949.

148 Pulitzer Prize Web site, Prizes for 1926, http://www.pulitzer.org/awards/1926.

149 Hopkins Named Professorships Web site, http://webapps.jhu.edu/namedprofessorships/professorshipdetail.cfm?professorshipID=345; *Model*, pgs. 267–268.

150 *Style* magazine, Nov. 1, 2008, "Baltimore Blues," Deborah Rudacille; *Model*, pg. 146.

151 Ibid.

152 Ibid.

153 Undated letter from Gertrude Stein to Lewellys F. Barker (ca. 1902); letters between Barker and H.W. Knower, secretary of *The American Journal of Anatomy*, April, 1902, courtesy of L. Randol Barker, M.D., Professor of Medicine, Johns Hopkins Bayview Medical Center.

154 *Model*, pgs. 38–39.

155 Ibid.; Hopkins Medical Archives on Thayer, http://www.medicalarchives.jhmi.edu/sgml/thayer.html.

156 *Model*, pgs. 38–39.

157 Ibid.; Named Professorships Web site, http://webapps.jhu.edu/namedprofessorships/professorshipdetail.cfm?professorshipID=273; *Proto* magazine, published by Massachusetters General Hospital, Spring 2010, "Paul Ehrlich and the Salvarsan Wars," http://protomag.com/assets/paul-ehrlich-and-the-salvarsan-wars .

158 *Model*, pg. 40; Lasker Syndicate Web site http://www.smokershistory.com/wiscsynd.htm; http://www.smokershistory.com/ASCC.htm.

159 Ibid., pg. 40; *Journal of Clinical Orthopaedics and Related Research*, June 8, 2010, "Biographical Sketch: William S. Baer (1872–1931), M.M. Manning, Ph.D.; Jason H. Calhoun, M.D.

160 *Model*, pg. 41.

161 Ibid., pgs. 47–48.

162 *Model*, pg. 41; *Model*, Vol. II, pgs. 73; 80- 81.

163 *Baltimore Sun*, July 14, 1943, "Dr. L.F. Barker Rites Slated Tomorrow."

164 *Model*, pgs. 41–42.

165 Ibid.; pgs. 51, 68, 72, 85, 98.

166 Cushing, *Osler*, op. cit., pgs. 650–651; *The Johns Hopkins Hospital and The Johns Hopkins University School of Medicine: A Chronicle*, Alan Mason Chesney, Vol. 1, Johns Hopkins University Press, pg. 404.

167 Cushing, *Osler*, op. cit., pgs. 683–684; *Dome*, July/Aug. 2005, "Seat of Authority," Anne Bennett Swingle.

168 *Model*, pgs. 44–47.

169 Barker, *Maryland Historical Magazine*, Vol. XXXVIII, No. 1, March 1943, p. 7.

170 *Model*, pgs. 45.

171 Ibid., pg. 46

172 Ibid.

173 *Model*, pg. 48

174 Bonner, Thomas Neveille, *Iconoclast: Abraham Flexner and a Life in Learning,* Johns Hopkins Press, 2002, op. cit., pg. 374.

175 *Journal of the American Medical Association*, May 5, 2004, "The Flexner Report and the Standardization of American Medical Education," http://jama.ama-assn.org/cgi/content/full/291/17/2139; the Carnegie Foundation, pdf of Flexner Report, http://www.carnegiefoundation.org/sites/default/files/elibrary/Carnegie_Flexner_Report.pdf

176 Bonner, *Iconoclast*, op. cit.; Flexner, James Thomas, *American Saga: The Story of Helen Thomas & Simon Flexner*, Little, Brown & Co., 1984.

177 *Model*, pg. 50.

178 *Max Brödel: The Man Who Put Art Into Medicine*, Ranice W. Crosby; John Cody, Springer-Verlag, N.Y., 1999, pgs., 126–127.

179 Ibid., pgs. 147–149.

180 Ibid.; Department of Art as Applied to Medicine Web site, http://www.hopkinsmedicine.org/medart/HistoryArchives.htm

181 Ibid.

182 Crosby/Cody, op. cit., pgs. 48–49, 151.

183 Crosby/Cody, op. cit, pg. 85.

184 Ibid.; Crosby/Cody, pg. 251.

185 Art As Applied to Medicine Web site, http://www.hopkinsmedicine.org/about/history/history7.html.

186 *Hopkins Medicine*, Winter 2008, Vol. 31, No. 2, "When Psychiatry Was Very Young," Janet Farrar Worthington.

187 Ibid.

188 Ibid.

189 *Hopkins Medical News*, Winter 2003, Vol. 26, No. 2, "Where a Mind Could Find Itself Again," Anne Bennett Swingle; *Model*, op. cit., pg. 63.

190 *Model*, pgs. 296–297.

191 *Hopkins Medicine*, Winter 2008, op. cit.

192 Ibid.

193 Ibid.

194 *Model*, pgs. 55–57; National First Ladies' Library Web site, "First Lady Biography: Harriet Lane," http://www.firstladies.org/biographies/firstladies.aspx?biography=16 .

195 Ibid.; NNDB Web site profile, "Harriet Lane," http://www.nndb.com/people/902/000127521/ .

196 Ibid.

197 *Model*, pgs. 55–57.

198 Ibid.

199 Ibid.; Hopkins Medical Archives Web site on the Harriet Lane Home, http://www.medicalarchives.jhmi.edu/harrietlane.html .

200 Hopkins Medical Archives page on Howland, http://www.medicalarchives.jhmi.edu/papers/howland.html; American Pediatrics Society's Web page on Howland Award, http://www.aps-spr.org/APS/Awards/Howland.htm .

201 Ibid.; Hopkins Medical Archives page on Park, http://www.medicalarchives.jhmi.edu/papers/park.html .

202 Ibid.; *Hopkins Medical News*, Fall/Winter 1988, Vol. 12, No. 2, "Remembering the Harriet Lane," F. Howell Wright, M.D.

203 Ibid., Leo Kanner page, http://www.medicalarchives.jhmi.edu/papers/kanner.html; *Model*, pgs.225–226.

204 *Model*, pgs. 190–191.

205 Ibid.

206 Ibid.

207 Ibid., pg. 193.

208 Ibid., pgs. 194–197; "Welch," Flexner, op. cit., pgs. 311–321.

209 Ibid., pgs. 320–321.

210 *Model*, pg. 192.

211 http://medicalarchives.jhmi.edu/sgml/barker.html; "The Choice: Lewellys F. Barker and The Full-Time Plan," Charles S. Bryan, M.D., *Annals of Internal Medicine*, Sept. 17, 2002, Vol. 137, No. 6; Economic History Services Web site (for estimate of current value of historic sums), http://www.measuringworth.com/uscompare/ . "The 20th Annual Baltimore Symphony Decorators' Show House, 1996, program, "Stratford-On-The-Green."

212 *Heritage of Excellence: The Johns Hopkins Medical Institutions, 1914–1947,* Thomas B. Turner, Johns Hopkins Press, 1974, pg. 227.

213 *Model,* pgs. 54–55.

214 Ibid.

215 *Hugh Young, A Surgeon's Autobiography,* pgs. 216–231.

216 Ibid.

217 Ibid.

218 Ibid., pgs. 216–217.

219 Ibid.

220 Ibid., pg. 225.

221 Ibid., pg. 225.

222 Ibid., pg. 231.

223 Ibid., Medical Archives page on Janeway, http://www.medicalarchives.jhmi.edu/sgml/janeway.html; "Welch," Flexner, pgs. 326; Turner, *Heritage of Excellence*, op. cit, pg. 28.

224 *Model*, pg. 199.

225 Barry, John M., *The Great Influenza: The epic story of the deadliest plague in history*. Penguin, 2005, pg. 181; Turner, op. cit., pg. 43.

226 Turner, op. cit., pgs. 32–33.

227 "John Miller Turpin Finney: First President of the American College of Surgeons," John L. Cameron, presidential address, 2009; *Military Times* Web site on Young and Thayer medals, "Hall of Valor," http://militarytimes.cm/citations-medals-awards/recipient.php?recipientid=18397 (Young); "Hall of Valor," http://militarytimes.cm/citations-medals-awards/recipient.php?recipientid=18244 (Thayer); Hopkins Medical Archives, Welch chronology, http://www.medicalarchives.jhmi.edu/welch/chronology.htm.

228 Cushing, *Osler*, op. cit, pg. 961.

229 *Model*, pg. 57; Cushing, *Osler*, op. cit., pgs. 1263–1265.

230 *Model*, pg. 57; Hopkins named professorship Web site (for identity of Joseph R. DeLamar), http://webapps.jhu.edu/namedprofessorships/professorshipdetail.cfm?professorshipID=128 .

231 *Model*, pg. 57.

232 Ibid., pgs. 219–229; Project Muse, "Dean Milton C. Winternitz at Yale," from *Perspectives in Biology and Medicine*, Summer 2003, Vol. 46, No.3, http://muse.ju.edu/journals/perspective_in_biology_and_medicine/v046/46.3spiro.pdf.

233 Flexner, "Welch," op. cit., pgs. 397–413.

234 *Model*, pgs. 64–69

235 Ibid.

236 *Model*, pg. 69; "Buildings of The Johns Hopkins Medical Institutions," Louise Cavagnaro, last updated on June 26, 2001, Chesney Medical Archives.

237 Web site, "Ron Schuler's Parlour Tricks," July 28, 2006, "Aida," http://rsparlourtricks.blogspot.com/2006/07/aida.html; Google Books, *Latinas in the United States*, Vicki Ruiz, pgs. 188–189; http://books.google.com/books?id=_62IjQ-XQScC&pg=PA189&lpg=PA189&dq=Ada+de+Acosta+Root+Breckinridge&source=bl&ots=WMcLpwlsSU&sig=D4bhidcuB_swEMRpI5J1SGPpS3E&hl=en&ei=LYnHTPjpBISClAeO49H6AQ&sa=X&oi=book_result&ct=result&resnum=2&sqi=2&ved=0CBkQ6AEwAQ# .

238 Ibid.; Library of Congress photo, flickr Web site, "Mrs. Oren Root," http://www.flickr.com/photos/library_of_congress/3294667187.

239 "Ron Schuler's Parlour Tricks" Web site, op. cit.

240 Ibid.

241 *Model*, pgs. 66–67; Named professorships Web site, "William Holland Wilmer Professorship in Ophthalmology, http://webapps.jhu.edu/namedprofessorships/professorshipdetail.cfm?professorshipID=228

242 *Model*, pgs. 66–67.

243 Flexner, "Welch," pgs. 416–417.

244 *Model*, op. cit. pgs. 66–67.

245 Ibid.

246 Flexner, "Welch," pg. 438.

247 Author's interview with Howard Jones, Aug. 5, 2008; National Endocrine and Metabolic Diseases Information Service Web site (on "Cushing Syndrome"), http://endocrine.niddk.nih.gov/pubs/cushings/cushings.htm.

248 Ibid.

249 Flexner, *Welch*, op. cit., pg. 445.

250 Ibid., pgs. 451–456.

251 Author's interview with Howard Jones, op. cit., Aug. 5, 2008.

252 *Johns Hopkins Magazine*, Vol. 31, Nos. 3 and 4, March and June, 1943, Thomas S. Cullen.

253 Turner, Thomas B., *Heritage of Excellence: The Johns Hopkins Medical Institutions, 1914–1947*, pg. 279.

254 April 1931 fundraising letter from Jesse B. Baetjer, president of the Women's Auxiliary Board, in Women's Board file in Hopkins Medical Archives.

255 *Model*, pgs.72- 73; Turner , op. cit, pg. 96.

256 *Model*, pg. 73.

257 Ibid., pg. 276.

258 Mencken, H.L., *The Baltimore Sun*, July 6–28, 1937, bound volume of typescripts in Chesney Medical Archives.

259 Ibid.

260 Turner, op. cit., pg. 246.

261 Turner, op. cit., pg. 397.

262 "Harvey Cushing," *American Journal of Roentgenology Online*, http://www.ajronline.org/cgi/content/full/194/2/296.

263 *Molecular Interventions*, April 2009, Vol. 9, No. 2, "John Jacob Abel: The Fifth Horseman," Rebecca J. Anderson.

264 *Johns Hopkins Magazine/Hopkins Medical News*, joint issue, June 1989, "Later Pioneers," Michael Bowman; Notable Names Data Base Web site, http://www.nndb.com/people/804/000165309/

265 Society of Laproscopic Surgeons, *Nezhat's History of Endoscopy*, http://laparoscopy.blogs.com/endoscopyhistory/chapter_15/

266 U.S. National Library of Medicine Web site, "The wonderful apparatus of John Jacob Abel called the 'artificial kidney,' G. Eknoyan, http://www.ncbi.nlm.nih.gov/pubmed/19573009.

267 *Model*, pgs. 273–274

268 *Johns Hopkins Magazine/Hopkins Medical News*, June 1989, op. cit.

269 Johns Hopkins Bloomberg School of Public Health Web site, http://www.jhsph.edu/school_at_a_glance/index.html.

270 *Model*, pg. 267.

271 *Johns Hopkins Magazine/Hopkins Medical News*, June 1989, op. cit.

272 *Molecular Interventions*, April 2009, op. cit.

273 *Hugh Young, A Surgeon's Autobiography*, Harcourt, pg. 261; *Journal of the American Medical Association*, Vol.87, No. 17, Oct. 23, 1926, "The Sterilization of Local and General Infections Experimental and Clinical Evidence of Results Obtained by Intravenous Injection of Mercurochrome-220 Soluble," Hugh H. Young, M.D.

274 *Model*, pg. 268; Spine Hall of Fame Web site, http://www.burtonreport.com/infspine/HallFameBios.htm; "Loose Cartilage from Intervertebral Disc Simulating Tumor of the Spinal Cord," Dandy, W.J., Arch Surg, 19:660–672, 1929).

275 *Fertility and Sterility*, Aug. 2005, Vol. 84, No. 2, "In Memoriam: Georgeanna Seegar Jones, M.D.: her legacy lives on," Marian D. Damewood, M.D., and Johns A. Rock, M.D., the Jones Institute Web site, http://www.jonesinstitutefoundation.org/downloads/GeorgeannaJones.pdf; National Library of Medicine, "Georgeanna Seegar Jones, M.D., 'Pioneer in Reproductive Medicine'," http://www.nlm.nih.gov/locallegends/Biographies/Jones_Georgeanna.html.

276 *Journal of Neurosurgery*, June 2010, "Walter E. Dandy's Contributions to Vascular Neurosurgery," Kretzer, R.M., Coon, A.L., Tamargo, R.J., http://www.ncbi.nlm.nih.gov/pubmed/20515365.

277 *Hugh Young: A Surgeon's Autobiography*, pg. 261.

278 "Deans and Directors," Hopkins Hospital Centennial press kit, 1989. On file with the author.

279 Ibid.

280 Ibid.

281 Ibid.

282 Ibid.; Chesney Archives Web page on Chesney, http://www.medicalarchives.jhmi.edu/sgml/chesney.html.

CHAPTER 2

1 *The Baltimore Sun*, March 8, 1999, "Augusta Townsend, 94, best-selling author of novel about Hopkins medical students," Edward Gunts.

2 Ibid.

3 *The Baltimore Sun*, Jan. 13, 1943, "Kelly Rites Set at 3 P.M. Tomorrow"; TeLinde, Richard W., presentation on Howard Atwood Kelly at the American Gynecological Society meeting, May 1954, reprinted in *Johns Hopkins Magazine*, Dec. 1954; University of Pennsylvania Archives, "Howard Atwood Kelly," Web site, http://www.archives.upenn.edu/people/1800s/kelly_howard_atwood.html.

4 The American Physiological Society Web site, "4th APS President (1905–1910)William H. Howell"; http://www.the-aps.org/about/pres/introwhh.htm; Nobel Prize Web site, http://nobelprize.org/nobel_prizes/medicine/laureates/1944/erlanger-bio.html; Chesney Medical Archives, http://www.medicalarchives.jhmi.edu/sgml/howell.html; Who Named It Web site, "William Henry Howell," http://www.whonamedit.com/doctor.cfm/442.html.

5 *Journal of the American College of Surgeons*, Nov. 2008, pgs. 327–332, "John Miller Turpin Finney: The First President of the American College of Surgeons," John L. Cameron; Turner, Thomas B., *Heritage of Excellence: The Johns Hopkins Medical Institutions, 1914–194 7*, Johns Hopkins Press, 1974, pgs. 111–112.

6 Ibid.; *Model*, pg. 60; *Hopkins Medical News*, Winter 2003, Vol. 26, No. 2, "Goodbye to All That," Anne Bennett Swingle.

7 Turner, op. cit, pg. 182.

8 Ibid., pg. 457.

9 Ibid., pgs. 457–458.

10 Ibid.; Answers.com "Vivien Thomas" biography, http://www.answers.com/topic/vivien-thomas.

11 Turner, op. cit., pgs. 463–464.

12 Ibid.

13 Answers.com Web site biography of Vivien Thomas, quoting from "Like Something the Lord Made," *Washingtonian* magazine, Aug. 1989, Katie McCabe.

14 Turner, op. cit, pg. 463; *Model*, pg. 268.

15 Turner, op. cit. pg. 464; *Model*, pgs. 268–270; National Heart, Lung and Blood Institute Web site, "Tetralogy of Fallot," http://www.nhlbi.nih.gov/health/dci/Diseases/tof/tof_what.html; Library of Biographical Information Web site, "History of Tetralogy of Fallot," most-cited sources, http://lib.bioinfo.pl/meid:120413; Medicine.net, definition of aortic value stenosis, http://www.medicinenet.com/aortic_stenosis/article.htm.

16 Ibid.

17 Ibid.

18 *Model*, pgs. 268–270; Answers.com biography of Vivien Thomas, op. cit.; Turner, op. cit. pg. 465.

19 Minetree, Harry, *Cooley: The Career of a Great Heart Surgeon*, Harper's Magazine Press, 1973, pgs. 89–91; author's interview with Cooley, July 31, 2008; *Sports Illustrated*, April 6, 1981, re: opening of Cooley athletic center at Hopkins (for height of Cooley); Academy of Achievement Web site profile of Cooley, http://www.achievement.org/autodoc/page/coo0bio-1.

20 Author's interview with Cooley, op. cit.

21 Ibid.; Turner, op. cit. pg. 461.

22 Author's interview with Cooley, op. cit. About.com Web site, definition of subclavian artery; pulmonary artery, http://biology.about.com/library/organs/heart/blsubclavianartery.htm; http://biology.about.com/library/organs/heart/blpulmartery.htm.

23 Stoney, William S. Stoney, *Pioneers of Cardiac Surgery*, Vanderbilt, 2008; author's interview with Cooley, op. cit.

24 Turner, op. cit., pgs. 464–470; Answers.com bio of Vivien Thomas, op. cit.

25 Chesney Medical Archives, 50th anniversary of Blue Baby operation, http://www.medicalarchives.jhmi.edu/page1.htm.

26 *Model*, pg. 269; Minetree, *Cooley*, op. cit., pg. 91.

27 *Texas Heart Institute Journal*, Vol. 36, 2009, Taussig-Bing Anomaly," Igor E. Konstantinov; *Texas Heart Institute Journal*, Vol. 37, 2010, "Helen Taussig," letter to the editor from Heinrich Taegtmeyer, M.D.; *The New York Times*, Nov. 13, 2010, "Richard Bing, Pioneering Heart Researcher, Dies at 101."

28 Turner, op. cit, pg. 454.

29 Ibid., 475; 492

30 Kastor, John A., *Governance of Teaching Hospitals: Turmoil at Penn and Hopkins*, Johns Hopkins Press, 2004, footnote, pg. 163; Turner, op. cit., pg. 473; *Model*, pgs. 75–76.

31 Turner, op. cit., pg. 475; *Model*, pg. 75.

32 Turner, op. cit., pgs. 471–498.

33 Turner, op. cit., pg. 487.

34 Ibid., 471–498; author's interview with Vernon Mountcastle, Oct. 10, 2006.

35 Turner, op. cit., pg. 476.

36 Ibid., pgs. 471–498; *Model*, pg. 258.

37 *Model*, pgs. 87;Turner, op. cit., pg. 499.

38 Cavagnaro, Louise, "A History of Segregation and Desegregation at The Johns Hopkins Medical Institutions," Feb. 18, 1992, Chesney Medical Archives; *Dome*, Sept. 2004, Vol. 55, No. 7, "The Way We Were," Louise Cavagnaro, with Anne Bennett Swingle.

39 "A History of Segregation," op. cit., pg. 5

40 Ibid., pg. 6.

41 *Dome*, Sept. 2004, op. cit.

42 Ibid.

43 Author's interview with Cavagnaro, May 28, 2009.

44 National Medical Association Web site, http://www.nmanet.org/index.php/nma_sub/history; *Journal of the American Medical Association*, Jully 16, 2008, Vol. 300, No. 3, "African American Physicians and Organized Medicine: 1846–1968," http://jama.ama-assn.org/cgi/content/full/300.3.306.

45 Dans, Peter, *Doctors in the Movies: Boil the Water and Just Say Aah,"* Medi-Ed Press, 2000, pgs. 149–150, citing Ludmerer, K.M., *Learning to Heal: The Development of Medical Education from the Turn of the Century to the Era of Managed Care*, Oxford University Press, 1999; National Medical Association Web site, op. cit.

46 "A History of Segregation," pgs. 12–13.

47 *Model*, pg. 123; *Hopkins Medical News*, Winter 1995, Vol. 18, No. 2, "Up From Bigotry," Randi Henderson.

48 Ibid.

49 Cushing, Harvey, *The Life of Sir William Osler*, Oxford University Press, 1940, pgs. 214–215.

50 Flexner, Simon and Flexner, James T., *William Henry Welch and the Heroic Age of American Medicine*, Johns Hopkins Press, 1993 edition, foreword by Turner, Thomas B., pg. xiii; Flexner, James T., *American Saga: The Story of Helen Thomas and Simon Flexner*, Little, Brown & Co., 1984, pgs. 241–242.

51 Valentine, William N., "Maxwell Mayer Wintrobe, 1901–1986, A Biographical Memoir," National Academy of Sciences, http://books.nap.edu/html/biomems/mwintrobe.pdf; *Hopkins Medical News*, Spring, 1995, Vol. 18, No. 3, Letter to the Editor, Kenneth Zierler.

52 Rudacille, Deborah, "Baltimore Blues," *Style* magazine, Nov. 1, 2008; *Hopkins Medical News*, Spring, 1995, Vol. 18, No. 3; Rengel, Marian, *Encyclopedia of Birth Control*, Oryx Press, Phoenix, Ariz., pg. 100, http://books.google.com/books?id=dx1Kz-ezUjsC&pg=PA100&lpg=PA100&dq=Alan+Guttmacher+%2B+1947+Lasker+Award&source=bl&ots=jrYjhlmuSu&sig=9_McDyMplkMLVxUmAMo9P73TFCs&hl=en&ei=NQbUTJzGCoP7lweW3OnaBQ&sa=X&oi=book_result&ct=result&resnum=5&ved=0CCcQ6AEwBA#v=onepage&q=Alan%20Guttmacher%20%2B%201947%20Lasker%20Award&f=false; Guttmacher Institute Web site, "Alan F. Guttmacher, 1898–1974)," http://www.guttmacher.org/about/alan-bio.html.

53 Rudacille, Deborah, "Baltimore Blues," *Style* magazine, Nov. 1, 2008, op. cit.; *Hopkins Medical News*, Spring, 1995, Vol. 18, No. 3, Letter to the Editor, Kenneth Zierler, op. cit.; Harrison, Tim, interview with Vernon Mountcastle, undated, Chesney Medical Archives, Mountcastle personal file, Item #41.

54 Harrison interview of Mountcastle, op. cit.

55 Turner, op. cit., pgs. 520–521; *Model*, pgs. 86–87.

56 Harrison interview of Mountcastle, op. cit.

57 *Hopkins Medicine*, Spring/Summer, 2005, Vol. 28, No. 3, "The Guy's and Us," Janet Farrar Worthington.

58 Ibid.

59 Ibid.

60 Ibid.

61 Ibid.

62 Turner, op. cit, pg. 466.

63 *Hopkins Medicine*, Spring/Summer 2005, op. cit.

64 Cushing, *Osler*, pgs. 361–362.

65 Turner, op. cit., pgs. 237–238.

66 *Model*, pg. 106.

67 *Hopkins Medicine*, Fall 2006, Vol. 30, No. 1, "When Reed Hall Is Your Home," Ramsey Flynn; Hopkins Medicine Web site, campus housing, "Reed Hall," http://www.hopkinsmedicine.org/som/students/life/housing/on_campus/; *Model*, op. cit. pg. 106.

68 Ibid.

69 *Hopkins Medicine*, Fall 2004, Vol. 28, No. 1, "Compound Bonding," Janet Farrar Worthington.

70 Ibid.

71 Ibid.

72 Ibid.

73 *Model*, pgs. 105–107.

74 Ibid.

75 Ibid.

76 Ibid.; Cavagnaro, Louise, "Buildings of The Johns Hopkins Medical Institutions," 2001, pg. 7 (date of "old" surgery building construction)

77 *Model*, pg. 106; *Model*, Vol. II, *A Pictorial History of Medicine at Johns Hopkins*, 1989, pgs. 11, 26–27, for original Hospital plan and photos of Octagon Ward.*Model*, pg. 106; Chesney Medical Archives page on W. Barry Wood papers, http://www.medicalarchives.jhmi.edu/sgml/woodwb.html.

78 *Model*, pg. 75.

79 Ibid., pgs. 75–76.

80 Ibid.

81 Ibid., pg. 80.

82 Ibid.

83 *Hopkins Medical News,* Fall 1985, Vol. 9, No. 3, "A Tale of Six Deans," Claudia Ewell; Turner, *Part of Medicine, Part of Me,* op. cit., pgs. 119–120, 81.

84 *Model*, pgs. 270–271.

85 Ibid.

86 Ibid.; *The Baltimore Sun*, Jan. 8, 2008, "William R. Milnor," obituary, Jacques Kelly; Biophysical Society Web site, "Biophysicists in Profile, Samuel Talbot….," http://www.biophysics.org/LinkClick.aspx?fileticket=hnX25N%2FryA0%3D&tabid=524.

87 *Model*, pgs. 271–272; Science Heroes Web site, "G. Guy Knickerbocker," http://www.scienceheroes.com/index.php?option=com_content&view=article&id=338&Itemid=284; James Jude, http://scienceheroes.com/index.php?option=com_content&view=article&id=337&Itemid=285; *Annals of Surgery*, April 2003, Henry T. Bahnson obituary, http://www.ncbi.nlm.nih.gov/pmc/articles/PMC1514482/.

88 *Model*, pg. 271; *Dome*, Sept. 2008, Vol. 59, No. 7, "Intensive Foresight," Neil A. Grauer; *Centuries of Caring: The Johns Hopkins Bayview Medical Center Story*, Neil A. Grauer, pg. 28; *The New York Times*, June 8, 2003, "Peter Safar, 'The Father of C.P.R.,' is Dead at 79,: http://www.nytimes.com/2003/08/06/us/peter-safar-the-father-of-cpr-is-dead-at-79.html.

89 *The New York Times*, June 8, 2003, op. cit., *Pittsburgh Post-Gazette*, March 31, 2002, "Peter Safar: A life devoted to cheating death," Anita Srikameswaran, http://www.post-gazette.com/lifestyle/20020331safar0331fnp2.asp; Science Heroes Web site, op. cit., "James Jude."

90 *Model*, pg. 270.

91 Ibid., pg. 272; *New England Journal of Medicine*, July 29, 2010, "In CPR, Less May Be Better," Myron L. Weisfeldt, M.D.

92 Author's interview with Mountcastle, Oct. 10, 2006; *Hopkins Medicine*, Winter 2007, Vol. 30, No. 2, "The Brain Voyager," Neil A. Grauer.

93 Ibid.

94 Ibid.

95 Ibid.

96 Ibid.

97 Ibid.

98 Ibid.

99 *Model*, pgs. 274–277.

100 Ibid.

101 Ibid.; National Foundation for Infantile Paralysis Web site, http://www.enotes. com/1950-medicine-health-american-decades/national-foundation-infantile-paralysis ; National Academy of Sciences, Maxcy memorial, http://books.nap.edu/html/biomems/kmaxcy.pdf.

102 *Model*, pg. 275; Hopkins Hospital Centennial time-line (for 1949 date of discovery of three polio viruses).

103 *Model*, pg. 276; Answers.com for definition of gamma globulin, http://www.answers.com/topic/gamma-globulin.

104 *Model,* pgs. 276–277; Cincinnati Children's Web site, on Albert Sabin, http://www.cincinnatichildrens.org/about/history/sabin.htm.

105 *Hopkins Medicine*, Fall, 2008, Vol. 32. No. 1, "Loss of a Legend," Neil A. Grauer; *Hopkins Medicine*, Winter 2009, Vol. 32, No. 2, letter to the editor, "More on the Loss of a Legend," Carol Bocchini, Joanna S. Amberger, and Ada Hamosh; phenotype definition, the Free Dictionary, http://www.thefreedictionary.com/phenotype.

106 *Hopkins Medicine*, Fall, 2008, op. cit.

107 Ibid.

108 Ibid.

109 Ibid.

110 Ibid.

111 Ibid.

112 Ibid.

113 Ibid.

114 *Model*, pg. 248.

115 Ibid; press release, Feb. 1, 2010, "A Statement From Johns Hopkins Medicine About HeLa Cells and Their Use"; Johns Hopkins Medicine, "Suggested Talking Points," Feb. 1, 2010; *Baltimore City Paper*, April 12, 2002, "Wonder Woman: The Life, Death, and Life After Death of Henrietta Lacks, Unwitting Heroine of Modern Medical Science," Van Smith; *Wired* magazine, Jan. 2010, Wired.com, "Henrietta Everlasting: 1950s Cells Still Alive, Helping Science, Erin Biba, http://www.wired.com/magazine/2010/01/st_henrietta/; *The Guardian*, London, June 23, 2010,"Henrietta Lacks: the Mother of Modern Medicine," Joanna Moorhead, http://www.guardian.co.uk/science/2010/jun/23/henrietta-lacks-cells-medical-advances; *The Washington Post*, "Book Review: 'The Immortal Life of Henrietta Lacks' by Rebecca Skloot,'" http://www.washingtonpost.com/wp-dyn/content/article/2010/01/29/AR2010012902147.html.

116 Ibid.; Answers.com on "informed consent" origin, http://www.answers.com/topic/informed-consent.

117 Ibid.; *Guardian*, op. cit.; *City Paper*, op. cit., press release, Feb. 1, 2010, op. cit., "Talking Points," op. cit.; *Johns Hopkins Magazine*, June 2010, "Immortal Cells, Enduring Issues," Dale Keiger.

118 Ibid.

119 Ibid.; *The Washington Post*, Jan. 31, 2010, op. cit.

120 Ibid.; *Guardian*, June 23, 2010, op. cit., "Talking Points," op. cit.

121 Ibid., *City Paper*, April 17, 2002, op. cit.; *The Washington Post*, Jan. 31, 2010, op. cit.; *Baltimore Sun* Web blog Read Street, " 'Henrietta Lacks' tops Amazon's Best Books of 2010," http://weblogs.baltimoresun.com/entertainment/books/blog/2010/11/henrietta_lacks_tops_amazons_b.html; press release, Feb. 1, 2010, op. cit; "Talking Points," op. cit.

122 *Model*, pgs. 106–107.

123 *Hopkins Medicine*, Fall 2006, Vol. 30, No. 1, "Triumph Amid the Tumult," Neil A. Grauer; *Model*, pg. 94.

124 Ibid.

125 Ibid.

126 *The Baltimore Sun*, Jan. 4, 1987, "Owens to Head Johns Hopkins Hospital"; June 8, 1988 press release (Owens personal file at Archives, Box I.)

127 Ibid.; *Model*, pg. 101.

128 Press release, Oct. 15, 1986, (Owens personal file at Archives, Box I); press release, June 8, 1988, op. cit.; *The Baltimore Sun*, Jan. 4, 1987, op. cit.; *The Daily Record*, Dec. 12, 1986, "JHH Plans $110 million, 88-bed Cancer Center," Owens Folder No. 2, personal file in Chesney Archives.

129 Press release, June 8, 1988, Owens Folder I , personal file in Chesney Archives.

130 Kastor, John A., *Governance of Teaching Hospitals: Turmoil at Penn and Hopkins*, Johns Hopkins Press, 2004, pgs. 185–186; *The Baltimore Sun*, Jan. 4, 1987, op. cit.

131 Press release, Jan. 26, 1988, Owens Folder No. 1, op. cit.

132 *The Evening Sun*, Baltimore, Oct. 13, 1978, "Scientists Help Unravel Mysteries of Genetics," Jon Franklin and Michael Himowitz; *The Baltimore Sun*, April 12, 1999, "After the Prize," Douglas Birch; faqs.org Web site, "Werner Arber Biography (1929 -)," http://www.faqs.org/health/bios/5/Werner-Arber.html.

133 *The Baltimore Sun,* April 11, 1999, "After the Prize," op. cit.,; Freeman, Elaine; Rodgers, Joann, *Uniquely Hopkins*, op. cit., pg. 21.

134 *Uniquely Hopkins*, op. cit.; *The Washington Post*, Oct. 13, 1978, "Americans, Swiss to Share Nobel Prize in Medicine," B.D. Cohen; *The Journal of Biological Chemistry*, Jan. 15, 2010, "The Characterization of Restriction Endonucleases: The Work of Hamilton Smith," Kresge, Nicole, Simoni Robert, Hill, Robert, http://www.jbc.org/content/285/3/e2.full.

135 *Uniquely Hopkins*, op. cit., *Model*, pg. 319; Nature Education Web site, "Restriction Enzymes," Leslie A. Pray, Ph.D., http://www.nature.com/scitable/topicpage/restriction-enzymes-545.

136 *Model*, pgs. 319–320; *The Baltimore Sun*, April 12, 1999, "After the Prize," op. cit., *The Nobel Prize for Medicine*, 1978 booklet produced by Hopkins, citing Nobel Prize citation.

137 *Model*, pgs. 90–91.

138 Ibid.; pg. 106; *Hopkins Medical News*, Winter 1995, Vol. 18, No. 2, "Master Builder," Jack Lessenberry; University of Michigan Department of Surgery Web site, Zuidema profile, http://surgery.med.umich.edu/portal/about/emeritus/faculty/zuidema_bio.shtml; Longmire, William P. Jr., *Alfred Blalock: His Life and Times*, (self-published, 1991), pgs. 244; 253–255; 260–266.

139 *Hopkins Medical News*, Winter 1995, op. cit.

140 Ibid.

141 *Model*, pg. 91.

142 *Hopkins Medical News*, Winter 1995, op. .cit.

143 Ibid.

144 Ibid.; *Model*, pg. 113.

145 Ibid.

146 *Hopkins Medical News*, Winter 1995, op. cit.

147 *Hopkins Medical News*, Winter 1995, op. cit.

148 Ibid.; University of Michigan Department of Surgery Web site, op. cit.

149 *Model*, pgs. 40, 67–68, 92, 212–213.

150 Ibid., pg. 212.

151 Ibid., pg. 93.

152 Ibid.

153 Ibid., pgs. 94–95.

154 Ibid.; http://www.hopkinsmedicine.org/neurology_neurosurgery

155 *Model*, pg. 96; Web site of the American Skin Association, "George W. Hambrick, Jr., M.D.," http://www.americanskin.org/about/hambrick.php.

156 *Hopkins Medicine*, Winter 2006, Vol. 29, No. 2, "The House that Sol Built," Kate Ledger.

157 *Hopkins Medicine*, Winter 2007, op. cit.

158 Ibid.; online "Free Dictionary," definition of neurotransmissions, http://encyclopedia.farlex.com/neurotransmissions.

159 *Hopkins Medicine*, Winter 2007, op. cit.

160 Ibid.

161 Ibid.

162 Ibid.

163 *Model*, pgs.171–172.

164 Ibid., pgs. 174–176.

165 Ibid., pgs. 177–178; *Hopkins Medical News*, Fall 1983, Vol. 8, No. 2, "Shakespeare and a Year Off."

166 *Model*, pgs. 177–179.

167 Ibid., pg. 177.

168 Ibid., pg. 178.

169 *Model*, pg. 124; *Hopkins Medicine*, Fall 2008, Vol. 32, No. 1, "Issues of Identity," Elaine K. Freeman.

170 Ibid.

171 *Hopkins Medicine*, Fall 2008, op. cit.

172 Ibid.

173 Ibid.

174 Ibid.

175 Ibid.

176 Ibid.

177 *Model*, pgs. 124–125.

178 Ibid.

179 *Hopkins Medicine*, Fall 2007, Vol. 31, No. 1, "Welcome to the Firm," Mike Field.

180 Ibid.

181 Ibid.; *Model*, pg. 125.

182 Ibid.; *Model*, pg. 125.

183 *Hopkins Medicine*, Fall 2007, op. cit.

184 Ibid.

185 *Model*, pg. 107.

186 Ibid., pgs. 131–132.

187 Ibid.; *Hopkins Medical News*, Fall 2001, Vol. 24, No. 4, "A Quarter Century of Stories from Hopkins Medicine."

188 Ibid.

189 Ibid.; pg. 35

190 *Model*, pg. 132; Cavagnaro, Louise, "Buildings of The Johns Hopkins Medical Institutions," updated last on June 26, 2001, from Chesney Archives.

191 Kastor, op. cit., pgs. 189.

192 Ibid., pg. 190.

193 *Model*, pg. 120

194 Ibid.; *Centuries of Caring: The Johns Hopkins Bayview Medical Center Story*, Johns Hopkins Bayview, 2004, Neil A. Grauer, pgs. 32–33.

195 *Model*, pgs. 120–121.

196 Ibid.

197 Ibid.

198 *The Baltimore Sun*, Oct. 13, 1978, "Laureates lead regular family lives," Charles W. Flowers.

199 *The News-American*, Oct. 12, 1978, "2 Hopkins Doctors Win Nobel Prize," Joann Rodgers; list of Hopkins Nobel Prize-winners, Web site, http://webapps.jhu.edu/jhuniverse/information_about_hopkins/facts_and_statistics/nobel_prize_winners/index.cfm.

200 *The Baltimore Sun*, Oct. 13, 1978, "2 at Hopkins share Nobel prize for research work in heredity," Albert Shelstedt, Jr.; *Hopkins Medical News*, Nov./Dec. 1978, Vol. 3, No. 5, "A Day of 'Holy Cows!' and 'I would like to get confirmation.' "

201 *Hopkins Medical News*, Nov./Dec. 1978, op. cit.

202 Ibid.

203 *The Baltimore Sun*, Oct. 13, 1978, op. cit; *Hopkins Medical News*, Nov./Dec. 1978, op. cit.; *The News-American,* Oct. 12, 1978, op. cit.

204 *The Baltimore Sun*, Oct. 13, 1978, op. cit,. *The News-American*, Oct. 12, 1978, op. cit.

205 *The News-American*, Oct. 16, 1978, editorial, "A Bright Day in Baltimore."

206 Ibid., "The Hopkins Family Honors Its Nobel Laureates."

207 Ibid.

208 *The Baltimore Sun*, Dec. 1, 1980, "4 hospitals preparing for a strike," Eileen Canzian; *The Baltimore Sun*, Dec. 2, 1980, "Strikers are arrested at Hopkins," Eileen Canzian.

209 *The Evening Sun*, Dec. 16, 2008, "Hopkins Hospital works out agreement," Norman Wilson.

210 *The Baltimore Sun*, Dec. 1, 1980, op. cit.

211 *The Evening Sun*, Dec. 16, 2080, op. cit.

212 *The Baltimore Sun*, Dec. 4, 1980, "Widening of strike postponed," Eileen Canzian and Will Englund; *The Baltimore Sun*, Dec. 17, 1980, "Hopkins workers approve pact, expected back to work today," Eileen Canzian.

213 *The Baltimore Sun*, Dec. 4, 1980, op. cit.

214 Ibid.; *Dome*, Special Strike Issue, Dec. 8, 1980, "…Hospital occupancy 94%...."

215 *The Baltimore Sun*, Dec. 14, 1980; Dec. 17, 1980, op. cit.

216 University of Texas athletics Web site, March 13, 2006 press release, "Dr. Denton Cooley named one of NCAA's 100 most influential student-athletes," http://www.texassports.com/sports/m-baskbl/spec-rel/031306aab.html; *Sports Illustrated*, April 6, 1981.

217 *Dome*, Oct. 2002, Vol. 53, No. 9, "The Cooley Center Gets into Shape," Mary Ellen Miller; *Hopkins Medical News*, Fall 2001, Vol. 24, No. 4, "A Quarter Century of Stories from Hopkins Medicine."

218 *Dome*, Jan. 1984, "New Orthopedic Center Opens at Hopkins," Karen L. Helsley.

219 Ibid.

220 *Hopkins Medical News*, Summer 1984, Vol. 8, No. 5, "The Clayton Heart Center."

221 Ibid.; also, same issue, "Names in Hopkins' History of the Heart"; Named Professorships Web site, E. Cowles Andrus Professorship in Cardiology, http://webapps.jhu.edu/namedprofessorships/professorshipdetail.cfm?professorshipID=106; MedicineNet.com (for definition of "etiology") http://www.medterms.com/script/main/art.asp?articlekey=3334 ; The Free Dictionary Web site (for definition of "psychophysiology") http://www.thefreedictionary.com/psychophysiological.

222 *Dome*, Sept. 1983, Vol. 33, No. 8, "Transplant Service Begins Procedures"; *Hopkins Medical News*, Summer 1984, Vol. 8, No. 5, "The Clayton Heart Center."

223 *Hopkins Medical News*, Summer 1984, Vol. 8, No. 5, "The Clayton Heart Center."

224 *Hopkins Medical News*, Fall/Winter 1988, Vol. 11, No. 2, "AIDS: Plague of the '80s," Janet Farrar Worthington.

225 Ibid.

226 Ibid.

227 Ibid.

228 Ibid.

229 Ibid.

230 *Dome*, July 2009, Vol. 60, No. 6, "A Quarter Century Strong," Neil A. Grauer.

231 Ibid.

232 Ibid.

233 Ibid.

234 Ibid.

235 Ibid.

236 Ibid.

237 *Model*, pg. 107.

238 Turner, Thomas B., *Part of Medicine, Part of Me: Musings of a Johns Hopkins Dean*, Waverly Press, 1981, pg. 120

239 *The Baltimore Sun*, May 28, 1989, editorial, "Baltimore's Hopkins."

240 *Model,* pgs. 257–258; Power Point presentation on Hopkins' association with the military.

241 *Johns Hopkins Magazine*, June 2001, "The Study of a Lifetime," Elaine F. Weiss; Freedman, Elaine; Rodgers, Joann, *Uniquely Hopkins*, 1989, pg. 33.

242 *Model*, pgs. 289–290.

243 *Johns Hopkins Gazette*, Feb. 8, 2010, Vol. 39, No. 21, "C. Lockard Conley, 94, pioneering hematologist," Neil A. Grauer.

244 Grauer, Neil A., *Centuries of Caring: The Johns Hopkins Bayview Medical Center Story*, pgs. 15, 21; The Alan Mson Chesney Medical Archives of The Johns Hopkins Medical Institutions, "Harold E. and Helen C. Harrison Collection," http://www.medicalarchives.jhmi.edu/papers/harrison.html.

245 *Model*, pg. 271.

246 Ibid.

247 *Model*, pg. 288.

248 *Model*, pgs. 318–319; *Uniquely Hopkins*, op. cit.

249 Named Professorships Web site, "Vincent L. Gott Professorship," http://webapps.jhu.edu/namedprofessorships/professorshipdetail.cfm?professorshipID=143; Free web encyclopedia (for definition of the Gott Shunt, http://www.jrank.org/health/pages/25550/Gott-shunt.html.

250 *Uniquely Hopkins*, op. cit., pg. 10.

251 *Hopkins Medical* News, Winter 1987, Vol. 10, No. 3, "Hopkins Firsts"; Email from David Hungerford, Nov. 16, 2010.

252 *Uniquely Hopkins*, op. cit., pg. 29; Opiates.com Web site, "Solomon Snyder," http://www.opioids.com/endogenous/solomon-snyder.html.

253 Email from Alex Haller, Oct. 15, 2010 (forwarded by Ekaterina Pesheva in Children's Center).

254 Ibid.; Chesney Medical Archives, Donner Collection, http://www.medicalarchives.jhmi.edu/papers/donner.html; Hopkins Swallowing Disorders Program Web site, http://www.hopkinsmedicine.org/Rehab/services/swallowing_disorders.html.

255 *Model*, pg. 272; Hopkins Medicine Website on African Americans at Hopkins, Levi Watkins, http://afam.nts.jhu.edu/people/Watkins/watkins.html.

256 *Hopkins Medical News*, Fall 2001, op. cit.

257 Named Professorships Web site, "Vincent L. Gott Professorship," op. cit., *Johns Hopkins Gazette*, Nov. 30,2009, "Kenneth L. Baughman, 63, former director of Cardiology," David March, http://gazette.jhu.edu/2009/11/30/kenneth-l-baughman-63-former-director-of-cardiology/.

258 E-mails from Richard Bennett, president of Johns Hopkins Bayview Medical Center, and William Greenough III, Sept. 13, 2011; Greenough Web page, with CV, on Hopkins Medicine Web site, http://www.hopkinsmedicine.org/burn/burn%20team/greenough/index.html; Rehydration Project Web site, re: Faisal Prize: http://rehydrate.org/dd/dd16.htm#page2.

259 *Uniquely Hopkins*, op. cit., pg. 34.

260 *Uniquely Hopkins*, op. cit., pg. 25; e-mail exchange with Bill Baumgartner, May 11, 2011.

261 *Dome*, Oct. 1987, Vol. 37, No. 6, photo montage, J. Pat Carter; *Newsweek*, Sept. 21, 1987.

262 "Deans and Directors," press kit for 1989 centennial of The Johns Hopkins Hospital. On file with the author.

263 Kastor, op. cit., pg. 171.

264 Kastor, op. cit., pg. 168.

265 Harvey, A. McGehee, Brieger, Gert H., Barams, Susan L., McKusick, Victor A., *Model*, pg. 101

266 "Deans and Directors" press kit, op. cit.; Chesney Archives, page on Chesney, http://www.medicalarchives.jhmi.edu/sgml/chesney.html.

267 "Deans and Directors" press kit, op. cit.; Chesney Archives, page on Bard, http://medicalarchives.jhmi.edu/sgml/bard.html.

268 "Deans and Directors," press kit., Chesney Archives, page on Turner, http://www.medicalarchives.jhmi.edu/sgml/turner.html.

269 "Deans and Directors," press kit, op. cit.; *Hopkins Medical News*, Summer 1990, Vol. 14, No. 1, "Richard Ross: A Tough Act to Follow," Janet Farrar Worthington.

CHAPTER 3

1 Ross, Richard S., "Governance of the Medical Institutions," in Ross papers in the Alan Mason Chesney Medical Archives, Johns Hopkins, "Governance" box, Folder 27.

2 Ibid.

3 Heyssel, Robert, Memo to Executive Committee, Sept. 29, 1989; in Ross papers "Governance" box, Folder 33.

4 Kastor, John A., M.D., *Governance of Teaching Hospitals: Turmoil at Penn and Hopkins*, Johns Hopkins University Press, 2004, pgs. 170–171; 227 (hereafter referred to as "Kastor"). As Kastor notes in his exceptional book, Ross essentially proposed that the School of Medicine be separated from the University and join the Hospital in a separate corporation, "The Johns Hopkins Medicine Institutions," which would have a CEO to whom both the medical school dean and Hospital president would report. That was not done, but a single Dean/CEO position for Hopkins Medicine was created.

5 Hopkins Centennial Press Kit; *Johns Hopkins Magazine/Hopkins Medical News*, joint issue, June 1989, pg. 104. On file with the author.

6 *The Baltimore Sun Magazine*, May 28, 1989.

7 Joann Rodgers recollections, Dec. 18, 2009.

8 *Dome*, May 1989, Vol. 39, No. 4, "Opening Day at the Hopkins"; Gore family photo.

9 *Johns Hopkins Medical News*, Fall 1989, Vol. 13, No. 2, "Opening Day."

10 Ibid.

11 Ibid.

12 *Hopkins Medical News*, Fall 1993, Vol. 17, No. 1, "Hillary at Hopkins"; *Baltimore Sun*, June 11, 1993, "First lady issues call to action at Hopkins; Mrs. Clinton pleads for more doctors in primary care," Michael Ollove.

13 Ibid.

14 *Dome*, Nov. 1993, Vol. 44, No. 6, "Diagnosing Health Care," Mary Ellen Miller.

15 *Hopkins Medical News,* Fall 1989, Vol. 13, No. 2, opt cit.; *Dome*, Nov. 1989, Vol. 39, No. 7, "Doors Open on a New Decade," Sharon Bondross.

16 *Dome*, Nov.1989, Vol. 39, No. 7, op. cit; press release, "Hopkins Forms Genetic Medicine Institute," Jan. 20, 1999.

17 *Dome,* Nov. 1989, Vol. 39, No. 7, op. cit.

18 *Hopkins Medical News*, Fall 1991, Vol. 15, No. 1, "Scientists Find Gene for Marfan Syndrome," Seema Kuar and Lynn Crawford Cook.

19 Johns Hopkins Medicine Web site on Smilow Center, www.hopkinsmedicine.or/geneticmedicine/Clinical_Resources/Smilow_Mafan/; *Hopkins Medicine,* Winter 2008, letter to editor from Hal Dietz, M.D., director of the Smilow Center.

20 *Dome*, Dec. 1993, Vol. 44, No. 6, "Colon Cancer Gene Found"; *Dome*, Jan. 1994, Vol. 45, No. 1, "The Ultimate Team Effort," Merrill Witty.

21 Author's interview with Vogelstein, Aug. 11, 2009.

22 *Hopkins Medicine*, Winter 2004, Vol. 27, No. 2 "Cites for More Eyes," Mary Ann Ayd.

23 *Hopkins Medicine*, Winter 2004, Vol. 27, No. 2, "Waterway to Stockholm," Mary Ann Ayd.

24 *Hopkins Medicine*, Fall/Winter 1988, "Researchers Face Serious Space Shortage," Sandy Budd.

25 *Dome*, Summer 1991, Vol. 4, No. 4, "The Ross Building: Home for a new brand of medical research."

26 *Dome,* Nov. 1988, Vol. 38, No. 8, "Rutland Research Building Going Up."

27 *Dome,* March 1990, Vol. 40, No. 2, "Johns Chosen Next Dean of Medical School"; "It's Now Officially the Ross Building"; *Hopkins Medical News*, Fall 1990, Vol. 14, No. 2, "Sustaining the Triple Threat"; *Hopkins Medical News*, Spring 1992, Vol. 15, No. 3, "First Ross Physician Scientist Award."

28 *Hopkins Medical News*, Spring 1992, Vol. 15, No. 3, op. cit. *Hopkins Medicine*, Fall, 2007, Vol. No., "Silver Bullet for Blake," Elaine Freeman.

29 Letter to Ross, Aug. 31, 2009, from Ed Miller and George Dover re: Ross Fund's current status.

30 *Dome,* Nov. 1989, Vol. 39, No. 7, "The Johns Hopkins Asthma & Allergy Center"; "Introducing Bayview."

31 *Dome*, Summer 1991, Vol. 1, No. 4, op. cit.; Grauer, Neil A., *Centuries of Caring: The Johns Hopkins Bayview Medical Center Story*, pg. 139.

32 *Dome,* Feb. 1990, Vol. 40, No. 1, "The Johns Hopkins Outpatient Center: It's good medicine," Sharon Bondroff.

33 Ibid.

34 *Dome*, Summer 1993, Vol. 44, No. 4, "Hopkins Heads Home," Elaine Weiss; *Hopkins Medicine in Brief, 2009*, pg. 20.

35 *HopkinsMedical News,* Fall 1991, Vol. 15, No. 1, "Hopkins Launches Geriatrics Center, Research Building," Carol Pearson.

36 Welch, Robert B., M.D., *The Wilmer Ophthalmological Institute, 1925–2000*, East Wind Publishing, Annapolis and Trappe, Md., 2000, pg. 291.

37 *Dome*, Dec. 1992, Vol. 42, No. 5, "Molding Medical Campus"; *Dome*, Oct. 1999, Vol. 50, No. 7, "Unveiling the Weinberg Building," Mary Ellen Miller.

38 Ibid.

39 *Dome*, Oct. 1999, Vol. 50, No. 7, "Unveiling the Weinberg Building," op. cit.; *Hopkins Medical News*, Fall 1999, "Do New Buildings *Truly* Make a Difference?" (Special supplement on openings of Weinberg and Bunting♦Blaustein buildings.)

40 Ibid.; *Dome*, Feb. 1990, Vol. 40, No. 1, "Heart Association Leadership: A Hopkins Tradition."

41 *Dome,* April/May 1990, Vol. 40, No. 3, "Henry Ciccarone Inspires a New Team"; Blumenthal, Roger, *The Hopkins Ciccarone Center: Facing Off Against Heart Disease*, 20th annual update, Dec. 2009.

42 *Dome*, March 1993, Vol. 44, No. 2, "Fueling the Fires of Research," Edith Nichols.

43 *Dome*, February 1991, Vol. 41, No. 1, "Hopkins Researchers Get to the Heart of the Matter."

44 *Dome*, Dec. 1992, Vol. 42, No. 5, "A Giant Step Found in Treating Sickle Cell"; Hopkins Children's Center Web site, www.hopkinschildrens.org/staffDetail.aspz?id=4218; *Hopkins Medicine*, Winter 2009, Vol. 32, No. 2, "Sweet Relief," Christen Brownlee.

45 *Dome,* Jan. 1993, Vol. 44, No. 1, "For Hopkins History Buffs"; "Curtain Call for Big Johns" (stamp).

46 *Dome,* August 1992, Vol. 42, No. 4, "Reshaping the American Physician."

47 Ibid.

48 Ibid.; *Hopkins Medical News*, Fall 1996, Vol.20, No.1, "A Class and a Curriculum Come of Age," Kate Ledger.

49 *Hopkins Medical News*, Spring 1992, Vol. 15, No. 3, "Fine-tuning the medical school curriculum," Carol Pearson; *Hopkins Medical News*, Fall 1996, Vol. 20, No. 1, op. cit.

50 *Hopkins Medical News*, Fall 1993, Vol. 17, No. 1, "For the First Time: A Generalist Rotation at the School of Medicine."

51 *Dome*, Sept. 1993, Vol. 44, No. 5, "Getting Into Hopkins," Mary Ellen Miller.

52 *Dome*, Sept. 1991, Vol. 41, No. 5, "*US News & World Report* Says We're the Best,"; *Hopkins Medical News*, Fall 1991, Vol. 15, No. 1, "Hopkins Ranked America's Top Hospital" (photo caption Heyssel quote).

53 *Dome*, Sept. 2005, Vol. 56, No. 6, "Miss Terrie Understands: Hard lessons ring familiar for counselor at Bayview's Center for Addiction and Pregnancy," Judy Minkove.

54 *Dome,* June 1992, Vol. 42, No. 3, "Reaching Out: The Office of Community Health," Wendy Brewer; *Hopkins Medical News,* Spring 1992, Vol. 15, No. 3, "Boosting the Health of the Neighborhood," Jan Shulman; *Dome*, Dec. 1998,Vol. 49, No. 10, "Lessons Learned: HEBAC."

55 *Dome*, May 1990, Vol. 40, No. 3, "To BE a Good Neighbor," Sharon Bondroff.

56 *Hopkins Medical News*, Spring/Summer 1998, Vol. 21, No. 3, "Women in Medicine Stand Tall," Gary Logan.

57 Ibid.

58 Ibid.

59 *Hopkins Medicine*, Spring/Summer 2009, Vol. 32, No. 3, "Sixty-Five Years of Tinkering," Kate Ledger, pgs. 36–37.

60 Ibid.; Alan Mason Chesney Medical Archives, The Richard J. Johns Collection, biographical sketch.

61 Ibid.

62 Author's interview with Elliott McVeigh, Sept. 7, 2009; *Hopkins Medicine*, Spring/Summer 2009, op. cit.

63 *Hopkins Medicine*, Spring/Summer 2009, op. cit.

64 Interview with Murray Sachs, conducted by Fredrik Nebeker of the Institute of Electrical and Electronics Engineers (IEEE) for the IEEE History Center, April 25, 2000.

65 Ibid.; Department of Biomedical Engineering Web site fact sheet, http://www.mbe.jhu.edu/welcome/bmefacts.htm.

66 *Dome*, Feb. 1990, Vol. 40, No. 1, "Hail to Our Chief"; Kastor, pg. 217; *The Washington Post*, Jan. 11, 1990.

67 Kastor, op. cit, pgs. 193–194.

68 Author's interview with Ross, July 23, 2009

69 Ibid.

70 *Dome*, Summer 1990, Vol. 40, No. 4, "Then/Now: A Dean for 15 Seasons."

71 Ibid.

72 Ibid.

73 Author's interview with Ross, July 23, 2009.

74 Ibid.

75 Author's interview with Ross, July 23, 2009.

76 Kastor, pgs. 193–194; *Hopkins Medical News*, Summer 1990, Vol. 14, No. 1, "Meet the New Dean."

77 Ibid.

78 Ibid.

79 Kastor, pg. 195.

80 Kastor, pg. 214.

81 Ibid., pgs. 196, 214.

82 *Hopkins Medical News,* Spring 1992, Vol. 15, No. 3, "On Leaving the Helm," Janet Farrar Worthington.

83 Ibid.; Kastor, pg. 186.

84 Ibid; pg. 178

85 Kastor, pg. 178; *Hopkins Medical News,* Spring 1992, op cit.

86 Kastor, pgs. 181–187

87 Kastor, pg. 178; *Hopkins Medical News*, Spring 1992, op. cit.

88 Kastor, pg. 173; author's interview with Ross, July 23, 2009.

89 Author's interview with Peterson, Oct. 22, 2008.

90 Ibid.

91 Kastor, pgs. 196–198; pg. 178

92 Ibid., pg. 194

93 Ibid., pgs. 170–172

94 Kastor, ,pgs. 214–229, 160

95 Author's interview with Ross, op. cit.

CHAPTER 4

1 http://en.wikipedia.org/wiki/North_American_blizzard_of_1996; "North American blizzard of 1996"; *Dome*, Jan. 1996 special issue.

2 *Dome*, Jan. 1996 special issue.

3 Kastor, op. cit, pgs. 159–277; *Dome*, July/August 1994, Vol. 45, No. 5 "Speaking with One Voice," Mary Ellen Miller; *Dome*, Dec. 1995, Vol. 46, No 7 "The Pressures Mount: It's a one-two punch. First came Manged care. No Medicaid Cutbacks...," Mary Ellen Miller; *Dome*, March 1996, Vol. 47, No. 2, "CEO Position Created for Johns Hopkins Medicine"; press release, Feb. 12, 1996, "Hopkins Names Miller Interim Dean"; press release, Dec. 18, 1996; "Peterson Named President of Hopkins Hospital"; *Dome,* Feb. 1997, Vol. 48, No. 1 "Miller' Charge," Mary Ellen Miller; "Peterson Officially Named JHHS/JHH President," Who/What page; press release, Jan. 15, 1997, "Miller Named Hopkins Medicine CEO/Dean"; press release, Feb. 21, 1997, "Peterson Named President of Hopkins Health System"; *Hopkins Medical News*, Spring/Summer 1997, Vol. 20, No. 3, "HMN Man of the Year"(Miller profile), Charles Salter, Jr.; "A Quarter-Century Climb to a Stormy Crest," (Interview with Peterson), Edith Nichols; *Change*, Feb. 27, 1997, Inside Hopkins: "Hiring Freeze to Cut Costs"; *Change* , June 12, 1997, "Pitching Medical Assistance"; *Change*, June 25, 1997, "Don't Get Carried Away by the Rankings," Charles Cummings, M.D.; *Change*, May 5, 1998, Inside Hopkins "More Lab Space"; "Upping the Ante"; "In Harm's Way: Buffeted by financial salvos from all quarters, Hopkins waits for the one shot that could cause further budget cuts"; *Change*, May 19, 1998, Inside Hopkins: "Belt Tightening: The Next Act."

4 *Dome*, April 1994, Vol. 44, No. 3, "Breast Cancer Services Now Under One Roof," Melissa Hendricks; "Introducing Johns Hopkins Bayview Medical Center"; *Dome*, Jan/Feb 1995, Vol. 46, No. 2, "Keeping Hopkins Safe," Mary Ellen Miller; "No Waiting, No Appointments: What's Up?" Mary Ellen Miller; *Dome,* May 1995, Vol. 46, No. 4, "Nathans Named JHU Interim President" (Who/What page);*Change*, Sept. 26, 2000, Vol. 4, No. 16, "Rebuilding a Community"; "What They're Saying About '24/7'"; *Dome,* May 1994, Vol. 45, No. 4, "Re-Engineering Wilmer," Mary Ellen Miller; "Planning the Cancer Center," Mary Ellen Miller; *Dome*, Sept./Oct. 1994, Vol. 45, No. 6, "Suburban Outposts," Elaine Weiss with Mary Ellen Miller; "Green Spring: The Hopkins Piece"; *Dome*, Jan/ Feb. 1995, Vol. 46, No. 1, op. cit.; *Dome,* Aug. 1995, Vol. 46, No. 5, "Hopkins Medicine Re-Examined," Mary Ellen Miller; "Hopkins, Other Area Hospitals Form Alliance"; "Executive Life" (Peterson profile), Mary Ellen Miller; *Dome*, Oct. 1995, Vol. 46, No. 6, "Osler 3 Overhauled," Mary Ellen Miller; "Of Hospitals and Hoteliers," (Marburgh refurbishing), Mary Ellen Miller; *Hopkins Medical News*, Fall 1995, Vol. 19, No. 1, "Holding Back the Sickle," Janet Farrar Worthington; "Gehrig-Ripken-Hopkins Triple Play," Kate Ledger; press release, Nov. 27, 1996, "Hopkins Establishes Comprehensive Transplant Center"; *Hopkins Medical News*, Fall 1996, Vol. 20, No. 1, "Onward to Information Future," Elaine Weiss: "By George, He's Got It" (Dover profile), Randi Henderson; *Dome*, Nov. 1996, Vol. 47, No. 7, "Brody's Back," Mary Ellen Miller; "Wilmer's New Pavilion Unveiled," Mary Ellen Miller; News Round-Up (Howard Couty; Suburban items); "Beyond The Cutting Edge," Marjorie Centofanti; *Dome*, Feb. 1997, Vol. 48, No. 1, op. cit; *Dome*, Nov. 1997, Vol. 48, No. 6, "Bricks and Mortar, Cranes and Girders," Mary Ellen Miller; *Dome*, Feb. 1998, Vol. 49, No. 1, "Examining Emergency Medicine," Patrick Gilbert; "The Coming Age of Rehab Medicine," Mary Ellen Miller; *Dome*, March 1998, Vol. 49, No. 2, "Through a Child's Eyes"; *Dome*, Jun 1998, Vol. 49, No. 5, "News from Town (Hurd) Hall"; "The Healing Power of Talk," Mary Ellen Miller; *Dome,* July/Aug. 1998, Vol. 49, No. 6, "Reconstructing Halsted," Mary Ann Ayd; "Eight Straight!" (Who/What page); *Dome,* Sept. 1998, Vol. 49, No. 7, "Welcome Class of '02," (Briefcase); press release, Oct. 26, 1998, "Hopkins to Build New Medical Center at White Marsh."

5 Author's interview with Peterson, Oct. 22, 2008.

6 Kastor, pgs. 160–161.

7 "What Hopkins Built, and Where," press kit for Johns Hopkins Medicine centennial, 1989. On file with the author.; *Model*, pgs. 8–9; Kastor, pgs. 160–161; *Hopkins Medical News*, Summer 1990, "A Tough Act to Follow," Janet Farrar Worthington.

8 Kastor, op. cit, pgs. 162–163.

9 Ibid; pgs. 168–170.

10 Ibid., pgs. 176–177 and 179–180; Johns Hopkins Medicine Publications Style Sheet.

11 Kastor, pgs. 179–180.

12 Kastor, pgs. 180, 196–199

13 Author's interview with Peterson, op. cit.; Kastor, pgs. 222, 233

14 Ibid.

15 Kastor, pg. 197.

16 Ibid., pgs. 197, 221; *The Baltimore Sun*, "Internal rift is testing Hopkins' will; medical school, hospital chiefs' clash runs deep," June 19, 1995, John Fairhall and David Folkenflik.

17 Kastor, pgs. 219–220.

18 Author's interview with Edward Miller, Sept. 29, 2009.

19 Author's interview with Chi Dang, June 18, 2009.

20 Kastor, pg. 219.

21 Kastor, pg. 221.

22 Kastor, pg. 275.

23 Kastor, pg. 217.

24 Kastor, pg. 218; *The Baltimore Sun*, Dec. 29, 1994, "JHU president resigns to head foundation," David Folkenflik.

25 Kastor, pg. 231; author's interview with Edward Miller, op. cit., Sept. 29, 2009.

26 Miller interview, op. cit.

27 Ibid.

28 *The Baltimore Sun*, June 15, 1995, "Boards seek stronger ties for 2 Hopkins institutions," John Fairhall and David Folkenflik; *Hopkins Medical News*, Spring/Summer 1997, Vol. 20, No. 3 "HMN Man of the Year," pg. 23.

29 Kastor, pgs. 223–224.

30 Kastor, pgs. 224–225.

31 Kastor, pg. 225.

32 Kastor, pgs. 226–227.

33 *Hopkins Medical News*, Winter 1997, Vol. 20, No. 2, inside cover, "Man at the Top."

34 Miller interview, op. cit.

35 Ibid.; Kastor, op. cit, pg. 230.

36 Press release, Feb. 12, 1996, "Hopkins Names Miller Interim Medical Dean."

37 Ibid.

38 *Dome*, Feb. 1997, Vol. 48, No. 1, "Miller's Charge"; *Hopkins Medicine*, Spring/Summer 2006, Vol. 29, No. 3, "Hi, Ed!"; *Hopkins Medicine*, Spring/Summer 2007, Vol. 30, No. 3, "Top Guy."

39 *Hopkins Medicine*, Spring/Summer 2007, Vol. 30, No. 3, "Top Guy."

40 *Hopkins Medicine*, Spring/Summer, 2007, Vol. 30, No. 3, "Top Guy"; *Hopkins* Medical *News*, Winter 1997, Vol. 20, No. 2, inside cover, "Man at the Top"; *Hopkins Medical* News, Spring/Summer 1997, Vol. 20, No. 3, "HMN Man of the Year," pg. 25; Kastor, pg. 237n.

41 *Hopkins Medical Magazine*, Spring/Summer 1997, Vol. 20, No. 3, "HMN Man of the Year," pgs. 26–27.

42 Kastor, op. cit., pgs. 227–230.

43 Ibid.

44 Ibid., pg. 236.

45 Ibid., pgs. 231–232.

46 Kastor, pg. 268 footnote.

47 Ibid., 232; *Model*, pg. 82.

48 Kastor, pg. 233.

49 Ibid.

50 Kastor, pgs. 235–236; press release, Jan. 15, 1997, "Miller Named Hopkins Medicine CEO/Dean"; *Hopkins Medical News*, Vol. 20, No. 2, "Man at the Top," inside cover; *Dome*, Feb. 1997, Vol. 49, No. 1, "Miller's Charge."

51 *The Baltimore Sun*, "Internal rift is testing Hopkins' will; medical school, hospital chiefs' clash runs deep," June 15, 1995, John Fairhall and David Folkenflik, op.cit.

52 Kastor, pg. 227.

53 *The Baltimore Sun*, op. cit., June 15, 1995.

54 Kastor, pgs. 199–200; *Dome*, Sept/Oct. 1994, Vol. 45, No. 6, "Suburban Outpost," Elaine Weiss with Mary Ellen Miller; *Dome*, June 2004, Vol. 55, No. 5, "Growing Green Spring," Anne Bennett Swingle.

55 Ibid.

56 *Dome*, Summer 1996, Vol. 47, No. 4, "Hopkins in the Burbs Booming," Martha Merrell and Kate Ledger; author's interview with Gill Wylie, Sept. 6, 2011.

57 *Dome*, Nov. 1997, Vol. 48, No. 6, "Bricks and Mortar, Cranes and Girders," Mary Ellen Miller; author's interview with Gill Wylie, op. cit.

58 Kastor, pgs. 200–201; *Dome*, June, 2004, Vol. 55, No. 5, "Growing Green Spring," op. cit.

59 Ibid.; e-mails from Linda Gilligan, vice president and COO of Johns Hopkins Community Physicians, Oct. 10, 2010; Jeff Richardson, M.D., chief of internal medicine for Johns Hopkins Community Physicians, Oct. 14, 2010.

60 *Dome*, Sept. 2004, Vol. 55, No. 7, "The Next Big Thing: Odenton," Anne Bennett Swingle; author's interview with Gill Wylie.

61 Kastor, pgs. 201–204.

62 *The Daily Record*, July 25, 1996; *Hopkins Medical News*, Fall 1996, Vol. 20, No. 1; "Suburban Hospital's Partnership NIH and Johns Hopkins," http://www.suburbanhospital.org/Cardiac/NIH.aspx; press release, July 2, 2009, "Suburban Hospital Healthcare System Joins Johns Hopkins Medicine."

63 *Dome*, July/August, 1998, Vol. 49, No. 6, "It Takes a Network," Mary Ellen Miller; Kastor, pgs. 252–258.

64 Press release, March 18, 1998, op. cit., Kastor, pgs., 252–254.

65 Kastor, pgs. 252–253, 258.

66 *Three Little Words and the difference they've made; Johns Hopkins Medicine: A ten-year report*, 1997, pg. 34.

67 Ibid.

68 Press release, March 18, 1998, "Howard Health System Unites With Johns Hopkins Medicine"; Kastor, pg. 256.

69 *Dome*. Jan. 1993, Vol. 44, No. 1, "Training the Modern M.D.," Helen Beilenson; "How Patients Rate Johns Hopkins," Edith Nichols; *Hopkins Medical News*, Fall 1998, Vol. 22, No. 1, "Bedside Manner on the Blink?"

70 Ibid, *Dome*, March 2008, Vol. 59, No. 2, "Delivering Bad News," Neil A. Grauer

71 *Hopkins Medical News*, Fall 1997, Vol. 21, No. 1, "Beyond Good and Evil," Kate Ledger

72 Ibid.

73 Ibid, Berman Institute Web Site, http://bioethicsinstitute.org/Web/page/869/sectionid/387/pagelevel/2/interview.asp

74 *Change*, May 22, 1997, Vol. 1, No. 9, "Priority Partners OK'd"; press release, May 15, 1997, "State Approves Priority Partners, Managed Care Organization Formed by Johns Hopkins, Maryland Community Health System."

75 *Change*, April 24, 1997, Vol. 1, No. 7, "Playing the Medicaid Game"; *Hopkins Medicine in Brief*, 2009, pg. 21; Gary Stephenson, Hopkins Medicine director of media relations, discussion, March 3, 2010; discussion with Ron Peterson, Sept. 23, 2010.

76 *Dome*, May 1996, Vol. 47, No. 3, "A Primary Care Network of Our Own."

77 Kastor, op. cit, pgs. 246–247; e-mails from Linda J. Gilligan, vice president and COO of JHCP, Oct. 7, 2010 (on latest patient figures).

78 *Change*, Aug. 13, 1997, Vol. 1, No. 13, "The Long Arm of Medical Care."

79 Ibid.

80 *Dome*, Feb. 1998, Vol. 49, No. 1, Who/What page; press release, Jan. 21, 1998, "Hopkins Medicine Names Patrician Brown Senior Director of Managed Care"; Press release, Jan. 27, 2010, University of Richmond, "Jepson School Names Patricia M.C. Brown its 2010 Leader-in-Residence."

81 *Dome*, Aug. 1995, Vol. 46, No. 5, "Marketing Hopkins Medicine Around the World"; *Time* magazine, Jan. 1, 1934, http://www.time.com/time/magazine/article/019171746701,00.html; *British Medical Journal*, http://bjo.bmj.com/content/20/8/493/.full.pdf.

82 *Dome*, Feb. 1998, Vol. 49, No. 1, "Johns Hopkins, Singapore Forming Medical Center of Excellence."

83 *Dome*, May 1997, Vol 49, No. 2, "The Out-of-Towners Get a Guide."

84 Sports Celebrity Entertainment Web site, http:www.nationalsportsagency.com/cripken.html; Press release, "Orioles Skybox Seats Available for 'Lou Gehrig's Disease Fund-Raiser Celebrating Ripken Retirement," Aug. 8, 2001.

85 Packard Center Web site, www.alscenter.org/about_rpc/our_story.html; "ALS' Key Adversary," *Hopkins Medicine Magazine*, Fall 2007, Vol. 31; No. 1; "Tom Watson hosts annual golf event to remember caddie, fight ALS," *New York Daily News*, July 5, 2008, http://www.nydailynews.com/fdcp?1268664197414.

86 *Dome*, Feb. 1998, Vol. 49, No. 1, "A Cause from the Heart."

87 Author's interview with Ted Dawson, scientific director of the Institute for Cell Engineering, June 10, 2011.

88 *Hopkins Medical News*, Winter 1996, Vol. 19, No. 2, "Down Genetics' Rocky Road [From Benchtop to Bedside]"

89 Ibid; press release, NIH News, "Gregory G. Germino, M.D., Named Deputy Director of NIH's National Instute of Diabetes and Digestive and Kidney Diseases," June 26, 2009, http://www.nih.gov/news/health/jun2009/niddk-26.htm.

90 Author's interview with Ted Dawson, June 10, 2011; Web site for Johns Hopkins Institute for NanoBio Technology, "EOC leader Gregg Semenza wins Canada Gairdner Award," http://inbt.jhu.edu/blog/2010/04/19/eoc-leader-gregg-semenza-wins-canada-gairdner-award/; Gairdner Award Web site, http://inbt.jhu.edu/blog/2010/04/19/eoc-leader-gregg-semenza-wins-canada-gairdner-award/; The Lancet, April 10, 2010, Vol. 375, Issue 9722, "The winners of the 2010 Gairdner Awards, http://www.thelancet.com/journals/lancet/article/PIIS0140-6736(10)60526-0/fulltext.

91 Hopkins Medical News, Winter 1997, Vol. 20, No. 2, "Gene Partners"; Center for Inherited Disease Research Web site, http://www.cidr.jhmi.edu.

92 Facts & Figures, 2004–2005, "Research Highlights, 1952–2003; Academy of Achievement Web biography, http://www.achievement.org/autodoc/printmember/gea0bio-1; press release, July 23, 2008, University of Pennsylvania, "John Gearhart, Stem Cell Pioneer, Named Penn's Institute for Regenerative Medicine Director and PIK Professor."

93 Hopkins Medical News, Winter 1994, Vol. 17, No. 2, "The Move Toward Gene Therapy."

94 Dome, March 1994, Vol. 45, No. 2, Who/What page; Hopkins Medical News, Spring 1995, Vol. 18, No. 3, "Remaking Emergency Medicine."

95 Dome, Feb. 1998, Vol. 49, No. 1, 'Examining Emergency Medicine."

96 Dome, Jan./Feb. 1995, Vol. 46, No. 1 "The 200th Heart."

97 Press release, Nov. 27, 1996, "Hopkins Establishes Comprehensive Transplant Center."

98 Ibid.; JHCTC Web site, http://www.hopkinsmedicine.org/transplant.

99 Press release, Aug. 16, 2001; "Hopkins Starts Paired Kidney Exchange Program"; press release, Mr. 4, 2004, " 'Triple Swap' Kidney Transplant Operation a Success"; press release, July 7, 2009, "Johns Hopkins Leads First 16-Patient, Multicenter 'Domino Donor' Kidney Transplant."

100 Dome, April 1994, Vol. 45, No. 3, "Breast Care Services Now Under One Roof."

101 Hopkins Medical News, Winter 1997, Vol. 20, No. 2, "Chemotherapy Straight into the Brain Tumor"; Hopkins Medicine magazine, Winter 2005, Vol. 28, No. 2, "In Spite of All Odds"; MIT News, Oct. 2, 1996, " Implant wafer approved for brain cancer treatment," http://web.mit.edu/newsoffice/1996/wafer-1002.html.

102 Hopkins Medicine magazine, Winter 2005, op. cit.

103 Hopkins Medical News, Spring/Summer 1997, Vol. 20, No. 3, "How Are We Doing, Dr. Cameron?"

104 Ibid.

105 Dome, May 1996, Vol. 47, No. 3, Who/What page, "Vogelstein, Kinzler Cited for Reports"; Hopkins Medical News, Winter 1998, Vol. 21, No. 2, "Colon Cancer Conversation."

106 Ibid.

107 Dome, Oct. 1997, Vol. 48, No. 5, "McKusick and Sommer Win Lasker Awards."

108 Ibid.

109 Ibid.

110 Press release, June 24, 1996, "Hopkins Nurse-Scientist is President-Elect of American Heart Association.

111 Dome, March 1994, Vol. 45, No. 2,"Tearing Down Fences," Mary Ellen Miller; Dome, Dec. 1998, Vol. 49, No. 10, "Lessons Learned: HEBCAC."

112 Ibid.; Dome, Dec. 2007, Vol. 58, No. 10, "To Catch a Thief"; Change, Sept. 12, 2000, Vol. 4, No. 15, "Mean Streets."

113 Change, Sept. 12, 2000, Vol. 4, No. 15, "Mean Streets."

114 Kastor, op. cit., pgs. 209–212; Hopkins Medicine magazine, Fall 2007, Vol. 31, No. 1, "The Stobo Touch."

115 Ibid.; press release, Sept. 18, 2008, "Dr. John Stobo appointed UC senior vice president for health sciences and services," http://www.universityofcalifornia.edu/news/article/18591.

116 Dome, Aug. 1995, Vol. 46, No. 5, "Benz Heads Department of Medicine" (Who/What page); press release, July 26, 2000, "Edward J. Benz, Jr., M.D., named President of Dana-Farber Cancer Institute," http://www.dana-farber.org/abo/news/press/072600.asp.

117 Press release, May 31, 1996, "Pediatric Researcher Named to Top Post at Hopkins Children's Center"; Hopkins International Physician Update, Spring 1996, "George Dover, M.D.: Interim Director: Distinguished Physician, Teacher, Researcher"; Pediatric News, Jan. 1997, "Former Pediatrics Chief at Johns Hopkins Dies."

118 Memorandum for the file, July 30, 1996, from Jo Martin, regarding Dover meeting with Jonathan Bor and Diana Sugg of The Baltimore Sun.

119 Children's Center News, Spring 2001, "Radiothon Triumphs Again"; Spring 2002, "Mix 106 FM Radiothon Tops $1 million; Spring, 2003, "Radiothon Sets New Record."

120 Children's Center News, Winter 1999, "Jos. A. Bank's continuous support for the Children's Center"; Spring 2001, "WLIF Ties Celebrities to Children's Center"; Fall 2001, "Flowers and Bunnies and Books, Oh My!"

121 Children's Center News, Winter 2004, "$20 Million Gift Lays Foundation for New Hospital."

122 Press release, June 9, 2008, "Johns Hopkins Children's Tower To Be Named for Charlotte Bloomberg."

123 Press relase, Oct. 19, 1996, "Johns Hopkins Selects New GYN/OB Director" [Fox]; http://womenshealth.jhmi.edu/birthcenter/team/fox.html; press release, Nov. 17, 1997, "Hopkins Med School Names New Biological Chemistry Director" [Hart]; press release, June 2, 1998, "Hopkins Medical Institutions Appoints New Neurology Director" [Griffin].

124 Press release, Oct. 19, 1996, op. cit.

125 http://womenshealth.jhmi.edu/birthcetner/team/fox.html, op. cit.

126 Press release, Nov. 17, 1997, "Hopkins Med School Names New Biological Chemistry Director," op. cit.; Web page, Department of Biological Chemistry, "Biological Chemistry Celebrated its 100th Anniversary on November 7, 2008," http://biolchem.bs.jhmi.edu/pages/index.html.

127 "Alumni News: Featuring Dr. Gerald W. Hart," Kansas State University alumni news release, 2008, http://www.ksu.edu/biology/alumnews.html; Web page, Department of Biological Chemistry, op.cit.; Thaindian News, "Study sugars role in cells working may pave way for new therapies," Oct. 21, 2008, http://www.thaindian.com/newsportal/india-news/studying-sugars-role-in-cells-working-may-pave-way-for-new-therapies_100109635.html "Sugar plays vital role in cell division," Feb. 6, 2010, http://www.thaindian.com/newsportal/health/sugar-plays-vital-role-in-cell-division_100315816.html.

128 Press release, June 2, 1998, "Hopkins Medical Institutes Appoints New Neurology Director"; "Dr. John W. Griffin" Web page, with its link to CV: http://www.hopkinsmedicine.org/neurology_neurosurgery/experts/team_member_profile/C1711EA977B5A71EF1DEF752A59530E5/John_Griffin.

129 Ibid.

130 Press release, April 4, 1997, "Hopkins Dean/CEO Miller Announces School of Medicine Appointments"; Kastor, op. cit., pgs. 249–251; Change, May 18, 1999, Vol. 3, No. 10, "Elias Zerhouni: His Last Word"; Change, Feb. 24, 2000, Vol. 4, No. 4, "Changes in the Dean's Office: Questions for Ed Miller."

131 Change, Feb. 24, 2000, Vol. 4, No. 4, op. cit; Dome, April 2002, Vol. 53, No. 4, "Zerhouni Awaits Senate Confirmation Hearing," Mary Ann Ayd; press release, April 20, 2009, "Former NIH Director Elias Zerhouni Rejoins Johns Hopkins Medicine as Senior Advisor."

132 Press release, Jan. 26, 2005, "Steven Thompson Promoted to Senior Vice President of Johns Hopkins Medicine."

133 Dome, April 2007, Vol. 58, No. 3, "Christine White: Career Diplomat," Anne Bennett Swingle.

134 Press release, March 20, 1998, "Johns Hopkins Hospital Appoints Karen Haller VP of Nursing, Patient Care."

135 Dome, Sept./Oct. 1994, Vol. 45, No. 6, "Move Over, Guys."

136 Ibid.

137 Hopkins Medical News, Winter 1995, Vol. 18, No. 2, "Up From Bigotry"; Hopkins Medical News, Winter 1998, Vol. 21, No. 2, "In a Sea of White Faces."

138 Ibid.

139 Change, Dec. 18, 1997, Vol. 1, No. 2, "Standing Tall; Task Force Turns the Key for Women in Medicine."

140 Ibid.

141 Change, Jan. 1997, Vol. 1, No. 1

"Reality Hits; Johns Hopkins Medicine LeadersOutlineSignificant Budget Cuts Needed to Keep the Institution Competitive."

142 Ibid.

143 Ibid.; Author's interview with Miller, Sept. 29, 2009.

144 Author's interview with Miller, op.cit.; *Dome*, June 1998, Vol.49, No. 5, "News from Town (Hurd) Hall."

145 *Dome*, June 1998, Vol. 49, No. 5, "News from Town (Hurd) Hall."

146 Ibid.

147 *Change*, May 19, 1998, Vol. 2, No. 10, "Belt Tightening: The Next Act"; Kastor, op. cit., pgs. 249–251; *Change*, April 23, 2006, Vol. 10, No. 7, "The Long-Distance Runner: Endurance and Discipline Have Been Ken Wilczek's Mainstay in Reshaping the Faculty Practice," Neil A. Grauer.

148 *Change*, May 19, 1998, Vol. 2, No. 10, "Belt Tightening: The Next Act, op. cit. *Dome*, June 1998, Vol. 49, No.5, "News from Town (Hurd) Hall," op. cit.; *Hopkins Medicine,* Spring/Summer 2007, Vol. 30, No. 2, "Top Guy."

149 *Dome*, June 1998, Vol 49, No. 5, "News from Town (Hurd) Hall," op. cit.

150 *Hopkins Medicine*, Spring/Summer 2007, Vol. 30, No. 2, "Top Guy," op. cit.; *Hopkins Medical News*, Winter 1995, Vol. 18, No. 2, "Circling the Dome: Next Stop $900 Million."

151 *Hopkins Medical News*, Winter 1995, op. cit.; *Change*, May 5, 1998, Vol. 2, No. 9, "Inside Hopkins: Upping the Ante."

152 *Change*, May 5, 1998, op. cit.; *Hopkins Medical News*, Winter 2001, Vol. 24, No. 2, "Oh, What a Capital Campaign Can Do!"

153 Ibid.

154 *Dome*, May 1996, Vol. 47, No. 3, "Global Radio"; *Change*, April 10, 1997, Vol. 1, No. 6, "Getting the Word Out, Wirelessly."

155 Press release, April 11, 1997, "Johns Hopkins Medicine Announces World Wide Web Site." Press release, Nov. 18, 1997, "Hopkins Grand Rounds Now on Internet."

156 Kastor, op. cit, pg. 182; *Centuries of Caring: The Johns Hopkins Bayview Medical Center Story*, Grauer, Neil A., pg. 58.

157 *Centuries of Caring*, op. cit, pg. 100.

158 *Dome*, Nov. 1996, Vol. 47, No. 7, "Wilmer Eye Care Pavilion Unveiled."

159 Ibid.; *Dome*, March 1998, Vol. 49, No. 2, "Through a Child's Eyes."

160 *Dome*, Feb. 1997, Vol. 48, No. 1, "Creating a More Livable Cardiology Care Unit."

161 *Dome*, Feb. 1998, Vol. 49, No. 1, "A Welcome Home."

162 *Hopkins Medical News,* Winter 1998, Vol. 21, No. 2, "Bricks and Mortar, Cranes and Girders."

163 *Hopkins Medical News*, Spring/Summer 1998, Vol. 21, No. 3, "They called him simply, 'The Prof'."

164 Ibid.

165 Press release, May 8, 1998, "A. McGehee Harvey, Hopkins Medical Luminary, Dies."

166 Ibid.; *Hopkins Medical News*, Spring 1989, Vol. 12, No. 3, "Mac Harvey: Portrait of a Hopkins Giant."

167 Ibid.; *Hopkins Medical News,* Fall 1998, Vol. 22 No.1,"Goodbye to a Sage of Medicine."

168 *Hopkins Medical News,* Fall 1998, Vol. 22 No.1,"Goodbye to a Sage of Medicine," op. cit.; *Hopkins Medical News*, Spring 1989, Vol. 12, No. 3, op. cit.

169 *Dome*, Sept. 2004, Vol. 55, No. 7, "The Way We Were"; *Hopkins Medical News*, Spring 1995, Vol. 18, No. 3, letter-to-the-editor by Kenneth Zierler, M.D., regarding previous issue's article, "Up From Bigotry." *Hopkins Medical News*, Spring 1989, Vol. 12, No. 3, op. cit.

CHAPTER 5

1 Letter, June 16, 1999, from Joann Rodgers to Severn Sandt, referring to telephone conversation "earlier today"; Sept. 6, 1999 letter from Phyliss McGrady, vice president, ABC News, to Edward Miller and Ronald Peterson; August 2000 "Dear Colleagues and Friends" letter signed by Ed Miller and Ron Peterson; *Dome*, Jan. 2000, Vol. 51, No. 1, "Anatomy of a Documentary."

2 Ibid.

3 Ibid.

4 *Dome*, April 1999, Vol. 50, No. 3, "Dissecting the Y2K Bug."

5 Ibid.

6 Author's interview with Reel, March 10, 2010.

7 "Getting to Yes," post-production Power Point presentation by Joann Rodgers, 2000; letter from Phyllis McGrady to Ed Miller and Ron Peterson, Sept. 6, 1999.

8 Ibid.

9 *Dome*, Jan. 2000, Vol. 51, No. 1, "Anatomy of a Documentary," op. cit.

10 Ibid., *Dome*, Nov. 2000, Vol. 51, No. 10, "Hopkins Health Heroes," by William Brody.

11 *Dome*, Sept. 2000, Vol. 51, No. 8, "Spotlight on Hopkins"

12 *Dome*, Oct. 2000, Vol. 51, No. 9; "Hopkins 24/7—the Aftermath." Joann Rogers undated presentation on impact of series.

13 "About the duPont Columbia Awards: Excellence in Broadcast Journalism," http://www.dupointawards.org/year/2002.

14 *Hopkins Medicine*, Spring/Summer 2004, Vol. 27, No. 3, "A Remedy of Errors," Mary Ann Ayd; *Safe Patients, Smart Hospitals: How One Doctor's Checklist Can Help Us Change Health Care from the Inside Out*, Peter Pronovost, Ph.D., M.D., and Eric Vohr, Hudson Street Press, N.Y., 2010, pgs. vii-xxii.

15 Ibid.

16 Ibid.

17 Ibid.

18 Ibid.

19 Ibid.

20 Ibid. Josie King Foundation Web site, http://www.josieking.org/page.cfm?pageID=1

21 *Hopkins Medicine,* Spring/Summer 2004, op. cit.

22 Ibid.

23 *Dome*, April 2002, Vol. 53, No. 4, "Blame It on the System," Mary Ellen Miller.

24 *Hopkins Medical News*, Fall 2001,Vol. 24, No. 4, "Dark Days at Johns Hopkins," Edith Nichols.

25 Ibid.; press release, July 14, 2001, "Volunteer Dies After Participation in Research Study"; "Johns Hopkins' Tragedy: Could Librarians Have Prevented a Death?" Eva Perkins, Information Today, Inc., Web site: http://newsbreaks.infotoday.com/nbreader.asp?ArticleID=17534, posted Aug. 7, 2001; *Johns Hopkins Magazine*, Feb. 2002, "Trials & Tribulations," Dale Keiger and Sue DePasquale.

26 *Johns Hopkins Magazine*, Feb. 2002, op. cit.

27 Press release, July 19, 2001, "Hopkins Responds to OHRP Suspension of Research"; press release, July 23, 2001, "Federal Office Allows Hopkins Research to Resume."

28 *Johns Hopkins Magazine*, Feb. 2002, "Trials & Tribulations," Dale Keiger and Sue DePasquale.

29 Ibid.

30 *Hopkins Medical News*, Fall 2001, Vol. 24, No. 4, Miller column, op. cit.

31 Press release, Nov. 16, 2001, "Hopkins Names Vice Dean for Clinical Investigation"; Daniel Ford profile on Hopkins Medicine Web site, http://www.hopkinsmedicine.org/about/leadership/biography/DABE674C76FAE96BC05D527524905A6B/Daniel_Ford.

32 Author's interview with Miller, Sept. 29, 2009.

33 Ibid.

34 Author's interview with Dang, June 18, 2009.

35 "Medical Errors: The Scope of the Problem," Agency for Healthcare Research and Quality, U.S. Dept. of Health and Human Services," reference to 1999 Institute of Medicine report, *To Err Is Human: Building A Safer Health System*, http://www.ahrg/gove/qual/errback.htm; *Dome*, April 2002, Vol. 53, No. 4, "Blame It on the System"; *Hopkins Medicine*, Spring/Summer 2004, Vol. 27, No. 3, "A Remedy of Errors."

36 *Hopkins Medicine*, Spring/Summer 2004, op. cit.

37 "Peter J. Pronovost, M.D., Ph.D., F.C.C.M.," http://hopkinsmedicine.org/anesthesiology/Team/summaries/Pronovost_Peter.html; *The New Yorker*, Dec. 10, 2007, "Annals of Medicine: The Checklist," Atul Gwande; *Dome*, April, 2002, op. cit.

38 *New Yorker*, Dec. 10, 2007, op. cit.; Pronovost Web site, op. cit.

39 Press release, Dec. 19, 2003, "JHM Reports Untimely Death of Child Cancer Patient."

40 Ibid.

41 Ibid.

42 *Dome*, March 2010, Vol. 61, No.2 "An A+ for Home Care," Judy F. Minkove.

43 *Hopkins Medicine*, Winter 2004, "Out of Time," op. cit.; *Hopkins Medicine,* Winter 2006, Vol. 29, No. 2, "Time Clocks in the Trenches," Katherine Millet.

44 *Hopkins Medicine*, Winter 2006, "Time Clocks in the Trenches," op cit.

45 *Hopkins Medicine*, Winter 2004, "Out of Time," op. cit.

46 Ibid.

47 Ibid.

48 Ibid.; author's interview with Ed Miller, Sept. 29, 2009.

49 Author's interview with Miller, Sept. 29, 2009.

50 Ibid.

51 *Hopkins Medicine*, Winter 2004, op. cit.; *Dome*, Oct. 2003, Vol. 54, No. 8, "Internal Medicine," Anne Bennett Swingle.

52 Ibid.

53 *Hopkins Medicine*, Winter 2004, op. cit.

54 Ibid.; Weisfeldt, "State of Medical Grand Rounds," Sept. 5, 2003, op. cit.

55 *Dome*, Jan. 2000, Vol. 51, No. 1, "Anatomy of a Documentary," op. cit.

56 Author's interview with Dang, June 18, 2009.

57 Ibid.

58 Press release, Jan. 20, 1999, "Hopkins Forms Genetic Medicine Institute."

59 Ibid.

60 Ibid.

61 Ibid., press release, June 28, 2000, "Aravinda Chakravarti Named Head of Genetic Medicine Institute"; *Hopkins Medicine*, Spring/Summer 2007, Vol. 30, No. 2, "Opening the Book of Life."

62 *Hopkins Medicine*, Spring/Summer 2007, Vol. 30, No. 2, "Opening the Book of Life."

63 Press release, Dec. 7, 2000, "Hopkins Opens Institute Focused on Fundamental Research."

64 Ibid.

65 Ibid.

66 "An Interview with Thomas J. Kelly," Sloan-Kettering Cancer Center Web site, http://www.mskcc.org/mskcc/print/50920.cfm.

67 Press release, Aug. 27, 2003, "Jeremy M. Berg, Ph.D., Named New Director of NIH's National Institute of General Medical Sciences," http://www.nigms.nih.gov/News/Results/Berg.htm; *Johns Hopkins Gazette*, May 24, 2010, "Society of Scholars."

68 *Dome*, April 2002, Vol. 53, No. 4; *Hopkins Medical News*, Spring/Summer 2002, Vol. 25, No. 3.

69 Press release, Oct. 25, 2003, "Desiderio to Head Hopkins' Institute for Basic Biomedical Sciences"; *Change*, Feb. 13, 2001, "Molecular Star Wars."

70 Press release, Jan. 30, 2001, "Hopkins Launches Cell Engineering Institute with $58.5 M. Gift."

71 Chi Dang Web page, http://www.hopkinsmedicine.org/hematology/faculty_staff/hang.html; neoplasia definition, http://medical-dictionary.thefreedictionary.com/Neoplastic+disease; oncogene definition, http://www.biochem.northwestern.edu/Holmgren/Glossary/Definitions/Def-O/oncogene.html; myc oncogene definition, http://www.ncbi.nlm.nih.gov/pubmed/11885563.

72 Ibid.; Press release, Jan. 30, 2001, "Hopkins Launches Cell Engineering Institute with $58.5 M. Gift," op. cit.; press release, Dec. 16, 2008, "University of Maryland School of Medicine Recruits Curt I. Civin, M.D., to Lead New Stem Cell Research Center," http://www.umm.edu/news/releases/civin.html.

73 *Hopkins Medicine*, Fall 2006, Vol. 30, No. 1, "The Secret Life of Curt Civin."

74 *Hopkins Medicine*, Winter 2006, Vol. 29, No. 2, "Stem Cell Dispute," pg. 2.

75 *Change*, Oct. 10, 2000, Vol. 4, No. 17, "Healing in All Forms."

76 *Dome*, March 2005, Vol. 56, No. 2, op. cit; *Dome*, Oct. 2005, Vol. 56, No. 8, "Acupuncture, Actually."

77 Press release, July 11, 2002, "Center for ALS Research at Hopkins Named for Robert Packard."

78 *Hopkins Medicine*, Fall 2007, Vol. 31. No. 1, "ALS' Key Adversary"; MedicineNet.com, "Definition of Neurotransmitter," http://www.medterms.cm/script/main/art.asp?articlekey=9973.

79 *Hopkins Medicine*, Fall 2007, op. cit.

80 Ibid; Marjorie Centofanti, editor of Hopkins Medicine publications, *ALS Alert* and *BrainWaves*, April 8, 2010; e-mail from Kathy Davis, administrative director of the Packard Center, Sept. 27, 2010.

81 Press release, April 23, 2002, "David Nagey, Hopkins' Diector of Perinatal Outreach, Dies at 51."

82 Press release, May 28, 2002, "Jeffery Williams, Hopkins' Neurosurgeon, Dies at 50."

83 Press release, Nov. 22, 2002, "Fredrick J. Montz, M.D. (1955–2002)."

84 Press release, May 9, 2003, "Johns Hopkins Receives $24 Million from Donald W. Reynolds Foundation to Study Sudden Cardiac Death."

85 Ibid.

86 Press release, Jan. 6, 1999, "Hopkins Establishes Johns Hopkins International L.L.C."

87 Ibid.

88 Ibid.

89 Press release, Jan. 26, 2005, "Steven Thompson Promoted to Senior Vice President of Johns Hopkins Medicine."

90 *Change*, May 24, 2005, Vol. 9, No. 9, "The Right-Hand Man," Patrick Gilbert; *Hopkins Medicine*, Spring/Summer 2005, Vol. 28, No. 3, "Minnesota Grit," Edith Nichols.

91 *Change*, May 24, 2005, op. cit.

92 Ibid.

93 Ibid.; *Change*, June 7, 2005, "Emerging from the Background," Patrick Gilbert.

94 Ibid.

95 Change, March 18, 2003, Vol. 7, No. 5, "Exporting Hopkins." *Hopkins Medicine*, Fall 2006, Vol. 30, No. 1, "Partnership Unraveled."

96 *Dome*, June 1999, Vol. 50, No. 4, "JHH Labor and Delivery Announces Redesigned Unit," Mary Ellen Miller; "Building Boom at Bayview," P. Susan Davis; *Centuries of Caring*, op. cit., pg. 114.

97 *Dome*, Oct. 2000, Vol. 51, No. 9, "At Howard County General, Major Surgery," Mary Ellen Miller.

98 *Dome*, Oct. 2000, Vol. 51, No. 9, "What's News: Land Exchange OK'd"; *Dome,* Nov. 2003, Vol. 54, No. 9, "Broadway Overlook," Anne Bennett Swingle; *Dome*, Vol. 60, No. 5, "Working on the Towline," Mary Ellen Miller; author's interview with Ron Peterson, Sept. 23, 2010.

99 *Dome*, Oct. 1999, Vol. 50, No. 7, "Homing In," Mary Ann Ayd.

100 Ibid.

101 Ibid.

102 *Dome*, Oct. 1999, Vol. 50, No. 7, "The Rejuvenation of Hopkins," Patrick Gilbert; *Dome*, Oct. 1995, Vol. 46, No. 6, "Unearthing Baltimore History"; *Dome*, Dec. 1995, Vol. 46, No. 7, "Archaeological Update."

103 *Dome*, Oct. 1999, Vol. 50, No. 7, "Unveiling the Weinberg Building," Mary Ellen Miller, op. cit.

104 Ibid.

105 Ibid.; interview with Elaine Freeman, Sept. 14, 2010.

106 *Hopkins Medical News*, Winter 2002 Vol. 25, No. 2, "Circling the Dome: $150 million to Johns Hopkins—Its Biggest Gift Ever," Anne Bennett Swingle; press release, Nov. 14, 2001, "Sidney Kimmel Gives $150 Million To Hopkins For Cancer Research and Patient Care"; press release, April 29, 2002, "Big Gift, Big Names, Big Celebration: Johns Hopkins Dedicates Sidney Kimmel Comprehensive Cancer Center."

107 Ibid.

108 Ibid.; *Dome*, Oct. 1999, op. cit.

109 *Dome*, Sept. 2003, Vol. 54, No. 7, "What's News: BRB to Open"; *Dome,* Oct. 2003, Vol. 54, No. 8, "Open for Business on Broadway"; *Dome*, April 2003, Vol. 54, No. 3, "What's News: CRB2 Ushered In"; Koch Industries Web site, http://www.kochind.com/files/KochDavid.pdf.

110 *Dome*, Oct. 2003, op. cit.; National Center for Biotechnology Information Web site, "Microarrays: Chipping Away at the Mysteries of Science and Medicine," http://www.ncbi.nlm.nih.gov/About/primer/microarrays.html; Children's Hospital Web site, "Introduction to Proteomics," http://www.childrenshospital.org/cfapps/research/data_admin/Site602/mainpageS602P0.html.

111 Ibid.

112 *Johns Hopkins Gazette*, March 3, 2003, "University to Buy Mt. Washington Campus," Dennis O'Shea; *Dome*, March 2006, Vol. 57, No. 2, "Mt. Washington: The New Destination," Anne Bennett Swingle.

113 Miller, Mark, *Mount Washington, Baltimore Suburb: A History Revealed through Pictures and Narrative*, GBS Publishers, a Division of Gordon Booksellers, Baltimore, pgs. 18–20; Foster, Stephen T., "Captain Belle Boyd, 'La Belle Rebelle,' May 9, 1843–June 11, 1900," http://www.geocities.com/Heartland/Meadows/9710/bboyd.html?20052; "Belle Boyd: Confederate Spy," http://www.angelfire.com/ga3/southernrebels/belle.html; About Women's History, "Belle Boyd," http://womenshistory.about.com/library/bio/blbio_belle_boyd.htm; "American Memory," "Belle Boyd to Abraham Lincoln, Tuesday, January 24, 1865," in Abraham Lincoln Papers at the Library of Congress; "Belle Boyd, Cleopatra of the Secession," http://www.civilwarhome.com/belleboyd.htm; "Belle Boyd," Wikipedia, http://en.wikipedia.org./wiki/Belle_Boyd.

114 *Dome*, Nov. 1999, Vol. 50, No. 8, "The Reitz Stuff," op. cit.

115 Ibid.; *Dome*, Sept. 2000, Vol. 51, No. 8, "Make It Easy," Mary Ellen Miller.

116 *Hopkins Medical News*, Fall 2002, op. cit., "Stents of a New Stripe," Anne Bennett Swingle; Angioplasty.org Web site, http://www.ptca.org/des.html.

117 *Dome*, Dec. 2003, Vol. 54, No. 10, "In the Moore Clinic with Wesla and Gloria," Anne Bennett Swingle; *Hopkins Medicine*, Winter 2006, Vol. 29, No. 2,"Beyond the Abyss," Ramsey Flynn.

118 The Moore Clinic for HIV Care Web site, http://www.hopkinsmedicine.org/gim/fellowship/moore_clinic.html.

119 *Dome,* December 2003, op. cit.; "More on Moore," Anne Bennett Swingle.

120 Epocrates Web site, Bartlett bio, http://www.epocrates.com/products/mrc/bio_bartlett.html; Medpage Today Web site, http://www.medpagetoday.com/HIVAIDS/HIVAIDS/17665; BioMed Central Web site, "History of HAART" reference in *Retrovirology* report on the 2006 International Meeting of the Institute of Human Virology, http://www.ncbi.nlm.nih.gov/pmc/articles/PMC1716971/; *Hopkins Medicine*, Winter 2006, op. cit.

121 Ibid.; Division of Infectious Diseases Web site, http://www.hopkinsmedicine.org/Medicine/id/DirectorStatement.html; Named Professorships Web site, http://webapps.jhu.edu/namedprofessorships/professorshipdetail.cfm?professorshipID=111 .

122 *Hopkins Medical News*, Spring/Summer 2001, Vol. 24, No. 3, "New Formula for Killing Inoperable Liver Lesions," Gary Logan; Johns Hopkins Medicine International Web site, "Interview with J.F. Geschwind: The director of cardiovascular and interventional radiology," http://www.ccghe.jhmi.edu/JHI/English/Doctors/Publications/IPU_Mar04_InterviewGeschwind.asp.

123 *Hopkins Medical News*, Winter 2002, Vol. 25, No. 2 "New [molecular] Test [developed at Hopkins] Detects Early Breast Cancer," Gary Logan.

124 Ibid.

125 *Hopkins Medical News*, Fall 2002, Vol. 26, No. 1, "Better Send It to Epstein," Janet Farrar Worthington.

126 *Hopkins Medical News*, Fall 2002, Vol. 26, No. 1, op. cit.

127 *Dome*, Nov. 1999, Vol. 50, No. 8, Who/What page; *Dome*, April 2000, Vol. 51, No. 5, "Wilmer at 75."

128 *Dome*, Nov. 1999, op. cit; Encyclopedia.com, http://www.encyclopedia.com/doc/1P3-74816062.html; *Reporter*, Vanderbilt Medical Center newspaper, March 4, 2015, "Remember J. Donald M. Gass, M.D., Nancy Humphrey; http://www.mc.vanderbilt.edu/reporter/index.html?ID=3805; Wilmer Institute Web site, faculty, William Richard Green, http://www.hopkinsmedicine.org/wilmer/employees/cvs/Green.html.

129 *Dome*, April 2000, Vol. 51, No. 5, "Wilmer at 75"; "A Sampler of Wilmer Research Firsts."

130 Ibid., "A Sampler of Wilmer Research Firsts."

131 Otolaryngology Web site, http://esgweb1.nts.jhu.edu/otolaryngology/faculty.html; *Dome*, Oct. 2002, Vol. 53, No. 9, "Music to Her Ears."

132 Ibid.

133 Ibid.

134 *Dome*, July 2002, Vol. 53, No. 7, "The Eureka Factor," Kate Ledger.

135 Ibid.

136 Ibid.; National Institute of Neurological Disorders and Stroke Web site, http://www.ninds.nih.gov/disorders/huntington/huntington.htm; Science Magazine Web site, previous issues, March 23, 2001, "Interference by Huntingtin and Atrophin-1 with CBP-Mediated Transcription Leading to Cellular Toxicity," Ross, et al. http://www.sciencemag.org/cgi/content/abstract/sci;291/5512/2423?maxtoshow=&hits=10&RESULTFORMAT=&fulltext=Christopher+Ross&searchid=1&FIRSTINDEX=0&issue=5512&resourcetype=HWCIT.

137 *Hopkins Medical News*, Winter 2000, "Have a Kidney," Anne Bennett Swingle.

138 Press release, Aug. 16, 2001, "Hopkins Starts Paired Kidney Exchange Program"; *Hopkins Medical News*, Fall, 2000, Vol. 26, No. 1, "A Door Opens in the World of Kidney Transplants," Gary Logan.

139 Press release, Aug. 1, 2003, "Johns Hopkins Surgeons Perform World's First 'Triple Swap' Kidney Transplant Operation"; *Dome* , Sept. 2003, Vol. 54, No. 7, "Triple-Swap Transplant," Anne Bennett Swingle; *The Baltimore Sun*, May 31, 2010, "Pushing the Boundaries," Arthur Hirsch.

140 *Hopkins Medical News*, Spring/Summer 2000, Vol. 23, No. 3, "Videotapes Offer Clues to Variations in Prostate Surgery," Janet Farrar Worthington.

141 Ibid.

142 Ibid.

143 Ibid.

144 Ibid.

145 Brady Urological Institute Web site, http://urology.jhu.edu/prostate/video1.php.

146 Alison Currie, Walsh assistant, telephone interview with author, June 2, 2010.

147 *Hopkins Medicine*, Fall 2007, Vol. 31, No. 1, "Hopkins Reader: Gray Matters and a Prostate Classic, Take Two," Neil A. Grauer; Alice Currie telephone interview, op. cit.

148 *Hopkins Medical News*, Fall 2000, Vol. 23, No. 3, "The HPV Warriors," Anne Bennett Swingle; *Hopkins Medicine*, Fall 2006, Vol. 30,No. 1, "The Making of a Phenom," Jon Jefferson; press release, May 14, 2003, "Hopkins Scientists Uncover Role of Faconi Anemia Genes in Pancreatic Cancer."

149 *Hopkins Medical News*, Spring/Summer 2001, Vol. 24, No. 3, "The Pathologist Who Struck Gold," Anne Bennett Swingle.

150 Ibid.

151 Ibid.

152 Ibid.; author's interview with Chi Dang, July 8, 2011.

153 *Hopkins Medical News*, Spring/Summer 2002, Vol. 25, No. 3, "Does Cardiac Bypass Induce a Cognitive Slump?" Marjorie Centofanti.

154 Press release, Aug. 3, 2009, "Is There Long-Term Brain Damage After Bypass Surgery?"

155 *Dome*, March 2002, Vol. 53, No. 3, "What's on *Your* Plate?"; *Johns Hopkins Magazine*, April 2008, Vol. 60, No. 2, "A Nibble of Prevention," Michael Anft.

156 Ibid.; *Pharmacentrical Field* magazine Web site, "The Balancing of Oxidants and Antioxidants," http://www.pharmafield.co.uk/article.aspx?issueID=9&articleID36.

157 *Johns Hopkins Magazine*, April 2008, op.cit.

158 Ibid.

159 *Dome*, June 2006, Vol. 57, No. 5, "Neighborhood Students Sample Science," Anne Bennett Swingle.

160 *Dome*, Jan. 2002, Vol. 53, No. 1, "Working to Stay Clean," Mary Ann Ayd.

161 Ibid.

162 *Dome*, Sept. 2000, op. cit., "Health on Wheels," Mary Ann Ayd.

163 Press release, Sept. 11, 2001, "National Emergency Preparations"; author interview with Elaine Freeman, Sept. 14, 2010.

164 *Dome*, Oct. 2001,Vol. 52, No. 8, "No Degrees of Separation," Mary Ann Ayd.

165 *Dome*, Oct. 2001, Vol. 52, No. 8, "Voices of Reassurance," Mary Ann Ayd.

166 *Dome*, Oct. 2001, Vol. 52, No. 8, "Needing To Do *Some*thing," Mary Ann Ayd.

167 *Dome*, Sept. 2002, Vol. 53, No. 8, "The Sum of All Parts," Mary Ann Ayd; http://www.hopkins-cepar.org/About/staff.html.

168 CEPAR Web site: http://www.hopkins-cepar.org/About/index.html.

169 Ibid., http://www.hopkins-cepar.org/ EMCAPS/EMCAPS.html; http://www. hopkins-cepar.org/index.html.

170 Press release, Oct. 12, 1999, "Judy Reitz Appointed Executive Vice President and Chief Operating Officer of The Johns Hopkins Hospital."

171 *Dome*, Nov. 1999, Vol. 50, No. 8, "The Reitz Stuff," Mary Ellen Miller.

172 Ibid.

173 *Centuries of Caring*, op. cit., pgs. 102, 108.

174 *Hopkins Medical News*, Spring/Summer 2000, Vol. 23, No. 2, "Who/What: New in the Dean's Office."

175 *Hopkins Medicine*, Winter 2005, Vol. 28, No. 2, "Medical School Reformer," Mary Ellen Miller.

176 *Hopkins Medical News*, Spring/Summer 2002, Vol. 25, No. 3, "Poised for Power," Anne Bennett Swingle.

177 Press release, Dec. 11, 2002, "Clements Named Director of Comparative Medicine at Johns Hopkins"; *Change*, Feb. 14, 2003, Vol. 7, No. 3, "The Changing Face of Comparative Medicine."; http://awic.nal.usda.gov/ nal_display/index.php?info_center=3&tax_ level=3&tax_subject=182&topic_ id=1118&level3_id=6735.

178 *Hopkins Medicine*, Winter 2005, Vol. 28, No. 2, "Medical School Reformer," Mary Ellen Miller.

179 Ibid.

180 Ibid.

181 Ibid.

182 Ibid.

183 Ibid.; Office of Academic Computing Web site, http://oac.med.jhmi.edu/oacnewsite/.

184 *Hopkins Medical News*, Spring/Summer 2000, "New In the Dean's Office," op. cit.

185 *Hopkins Medicine*, Winter 2007, Vol. 30, No. 2, "Hot Dang," Neil A. Grauer.

186 Ibid.

187 "Farewell to Chi V. Dang, M.D., Ph.D.," e-mail June 21, 2011.

188 Press release, Nov. 16, 2001, "Hopkins Names Vice Dean for Clinical Investigation"; Daniel Ford profile on Hopkins Medicine Web site, http://www.hopkinsmedicine.org/about/ leadership/biography DABE674 C76FAE96BC05D527524905A6B/ Daniel_Ford.

189 Ibid.

190 Press release, Aug. 16, 1999, "Weiss Named Hopkins Associate Dean of Admissions and Academic Affairs."

191 *Hopkins Medical News*, Spring/Summer, Vol. 25, No. 3; *Hopkins Medical News*, Winter 2003, Vol. 26, No. 2, letter from Henry Seidel on longstanding "secret grading" system; *The Hopkins Internist*, Vol. 5, No. 1, May 2001, "Interview: James Weiss, Associate Dean of Admissions."

192 *Dome*, Sept. 2003, Vol. 54, No. 7, "What's News: Class of '07"; *Dome*, Sept. 2005, Vol. 56, No. 6, "Welcome to the Class of '09,"Lydia Lewis Bloch.

193 Press release, Dec. 1, 1999, "Roderer Named Director of Hopkins Welch Medical Library."

194 Ibid.

195 Ibid.

196 *Hopkins Medicine*, Fall 2010, Vol. 33, No. 3, "At 81, the Welch Bids Goodbye to Books," Geoff Brown.

197 Ibid.

198 *Hopkins Medical News*, Fall 2000, Vol. 23, No. 3, "The Booming Business of Keeping M.D.s Current," Mat Edelson.

199 Ibid.

200 Ibid.; *Change*, March 2007, Vol. 11, No. 3, "CPA News: "CME's Changing World," Neil A. Grauer; Office of Continuing Medical Education Web site, "message from the Associate Dean," Todd Dorman, http:// www.hopkinscme.edu/About/MsgDean.aspx .

201 *Dome*, Feb. 1999, Vol. 50, No. 1, "Now at a Bookstore Near You," Anne Bennett Swingle.

202 Ibid., also "You Guide to the Medicine Chest," sidebar.

203 *Hopkins Medical News*, Fall 2002, Vol. 26, No. 1, "Going National with Diabetes," Anne Bennett Swingle.

204 Ibid.

205 Press release, Nov. 16, 1999, "Noted Hopkins Molecular Biologist, Daniel Nathans, Dies"; *Dome*, Dec. 1999, Vol. 50, No. 9, "In Memoriam," Marjorie Centofanti; *Hopkins Medical News,* Winter 2000, Vol. 23, No. 2, "Farewell to a Quiet Man."

206 *Hopkins Medical News*, Winter 2001, Vol. 24, No. 2, "Library Notes."

207 *Hopkins Medical News*, Spring/Summer 2000, Vol. 23, No. 2, "'Who Will Look After Us Now?'" http://www.giving.jhu.edu/ kfw.

208 Ibid.; press release, Feb. 24, 2000, "Carol Johns, Hopkins Lung Specialist, Dies."

209 Ibid.

210 Ibid.

211 *Hopkins Medical News*, Spring/Summer 2000, op. cit.

212 Press release, May 21, 2001, "Russell Nelson, Former Hopkins Hospital President, Dies at 88"; press release, June 13, 2001, "Obituary: Robert Heyssel, Hopkins Hospital President Emeritus."

213 Ibid.

214 Ibid.

215 Ibid.

216 Ibid.

217 *Hopkins Medical News*, Spring/Summer 2001, Vol. 24, No. 3, "The Man Who Knew Welch," Anne Bennett Swingle; press release, July 23, 2002, "Owsei Temkin, Renowned Historian of Medicine, Dies."

218 Ibid.

219 Press release, July 23, 2002, op. cit.

220 Press release, July 16, 2003, "Benjamin Baker, Johns Hopkins 'Renaissance' Physician, Dies at 101"; *Hopkins Medical News*, Spring/Summer 2002, Vol. 25, No. 3, "He Led a Charmed Life, but at 100, the Devil Is in the Details," Anne Bennett Swingle; *Johns Hopkins Magazine*, April 2002, Vol. 54, No. 2, "Benjamin Maker, Med '27, One Hundred Hears Later, 'Here I Am'," Jeanne Johnson.

221 Ibid., *Hopkins Medical News*, Spring/ Summer 2002, op. cit.

222 Ibid.

223 Ibid.

224 Press release, July 16, 2003, op. cit.; Encyclopedia Britannica on-line, http:// www.britannica.com/EBchecked/ topic/50684/ballistocardiogram.

225 Ibid., press release, July 16, 2003.

226 *Dome*, Feb. 2003, Vol. 54, No. 1, "Almost a Miracle," Mary Ann Ayd.

227 Ibid.

228 Ibid.; GlaxonSmithKline web site, http:// www.gsk.com/about/ataglance.htm; Organization of American Historians web site, http://www.oah.org/activities/awards/ barnouw/winners.html Congressional Black Caucus web site, http://www. cbcfinc.org/2003-archive/284-cbcf-and-glaxosmithkline-unveil-new-scholarship-program-to-increase-minority-presence-in-the-sciences.html; standard press release from Rep. James E. Clyburn Office regarding Vivien Thomas Scholarship, http://clyburn. house.gov/press/030429thomasscholarship. html; *Dome*, Feb. 2003, op. cit.

229 *The Washington Post*, Nov. 11, 2007, "Dentist Had Hankering for Show Business," Mat Schdel (obituary of STLM co-producer, Irving Sorkin), http://www.washingtonpost. com/wp-dyn/content/article/2007/11/10/ AR2007111001586.html; *Los Angeles Times,* Oct. 25, 2007, "Irving Sorkin, 88; dentist saw Hollywood dream come true as award-winning producer," Dennis McLellan, http:// articles.latimes.com/2007/oct/25/local/me-sorkin25; *Hopkins Medicine*, Spring/Summer 2004, Vol. 27, No. 3, "Something the Lord Made," Mary Ann Ayd; *Dome*, May 2004, Vol. 55, No. 4; *Dome*, May 2005, Vol. 56, No. 4.

230 *Dome*, May 2004, Vol. 55, No. 4, "Hopkins Meets HBO," Anne Bennett Swingle.

231 Ibid.; *Hopkins Medicine*, Spring/Summer 2004,Vol. 27, No. 3, "Something the Lord Made," Mary Ann Ayd. http://articles. latimes.com/2007/oct/25/local/me-sorkin25. http://articles.latimes.com/2007/oct/25/ local/me-sorkin25.

232 Ibid.

233 *The Washington Post*, Nov. 11, 2007, "Dentist Had Hankering for Show Business," op. cit.; *Los Angeles Times*, Oct. 25, 2007, "Irving Sorkin, 88; dentist saw Hollywood dream come true as award-winning producer," op. cit.; *Dome*, May 2005, Vol. 56, No. 4, "Who/What, STLM Wins Peabody."

234 Press release, May 26, 2004, "Hopkins Launches Vivien Thomas Fund to Increase Diversity."

235 *Time*, Aug. 20, 2001, http://www.time.com/ time/magazine/article/0,9171,1000602,00. html.

236 David Sidransky profile, Hopkins Medicine Web site, http://www.hopkinskimmelcancercenter. org/index.cfm/cID/1686/mpage/expertdata. cfm/expID/15; Sidransky profile, Forbes. com, http://people.forbes.com/profile/david-sidransky/2906; Sidranksky biography, Spoke. com, http://www.spoke.com/info/p6SWRee/ DavidSidransky ; press release, Jan. 17, 2008, "Alfacell Appoints David Sidranksy Chairman of the Board of Directors, http:// salesandmarketingnetwork.com/news_ release_print.php?ID=2022750.

237 *Time*, Aug. 20, 2001, op. cit. http:// www.time.com/time/magazine/ article/0,9171,1000595,00.html.

238 Sidransky profile on Kimmel Cancer Center Web site, http://www. hopkinskimmelcancercenter.org/index.cfm/ cID/1686/mpage/expertdata.cfm/expID/15.

239 *Hopkins Medical News*, Winter, 1997, Vol. 20, No, 2, "From the Lab to the Limelight," Sandra Salmons; *Time*, Aug. 20, 2001, op. cit., http://www.time.com/time/magazine/ article/0,9171,1000 589,00.html.

240 Academy of Achievement, Benjamin S. Carson biography, http://www.achievement. org/autodoc/page/car1bio-1; The History of African Americans @ Johns Hopkins University Web site, http://afam.nts.jhu.edu/ people/Carson/carson.html; "The Healer Beyond the Operating Room," Antoinette Rainey, The History of African Americans @ Johns Hopkins Web site, op. cit.; Biography. com Web site, "Ben Carson Biography," http://www.biography.com/articles/Ben-Carson-475422?print; The Carson Scholars Fund Web site, http://carsonscholars.org/.

241 Ibid.; Biography.com, op. cit.

242 Ibid.; Carson c.v., http://www. hopkinsmedicine.org/neurology_ neurosurgery/cv/benjamin_carson.pdf.

243 *Hopkins Medical News*, Winter 1997, Vol. 20, No. 2, op. cit.

244 Carson Scholars Fund Web site, op. cit.

245 Ibid.

246 Press release, Oct. 24, 2001, "Two Hopkins Faculty Members Receive 'Genius' Awards."

247 Ibid.

248 *Hopkins Medical News*, Spring/Summer 2002, Vol. 25, No. 3, "The Winning Worm," Anne Bennett Swingle.

249 Ibid.; Seydoux c.v., http://www.mbg.jhmi. edu/Pages/cv/CVSeydoux.pdf.

250 Press release, Oct. 24, 2001.

251 *Hopkins Medical News*, Spring/Summer 2002, op. cit.

252 Press release, May 9, 2002, "McKusick, 'Father of Genetic Medicine,' To Get National Medal of Science."

253 Ibid.

254 *Hopkins Medical News*, Fall 2002, Vol. 26, No. 1, "Two Days in June, Two Big Awards"; McKusick c.v., circa 2007, http://www. socgen.ucla.edu/hgp/files/McKusickCV2007. pdf.

255 *The Globe and Mail*, Toronto, Canada, Oct. 9, 2003, "Quote of the Day"; *The New York Times*, Oct. 9, 2003, "Two Americans Win Nobel for Chemistry," Kenneth Chang.

256 *Hopkins Medical News*, Fall 1997, Vol. 21, No. 1, "Aquaporin's Secrets," Marjorie Centofanti; press release, Oct. 8, 2003, The Royal Swedish Academy of Sciences, "Molecular channels let us enter the chemistry of the cell."

257 Ibid., author's interview with Agre, Aug. 19, 2009.

258 Author's interview, with Agre, op. cit.

259 Press release, Oct. 8, 2003, op. cit.; *The New York Times*, Oct. 9, 2003, op. cit.

260 Press release, Oct. 8, 2003, op. cit.; *Dome*, November 2003, Vol. 54, No. 9, "A Day Like No Other," Anne Bennett Swingle.

261 *Hopkins Medicine*, Winter 2004, Vol. 27, No., 2 "Waterway to Stockholm," Mary Ann Ayd.

262 Ibid.

263 Ibid.; Press release, Feb. 16, 2009, Bloomberg School of Public Health, "Peter Agre to Begin Term as President of AAAS."

264 Ibid., *Hopkins Medical News*, Fall 1997, op. cit.

265 Ibid.

266 Ibid.; *The New York Times*, Oct. 9, 2003, op. cit.

267 Ibid.

268 Author's interview with Agre, op. cit.

269 *Hopkins Medicine*, Spring/Summer 2004, Vol. 27, No. 3, "After the Nobel," Mary Ann Ayd.

270 Ibid.

271 Author's interview with Agre, op. cit.

272 Ibid.

273 Press release, Feb.16, 2009, Bloomberg School of Public Health, op. cit.; author's interview with Agre, op. cit.

274 Ibid; *Hopkins Medicine*, Spring/Summer 2009, "When Suitors Come Calling," Ramsey Flynn.

275 *Hopkins Medicine*, Spring/Summer 2007, Vol. 30, No. 2, "Pushing Every Day: Ron Peterson," Edith Nichols.

276 Ibid.

277 *Dome*, Oct. 1999, Vol. 50, No. 7, "The Rejuvenation of Hopkins," Patrick Gilbert; author's interview with Peterson, Oct. 22, 2008.

278 Author's interview with Miller, Sept. 29, 2009.

279 Ibid.

280 Ibid.

281 *Change*, June 24, 2003, "First Shot Out of the Gate."

282 Ibid.

283 Ibid.

284 Press release, May 6, 2002, "Johns Hopkins Sets $1 Billion Campaign Goal"; *Hopkins Medical News*, Fall 2002, Vol. 26, No. 1, "Seeking Cures, Saving Lives," Marlene England.

285 "Knowledge for the World" Web site, http:// www.giving.jhu.edu/kfw; Fund for Johns Hopkins Medicine campaign total listing as of Dec. 31, 2008, courtesy of Kathy White, senior administrative coordinator to Steve Rum, Fund for Johns Hopkins Medicine, e-mail, June 11, 2010.

CHAPTER 6

1 *Change*, Dec. 2, 2005, Vol. 9, No. 18, "Upping the Ante."

2 *Change*, Nov. 11, 2005, Vol. 9, No. 17, "Manna for Teaching."

3 Author's interview with McVeigh, July 7, 2009.

4 CBID Web site, op. cit., http://cbid.bme. jhu.edu/news-and-events/index.php; *The New York Times*, Jan. 4, 2009, "Getting Up and Around, Even in Intensive Care," Gina Kolata; "So the Medicine Goes Down," Gina Kolata.

5 Press release, Sept. 21, 2004, "Arthur B. Modell to Head Board of Governors for New Johns Hopkins Heart Institute"; *Change*, Oct. 14, 2004, Vol. 8, No. 15, "Launching the Heart Institute"; *Hopkins Medicine*, Winter, 2005, Vol. 28, No. 2, "New Beat, All Heart," Mary Ellen Miller; *Dome*, Oct. 2004, Vol. 55, No. 8, "A New Cardiac Enterprise," Anne Bennett Swingle.

6 Ibid.; *Hopkins Medicine*, Fall 2006, Vol. 30, No. 1, "At the Heart of It All."

7 Ibid.; *Hopkins Medicine* , Fall 2006, op. cit.; *Hopkins Medicine*, Spring/Summer 2005, Vol. 28, No. 2, "Change of Heart," Mary Ann Ayd; Merck Manuals Medical Library Online, http://www.merck.com/mmhe/sec03/ ch026/ch026d.html.

8 *Dome*, Feb. 2009, Vol. 60, No. 1, "Keys to a Woman's Heart," Judy Minkove.

9 *Cardiovascular Report*, Fall, 2008, "Totally Wireless."

10 *Dome*, Nov. 2006, Vol. 57, No. 9, "Into the Wild Blue Yonder," Sarah Richards.

11 Ibid.; *Centuries of Caring: The Johns Hopkins Bayview Medical Center Story*, op. cit., pgs. 25, 28, 31

12 *Hopkins Medicine*, Spring/Summer 2002, Vol. 25, No. 3, "The face of frailty," Mat Edelson; *Dome*, May 2008, Vol. 59, No. 4, "Biology of Frailty," Neil A. Grauer.

13 *Hopkins Medicine*, Spring/Summer 2002, op. cit.; NIH "Clinical Trials.gov" Web site, http://clinicaltrials.gov/ct2/show/ NCT00005133.

14 Ibid.

15 *Hopkins Medicine*, Spring/Summer 2009, "Synergy at Last," Marjorie Centofanti; *Hopkins Medicine*, Spring/Summer 2005, Vol. 28, No. 3, "Consultation with Constantine Lyketsos"; Department of Psychiatry Web site, http://www.hopkinsmedicine.org/ neurology_neurosurgery/experts/team_ member_profile/9DB533E101255B7CE3cc61 4A64B69919/Constantine_Lyketsos.

16 *Hopkins Medicine*, Fall 2007, Vol. 31, No. 1, "A Silver Bullet for Blake," Elaine Freeman.

17 *Hopkins Medicine*, Winter 2009, op. cit.; McKusick-Nathans Institute of Genetic Medicine newsletter *Genetic News* Vol. 1, Issue 1. March, 2009, "Individualized Medicine Comes to Johns Hopkins," Maryalice Yakutchik.

18 *Dome*, May 2005, Vol. 56, No. 2, "The Gene Gurus," Judith Minkove.

19 Ibid.

20 *Hopkins Medicine*, Winter 2009, Vol. 32, No. 2, "Sweet Relief," Christen Brownlee.

21 Ibid.

22 Ibid., "So-Long to Sickled Cells."

23 Ibid.

24 Ibid.

25 *The Baltimore Sun*, June 27, 2010, "Gateway to the Brain," Meredith Cohn.

26 Ibid.

27 Ibid.

28 *Hopkins Medicine*, Winter 2009, "Special Delivery," Ramsey Flynn.

29 *Dome*, Oct. 2009, Vol. 60, No. 8, "The Wilmer Way," Neil A. Grauer.

30 Ibid.

31 Press release, June 18, 2009, "Hopkins Children's on *US News & World Report* List of Best Children's Hospitals."

32 *The Johns Hopkins Heart Institute Cardiovascular Report*, Summer 2007, "Getting to Loeys-Dietz in Time," Mary Ann Ayd; National Institutes of Health Web site, "Harry C. Dietz—Curing Connective Tissue Diseases," http://recovery.nih.gov/stories/dietz.php.

33 Ibid.; Web site of the Archives of Disease in Childhood, June 20, 2010, http://adc.bmj.com/content/92/2/119.full.pdf.

34 *Hopkins Medicine*, Fall 2007, Vol. 31, No. 1, "The Latest Big Fish Story," Neil A. Grauer.

35 *Hopkins Medicine*, Fall 2004, Vol. 28, No. 1, "Alternative to Angiography?" Michael Levin-Epstein; *Dome*, July 2006, Vol. 57, No. 6, "No. 1 in Quality," Anne Bennett Swingle; "In Radiology, Film Clerks Go Digital," Deborah Rudacille; *Hopkins Medicine*, Fall 2007, Vol. 31, No. 1, op. cit.; press release, "Johns Hopkins Housing and Testing Only 256-Slice CT Scanner in North America"; press release, Nov. 26, 2007, "Johns Hopkins Installs First 320-Slice CT Scanner in North America."

36 Press release, Nov. 26, 2007, op. cit.

37 *Hopkins Medicine*, Fall 2004, Vol. 28, No. 1, "Incurable, not Untreatable," Mary Ann Ayd; Dept. of Rheumatology Web site, Livia Casciol-Rosen page, http://www.hopkinsmedicine.org/rheumatology/research/faculty_staff/livia_casciola_rosen.

38 *Hopkins Medicine*, Spring/Summer 2007, Vol. 30, No. 3, "Just say 'ah,'" Ramsey Flynn.

39 *Dome*, July 2009, Vol. 60, No. 6, "A Swap for a Co-Worker," Mary Ellen Miller.

40 Press release, April 8, 2008, "Hopkins Performs Historic 'Six-Way Domino' Kidney Transplant.

41 Ibid.

42 *Hopkins Medicine*, Spring/Summer 2009, "An Easier Kidney Donation," op. cit.

43 Press release, Feb. 16, 2009, "Hopkins Leads First 12-pateint, Multicenter 'Domino Donor' Kidney Transplant."

44 Ibid., *Hopkins Medicine*, Spring/Summer 2009, "Kidney Swap Central," op. cit.

45 *Hopkins Medicine*, Spring/Summer 2009, op. cit.

46 Ibid.

47 *Dome*, July 2009, Vol. 60, No. 6, "1,000th Kidney Transplant."

48 *Hopkins Medicine*, Spring/Summer 2009, Vol. 32, No. 3, "Kidney Swap Central"; "An Easier Kidney Donation," Ramsey Flynn; press release, Feb. 2, 2009, "Hopkins Transplant Surgeons Remove Healthy Kidney Through Donor's Vagina"; press release, Feb. 16, 2009, "Hopkins Leads First 12-patient, Multicenter 'Domino Donor' Kidney Transplant"; press release, July 7, 2009; *Dome*, July 2009, Vol. 60, No. 6, "A Swap for a Co-Worker," Mary Ellen Miller.

49 Press release, Feb. 16, 2009, op. cit.; press release, July 7, 2009, "Johns Hopkins Leads First 16-Patient, Multicenter 'Domino Donor' Kidney Transplant."

50 *Dome*, July 2009, Vol. 60, No. 6, "1,000th Kidney Transplant"; press release, March 23, 2010, "Johns Hopkins Reaches Milestone in Pioneering 'Incompatible Donor' Kidney Transplants."

51 Press release, Oct. 5, 2009, "Johns Hopkins First in R&D Expenditures for 30th Year."

52 Dome, Sept. 2011, Vol. 62, No. 7, "Catching up with Landon King," Linell Smith; press release, July 21, 2011.

53 Press release, Oct. 12, 2005, "Institute Taps Computer Power to Advance Medical Research."

54 Ibid.

55 *Johns Hopkins Gazette*, Sept. 4, 2009, op. cit.; *Johns Hopkins Gazette*, Dec. 2, 2009, op. cit.; Institute for Computational Medicine Web site, http://www.icm.jhu.edu/cancer_research/; http://www.icm.jhu.edu/cardio_research/; http://www.icm.jhu.edu/brain_research/.

56 *Johns Hopkins Gazette*, March 19, 2007, Vol. 36, No. 25, "JHM Launches Unique Brain Science Institute," Greg Rienzi; Brain Science Institute Web site, Frequently Asked Questions, http://www.brainscienceinstitute.org/index.php/about/faq/.

57 Johns Hopkins Medicine Web site, Brain Science Institute, http://www.hopkinsmedicine.org/brainscience; Brain Science Institute Web site, Frequently Asked Questions, op. cit.

58 Ibid.

59 Ibid.; Radiological Society of North America Web site, "Brain Imaging Discoveries Translated into Practice at New Center," July 2009, http://www.rsna.org/Publications/rsnanews/July–2009/Brain_Imaging_feature.cfm; author's interview with Chi Van Dang, July 8, 2011.

60 Press release, Sept. 18, 2007, "Johns Hopkins Joins National Consortium to Speed Research from Clinic to Community."

61 *Johns Hopkins Gazette*, Aug. 16, 2010, "Johns Hopkins establishes new clinical research network," Gary Stephenson.

62 Ibid.

63 Press release, June 28, 2004, "Next Up: All There Is to Know About Epigenetics; Hopkins to Found New Center"; *Hopkins Medicine*, Fall 2006, Vol. 30, No. 1, "Andy Feinberg, Chief Explainer," Gregory Mone; press release, Oct. 4, 2006, "Eight New Research Centers Opened at Johns Hopkins."

64 *Hopkins Medicine*, Spring/Summer 2010, Vol. 33, No. 2, "Nature Meets Nurture," Christen Brownlee.

65 Ibid.

66 Press release, Oct. 4, 2006, op. cit; author's interview with Stephen Desiderio, June 28, 2010, op. cit.; HiT Center Web site, http://www.hopkinsmedicine.org/institute_basic_biomedical_sciences/research/research_centers/high_throughput_biology_hit.

67 Ibid.; IBBS Web site, "Research Centers," http://www.hopkinsmedicine.org/institute_basic_biomedical_sciences/research/research_centers/.

68 Author's interview with Desiderio.

69 Ibid.

70 Ibid.; IBBS Research Centers Web site, op. cit.

71 Ibid.

72 Press release, Oct. 4, 2006, op. cit.; MyOptumHealth Web site definition of malabsorption, http://www.myoptumhealth.com/portal/ADAM/item/Malabsorption; author's interview with Chi Van Dang, July 8, 2011, op. cit.

73 *Hopkins Medicine*, Winter 2008, Vol. 31, No. 2, "High Time for HiCy?" Elaine Freeman.

74 Ibid.

75 Ibid.

76 Press release, Feb. 9, 2005, "'Broken Heart Syndrome: Real, Potentially Deadly But Recovery Quick"; *Hopkins Medicine*, Winter 2008.

77 Ibid.

78 *Hopkins Medicine*, Winter 2008, op. cit.

79 Author's interview with Ted Dawson, scientific director of the Institute for Cell Engineering, June 10, 2011.

80 Author's interview with Ted Dawson, June 10, 2011, op. cit., dendritic cell therapy for cancer Web site, http://www.dendritic.info/; International Society for Cellular Therapy Web site, http://www.celltherapysociety.org/files/PDF/Resources/OnLine_Dendritic_Education_Brochure.pdf.

81 Author's interview with Ted Dawson, June 10, 2011.

82 *Hopkins Medicine*, Winter 2004, Vol. 27, No. 2, "To Break TB," Robert Roper and Mary Ann Ayd.

83 Ibid.

84 Ibid.

85 Ibid.; Center for Tuberculosis Research Web site, http://www.hopkinsmedicine.org/DOM/TB_Lab/faculty/chaisson.html; http://www.hopkinsmedicine.org/DOM/TB_Lab/; *Nature Medicine* magazine, Jan. 2005, Vol. 11, No. 1, profile of Richard Chaisson, Bruce Diamond.

86 Press release, July 15, 2004, "Johns Hopkins-Led Research Group Receives $44.7 million Gates Foundation Grant to Evaluate New Strategies to Fight HIV-Related Tuberculosis."

87 Press release, Oct. 14, 2008, "Consortium to Respond Effectively to the Aids/Tuberculosis Epidemic Gets $32 Boost from Bill & Melinda Gates Foundation"; Sanjay Main page on the Hopkins Children's Center Web site, http://www.hopkinschildrens.org;staffDetail.aspx?id=2330.

88 Press release, Nov. 3, 2005, "Johns Hopkins Celebrates Its First Century of Neuroscience."

89 Ibid.

90 Press release, Nov. 7, 2005, "Nov. 11 Event Celebrates a Century of Brain Science at Johns Hopkins."

91 Press release, Nov. 3, 2005, op. cit.; *Hopkins Medicine*, Winter 2006, Vol. 29, No. 2, "The House That Sol Built," Kate Ledger.

92 *Hopkins Medicine*, Winter 2007, op. cit.

93 Ibid.; Netdoctor Web site on Huntington's Disease name, http://www.netdoctor.co.uk/diseases/facts/huntingtons.htm.

94 *Hopkins Medicine*, Winter 2007, op. cit.

95 Ibid.; Stanford University's Hopes, Huntington's Outreach Project Web site, statistics, http://hopes.stanford.edu/causes/popgen/aj5.html.

96 *Hopkins Medicine*, Spring/Summer 2008, Vol. 31, No.3, "When Alzheimer's Doesn't Mean Dementia," Ramsey Flynn; author's interview with Chi Van Dang, July 8, 2011, op. cit.; http://pathology.jhu.edu/department/giving.cfm "The Donald L. Price Research Fund"; http://alzheimers.about.com/od/tratmentofalzheimers/p/aricpet.htm.

97 *Hopkins Medicine*, Spring/Summer 2008, op. cit.

98 Ibid.

99 Press release, Jan. 13, 2006, "Five Kimmel Cancer Researchers Called Best in Their Field/Kimmel Cancer Center Called Research Powerhouse."

100 Ibid.

101 Author's interview with Vogelstein, August 11, 2009.

102 Ibid.

103 Ibid.

104 Ibid.

105 Author's interview with Chi Dang, July 8, 2011.

106 Ibid.

107 Ibid.

108 Author's interview with Vogelstein, August 11, 2009

109 Ibid.

110 *Hopkins Medicine*, Spring/Summer 2005, Vol. 28, No. 3, "Goldman Family Turns Tide"; *Hopkins Medicine*, Winter 2009, Vol. 32, No. 2, "A Transformative Sequence of Events," Marlene England.

111 *Hopkins Medicine*, Spring/Summer 2005, op. cit.

112 Ibid.; *Hopkins Medicine*, Winter 2009, op. cit.

113 *Hopkins Medicine*, Winter 2009, op. cit.

114 Press release, June 11, 2010, "$20 Million to Johns Hopkins for Pancreas Cancer Research"; *Johns Hopkins Gazette*, June 21, 210, "Kimmel Center receives $20 mill for pancreas cancer research, care."

115 *Hopkins Medicine*, Spring/Summer 2006, "Taming the Beast," op. cit.

116 Author's interview with Chi Dang, July 8, 2011; *Hopkins Medicine*, Winter 2009, Vol. 32, No. 2, "The Vision Nobel," Neil A. Grauer.

117 *Hopkins Medicine*, Spring/Summer 2008, Vol. 31, No. 3, "Music on the Mind," Nick Zagorski; *Dome*, Feb. 2008, Vol. 59, No. 1, "CSI: Beethoven," Neil A. Grauer.

118 *Hopkins Medicine*, Fall 2004, Vol. 28, No. 1, "Patz Meets Palmer (Gets Medal, Too)," Neil A. Grauer.

119 Press release, June 18, 2004, "World Renowned Hopkins Eye Specialist to Receive Presidential Award."

120 Press release, Feb. 15, 2005, "Johns Hopkins Scientists Receive Presidential Medals"; *Hopkins Medicine*, Spring/Summer 2005, Vol. 28, No. 3, "Lab Klutz with Class," Neil A. Grauer.

121 Ibid.

122 Press release, Feb. 15, 2008, "Benjamin Carson, M.D., Awarded the Ford's Theatre Lincoln Medal"; Press release, June 20, 2008, "Hopkins Surgeon Ben Carson Receives Medal of Freedom"; *Dome*, July 2008, Vol. 59, No. 6, In the News, "Medal for Mettle"; *Hopkins Medicine*, Fall 2008, Vol. 32, No. 1, "In the Limelight," Neil A. Grauer.

123 *Hopkins Medicine*, Fall 2008, op. cit.

124 Ibid.

125 Transcript of video of Carson tribute dinner, July 10, 2008.

126 Ibid.

127 MacArthur Foundation Web site, descriptions of grant, http://www.macfound.org/site/c.lkLXJ8MQKrH/b.855245/k.588/About_the_Foundation.htm; http://www.macfound.org/site/apps/nlnet/content3.aspx?c=lkLXJ8MQKrH&b=4513915&ct=5984655.

128 *Dome*, Oct. 2007, No. 58, No. 8, "Cooper Named MacArthur Fellow," Mary Ellen Miller; *Dome*, Nov. 2007, Vol. 58, No. 9, "Q&A: Questions for Lisa Cooper," Mary Ellen Miller; MacArthur Foundation Web site profile of Cooper, http://www.macfound.org/site/c.lkLXJ8MQKrH/b.2913825/apps/nl/content2.asp?content_id=%7BF6E54349-A2BF-4FCD-8041-0DA09A56B5B5%7D&notoc=1.

129 *Dome*, Nov. 2007, op. cit.

130 Ibid.

131 Ibid.

132 Ibid.; MacArthur Foundation Web site profile of Cooper, op. cit.; press release, Sept. 25, 2007, "MacArthur 'Genius' Award Honors Expert on Minority Health at Johns Hopkins."

133 *Time* magazine, April 30, 2009, "Peter Pronovost," Kathleen Kingsbury; press release, Sept. 23, 2008, "Two Johns Hopkins Professors Receive 'Genius' Grants"; MacArthur Foundation award citation, http://www.macfound.org/site/c.lkLXJ8MQKrH/b.4537281/k.9A40/Peter_Pronovost.htm.

134 *The Baltimore Sun*, Oct. 5, 2010, "State joins anti-infection effort," Andrea K. Walker.

135 *The New York Times*, Oct. 13, 2009, "A Conversation with Carol W. Greider On Winning the Nobel Prize in Science," Claudia Dreifus.

136 Ibid.; CNN report, "A day in the 'normal' life of a Nobel Prize winner," Val Willingham, http://articles.cnn.com/2009-12-07/health/nobel.prize.mom.telomeres_1_carol-greider-telomerase-chromosomes?_s=PM:HEALTH.

137 Ibid.; *Dome*, Oct. 2006, "A Triumph for Basic Science: A Lasker Award—a.k.a. the "American Nobel"—goes to one of our own," Judy Minkove; *The Baltimore Sun*, Oct. 5, 2009, "Hopkins professor among Nobel winners in medicine," Karl Ritter and Matt Moore, Associated Press; *Hopkins Medical News*, Winter 2001, Vol. 24, No. 2, "The Flier," Rebecca Skloot.

138 *Dome*, Oct. 2006, op. cit.; *Dome*, Nov. 2009, "Prized Curiosity: Carol Greider's inquisitive nature and tenacity helped her win a Nobel, despite a learning disability," Judy Minkove; *The Washington Post*, Oct. 20, 2009, "Success is in her DNA," Liza Mundy; *The Baltimore Sun*, Oct. 5, 2009 Associated Press story, op. cit.

139 *Dome*, Nov. 2009, op. cit.

140 *The Washington Post*, Oct. 20, 2009, op. cit; *The New York Times*, Oct. 13, 2009, op. cit.

141 *The Washington Post*, Oct. 20, 2009, op. cit.

142 *Hopkins Medical News*, Winter 2001, op. cit.; e-mail from Ed Miller and Ron Daniels to "students, faculty and staff," announcing Greider's Nobel, Oct. 5, 2009; press release, Dec. 22, 2005, "A Little Telomerase Isn't Enough."

143 Miller and Daniels e-mail; press release, Oct. 5, 1109, "'Telomere Expert Carol Greider Sharels 2009 Nobel Prize in Physiology or Medicine'"; *The New York Times*, Oct. 13, 2009, op. cit.

144 *Hopkins Medical News*, Winter 2001, op. cit.

145 *The Baltimore Sun*, Oct. 5, 2009 Associated Press story, op. cit.

146 Ibid.

147 Carol Greider press conference, Oct. 5, 2009.

148 *Dome*, Feb. 2009, Vol. 60, No. 1, "Actor Captures Ben Carson's Personna," Kim Hoppe; *Hopkins Medicine*, Fall 2008, Vol. 32, No. 1, "In the Limelight," Neil A. Grauer.

149 *Dome*, Feb. 2007, Vol. 58, No. 1, "24/7 Reprised"; *Dome*, April 2007, Vol. 58, No. 3, "Behind the Scences, 24/7," Anne Bennett Swingle; *Dome*, May 2007, Vol. 58, No. 4, "Behind the Scenes, 24/7," Anne Bennett Swingle; *Dome*, Sept. 2007, Vol. 58, No. 7, "Appreciation from Behind the Camera" (Letter to the Editor from director of photography Richard Chisolm; *Hopkins Medicine*, Fall, 2008, Vol. 32, No. 1, "Made for Prime Time," Stephanie Shapiro; press release, April 2, 2009, "ABC Documentary 'Hopkins' Wins Prestigious Peabody Award."

150 *Dome*, Feb. 2007, op. cit.; "Hopkins" series episodes descriptions, http://hopkins.portfolio.crushlovely.com/episodes/1.

151 *Dome*, Feb. 2007, op. cit.; "Hopkins" series episodes descriptions; *Dome*, April 2007, op. cit.

152 Ibid.; *Dome* May 2007, op. cit.

153 Ibid.

154 "Hopkins" series episodes Web site, op. cit.

155 *Dome*, June 2008, op. cit.

156 *Dome*, Dec. 2004, Vol. 55, No. 10, "Campus Transformation," Anne Bennett Swingle.

157 Ibid.

158 Ibid.

159 *Change*, July 21, 2004, Vol. 8, No. 12, "On the Path of Construction"; press release, Oct. 14, 2005, "Hopkins Medical Campus Takes First Step in Massive Campus Redevelopment."

160 *Change*, July 21, 2004, op. cit.

161 *Hopkins Medicine*, Winter 2006, Vol. 29, No. 2, "Buildings Count in the Cancer War," Marlene England; *Dome*, May 2006, Vol. 57, No. 4, "Cancer Collaboration2," Judith Minkove.

162 *Dome*, June 2006, Vol. 57, No. 5, "Pediatric Medical Office Building," Anne Bennett Swingle; *Dome*, April 2007, Vol. 58, No. 3, "Rubenstein Building."

163 Ibid., sidebar, "What's Next."

164 *Hopkins Medicine*, Winter 2006, Vol. 29, No. 2, "Towering Inflation."

165 *Dome*, May 2006, Vol. 57, No. 4; "Breaking Ground"; *Dome*, Oct. 2006, Vol. 57, No. 8, "Future Docs' New Digs"; *Hopkins Medicine*, Spring/Summer 2006, Vol. 29, No. 3, "Armstrong Acts Now," Marlene England.

166 *Hopkins Medicine*, Fall 2007, Vol. 31, No. 1, "Dreaming Big" and "A Symbol of the Human Spirit," Marlene England; *Dome*, April 2007, Vol. 58, No. 3, "Eye Edifice"; author's interview with Peter McDonnell, July 13, 2009; "Fast Facts" on The Hackerman-Patz Patient and Family Pavilion, http://www.hopkinsmedicine.org/kimmel_cancer_center/our_center/facilities/hackerman_patz/fast_facts.html.

167 *Dome*, May 2006, Vol. 57, No. 4, "Biotech Park's First Building Gets Off the Ground," Anne Bennett Swingle; *Johns Hopkins Gazette*, April 17, 2006, Vol. 35, No. 3, "A New Era Begins in East Baltimore," Greg Rienzi; Forest City Science + Technology Group Web site, http://www.forestcityscience.net/hopkins/new_east_baltimore.shtml.

168 *Dome*, May 2006, op. cit.; Johns Hopkins Gazette, April 14, 2008, Vol. 37, No. 30, "New Era Begins in East. Balto.," Greg Rienzi; *Hopkins Medicine*, Spring/Summer 2006, Vol. 29, No. 3, "The Rangos Building Rises," Linell Smith.

169 *Johns Hopkins Gazette*, April 14, 2008, op. cit.

170 *Dome*, Feb. 2009, Vol. 60, No. 1, "A Place to Hang Their Coats—and Troubles," Janet Anderson.

171 Ibid.; *Dome*, Sept. 2005, Vol. 56, No. 6, "Extreme Makeover," op. cit.; *Dome,* Nov. 2007, Vol. 58, No. 9, Briefcase, "Patient Refuge."

172 Press release, April 30, 2007, "Major Gift to Johns Hopkins Medicine Honors U.A.E. Sheikh Zayed Bin Sultan Al Nahyan"; *Dome*, June 2007, Vol. 58, No. 5, "In Fund Raising, an Oasis," Anne Bennett Swingle; *Hopkins Medicine*, Fall 2007, Vol. 31, No. 1, "A Royal Gift of Tower-ing Impact," Lindsay Roylance.

173 Ibid.

174 Press release, April 30, 2007, op. cit.

175 Ibid.

176 *Dome*, July 2008, Vol. 59, No. 6, "To Mom, With Love."

177 *New York Daily News*, Dec. 31, 2009, "Mama Bloomberg Will Stay in Mass. Home for son's 3rd inauguration as New York mayor."

178 "Mayor Michael R. Bloomberg, A Biography," New York City Web site, http://www.carnaval.com/cityguides/newyork/Mayor/mayorB.htm; Bloomberg, Michael, Bloomberg on Bloomberg, John Wiley & Sons, 1997, pgs. 12–13; e-mail from Dennis O'Shea, executive director of communications and public affairs, The Johns Hopkins University, April 26, 2011.

179 Hopkins Medicine, Spring/Summer 2008, Vol. 31, No. 3, "Of Staggering Scale," Mary Ellen Miller.

180 *Hopkins Medicine*, Spring/Summer, 2007, Vol. 30, No. 2, "No Time to Rest"; *Hopkins Medicine*, Winter 2006, Vol. 29, No. 2, "What's Next"; Hopkins Children's Center Web site, http://www.hopkinschildrens.org/new-hospital.aspx.

181 *Dome*, June 2006, Vol. 57, No. 5, "Welcome to Your New Workplace"; *Hopkins Medicine*, Spring/Summer 2008, Vol. 31, No. 3, op. cit.

182 Ibid. *Dome*, April 2008, Vol. 59, No. 3, "Towers Rising," Mary Ellen Miller.

183 *Hopkins Medicine*, Spring/Summer 2008, op. cit.

184 *Dome*, Sept. 2005, Vol. 56, No. 6, "The Master Planners"; "Extreme Makeover," Anne Bennett Swingle.

185 Ibid.

186 Ibid.

187 *Dome*, June 2006, Vol. 57, No. 5, op. cit.; *Dome*, April 2008, Vol. 59, No. 3, op. cit.

188 *Hopkins Medicine*, Fall 2007, op. cit.; press release, Jan. 1, 2010, "Robert H. Smith Tribute"; press release, Oct. 20, 2009, "New Anne and Mike Armstrong Medical Education Building Dedicated"; *Sightline*, Fall 2009, "Dedication to a Cure"; press release, June 8, 2009, "Johns Hopkins Holds Ribbon-Cutting Ceremony for New Wilmer Eye Institute Building."

189 *Sightline*, Fall 2009, op. cit.; *Hopkins Medicine*, Spring/Summer 2007, Vol. 30, No. 2, op. cit.; American Society of Retinal Specialists Web site, http://www.amdawareness.org/asrs/?s_cid=0001&s_src=googleppc&gclid=CNr21OW3-KICFRNO5QodUlvXlg.

190 Johns Hopkins Named Professorships Web page re: Bendann/Iliff Professoship in Ophthalmology, http://webapps.jhu.edu/namedprofessorships/professorshipdetail.cfm?professorshipID=112; press release, May 16, 2006, "Rebecca Henry Named Scott Bendann Faculty Chair in Classical Music."

191 *Hopkins Medicine*, Fall 2009, Vol. 33, No. 1, "Marathon Munificence," Neil A. Grauer.

192 Ibid.

193 *The Baltimore Sun*, Oct. 24, 2009, "Building a curriculum," Edward Gunts.

194 Ibid.

195 *Hopkins Medicine*, Spring/Summer 2009, Vol. 32, No. 3, "Designed for the Needs of the Next Generation" Marlene England; *Dome*, Sept. 2009, Vol. 60, No. 7, "A 21st Century Approach to Medical Education"; press release, March 27, 2008, "Actor-Robots 'Staff' Part of New $5 Million Simulation Center"; *Change*, Sept. 21, 2009, "Anatomy of the New Curriculum," Linell Smith.

196 Ibid.; *Dome*, June 2006, Vol. 57, No. 5, What's News, "College Names."

197 Ibid.

198 Press release, Oct. 20, 2009, "New Anne and Mike Armstrong Medical Education Building Dedicated."

199 *Dome,* Sept. 2009, Vol. 60, No. 7, "A 21st Century Approach to Medical Education"; *Change*, Sept. 21, 2009, "Curricular Thinking," Linell Smith; *The Washington Post*, Nov. 10, 2009, "Med schools offer doses of new reality," Sarah Lovenheim.

200 Ibid., *Change*, Sept. 21, 2009, "Anatomy of the New Curriculum"; "Curricular Thinking."

201 Ibid.

202 Ibid., *Dome*, Sept. 2009, Vol. 60, No. 7, "A 21st Century Approach to Medical Education," op. cit.

203 *Change*, Sept. 21, 2009, op. cit.

204 Ibid., "Anatomy of the New Curriculum," op. cit.

205 Ibid.

206 Ibid.

207 Ibid.

208 *Hopkins Medicine*, Fall, 2009, Vol. 33, No. 1, "Taking Root," Mat Edelson and Linell Smith.

209 Ibid.

210 Author's interview with David Nichols, July 7, 2009.

211 *Hopkins Medicine*, Fall 2009, Vol. 33, No. 1, "What Patients Have to Teach," Linell Smith.

212 *Dome*, Dec. 2006, Vol. 57, No. 10, "An Iconic Painting Turns 100," Anne Bennett Swingle.

213 Ibid.

214 Author's interview with Jay Corey, Sept. 7, 2010.

215 Ibid.; transcript of Ed Miller "roast" video, April 2007.

216 Transcript of Ed Miller "roast" video, op. cit.

217 Ibid.

218 *Three Little Words and the difference they've made; Johns Hopkins Medicine: A ten-year report*, 1997.

219 Press release, July 13, 2007, "The Johns Hopkins Hospital Tops *US News & World Report* Honor Roll 17th Year in a Row"; *Dome*, Dec. 2007, Vol. 58, No. 10, "Top Docs" item in Who/What page.

220 *Dome*, Oct. 2004, Vol. 55, No. 8, "A Bayview, a Quartet of Construction Projects," Neil A. Grauer; NIH Office of Research Facilities Web site, http://orf.od.nih.gov/AboutORF/Buildings/BRC.htm; NIH News Briefs Web site, http://www.nia.nih.gov/NewsAndEvents/SOAR/v2n1/News.

221 Photo cutlines from Feb. 2009 to July 2009 issues of *The Banner*, provided by Kristin Meyers, physician liaison, Hopkins Bayview Medical Center, July 20, 2010.

222 *Dome*, April 2009, Vol. 60, No. 3, "Bayviewland."

223 *Dome*, July 2009, Vol. 60, No. 6, "Making Room in Howard County," Sharon Sopp; *Dome*, Nov. 2007, Vol. 58, No. 9, "Preserving Strong Ties"; *Dome*, Dec. 2005, Vol. 56, No.10, "Smart Space in Howard County," Anne Bennett Swingle; *Dome* Sept. 2005, Vol. 56, No. 6, What's News, "HCGH to Expand."

224 HCGH Web site, http://www.hcgh.org/content/RehabilitationServices.htm; http://www.hcgh.org/content/bolduclundy.htm; *Dome*, July 2009, op. cit.

225 *Dome*, Dec. 2005, op. cit.; HCGH press release, July 10, 2009, "HCGH Opens New Patient Pavilion."

226 *Dome*, June 2004, Vol. 55, No. 5, "Growing Green Spring," Anne Bennett Swingle; Hopkins Medicine in Brief, 2009, "Johns Hopkins Health Care and Surgery Center."

227 *Dome*, Sept. 2004, Vol. 55, No. 7, "White Marsh II Grand Opening"; *Dome*, April 2007, Vol. 58, No. 3, "Surgery Center to Open at White Marsh."

228 *Dome*, April 2007, op. cit.; Hopkins Medicine in Brief, 2009, op. cit.

229 *Johns Hopkins Gazette*, March 26, 2007, "JHM Inks Alliance with Anne Arundel Health System"; *Dome*, April 2007, Vol. 58, No. 3, What's News, "Arundel Alliance."

230 Ibid.

231 Press release, July 20, 2007, "Great Baltimore Medical Center (GBMC) Collaboration Expands Care Options for Patients in the Region"; *Dome*, Sept. 200, Vol. 58, No. 7, What's News, "GBMC Partners"; *The Baltimore Sun*, July 21, 2007, "Hopkins and GBMC form partnership," Trisha Bishop.

232 *The Baltimore Sun*, July 21, 2007, op. cit.; GBMC publication, *Greater Living*, Spring 2010, "Johns Hopkins Cardiology at GBMC."

233 Press release, July 20, 2007, op. cit.

234 *Hopkins Medical News*, Fall 1996, Vol. 20, No. 1, "Married, but Living Apart," Kate Ledger; press release, April 24, 2009, "Suburban Hospital Healthcare System to Join Johns Hopkins Medicine"; press release, July 2, 2009, "Suburban Hospital Healthcare System Joins Johns Hopkins Medicine."

235 Press release, July 2, 2009, op. cit.

236 *Dome*, July 2010, Vol. 51, No. 5, What's News, "Welcome Sibley Memorial Hospital"; E-mail from Ed Miller and Ron Peterson to Hopkins Medicine community, May 26, 2010, "Sibley Memorial Hospital to Join Johns Hopkins Medicine"; press release, May 27, 2010, "Sibley Memorial Hospital to Join Johns Hopkins Medicine."

237 Ibid.

238 E-mail, July 20, 2010, from Ed Miller and Ron Peterson, "A New Addition to the JHM Family: All Children's Hospital in Florida"; *The Baltimore Sun*, July 21, 2010, "Hopkins to buy Florida children's hospital," Andrea K. Walker.

239 *The Baltimore Sun*, July 21, 2010, op. cit.

240 E-mail, July 20, 2010, from Ed Miller and Ron Peterson, "A New Addition to the JHM Family: All Children's Hospital in Florida"; *The Baltimore Sun*, July 21, 2010, "Hopkins to buy Florida children's hospital," Andrea K. Walker.

241 *St. Petersburg Times*, July 21, 2010, "All Children's Hospital to join forces with Johns Hopkins health system," Richard Martin and Kris Hundley; http://www.hopkinsbayview.org/news/2007/070328jonellen.html.

242 E-mail, July 20, 2010, op. cit, "All Children's Hospital Integration: Frequently Asked Questions" attachment; *Tampa Tribune*, July 21, 2010, op. cit.

243 Johns Hopkins Medicine International 2009 Annual Report: Making an Impact, pgs. 22–31.

244 Author's interview with Nichols, July 7, 2009.

245 Johns Hopkins Medicine International 2009 Annual Report, op. cit., pgs. 25–29

246 Ibid., pg. 25

247 *Three Little Words and the difference they've made; Johns Hopkins Medicine: A ten-year report*, 1997, pg. 47

248 Johns Hopkins International 2009 Annual Report, op. cit, pgs. 30–31.

249 Ibid., pgs. 20; 22; 34–35.

250 *Dome*, Sept. 2006, Vol. 57, No. 7, "China Revisted," Anne Bennett Swingle; Dept. of Geriatric Medicine and Gerontology Web site, http://www.hopkinsmedicine.org/geriatrics/faculty/Leng.html.

251 Ibid.

252 *Hopkins Medicine*, Spring/Summer 2010, Vol. , No., "Big Step in the Middle East," Edward D. Miller; "Partnership of Global Proportions," Marlene England; press release, Jan. 11, 2010, "Leading Ophthalmological Centers in the United States and Saudi Arabia Announce Affiliation."

253 *Hopkins Medicine*, Spring/Summer 2010, "Partnership of Global Proportions," op. cit.

254 Ibid., "A Big Step in the Middle East"; Wilmer Web site, http://www.hopkinsmedicine.org/wilmer/employees/cvs/Behrens_e.html.

255 Ibid., "A Big Step in the Middle East."

256 Ibid.

257 *Johns Hopkins Gazette*, Nov. 8, 2010, Vol. 40, No. 10, "Malaysia-bound," Natalia Abel; Malaysia Tourist Web site, "Fast Facts," http://www.tourism.gov.my/en/about/facts.asp.

258 *Johns Hopkins Gazette*, Nov. 8, 2010, op. cit.

259 Ibid.

260 Ibid.

261 *Dome*, March 2005, Vol. 56, No. 2, "Critical Condition," Anne Bennett Swingle.

262 Ibid.

263 *Dome*, July 2009, Vol. 60, No. 6, "Stepping Aside," Jamie Manfuso.

264 *Dome*, Nov. 2005, Vol. 56, No. 9, "Rx for Patient Safety," Deborah Rudacille.

265 *Hopkins Medicine*, Winter 2005, Vol. 28, No. 2, "Infection Patrol," Patrick Gilbert

266 Press release, Jan. 14, 2008, "Rubber Gloves: 'Born'—and Now Banished—at Johns Hopkins."

267 *Dome*, July 2009, Vol. 60, No. 6, op. cit.

268 Press release, May 26, 2011, "Johns Hopkins Establishes Armstrong Institute for Patient Safety"; *Dome*, June 2011, Vol. 62, No. 6, "Pronovost to head new patient safety institute."

269 Press release, May 26, 2011, op. cit.

270 Ibid.

271 Press release, March 14, 2011, "Francis B. Burch Jr. Named New Chairman of Johns Hopkins Medicine."

272 *Hopkins Medicine*, Spring/Summer 2005, Vol. 28, No. 3, "Osler Lives On," Mary Ann Ayd.

273 Ibid.

274 *Dome*, Feb. 2007, Vol. 58, No. 1, "Celebrating a Century: Social Work Turns 100," Anne Bennett Swingle; *Dome*, March 2007, Vol. 58, No. 2, "True Champions," Anne Bennett Swingle; *Dome*, Marc. 2009, Vol. 60, No. 2, "No Stone Unturned," Judith Minkove

275 *Hopkins Medicine*, Spring/Summer 2008, Vol. 31, No. 3, "And the Survey Says…."; *Dome*, April 2009, Vol. 60, No. 3, "Making Amends," Karen Blum.

276 *Dome*, April 2009, "Making Amends," op. cit.; author's interview with Miller, Sept. 29, 2009.

277 Press release, March 8, 2010, "Johns Hopkins Hospital Earns 2010 'Hospital of Choice' Award"; e-mail, Oct. 18, 2011 from Adam Benash of National Research Corp., Lincoln, Nebraska.

278 *Dome*, Nov. 2005, Vol. 56, No. 9, "Doing Our Part," Anne Bennett Swingle; PatriotLife (publication of USFHP), Fall 2008, "Johns Hopkins US Family Health Plan Hits 98th Percentile in Member Satisfaction Again"; Fall 2009, "Independent Survey Shows Hopkins USFHP Member Satisfaction in 98th Percentile"; USFHP 2010 patient number, from Jason Teves, Aug. 5, 2010.

279 *Dome*, July 2007, Vol. 58, No. 6, "Top Priority," Judith Minkove; *The Wall Street Journal*, Dec. 4, 2009, "Health Reform Could Harm Medicaid Patients," Edward Miller.

280 *Dome*, July 2007, Vol. 58, No. 6, op. cit.

281 *The Wall Street Journal*, Dec. 4, 2009, op. cit.

282 Ibid.; *Dome*, July 2007, op. cit.

283 *The Wall Street Journal*, Dec. 4, 2009, op. cit.

284 Ibid; *The Baltimore Sun*, March 17, 2010, "Hopkins supports health care reform," Edward Miller.

285 *Dome*, May 2005, Vol. 56, No. 4, "At the Helm of Home Care," Neil A. Grauer

286 Ibid.; press release, April 4, 2005, "Daniel B. Smith Named President of Johns Hopkins Home Care Group."

287 Ibid.

288 JHHCG Fiscal Year 2010 patient visit figure provided by Cathy Rogers in Finance; e-mail from Daniel Smith, Sept. 17, 2010.

289 *Dome*, July 2008, Vol. 59, No. 6, "Questions for Barbara Cook," Linell Smith; *Hopkins Medicine*, Spring/Summer 2008, Vol. 32, No. 3, "Passionate About Patient Care," Neil A. Grauer.

290 Dome, July 2008, op. cit.; e-mail from Linda Gilligan, COO of JHCP, Oct. 7, 2010.

291 *Hopkins Medicine*, Spring/Summer 2009, op. cit.

292 *Dome*, June 2006, Vol. 57, No. 5, "Mixing It Up," Jamie Manfuso and Mary Ellen Miller.

293 *Hopkins Medicine*, Spring/Summer 2008, Vol. 31, No. 3, "Elusive Equity," Neil A. Grauer.

294 Author's interview with Ed Miller, Sept. 29, 2009.

295 Ibid.

296 *Dome*, Dec. 2005, Vol. 56, No. 10, "This Time, It Was All About Women," Deborah Rudacille; *Hopkins Medicine*, Fall 2005, Vol. 29, No. 1, "Catch Up Time," Mary Ann Ayd.

297 *Dome*, April 2010, Vol. 61, No. 3, "155 and Counting," Linell Smith.

298 *Johns Hopkins Gazette*, March 22, 2010, Vol. 39, No. 6, "SoM to host 'A Tribute to 150+ Women Professors' celebration," Stephanie Desmon.

299 "A Women's Journey" Web site, http://www. hopkinsmedicine.org/awomansjourney; e-mail, Nov. 20, 2009, update on attendance at 2009 "A Women's Journey," Rebecca DeMattos; *Dome*, Oct. 2009, Vol. 60, No. 8, "Voyage to Better Health," Judy F. Minkove.

300 *Dome*, June 2006, Vol. 57, No. 5, "Mixing It Up," op. cit.

301 *Hopkins Medicine*, Winter 2008, Vol. 31, No. 2, "The Right Direction," Ramsey Flynn; *Dome*, April 2007, Vol. 58, No. 3, "Diverse Scholars," Anne Bennett Swingle.

302 *Dome*, May 2009, Vol. 60, No. 4, "Diversity Scholars," Mary Ellen Miller.

303 Ibid.

304 Ibid.

305 *Dome*, Sept. 2009, Vol. 60, No. 7, "REACH-ing New Heights," Judy F. Minkove.

306 Ibid.

307 E-mail, Dec. 4, 2008, "New Dean for Diversity Appointed," Ed Miller and Janice Clemens; *Dome*, Feb. 2009, Vol. 60, No. 1, "Associate Dean" (Who/What page); *Dome*, March 2009, Vol. 60, No. 2, "Disparities and Diversity," Jamie Manfuso.

308 http://www.hopkinsmedicine.org/diversity/.

309 *Dome*, Nov. 2008, Vol. 59, No. 7, "Opportunity First" (What's News); *Dome*, March 2008, Vol. 59, No. 7, "Dedicated to Diversity" (Who/What page); *Dome*, Feb. 2009, Vol. 60, No. 1, "Minorities Matter."

310 *Johns Hopkins Magazine*, April 1999, "Aiming High," Melissa Hendricks; "Short Support" Web site, ABC News interview, "One of a Kind Surgeon: Chat with Michael Ain of Johns Hopkins Hospital," Sept. 14, 2000, http://www.shortsupport.org/News/0238. html.

311 Ibid.; Emedicine Web site for skeletal dysplasia definition, http://emedicine. medscape.com/article/943343-overview.

312 *Johns Hopkins Magazine*, April 1999, op. cit.

313 Ibid.

314 *Hopkins Medicine*, Winter 2007, Vol. 30, No. 2, "The Alfredo Story," David Dudley; *The New York Times*, May 13, 2008, "A Surgeon's Path from Migrant Fields to Operating Room," Claudia Dreifus, http://www. nytimes.com/2008/05/13/science/13conv. html?_r=2&pagewanted=print.

315 *Hopkins Medicine*, Winter 2007, op. cit.

316 Ibid.

317 Ibid.; *The New York Times*, May 13, 2008, op. cit.; *Dome*, Sept. 2006, Vol. 57, No. 7, "The Remarkable Journey of Doctor Q.", Deborah Rudacille; Quiñones CV, http:// www.hopkinsmedicine.org/neurology_ neurosurgery/cv/alfredo_quinones.pdf.

318 Ibid.

319 Ibid.

320 Hopkins Medicine Fall 2009, Vol. 33, No. 1, *The Long Way Her*e, Ramsey Flynn.

321 *Change*, Nov. 10, 1998, Vol. 2, No. 19, "Couple Recognition."

322 *Hopkins Medicine*, Spring/Summer 2004, Vol. 27, No. 3, "With This Ring," Mary Ann Ayd.

323 *Dome*, March 2005, Vol. 56, No. 2, "Tom Koenig: Up to the Match," Sarah Richards.

324 Ibid.

325 Ibid.

326 Ibid.

327 *Dome*, May 2006, Vol. 57, No. 4, "35 Years and Counting," Deborah Rudacille.

328 Ibid.

329 *Dome*, Feb. 2005, Vol. 56, No. 1, "Jesse!" Neil A. Grauer.

330 Ibid.

331 *Hopkins Medicine*, Winter 2006, Vol. 29, No. 2, "Fast Times," Janet Farrar Worthington.

332 *Hopkins Medicine*, Winter 2006, Vol. 29, No. 2, "Annals of Hopkins: Fast Times," Janet Farrar Worthington.

333 Ibid.; History of the Pithotomy Club, Robert H. Harrell, M.D., http://www. pithotomy.com/history.html.

334 *Dome*, Dec. 2009, Vol. 60, No. 10, "Without Charge," Mary Ellen Miller; *Dome*, Oct. 2004, Vol. 55, No. 8, "Fish Stories," Anne Bennett Swingle; *Dome*, Oct. 2005, Vol. 56, No. 8, "A School Partnership Bears Fruit, " Anne Bennett Swingle; *Dome*, March 209, Vol. 60, No. 2, "Down But Not Out," Judy F. Minkove.

335 *Dome*, Dec. 2009, op. cit.

336 *Dome*, April 2004, Vol. 55, No. 3, "Se Habla Español?" Lindsay Roylance; *Dome*, Nov. 2005, Vol. 56, No. 9, "Hola Hopkins!", Lydia Levis Bloch; *Dome*, May 2008, Vol. 59, No. 4, "Translating Health," Linell Smith; "The Interpreters," Linell Smith; "Se Hable Ingles?"

337 Ibid.; *Change*, June 5, 2009, "CPA News: A Question of Access," Linell Smith; *Inside Hopkins*, Sept. 1, 2010, "Barbara Cook to Receive Baltimore City Health Department Award."

338 *Dome*, Oct. 2005, op. cit.; Franklin Square Hospital Web site, "Second-Year Residents" (for Samyra Sealy), http://www. franklinsquare.org/body.cfm?id=947; *Dome*, March 2009.

339 *Hopkins Medicine*, Spring/Summer 2010, Vol. 33, No. 2, "Being Father Teresa," Ramsey Flynn.

340 Ibid.; official Web site of Rick Hodes, http:// rickhodes.org/.

341 Ibid.

342 Ibid.; HarperCollins Web site on Marilyn Berger, http://www.harpercollins.com/ author/microsite/About.aspx?authorid=35327 .

343 Author's interview with Miller, Sept. 29, 2009.

344 *Dome*, Nov. 2008, Vol. 59, No. 9, "Hopkinomics," Mary Ellen Miller.

345 *Dome*, June 2009, Vol. 60, No. 5, "Finding Financial Opportunities," Mat Edelson.

346 Bloomberg School of Public Health Web site on Klag, http://webapps.jhu.edu/jhuniverse/ information_about_hopkins/about_jhu/ principal_administrative_officers_and_ deans/michael_j_klag/; *Dome*, Sept. 2005, Vol. 56, No. 6, Who/What page, "Two Vice Deans"; *Hopkins Medicine*, Fall 2005, Vol. 29, No. 1, "A Ford in Researchers' Future," Neil A. Grauer.

347 *Hopkins Medicine*, Fall 2005, op. cit.

348 Ibid.

349 *Hopkins Medicine*, Fall 2009, Vol. 33, No. 1, "What She's Thinking," Neil A. Grauer.

350 Ibid.

351 Ibid.

352 Ibid.

353 *Dome*, Sept. 2005, op. cit.; "Dear Colleagues" letter on Pasternak departure, July 2005; Inova Health System press release, http://newsroom.inova.org/article_ display.cfm?article_id=5051; *Hopkins Medicine*, Fall 2005, op. cit., "Hopkins With Grass," Neil A. Grauer.

354 *Hopkins Medicine*, Fall 2005, "Hopkins With Grass," op. cit.; press release, July 27, 2005, "Hellmann Appointed New Vice Dean for Hopkins Bayview."

355 *Dome*, Oct. 2008, Vol. 59, No. 8, "Greg Schaffer Retires," Neil A. Grauer.

356 Ibid.; Tawam Hospital press release, Feb. 1, 2010, http://daraldaleel.com/13804/pressrelease/new-chief-executive-officer-appointed-at-tawam-hospital.html.

357 *Dome*, Nov. 2006, Vol. 57, No. 9, "Rich Bennett: Physician/EVP," Neil A. Grauer.

358 *Dome*, May 2009, Vol. 60, No. 4, "Bennett Named Bayview President," Neil A. Grauer.

359 *Dome*, June 2005, Vol. 56, No. 5, "She Made the Dome Shine Bright," Joann Rodgers and Edith Nichols.

360 Ibid.

361 Ibid.

362 Press release, Dec. 22, 2005, "Johns Hopkins Medicine Names D. J. Haldeman as Marketing and Communications Vice President"; *Hopkins Medicine*, Spring/Summer 2006, Vol. 29, No. 3, "The Maven of Scientific Marketing," Edith Nichols.

363 Press release, Dec. 22, 2005, op. cit.

364 Email from Dalal Haldeman, Oct. 24, 2011.

365 *Dome*, April 2008, Vol. 59, No. 3, "Brody to Retire," Neil A. Grauer.

366 Ibid.

367 Ibid.

368 *Dome*, Dec. 2008, Vol. 59, No. 10, "Daniels Named University President"; *Hopkins Medicine*, Winter 2009, Vol 32, No. 2, "Changing of the Guard," Ramsey Flynn.

369 Ibid.

370 *Hopkins Medicine*, Winter 2009, op. cit.

371 Press release, Aug. 21, 2009, "Lloyd Minor Named Provost"; *Dome*, Sept. 2009, Vol. 60, No. 7, "Major Appointment for Minor."

372 Ibid.

373 Ibid.; *Hopkins Medicine*, Vol. 33, No. 1, "Minor's Major Appointment," Neil A. Grauer.

374 *Hopkins Medicine*, Winter 2009, op. cit.; university-wide e-mail, Sept. 10, 2009, "Inauguration of University President Ronald Daniels (describing plans for "a fun run around the Homewood campus," Jerome D. Schnydman).

375 University-wide e-mail, Oct. 12, 2009, "A Message from JHU President Daniels."

376 University-wide e-mail, Oct. 22, 2009, "A Message from JHU President Daniels."

377 Ibid.

378 *Inside Hopkins*, July 16, 2010, "America's Best: Hopkins earns #1 spot for 20th year"; *Inside Hopkins*, July 21, 2011, "America's Best: Hopkins earns #1 spot for 21st year."

379 *Dome*, August 1992, Vol. 42, No. 4, "The Stress of Being Top Dog," Anne Childress.

380 *Change*, June 25, 1997, Vol. 1, No. 11, CPA News page, "Don't Get Carried Away by the Rankings," Charles Cummings, M.D.

381 Ibid.

382 *Inside Hopkins*, July 21, 2011, , op. cit.

383 *Inside Hopkins*, July 16, 2010, op. cit.

384 *Dome*, Sept. 1991, Vol. 41, No. 5, "US News & World Report."

385 Press release, April 2, 2004, "*US News & World Report* Ranks Hopkins in Top 3 Medical Schools"; *Hopkins Medicine*, Spring/Summer 2004, Vol. 27, No. 3, "Three's Company"; *Hopkins Medicine*, Spring/Summer 2005, Vol. 28, No. 3, "Back in the Saddle Again"; *Change*, June 25, 1997, op. cit.

386 *Hopkins Medicine*, Fall 2008, Vol. 32, No. 1, "Loss of a Legend," Neil A. Grauer.

387 Ibid.

388 Ibid.

389 Ibid.

390 Ibid.

391 *Hopkins Medicine*, Spring/Summer 2006, Vol. 29, No. 3, "Lock Conley Looks Back and Blushes," Neil A. Grauer.

392 Ibid.

393 *Hopkins Medicine*, Spring/Summer 2010, Vol. 33, No. 2, "Groundbreaking Geneticist," Neil A. Grauer.

394 Press release, Feb. 19, 2010, "Eminent Pediatrician and Geneticist Barton Childs Dies at Age 93."

395 Press release, March 25, 2010, "Beloved Hopkins Pediatrician and Educator Henry Seidel, M. D., 87, Dies."

396 Ibid.

397 Ibid.

398 *Change*, Nov. 13, 2009, "Hopkins' Stake in the Debate," Barry Rascovar.

399 Ibid.

400 Author's interview with Beth Felder, director of federal affairs for Hopkins, Sept. 9, 2010.

401 Ibid.

402 Ibid.

403 *Change*, Nov. 13, 2009, op. cit.; transcript of Miller speech to the National Press Club, June 21, 2010, http://www.hopkinsmedicine.org/news/stories/national_press_club.html.

404 Miller speech to the National Press Club, op. cit.

405 *Change*, Nov. 13, 2009, op. cit.

406 *The Wall Street Journal*, Dec. 4, 2009, "Health Reform Could Harm Medicaid Patients," Edward Miller.

407 *Dome*, Sept. 2011, Vol. 62, No. 7, "The Future of Johns Hopkins Medicine," Dean/CEO Edward D. Miller, M.D.; e-mail, Jan. 5, 2011 from Edward Miller and Ronald R. Peterson, "Steve Thompson Appointed New CEO of JHI"; e-mail, Feb. 17, 2011 from Edward Miller and Ronald R. Peterson, "Positioning Johns Hopkins Medicine for the future: Announcing a New Community Division."

408 Email, Feb. 17, 2011, from Edward Miller op. cit.

409 *Dome*, Sept. 2011, op. cit.

410 Ibid.

411 *Model*, pg. 216.

412 Hopkins Medicine, Summer 2009, Vol. 32, No. 3, "When Suitors Come Calling," Ramsey Flynn; author's interview with Miller, Sept. 29, 2009.

413 *Hopkins Medicine*, Summer 2009, op. cit.

414 Author's interview with Miller, Sept. 29, 2009, op. cit.

415 Ibid.

416 *Pittsburgh Post-Gazette*, Aug. 12, 2010, "UPMC hand transplant main surgeon leaving Oct. 1," Sean D. Hammill; *Pittsburgh Tribune-Review*, Aug. 11, 2010, "Pioneering hand transplant surgeon to leave UPMC," Luis Labregas.

417 Author's interview with Peterson, Oct. 22, 2008.

418 Ibid.

419 *Three Little Words and the difference they've made; Johns Hopkins Medicine: A ten-year report*, 1997, pg. 17

420 *The Baltimore Sun*, Oct. 6, 2009, "Low-key laureate," Kelly Brewington.

421 *The Daily Record*, Oct. 5, 2009, "Greider's Nobel means attention, and money, for Hopkins," Danielle Ulman.

422 Author's interview with Miller, Sept. 29, 2009.

423 Ibid.

424 *Three Little Words and the difference they've made; Johns Hopkins Medicine: A ten-year report*, 1997, pg. 24.

425 Ibid., pg. 29

426 Ibid.

427 Ibid.

428 Press release, March 10, 2003, "Amey Appointed Associate Dean for Research Administration at Hopkins Medical School."

429 Ibid.

430 Ibid.

431 PowerPoint presentation by Chi Dang, Sept. 14, 2010 Town Hall meeting, http://webcast.jhu.edu/mediasite/Viewer/?peid=2cbb43263dcc440ea075d8804361b43e.

432 Power Point presentation by Chi Dang, Sept. 14, 2010, op. cit.

433 Ibid.

434 *Johns Hopkins Medical News*, Spring/Summer 1997, Vol. 20, No. 3, "The Entrepreneurial Road to Save Research," Edward Miller.

435 Ibid.

436 *Hopkins Medicine*, Winter 2006,Vol. 29, No. 2, "Marketplace Catch-Up," Edward Miller.

437 Press release, Oct. 9, 2006, "Blakeslee Named Acting Director of Technology Transfer," http://www.jhu.edu/news/univ06/oct06/blakeslee.html; Technology Transfer Web site, Overview, http://www.techtransfer.jhu.edu/commercialOpportunities/index.html.

438 Author's interview with Miller, Sept. 29, 2009; *Hopkins Medicine*, Winter 2009, Vol. 23, No. 2, "Reaping the Fruits of Research," Nick Zagorski.

439 Ibid.; Chi Van Dang presentation at Hopkins Medicine Town Hall meeting, Sept. 14, 2010; definition of material transfer agreements, Web site of Harvard Office of Technology Development, http://www.techtransfer. harvard.edu/resources/agreements/materialtransfer/.

440 Ibid.; author's interview with Miller, op. cit; Gatorade origin, Snopes.com Web site: http://www.snopes.com/food/origins/gatorade.asp.

441 Press release, April 30, 2010, "Johns Hopkins Technology Transfer: There's An App For That"; definition of app, Webopedia Web site, http://www.webopedia.com/TERM/M/mobile_application.html.

442 Author's interview with Miller, Sept. 29, 2009.

443 Ibid.

444 Author's interview with Peterson, Oct. 22, 2008, op. cit.

445 Author's interview with Miller, Sept. 29, 2009; author's interview with Peterson, Oct. 22, 2008.

446 Author's interview with Miller, Sept. 29, 2009.

447 Ibid.; author's personal experience.

448 Author's interview with Miller, op. cit.

449 Author's interview with Peterson, op. cit.

450 Ibid.

451 *Model*, pg. 3.

452 Ibid.

453 Ibid.

APPENDICES

Appendix A
1999 – 2010: Leadership Changes Across the Board

1 Author's interview with Miller, Sept. 29, 2009.

2 *Hopkins Medicine*, Spring/Summer 2007, Vol. 30, No. 2, "Top Guy," David Dudley.

3 *Dome*, June 1999, Vol. 50, No. 4, Who/What page, "Another Johns at Hopkins."

4 Ibid.

5 Press release, Nov. 2, 2004, "Ulatowski Named New Director of Anesthesiology and Critical Care Medicine at Johns Hopkins."

6 Ibid.; Anesthesiology and Critical Care Department Web site, http://www.hopkinsmedicine.org/anesthesiology/Team/faculty.cfm; http://www.hopkinsmedicine.org/anesthesiology/Team/fellow_roster.cfm; http://www.hopkinsmedicine.org/anesthesiology/Team/resident_roster.cfm; http://www.hopkinsmedicine.org/anesthesiology/Patient_Care/directions.cfm.

7 Hopkins Medicine Named Professorships Web site, http://webapps.jhu.edu/namedprofessorships/professorshipdetail.cfm?professorshipID=177; *Model*, pg. 313.

8 Ibid.; Pharmacology Department's faculty list, http://www.hopkinsmedicine.org/pharmacology/index.html.

9 Press release, Dana-Farber Cancer Institute, July 26, 2000, "Edward J. Benz, Jr., M.D., named President of Dana-Farber Cancer Institute."

10 Press release, Aug. 27, 2001, "Weisfeldt To Head Hopkins' Department of Medicine."

11 Ibid.

12 Weisfeldt, M.D., "State of the Department," Medical Grand Rounds, Sept. 5, 2003, Hurd Hall; "Chairman's Welcome Letter," Dept. of Medicine Web site, http://www.hopkinsmedicine.org/Medicine/admin/welcome.html; *The Johns Hopkins Gazette*, March 3, 2008, Vol. 37, No. 24, "Minority Recruitment Recognized," David March.

13 *Hopkins Medical News*, Spring/Summer 2003, Vol. 26, No. 3, "Face Time/David Hellmann."

14 Press release, July 27, 2005, "Hellmann Appointed New Vice Dean for Hopkins Bayview."

15 *Centuries of Caring: The Johns Hopkins Bayview Medical Center Story*, (2004) Neil A. Grauer, pgs. 120–121.

16 Press release, July 27, 2005, "Hellmann Appointed New Vice Dean for Hopkins Bayview."

17 Press release, May 31, 2000, "Hopkins Appoints New Head of Cell Biology and Anatomy"; press release, Salk Institute, April 24, 1997, "An Enriched Environment Stimulates An Increase In The Number of Nerve Cells in Brains of Older Mice" [noting Pollard as president of Institute then]; http://en.wikipedia.org/wiki/Thomas_D._Pollard.

18 *Hopkins Medical News*, Spring 2000, Vol. 23, No. 2, "Four Students and a Cadaver," Anne Bennett Swingle; Hopkins Medical News, Fall 2002, Vol. 26, No. 1, "Anatomy Transformed," Anne Bennett Swingle.

19 Press release, Oct. 26, 2000, "Hopkins Names Orthopedics Chief"; press release, Feb. 27, 1998, "Richard Stauffer, Head Orthopaedic Surgeon At Hopkins, Dies."

20 *Centuries of Caring*, op. cit, pgs. 109; 111; Bayview News, Summer 2006, Vol. 22, No. 3.

21 *Hopkins Medical News*, Spring/Summer 2003, Vol. 26, No. 3; *Hopkins Medicine*, Winter 2008, Vol. 31, No. 2, "Joint Solutions," Ramsey Flynn.

22 Ibid.; *Johns Hopkins Bayview News*, Spring, 2007, Vol. 23, No. 2, "Restoring Hope," Neil A. Grauer.

23 Press release, Nov. 23, 2000, "Henry Brem Named New Director of Hopkins Neurosurgery."

24 Ibid.; *Dome*, July 2002, Vol. 53, No. 7, "Nerve Central," Mary Ann Ayd.

25 Neurological Surgery Training Program Web site, http://www.hopkinsmedicine.org/se/util/display_mod.cfm?MODULE=/se-server/mod/modules/semod_printpage/mod_default.cfm&PageURL=/neurology_neurosurgery/education/residencies/neurosurgery_residency/&VersionObject=8EA77C99FF629D27EA7B5CAB1D3EF149&Template=C3DCB6B2FEDE5BD04FFD54F6429A4331&PageStyleSheet=1DD8B8EA2404DA3607707FF796EDC73B; Neurology and Neurosurgery at Johns Hopkins Web site, http://www.hopkinsmedicine.org/neurology_neurosurgery; Justin McArthur Web site, http://www.hopkinsmedicine.org/neurology_neurosurgery/specialty_areas/project_restore/profiles/team_member_profile/F00456462E2B11FA0068C52042B30687/;

Justin McArthur Web site, http://www.hopkinsmedicine.org/neurology_neurosurgery/specialty_areas/project_restore/profiles/team_member_profile/F00456462E2B11FA0068C52042B30687/Justin_McArthur.

26 Ibid.; About.com, functions of the brain Web site, http://psychology.about.com/od/biopsychology/ss/brainstructure_6.htm.

27 Neurological Surgery Training Program Web site, op. cit; Free Medical Dictionary Web site, http://medical-dictionary.thefreedictionary.com/vasospasm.

28 *Model*, *The New York Times*, July 19, 2005, http://query.nytimes.com/gst/fullpage.html?res=9503E4DC1F30F93AA25754C0A9639C8B63&sec=&spon=&pagewanted=all; Johns Hopkins Named Professorships, http://webapps.jhu.edu/namedprofessorships/professorshipdetail.cfm?professorshipID=192.

29 Princeton Dermatology Associates, op. cit., Hopkins Dermatology Web site, including Sewon Kang's CV, http://www.hopkinsmedicine.org/dermatology/about_us/message_chairman.html.

30 Hopkins Dermatology Web site.

31 Press release, Sept. 28, 2001, "Brooks Jackson Named New Director of Hopkins Pathology."

32 http://pathology.jhu.edu; http://medical-dictionary.thefreedictionary.com/cytopathology; http://medical-dictionary.thefreedictionary.com/Flow+cytometry.

33 Press release, Nov. 16, 2001, "International Health Expert to Lead Hopkins' Department of History of…Medicine"; *Model*, pgs. 65, 89.

34 Press release, Nov. 16, 2001, op. cit.

35 Faculty in the Department of the History of Medicine, http://www.hopkinsmedicine.org/histmed/people/faculty; The Department of History of Science and Technology, http://host.jhu.edu/about.html; Program in the History of Science, Medicine, and Technology, http://web.jhu.edu/hsmt; Bulletin of the History of Medicine, http://www.press.jhu.edu/journals/bulletin_of_the_history_of_medicine/; Institute of the History of Medicine, http://www.welch.jhu.edu/ihm/iohmlibrary.html.

36 Press release, Feb. 15, 2002, "DePaulo New Director of Psychiatry at Hopkins."

37 "Chairman J. Raymond DePaulo, Jr. M.D.," http://www.hopkinsmedicine.org/psychiatry/about_us/chairman.html.

38 Press release, Aug. 8, 2007, "Hopkins Team Develops First Mouse Model of Schizophrenia"; Wired, Dec. 26, 2007, "The Top 10 New Organisms of 2007," http://www.wired.com/science/discoveries/news/2007/12/YE_10_organisms.

39 http://www.hopkinsmedicine.org/psychiatry/about_us/index.html

40 Press release, Dec. 11, 2002, "Clements Named Director of Comparative Medicine at Johns Hopkins."

41 Ibid.

42 M. Christine Zink, http://www.
hopkinsmedicine.org/neurology_neurosurgery/
research/JHU_NIMH/researchers/profiles/
czink.html; http://medical-dictionary.
thefreedictionary.com/phagocyte.

43 *Hopkins Medicine*, Winter 2009, Vol. 32, No.
2, "One Medicine," Neil A. Grauer.

44 Barbara de Lateur became the first woman
to head a clinical department when she was
named director of Physical Medicine and
Rehabilitation in 1995; Ranice Crosby headed
the Department of Art as Applied to Medicine
from 1943 to 1983.

45 *Hopkins Medical News*, Fall 1995, Vol. 19,
No. 1, "A Study in Harmony," Janet Farrar
Worthington; *Hopkins Medical News*, Winter
1994, Vol. 17, No.2, "Women in the Promised
Land," Randi Henderson.

46 Hopkins Dept. of Surgery Web site: http://
www.hopkinsmedicine.org/surgery/div/; http://
www.hopkinsmedicine.org/surgery/faculty/List
; http://www.hopkinsmedicine.org/surgery/
research/.

47 Press release, July 28, 2003, "Greider Named
Director of Molecular Biology at Hopkins";
Dome, Vol. 60, No. 9, Nov. 2009, "Prized
Curiosity," Judy Minkov; *The Baltimore Sun*,
Oct. 6, 2009, "Low-key laureate," Kelly
Brewington.

48 *Hopkins Medical News*, Winter 2001, op. cit;
Department of Molecular Biology & Genetics
Web site, http://www.mbg.jhmi.edu/Pages/
about/index.aspx; http://www.mbg.jhmi.edu/
Pages/people/faculty_list.aspx; http://www.
mbg.jhmi.edu/Pages/people/faculty_list_
secondary.aspx.

49 Ibid.

50 Press release, May 27, 2003, "Alumnus Peter
McDonnell to Head Hopkins' Wilmer Eye
Institute"; *Dome*, Oct. 2009, Vol. 60, No. 8,
"The Wilmer Way," Neil A. Grauer

51 Press release, May 27, 2003.

52 Author's interview with McDonnell, July 13,
2009; pamphlet, Dedicationi of the Robert
H. and Clarice Smith Building and Maurice
Bendann Surgical Pavilion, October 16, 2009.

53 Wilmer Web site: http://www.
hopkinsmedicine.org/wilmer/services/.

54 Ibid.

55 Ibid.

56 Press release, Dec. 2, 2003, "Lloyd Minor
Named Otolaryngology Chief at Hopkins";
Johns Hopkins International Web site,
http://www.hopkinsglobal.net/jhi/english/
doctors/cummingsendowedchair.asp; *Dome*,
March 2006, Vol. 57, No. 2, "Hopkins USA,
Reenergized," Lyndia Levis Bloch; Named
Professorships Web site, http://webapps.jhu.
edu/namedprofessorships/professorshipdetail.
cfm?professorshipID=351.

57 Johns Hopkins International Web site, op. cit.

58 Otology and Neurotolgy Web sites, http://
www.wrongdiagnosis.com/specialists/
otology-neurotology.htm; http://www.
editorialmanager.com/on/.

59 Otolaryngology Web site, http://esgweb1.nts.
jhu.edu/otolaryngology/faculty.html; *Dome*,
Oct. 2002, Vol. 53, No. 9, "Music to Her
Ears."

60 "Dear Colleagues" e-mail from Edward
Miller, Nov. 2011; Eisele profile, University
of California Helen Diller Family
Comprehensive Cancer Center, http://cancer.
ucsf.edu/people/eisele_david.php.

61 Ibid.

62 *Model*, pg. 90; press release, July 2, 2003,
"DeWeese First Director of Hopkins' New
Radiation Oncology Department."

63 Dept. of Radiation Oncology and Molecular
Radiation Sciences Web site, http://www.
radonc.jhmi.edu/faculty/dr_deweese.html.

64 Dept. of Radiation Oncology and Molecular
Radiation Sciences Web site, http://www.
radonc.jhmi.edu/education/faculty.html
; http://www.radonc.jhmi.edu/index.asp
; http://mrs.radonc.jhmi.edu/index.html
; http://www.radonc.jhmi.edu/faculty/
dr_deweese.html.

65 *Hopkins Medical News*, Spring/Summer 2001,
Vol. 24, No. 3.

66 Ibid.

67 Rheumatology Division Web site, http://www.
hopkinsmedicine.org/rheumatology/clinics/
faculty_staff/fredrick_wigley.html; http://
www.hopkinsmedicine.org/rheumatology/
research/faculty_staff/antony_rosen. http://
scleroderma.jhmi.edu/scleroderma-center/
differences.html

68 Rheumatology Division Web site, op.
cit., http://www.hopkinsmedicine.org/
rheumatology/index.html; Press release, July
16, 2009, "The Johns Hopkins Hosital Tops
US News & World Report 'Honor Roll' 19th
Year In a Row."

69 *Hopkins Medicine*, Winter 2004, Vol. 27,
No. 2, "Top Cardiologist, a True Success
Story"; *Johns Hopkins Gazette*, Nov. 30, 2009,
"Kenneth L. Baughman, 63, former director
of Cardiology," David March.

70 Ibid.; Heartwire, May 9, 2007, "Marbán
tomove to Cedars-Sinai," Lisa Nainggolan,
http://www.theheart.org/article/789629.do.

71 *Hopkins Medicine*, Spring/Summer 2009, Vol.
32, No. 3; http://www.thefreedictionary.com/
electrophysiological.

72 *Dome*, May 2009, Vol. 60, No. 4, Who/What
page, "New Cardiology Clinical Director."

73 *Dome*, Winter 2004, Vol. 27, No. 2, "Duly
Noted"; Thomas Jefferson University, Dept.
of Surgery Web site, http://www.jefferson.
edu/surgery/faculty_profile.cfm?key=cjy001;
Hellenic Medical Society of New York, 2006
event program, pg. 3, http://www.hmsny.org/
HMS-jrnl-06-Edit-web.pdf.

74 E-mail from John Hundt, administrator of the
Department of Surgery, Sept. 27, 2010; Web
site on named professorships, Handelsman
Professorship, http://webapps.jhu.edu/
namedprofessorships/professorshipdetail.
cfm?professorshipID=348; Choti profile on
Hopkins Medicine Web site, http://www.
hopkinsmedicine.org/surgery/faculty/Choti.

75 E-mail from John Hundt, Sept. 27, 2010,
op. cit.; *Model*, pg. 91 (on Ravitch); Hopkins
Medicine Web site on Eckhauser, http://
www.hopkinsmedicine.org/surgery/faculty/
Eckhauser; named professorship Web
site on Efron, http://webapps.jhu.edu/
namedprofessorships/professorshipdetail.
cfm?professorshipID=201.

76 E-mail from John Hundt, Sept. 27, 2010,
op. cit.

77 Press release, Oct. 23, 2003, "Eileen
Vining to Direct Pediatric Epilepsy
Program"; John Freeman profile on
Neurology & Neurosurgery Web site,
http://www.neuro.jhmi.edu/profiles/
freeman.html.

78 Ibid.; Neurology & Neurosurgery Web
site on "History of Pediatric Epilepsy at
The Johns Hopkins John M. Freeman
Pediatric Epilepsy Center," http://
www.hopkinsmedicine.org/neurology_
neurosurgery/history/pediatric_epilepsy_
history.html.

79 Press release, Oct. 23, 2003, op. cit.; vagus
nerve definition, MedicineNet.Com http://
www.medterms.com/script/main/art.
asp?articlekey=7631

80 Press release, April 29, 2003, "Linda Fried
Named Geriatrics Head at Johns Hopkins."

81 Press release, April 29, 2003, op. cit.

82 Press release, Jan. 16, 2008, Columbia
University, http://www.columbia.edu/cu/
president/docs/communications/2007-
2008/080116-mailman-dean-
announcement.html; *Dome*, May 2010,
Vol. 61, No. 4, Who/What page.

83 Press release, Nov. 2, 2004, "Ulatowski
Named New Director of Anesthesiology
and Critical Care Medicine at Johns
Hopkins"; press release, Nov. 12, 2004,
"Partin Named New Director of Urology
at Johns Hopkins."

84 Ibid.

85 Ibid.

86 Ibid.

87 Ibid.

88 Brady Institute Web site, History section,
"A Visit with Dr. Alan Partin," op. cit.

89 Ibid., home page, http://urology.jhu.edu/
index.html.

90 Ibid.; research laboratories, http://
urology.jhu.edu/research/laboratories.php
; Answers.com Web site for definition of
biomarkers, http://www.answers.com/
topic/biomarker.

91 Press release, "Jonathan S. Lewin, M.D., to
Head Hopkins Department of Radiology."

92 Ibid.

93 Web sites for radiology at Hopkins
Hospital, http://www.hopkinsradiology.
org/Radiology%20Divisions/index.
html; http://www.hopkinsradiology.org/
Radiology%20Faculty/index.html; and
Johns Hopkins Bayview, http://www.
hopkinsbayview.org/imaging/services.

94 Ibid., http://www.hopkinsradiology.org/
Radiology%20Divisions/index.html, op.
cit.

95 Press release, May 3, 2004, "Brancati
Named New Director of Hopkins General
Internal Medicine."

96 Press release, May 3, 2004, op. cit.

97 Division of General Internal Medicine
Web site, op. cit.

98 Ibid.

99 Ibid.; hospitalist definition, Medterms Web site, http://www.medterms.com/script/main/art.asp?articlekey=8384.

100 Division of General Internal Medicine Web site, op. cit., http://www.hopkinsmedicine.org/gim/clinical/; http://www.hopkinsmedicine.org/gim/clinical/services.html.

101 Press release, June 25, 2004, "Jeffrey Palmer Named New Chair of Physical Medicine and Rehabilitation at Johns Hopkins"; Barbara J. de Lateur profile on department Web site, http://www.hopkinsmedicine.org/rehab/faculty/delateur.html; http://www.hopkinsmedicine.org/Rehab/News/Delateurdistinguished.html.

102 Press release, June 25, 2004, op. cit.

103 Department of Physical Medicine and Rehabilitation Web site, http://www.hopkinsmedicine.org/Rehab/index.html.

104 Department of Physical Medicine and Rehabilitation Web site, op. cit.

105 Ibid.

106 Centuries of Caring, op. cit.; Johns Hopkins Magazine, Sept. 2007, "After the Fire," Maria Blackburn.

107 Hopkins Medicine, Winter 2004, op. cit.

108 Ibid.; Centuries of Caring, op. cit, pgs. 87–88; Johns Hopkins Magazine, Sept. 2007, op. cit.

109 Press release, Aug. 29, 2003 (Bayview Medical Center), "Robert J. Spence, M.D., Named Director of the Baltimore Regional Burn Center at the Johns Hopkins Bayview Medical Center"; Johns Hopkins Magazine, Sept. 2007, op. cit.

110 "Dear Colleagues" letter from Julie Freischlag, director of surgery, July 7, 2005, "New Chief of Hopkins Burn Services"; Good Samaritan Hospital Web site, "Burn Reconstruction Center," http://www.goodsam-md.org/body.cfm?id=557195&fr=true&utm_source=GSHBurnBalt&utm_medium=SEM&utm_campaign=GSHBurnBalt_SEM.

111 "Dear Colleagues" letter from Freischlag, op. cit.

112 Burn Center Web site, http://www.hopkinsmedicine.org/burn/index.html; Johns Hopkins Magazine, Sept. 2007, op. cit.

113 Ibid., http://www.hopkinsmedicine.org/burn/community-prevention/; http://www.hopkinsmedicine.org/burn/military%20training/.

114 Model, op.cit., pg. 97; American Society of Hematology Web site, http://www.hematology.org/About-ASH/ (definition of hematology); http://www.hematology.org/Publications/Hematologist/2007/1169.aspx (on Maxwell M. Wintrobe);Osleriana Archives Web site (on Maxwell M. Wintrobe), http://www.asksam.com/cgi-bin/as_web6.exe?Command=DocName&File=Osleriana&Name=2006-19%20Maxwell%20M.%20Wintrobe%2C%20The%20Hematologist%20from%20Halifax.

115 Hopkins Medicine, Spring 2005, Vol. 28, No. 3, "To the Manner Born," Neil A. Grauer; University of Pennsylvania Gazette, March/April 2008, obituaries, "Dr. Isadore Brodsky, C'51 M'55," http://www.upenn.edu/gazette/0308/obits.html.

116 Hopkins Medicine, Spring 2005, op. cit., first draft, March 2005; Hopkins Medicine, Winter 2006, Vol. 29, No. 22, "The Brodsky Approach Saves a Life," Michael Levin-Epstein.

117 Oslerliana Archives Web site, "Maxwell M. Wintrobe, The Hematologit from Halifax," Marvin J. Stone.

118 Hopkins Medicine, Spring/Summer 2006, Vol. 29, No.3, "Lock Conley looks back and blushes," Neil A. Grauer.

119 Hematology Web site on Jerry Spivak and Center for the Crhoic Myeloproliferftive Disorders, http://www.hopkinsmedicine.org/hematology/faculty_clinical_staff/spivak_j.html; http://www.mpdhopkins.org/.

120 Division of Hematology Web site, http://www.hopkinsmedicine.org/hematology/.

121 Hopkins Medicine, Fall 2005, Vol. 29, No., 1, "Computer Geek Division Chiefs," Neil A. Grauer.

122 Kalloo profile on named professorships Web page, http://webapps.jhu.edu/namedprofessorships/professorshipdetail.cfm?professorshipID=196; Medscape Web site (definition of Endoscopic cryotherapy, http://www.medscape.com/viewarticle/407958; Patentstorm Web site for definition of winged biliary stent, http://www.patentstorm.us/patents/5776160/description.html.

123 Ibid.; Hopkins Medicine, Fall 2005, op. cit.

124 Ibid.; Hopkins GI Web site, http://www.hopkins-gi.org/JHGI_Home.aspx?SS=&CurrentUDV=31.

125 Ibid.

126 Hopkins Medicine, Fall 2005, "Computer Geek Division Chiefs," op. cit.

127 Nephrology Division Web site, Scheel profile, http://www.hopkinsmedicine.org/nephrology.

128 Ibid.; definition of retroperitoneal space, Medcyclopaedia Web site, http://www.medcyclopaedia.com/library/topics/volume_ii/r/retroperitoneal_space.aspx; definition of peritoneum, MedicineNet.com, http://www.medterms.com/script/main/art.asp?articlekey=4842; definition of fibrosis, Wisegeek.com, http://www.wisegeek.com/what-is-fibrosis.htm.

129 Nephrology Division Web site, op. cit.; definition of kidney tubule, Astrograhics Web site, with image of one, http://www.astrographics.com/GalleryPrintsIndex/GP2079.html.

130 Hopkins Medicine, Fall 2005, op. cit.

131 Ibid.

132 Dome, May 2006, Vol. 57, No. 4, "Bridging Molecules and Mind," Deborah Rudacille.

133 Press release, Feb. 14, 2006, "New Director of Neurosciences at Johns Hopkins Medicine."

134 Ibid.

135 Ibid.

136 Ibid.; definition of synaptic plasticity, encyclopedia.com, http://www.encyclopedia.com/doc/1O6-synapticplasticity.html; Laboratory of Richard L. Huganir, Ph.D. Web site, http://www.bs.jhmi.edu/neuroscience/huganir/index.htm; Neuroscience Department Web site, "Message from the Director," http://neuroscience.jhu.edu/welcome.php : Dome, May 2006, op. cit.

137 Neuroscience Department Web site, "Overview," http://neuroscience.jhu.edu/.

138 Ibid.

139 Ibid.

140 Hopkins Medicine, Fall 2006, Vol. 30, No. 1, "Follow That Fish," Neil A. Grauer.

141 Ibid.; Cystic Fibrosis Foundation Web site, http://www.cff.org/AboutCF/index.cfm?dspPrintReady=Y.

142 Hopkins Medicine, Fall 2006, op. cit.; Guggino Web page, http://guggino.org/about.htm.

143 Press release, May 15, 2006, "New Director of Physiology at Johns Hopkins Medicine"; Dept. of Physiology Web site, op. cit.

144 Department of Physiology Web site, "Mission Statement," http://physiology.bs.jhmi.edu/pages/about/mission.aspx.

145 Hopkins' Human Genetics Training Program Web site, http://humangenetics.jhmi.edu/index.php/faculty/william-guggino.html ; the NIH's National Kidney and Urologic Diseases Information Clearinghouse Web site, http://kidney.niddk.nih.gov/kudiseases/pubs/polycystic/.

146 Partners in Discovery, the Cystric Fibrosis Center newsletter, Fall 2006, Vol. 3, No. 2, "Advancing the Knowledge."

147 Ibid.; Cystic Fibrosis Foundation Web site, op. cit.

148 Hopkins Medicine, Fall 2006, Vol. 30, No. 1, "Director with a Latin Beat," Neil A. Grauer.

149 Press release, July 17, 2006, "Johns Hopkins Appoints New Chair of Biophysics and Biophysical Chemistry."

150 Ibid.

151 Model, pg. 98; Dept. of Biophysics and Biophysical Chemistry Web site, http://biophysics.med.jhmi.edu/BIOPHYS/Pfaculty.html; http://nobelprize.org/nobel_prizes/medicine/laureates/2004/axel-autobio.html ; http://www.hhmi.org/biointeractive/cancer/vogelstein_career.html.

152 MedicineNet.com (definition of X-ray crystallography), http://www.medterms.com/script/main/art.asp?articlekey=12381.

153 Answers.com, http://www.answers.com/topic/calorimeter; YourDictionary.com, http://examples.yourdictionary.com/calorimetry (definition of microcalorimetry).

154 Stowers Institute for Medical Research Web site, http://research.stowers-institute.org/microscopy/external/Technology/FCS/index.htm; Answers.com, http://www.answers.com/topic/spectroscopy (definition of fluorescence spectroscopy).

155 Telephone conversation with Chester W. Schmidt, Jr., Sept. 29, 2010; press release, May 7, 2007, "Lyketsos, Head of Johns Hopkins Bayview Psychiatry, Accepts Endowed Chair."

156 Telephone conversation with Schmidt, op. cit.

157 Academic Psychiatry, March 2001, Vol. 25, "Commentary: Research Training During Psychiatric Residency, A Personal Reflection," Constantine G. Lyketsos, M.D., M.H.S.

158 Memory and Alzheimer's Treatment Center Web site, http://www.hopkinsmedicine.org/psychiatry/specialty_areas/memory_center.

159 Hopkins Bayview Psychiatry Department Web site, http://www.hopkinsmedicine.org/psychiatry/bayview; http://www.hopkinsmedicine.org/psychiatry/expert_team/faculty/.

160 Ibid.

161 *Hopkins Medicine*, Fall 2006, Vol. 30, No. 1, "From Uganda to Chief," Neil A. Grauer; press release, May 31, 2006, "Johns Hopkins Appoints New Chair of Infectious Diseases – Hepatitis Expert at Hopkins Since 1990."

162 Ibid.

163 Ibid.

164 Centers for Disease Control and Prevention Web site, http://www.cdc.gov/mrsa/index.html; WebMD Web site, http://www.webmd.com/skin-problems-and-treatments/understanding-mrsa-methicillin-resistant-staphylococcus-aureus.

165 "Dear Colleagues" e-mail from Edward Miller, Oct. 15, 2008; Neurology Department Web site, http://www.hopkinsmedicine.org/neurology_neurosurgery/experts/team_member_profile/F00456462E2B11FA0068C52042B30687/Justin_McArthur.

166 Ibid.

167 Ibid.

168 Dept. of Neurology Web site, McArthur profile, op. cit.

169 Ibid., http://www.hopkinsneuro.org/research/JHU_NIMH/researcher.cfm?i=38; http://www.neuro.jhmi.edu/nromusc/experts.htm; *Brain Wave*s, Winter/Spring 2003, Vol. 16 No. 1, "Balanced on an Antiretroviral Tightrope."

170 Neurology Dept. Web site, op. cit., http://www.hopkinsmedicine.org/bin/y/h/2009%20Annual%20Letter%20NRO.pdf; the American Recovery and Reinvestment Act Web site (for date of enactment), http://www.recovery.gov/About/Pages/The_Act.aspx.

171 Department of Neurology Annual Report, http://www.hopkinsmedicine.org/bin/g/u/neurology-annual-report.pdf.

172 Inside Hopkins, March 1, 2007, "Dear Colleagues" letter from Edward Miller.

173 Ibid.; Free Dictionary on-line (peroxisomal disorders definition), http://medical-dictionary.thefreedictionary.com/Peroxisomal+Disorders; *Hopkins Medicine*, Spring/Summer 2007, Vol. 30, No. 2, "Opening the Book of Life."

174 *Inside Hopkins*, March 1, 2007, op. cit.

175 McKusick-Nathans Institute of Genetic Medicine Web site, http://www.hopkinsmedicine.org/genetic medicine/People/Faculty.html; *Hopkins Medicine*, Spring/Summer 2007, op. cit.

176 *Hopkins Medicine*, Winter 2008, Vol. 31, No. 2, "Building Bridges," Neil A. Grauer.

177 Ibid.

178 Ibid.

179 Dept. of Biomedical Engineering Web site, http://www.bme.jhu.edu/welcome/welcome.htm.

180 Ibid.; press relase, Sept. 25, 2007, "Dr. Elliot McVeigh Named New Biomedical Engineering Director at Hopkins."

181 Ibid.

182 *Dome*, June 2007, Vol. 58, No. 5, "Honoring Abeloff," Anne Bennett Swingle; press release, Sept. 14, 2007, "Chief Oncologist Martin Abeloff Dies of Leukemia."

183 *Dome*, June 2007, op. cit.; transcript of Abeloff tribute CD, May 3, 2007.

184 Ibid.

185 Press release, Sept. 14, 2007, op. cit.

186 Press release, Nov. 4, 2008, "Johns Hopkins Prostate Cancer Specialist William Nelson to Head Institution Cancer Center."

187 Ibid.

188 Ibid.

189 Ibid.

190 Ibid.

191 Press release, March 31, 2009, "Johns Hopkins Appoints New Clinical Director of Cardiology."

192 Ibid., *Dome*, May 2009, Vol. 60, No. 4, "Who/What: New Cardiology Clinical Director."

193 Press release, March 31, 2009.

194 Ibid.; Heart & Vascular Institute Web site, http://www.hopkinsmedicine.org/heart_vascular_institute/experts.

195 Press release, April 20, 2009, "Johns Hopkins Names New Director of Pediatric Cardiac Surgery."

196 *Hopkins Medicine*, Winter 2006, Vol. 29, No. 2, "Let's Meet … Luca Vricella," Mary Ellen Miller.

197 Ibid.

198 Ibid.

199 Ibid.; kilos to pounds converter Web site, http://manuelsweb.com/kg_lbs.htm.

200 Press release, March 16, 2009, "Johns Hopkins Medicine International Launches New Cardiac Surgery Collaboration in Italy."

201 American Board of Thoracic Surgery Web site, http://www.abts.org/sections/About%20Us/article1.html.

202 *Johns Hopkins Gazette*, Dec. 13, 2010, "Duke Cameron is named next director of Cardiac Surgery," David March.

203 Ibid.

204 Ibid.; Hopkins Cardiac Surgery Web site, http://www.hopkinsmedicine.org/heart_vascular_institute/learn_more/cardiac_surgery.html

205 Hopkins Cardiac Surgery Web site, op. cit.; American Heart Association Web site (definition of atrial fibrillation), http://www.americanheart.org/presenter.jhtml?identifier=4451.

206 Hopkins Cardiac Surgery Web site, op. cit.

207 Hopkins Pediatric Oncology Web site, Donald Small profile, http://www.hopkinsmedicine.org/kimmel_cancer_center/centers/pediatric_oncology/our_experts/Physicians/Donald_Small.html.

208 Ibid.

209 Ibid.

210 Press release, July 9, 2009, "Donald Small to Lead Pediatric Oncology Division."

211 Pediatric Oncology Division Web site, types of cancer, http://www.hopkinsmedicine.org/kimmel_cancer_center/centers/pediatric_oncology/cancer_types/.

212 Ibid.

213 Hopkins Plastic and Reconstructive Surgery Web site, http://www.hopkinsmedicine.org/surgery/div/Plastics.html ; The John Staige Davis Society of Plastic Surgeons of Maryland Web site, "History," http://www.johnstaigedavissociety.org/pages/history; *Journal of Plastic & Reconstructive Surgery*, Sept. 1978, Vol. 62, No. 3, "The Life of John Staige Davis, M.D.," W. Bowdoin Davis, M.D., http://s3.amazonaws.com/webgen_einsteinwebsites/public/assets/16532/Life_of_JSD.PDF.

214 Hopkins Plastic and Reconstructive Surgery Web site, op. cit., http://www.hopkinsmedicine.org/surgery/ConditionsDiseases/Div/Plastic/index.htm.

215 *Pittsburgh Post-Gazette*, Aug. 12, 2010, "UPMC hand transplant main surgeon leaving Oct. 1," Sean D. Hamill, http://www.post-gazette.com/ng/10224/107928-114.stm.

216 The John Staige Davis Society of Plastic Surgeons of Maryland, op. cit.

217 *Journal of Plastic & Reconstructive Surgery*, Sept. 1978, op. cit.

218 Ibid.

219 Ibid.; The Johns Staige Davis Society of Plastic Surgeons of Maryland, op. cit.

220 Hopkins Plastic and Reconstructive Surgery Web site.

221 Pittsburgh Post-Gazette, Aug. 12, 2010, op. cit; Pittsburgh Tribune-Review, Aug. 11, 2010, http://www.pittsburghlive.com/x/pittsburghtrib/news/print_694359.html.

222 Ibid.

223 Press release, Nov. 30, 2010, "Johns Hopkins Alumnus W.P. Andrew Lee to Head Department of Plastic Surgery," http://www.hopkinsmedicine.org/news/media/releases/johns_hopkins_alumnus_w_p_andrew_lee_to_head_department_of_plastic_surgery.

224 *Dome*, September 2011, Vol. 62, No. 7, Who/What page; Johns Hopkins Medicine Special Reports, "What You Should Know About Weight-Loss Surgery," http://www.johnshopkinshealthalerts.com/special_reports/W_Loss_Surgery_reports/W_L_Surgery_landing.html.

225 E-mail, July 18, 2011, from Sara Baker, regarding appointments at Hopkins Bayview; Duncan CV, http://www.hopkinsbayview.org/bin/s/k/DuncanCV.pdf.

226 Bayview Surgery Division Web site, http://www.hopkinsbayview.org/surgery/index.html.

Credits and Sources

a = above

al = above left

ar = above right

b = below

bl = below left

br = below right

c = center

cr = center right

cl = center left

l = left

r = right

Photographers

Alman & Co., 193 ar
Bachrach, Inc., 35 ar; 48 al; 74 al
Bachrach, Louis Fabian, 12 a
Barber, Doug, 123 br
Bishop, Jennifer, 230 al
Boam, Maxwell, vi; 20 al
Bohrer, White House, 265 al
Bower, Elizabeth, 278 al
Burke, Robert, 136 cl
Busey, N.H., 27 c
Canner, Larry, 299 ar
Chellappa, Mohan, 274 cl
Christofersen, Jon, 246 al; 272–273, a
Ciesielski, Mike, 303 cr
Cole, Bernard, 104 a
Colwell, David, 162 al; 169 cr; 247 b; 257 ar
Corey, Jay, 283 ar
Coupon, William, 143 ar
Crocetti, Michael, 213 ar
Denison, Bill, 114 al; 138 al; 138 br; 141 ar;
    148 bl; 175 c; 207 ar; 229 al; 247 ar; 304 cl
Dorfman, Elena, 247 a
Eccles, Andrew, 270 al
Ekstromer, Jonas, 231 cr; 268 bl
Fetters, Paul, 250 cl
Fitzhugh, Susie, 109 ar
Freeman, Charles, 99 a
Friedman, Sy, 66 a
Gardner, Alexander, 5 ar
Gey, George, 89 br
Green, Bob, HBO Films, 227 ar; 227 cr
Greystone Studios, Inc. 96 al
Grief, Leonard J. Jr., 225 cr
Gummerson, 149 br
Hamilton, William C. 80 al
Harris and Ewing, 50 al
Harting, Christopher, 191 ar
Hartlove, Chris, 133 cr; 179 ar; 196 a; 216 bl;
    254 al; 281 a; 295 a; 302 al; 326 a;
Howell, Joe, 305 br
Howard, Peter C., 93 ar; 102 al; 111 ar; 122 c;
    123 ar; 128 cl; 133 ar; 147 ar; 171 ar; 177 ar;
    139 r; 147 ar; 160 cl; 229 ar
Ingrodi, Shanna, 245 b
Kareem, Zuhair, 164 br;
Karsh, Yousuf, 63 ar; 224 bl
Kirk, Will, 166 ar; 220 ar;
Klosicki, Bill, *Bio-Medical Photographer
    Bayview Medical Center*, 130 c
Kurniawan, David, 284 b
Levin, Aaron, x c; 11 a; 64 al
Liu, Dianni, 144 c
Logan, Gary, 301 cr

McDaniel, Steve, 124 a
McDonald, Doug, 150 al
McGovern, Michael, 145 ar
Melio, Jose, 289 cr
Mesny, 17 br
Meyers, Jill, 285 ar
Mitchell, James F., 32 al
Morton, Tadder, 65 ar
Mottar, Robert, 57 al; 70 al
Murphy, Paula, 242 c
Myra Photography, 135 ar
Osorio, Carlos, 126 al
Phillips, Robert, 90 al; 101 a
Potter, Robert D., 65 br; 103 ar
Push, Steve, 82 l
Rahba, Al, 289 cr
Riggins, Rich, 194 al; 213 br;
Rivard, Susan, *Johns Hopkins News-Letter*, 6 al
Rodriguez, Vince, 159 a; 167 ar; 210 bl;
Rubino, Joseph, 134 al
Schaefer, J.H. & Son, 94 al
Smith, Robert J., 125 cr; 129 b
Soloman, Ron, 274 al
Stockfield, Bob, 86 al; 129 ar
Tadder Associates, 65 ar
Tate, F., 78 b
Teves, Jason, 296 ar
Thompson, Richard C., 71 ar; 76 al; 77 ar
Van Rensselaer, Jay, 313 ar
Vasquez, Claudio, ii c
Wachter, Jerry, 161 ar
Weller, Keith, 85 a; 98 al; 103 b; 127 ar; 131
    ar; 133 b; 136 ar; 153 cr; 154 a; 155 ar; cr;
    156 al; 157 b; 160 ar; 163 ar; 164 al; 165
    ar; 166 bl; 168 ar; 170 ar; c; bl; 173 ar; 174
    al; 178 c; 180 a; 181 ar; 182 c; 187 a; 189 a;
    192 b; 195 br; 197 a; br; 198 a; 199 ar; br;
    200 ar; 201 a; 203 bl; ar; 205 ar; 206 al;
    209 ar; 210 al; 211 ar; 212 cl; 215 cr; 216 al;
    217 ar; 218 br; 219 ar; 221 ar; bl; 230 al; bl;
    232 ar; 233 ar; 234 al; 235 bl; 236 ar; 238
    c; 240 al; 241 cr; 242 ar; 243 al; br; 244 al;
    245 ar; 246 b; 248 al; cl; b; 249 ar; 250 ar;
    251 ar; c; bl; 252 br; 253 ar; 254 bl; 255 ar;
    bl; 256 cl; 258 al; 259 ar; 260 cr; 261 ar; bl;
    262–263 c; 264 al; 266 al; 267 al; br; 269
    al; 272 bl; 273 cr; 274 al; 275 a; 276 al; 277
    ar; 279 ar; 280 bl; 281 al; 282 b; 283 b; 285
    cr; 287 a; 287 br; 290 al; 293 br; 294 cl;
    296 al; 297 ar; 298 a; 300 cl; 305 a; 307 ar;
    308 cl; 308 cr; 309 ar; br; 310 al; br; 311 ar;
    bl; 312 al; 314 ar; 315 ar; bl; 317 ar; 318 al;
    321 ar; 323 a; cr; 324 al; 329 a; 331 c

Artists

Abrams, Herbert E., 57 br; 97 ar
Broedel, Max, 42 lc
Carey, May Lewis, 88 al
Cooper, Henry, 92 al
Corner, Thomas, x c; 15 ar; 36 bl
Draper, William, 46 al
Gee, Bob, 64 al
Haupt, Eric Guido, 33 c
Ingram, Wayne, 176 al
Parada, Roberto, 239 ar
Poynton, Sarah, 271 ar
Salisbury, Frank O., 53 br
Sargent, John Singer, 11 a
Stevens, Bradley, 224 ar
Wyeth, Jamie, 104 bl

Sources

Courtesy of The Alan Mason Chesney Medical Archives of The Johns Hopkins Medical Institutions

x c; 2 a; 7 ar; 11 a; 12 al; 14 al; 15 ar; 16 a; 17 br; 18 bl; 19 ar; 21 a; 23 a; cr; 26 al; 27 cr; 28 b; 29 ar; 30–31 c; 32 al; 33 cr; 34 al; cr; 36 bl; 37 ar; 38 al; 39 cr; 40 al; 42 a; lc; 43 ar; 44 al; 45 ar; br; 46 al; bl; 49 br; 52 al; 53 br; 55 ar; 56 bl; 57 al; br; 61 ar; 63 al; 64 al; 65 ar; br; 66 a; 68 al; br; 69 ar; 70 al; 71 ar; 74 b; 75 ar; 76 al; bl; 77 ar; 78 b; 79 ar; 80 al; 81 ar; 82 bl; 84 al; 86 al; 87 br; 88 al; 89 c; br; 90 al; br; 92 al; 94 al; 95 ar; 96 al; 97 ar; 99 a; 101 a; 103 a; 104 a; bl; 105 cr; 106 a; 107 cr; 109 ar; br; 110 al; bl; 115 ar; 117 a; 122 c; 127 a; 129 ar; bl; 132 al; 134 al; 140 al; 143 ar; 147 ar; 149 ar; 165 ar; 176 al; 193 ar; 209 ar; 213 bl; 222 cl; 224 ar; bl; 225 cr; 282 al; 406 c

Courtesy of The Ferdinand Hamburger Jr. Archives, The Johns Hopkins University

3 ar; 76 b; 148 bl; 325 ar

Courtesy of the Johns Hopkins Medicine Office of Marketing and Communications

ii c; vi c; 20 al; 62 bl; 62 br; 72 al; 81 ar; 83 ar; 85 a; 93 ar; 98 al; 102 al; 103 bl; 109 ar; br; 111 ar; 114 a; 116 al; br; 122 c; 123 ar; 124 a; 125 br; 127 ar; 128 cl; 129 al; bl; 130 c; 131 cr; 133 a; c; bl; 134–135 b; 135 ar; 136 ar; cl; 137 ar; 138 al; br; 139 cr; 140 al; 141 ar; 143 ar; 144 c; 145 ar; 146 a; 147 ar; 148 bl; 149 br; 150 al; 153 br; 154 a; 155 ar; cr; 156 al; 157 b; 159 a; 160 ar; bl; 161 ar; 162 al; 163 ar; 164 al; br; 165 ar; 166 bl; 167 al; br; 168 a; c; 169 ar; br; 170 ar; cl; bl; 171 ar; 173 ar; 174 al; 175 ar; c; 177 ar; 178 c; 179 ar; 180 a; 181 ar; 182 c; 185 cr; 187 a; 189 a; 191 ar; 192 b; 194 al; 195 br; 196 a; 197 a; br; 198 a; 199 ar; bl; bc; br; 200 al; 201 ar; 202 bl; 203 ar; 204 al; 205 ar; 206 al; 207 a; 208 cl; 209 ar; 210 al; bl; 211 ar; 212 cl; 213 al; cr; br; 214, ar; 215, cr; 216 al; bl; 217 ar; 218 br; 219 ar; 220 ar; 221 ar; bl; 222 cl; 223 ar; 225 cr; 227, ar; cr; 228 al; 229 ar; 230 al; bl; 233 ar; 234 al; 235 b; 236 ar; 238 c; 239 ar; 240 al; 241 ac; ar; br; 242 ar; c; 243 al; br; 244 al; 245 ar; bl; 246 al; bl; 247 ar; bl; 248 cl; bc; 249 ar; 250 ar; cl; 251 ar; cr; bl; 252 br; 253 ar; 254 al; 255 ar; bl; 256 cl; 257 ar; 258 al; 259 ar; 260 cl; 261 ar; bl; 262–263 c; 264 al; 265 al; cr; bl; 266 al; 267 al; br; 269 al, ar; 270 al; 271 ar; 272–273 a; 272 bl; 273 br; 274 al; cl; 275 ac; bl; 276 al; 277 ar; 278 al; 279 ar; 280 al; bl; 281 ac; 282–283 b; 282 al; 283 ar; 284 bc; 285 ar; cr; 286 al; 287 al; br; 288 ar; 289 ar; cr; 290 al; 291 bl; 292 al; 293 ar; 294 cl; 295 ac; 296 ar; cl; 297 al; 298 ac; 299 ar; 300 cl; 301 cr; 302 al; bl; bc; br; 303 br; 304 cl; 305 ac; br; 306 al; 307 ar, cr; 308 cl; cr; 309 ar, br; 310 al; br; 311 ar; bl; 312 al; 313 ar; 314 ar; 315 ar, bc; 317 ar; 318 al; 319 cr; 321 ac; 323 ac; cr; 324 al; 326 ac; 329 ac; 331 c

Courtesy of the Library of Congress

5 al; c; 8 al; 10 al; 38 bc; 54 al; 235 ar

Courtesy of the National Library of Medicine

9 r; 24 al; 25 ar; 50 al; br; 51 ar: 74 ar

Other Sources

1   ar, *Popular Science Monthly* Volume 63
4   bl, courtesy of www.medicalantiques.com
5   ar, courtesy of *The Rail Splitter*, New York, NY
13   b, *Time and the Physician*, Lewellys Barker, G. P. Putnam's Sons, New York, 1942
22   cr, used with courtesy from the American Physiological Society's picture archives
35   ar, courtesy of The Bancroft Library, University of California, Berkeley. Photo by Bachrach
41   ar, courtesy of Smith College. Sophia Smith Collection. Florence Rena Sabin Papers
47   ar, *Time and the Physician*, Lewellys Barker, G. P. Putnam's Sons, New York, 1942
48   al, courtesy of Smith College. Sophia Smith Collection. Florence Rena Sabin Papers
49   al, courtesy of Culver Pictures, Inc., NY
60   a, courtesy of the Margaret Herrick Library, the Motion Picture Academy of Arts and Sciences Foundation, Los Angeles, CA; reproduction rights courtesy of Universal Studios, Licensing LLC,© 1945 Universal Pictures, Los Angeles, CA
62   al, courtesy of Ranice Crosby
63   ar, courtesy of the Estate of Yousuf Karsh
63   br, © Bettmann/Corbis
65   ar, © Tadder Baltimore
67   cr, © Bettmann / Corbis
69   br, courtesy of the Franklin D. Roosevelt Library, Hyde Park, N.Y.
73   ar, courtesy of the University of Pennsylvania Art Collection, Philadelphia, PA
73   br, courtesy of University of Utah, J. Willard Marriott Library, Special Collections, Salt Lake City, Utah
83   br, Smithsoninan Postal Museum Web site, http://arago.si.edu/index.asp?con=1&cmd=1&mode=&tid=2045345
89   ar, courtesy of the Crown Publishing Group, a division of Random House, Inc.
91   br, courtesy of the University of Maryland, *News-American* Photograph Collection, Marylandia Room, Hornbake Library, College Park, MD; the Hearst Corporation, N.Y.
113 a, courtesy of the Division of Cardiac Surgery, The Johns Hopkins Hospital and Medical School
126 al, AP Photo/Carlos Osorio
147 cr, The Sheridan Libraries/The Milton S. Eisenhower Library Audiovisual Services
149 ar, courtesy of the Johns Hopkins University Office of News and Information
152 ar, courtesy of the Johns Hopkins University Office of News and Information
165 br, courtesy of Alfred Sommer
166 ar, courtesy of the Johns Hopkins University Office of News and Information
175 ar, courtesy of Johns Hopkins Bayview Biomedical Media Services
183 ar, cr, br courtesy of ABC News
184 bl, courtesy of the Josie King Foundation / Crosskeys Media, LLC
204 al, courtesy of the Kimmel Cancer Center
218 al, courtesy of the White House
224 bl, courtesy of the Estate of Yousuf Karsh
226 al, courtesy of Spark Media, Inc.
227 ar, Bob Green, HBO Films
227 cr, Bob Green, HBO Films
231 cr, AP Photo / Jonas Ekstromer
235 ar, Library of Congress, Geography and Map Division
265 cr, courtesy of the National Medals Foundation
265 bl, courtesy of the National Medals Foundation
268 bl, AP photo/Scanpix Sweden / Jonas Ekstromer
270 al, "Gifted Hands: The Dr. Ben Carson Story" courtesy Sony Pictures Television
271 ar, courtesy of the Department of Art as Applied to Medicine
302 bl, courtesy of Alfredo Quiñones-Hinojosa
302 bc, courtesy of Alfredo Quiñones-Hinojosa
302 br, courtesy of Alfredo Quiñones-Hinojosa
307 cr, courtesy of Rick Hodes

Index

Numerals in bold italics refer to illustrations or photo caption pages.

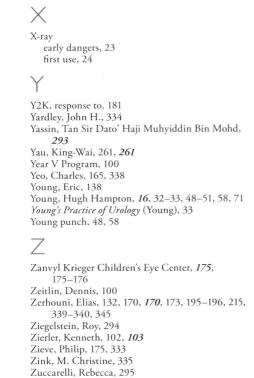